RADICAL POLYMERIZATION IN DISPERSE SYSTEMS

ELLIS HORWOOD SERIES IN POLYMER SCIENCE AND TECHNOLOGY

Series Editors: T. J. Kemp, Department of Chemistry, University of Warwick
J. F. Kennedy, Director of the Research Laboratory for the Chemistry of Bioactive Carbohydrates and Proteins, Department of Chemistry, University of Birmingham, and Professor of Applied Chemistry, North East Wales Institute of Higher Education.
Consulting Editor: Ellis Horwood MBE

This series, which covers both natural and synthetic macromolecules, reflects knowledge and experience from research, development and manufacture within both industry and academia. It deals with the general characterization and properties of materials from chemical and engineering viewpoints and will include monographs highlighting polymers of wide economic and industrial significance as well as of particular fields of application.

Barashkov, N. & Gunder, O.	FLUORESCENT POLYMERS
Bartoň, J. & Capek, I.	RADICAL POLYMERIZATION IN DISPERSE SYSTEMS
Bershtein, V. & Egorov, V.	DIFFERENTIAL SCANNING CALORIMETRY OF POLYMERS: Physics, Chemistry, Analysis, Technology
Blažej, A. & Košík, M.	PHYTOMASS: A Raw Material for Chemistry and Biotechnology
Bodor, G.	STRUCTURAL INVESTIGATION OF POLYMERS
Bryk, M.T.	DEGRADATION OF FILLED POLYMERS: High-Temperature and Thermal-Oxidative Processes
Challa, G.	POLYMER CHEMISTRY: An Introduction
Crompton, T.R.	POLYMER LABORATORY INSTRUMENTATION: Volumes 1 & 2
De, S.K. & Bhowmick, A.K.	THERMOPLASTIC ELASTOMERS FROM RUBBER-PLASTIC BLENDS
Gorbatkina, Y.A.	ADHESIVE STRENGTH OF FIBRE-POLYMER SYSTEMS
Hearle, J.W.S.	POLYMERS AND THEIR PROPERTIES: Vol. 1
Hearle, J.W.S.	POLYMER MATERIALS: Vol. 1: Structure, Properties and Performance Vol. 2: Complex Systems
Hongu, T. & Phillips, G. O.	NEW FIBERS
Kennedy, J.F. et al.	CELLULOSE: Structural and Functional Aspects
Kennedy, J.F: et al.	CELLULOSE SOURCES AND EXPLOITATION: Industrial Utilization, Biotechnology and Physico-Chemical Properties
Kennedy, J.F. et al.	LIGNOCELLULOSICS: Science, Technology, Development and Use
Kennedy, J.F. et al.	CELLULOSICS: Pulp, Fibre and Environmental Aspects
Kennedy, J.F. et al.	CELLULOSICS: Chemical, Biochemical and Material Aspects
Lazár, M. et al.	CHEMICAL REACTIONS OF NATURAL AND SYNTHETIC POLYMERS
Nachinkin, O.I.	POLYMERIC MICROFILTERS
Nevell, T.J. & Zeronian, S.H.	CELLULOSE CHEMISTRY AND ITS APPLICATIONS
Saifullin, R.S.	PHYSICAL CHEMISTRY OF INORGANIC POLYMERIC AND COMPOSITE MATERIALS
Stepek, J. et al.	POLYMERS AS MATERIALS FOR PACKAGING
Svec, P. et al.	STYRENE-BASED PLASTICS AND THEIR MODIFICATIONS
Ulberg, Z.R. & Deinega, Y.F.	ELECTROPHORETIC COMPOSITE COATINGS
Vakula, V.L. & Pritykin, L.M.	POLYMER ADHESION: Physico-Chemical Principles

ELLIS HORWOOD PTR PRENTICE HALL SERIES IN POLYMER SCIENCE AND TECHNOLOGY

Bartenev, G.M.	MECHANICAL STRENGTH AND FAILURE OF POLYMERS
Clarson, S.J.	SILOXANE POLYMERS
Galiatsatos, V.	MOLECULAR CHARACTERIZATION OF NETWORKS
Shlyapnikov, Y.A.	ANTIOXIDATIVE STABILIZATION OF POLYMERS
Vergnaud, J.M.	LIQUID TRANSPORT PROCESS IN POLYMERIC MATERIALS
Wirpsza, Z.	POLYURETHANES: Chemistry, Technology and Applications

RADICAL POLYMERIZATION IN DISPERSE SYSTEMS

J. BARTOŇ
and
I. CAPEK
both at the
Polymer Institute
Slovak Academy of Sciences
Bratislava

Translation Editor:
Professor T. J. KEMP
University of Warwick

ELLIS HORWOOD
NEW YORK LONDON TORONTO SYDNEY TOKYO SINGAPORE

First published 1994
in coedition between
Ellis Horwood Limited
Market Cross House, Cooper Street
Chichester
West Sussex, PO19 1EB
A division of
Simon & Schuster International Group

and
VEDA
Publishing House of Slovak Academy of Sciences
Klemensova 19, 814 30 Bratislava, Slovakia
ISBN 80-224-0408-X

© 1994 J. Barton, I. Capek / Ellis Horwood Limited
Translation © A. Rebrová

All rights reserved. No part of this publication may be reproduced, stored in a retrieval system, or transmitted, in any form, or by any means, electronic, mechanical, photocopying, recording or otherwise, without prior permission, in writing, from the copyright owners.

Printed and bound in Slovakia

Library of Congress Cataloging-in Publication Data
Available from the publisher
British Library Cataloguing in Publication Data
A catalogue record for this book is available from the British Library

ISBN   0-13-752353-X   (hbk)

Table of contents

Preface 9

1 Introduction 11
  References 14

2 Methods of preparation and classification of dispersed systems 16
  References 21

3 Polymerization and copolymerization in aqueous emulsions 23
  3.1 Classical and more recent theories of emulsion polymerization 23
    3.1.1 Harkins' theory 23
    3.1.2 Smith–Ewart's theory 25
    3.1.3 Gardon's theory 30
    3.1.4 Medvedev's theory 36
    3.1.5 Theory of homogeneous nucleation 38
    3.1.6 Polymerization in emulsified monomer droplets 45
  References 47
  3.2 Contemporary ideas about kinetic processes and mechanism
      of emulsion polymerization 49
    3.2.1 Some microscopic kinetic processes in emulsion polymerization 49
    3.2.2 Growth and deactivation of polymer particles 56
    3.2.3 Mathematical modelling of emulsion polymerization 64
    3.2.4 Site of initiation-partitioned polymerization approach 71
  References 76
  3.3 Characterization of micellar systems and principal components
      of emulsion polymerization systems 78
    3.3.1 Micellar systems: oil/water 78
    3.3.2 Initiating systems 104
  References 118
  3.4 Kinetics and mechanism of the radical polymerization of unsaturated
      monomers in micellar systems 123
    3.4.1 Styrene and its derivatives 123

    3.4.2 Acrylic and methacrylic alkyl esters 140
    3.4.3 Carboxylated monomers 160
  References 167
 3.5 Modification of polymer dispersions 170
    3.5.1 Hydrolysis 171
    3.5.2 The Hofmann reaction 172
    3.5.3 The Mannich reaction 172
    3.5.4 Hydroxymethylation of amide groups 173
    3.5.5 Reactions of double bonds 174
    3.5.6 Utilization of flexibility in the procedure for polymerization 174
  References 177
 3.6 Colloidal stability of polymer latices 178
  References 185

4 Inverse emulsion (microemulsion) polymerization 186
 4.1 Characterization of inverse emulsion and microemulsion systems 186
  References 192
 4.2 Kinetics and mechanism of radical polymerization in inverse micellar
    systems (water-in-oil emulsions and microemulsions) 192
  References 209
 4.3 Special cases of radical polymerization in water-in-oil micellar systems 210
  References 214

5 Dispersion polymerization in nonaqueous media 216
 5.1 Basic principles of formation and stabilization of nonaqueous
    polymer dispersions 216
  References 224
 5.2 Dispersants (steric stabilizers) and disperse media of nonaqueous
    dispersions of polymers 225
  References 233
 5.3 Kinetics and mechanism of radical dispersion polymerization
    in nonaqueous media 234
  References 244
 5.4 Properties and applications of nonaqueous polymer dispersions 244
  References 248

6 Special cases of polymerization in disperse systems 249
 6.1 Synthesis of polymer dispersions with a heterogeneous
    particle structure 249
  References 265
 6.2 Emulsifier-free emulsion polymerization 267
  References 272
 6.3 Polymerization of monomers with emulsifying properties
    (emulsifying monomers, surface-active monomers,
    internal emulsifiers, surfomers) 273
  References 277

Table of contents

6.4 Emulsion polymerization in the presence of compounds of the initiator-emulsifier type (surface-active initiators) 278
References 281
6.5 Polymerization in mono- and multilayers. Other special cases of emulsion polymerization 281
References 288

7 Suspension polymerization 290
7.1 Mechanism for the formation and stabilization of polymer particles in a suspension 290
References 297
7.2 Suspension polymerization and copolymerization of unsaturated monomers 298
    7.2.1 Styrene 298
    7.2.2 Vinyl chloride 302
    7.2.3 Other lipophilic monomers 312
    7.2.4 Unconventional suspension polymerizations 314
References 316

8 Methods of characterization of polymer dispersions 319
8.1 Methods of particle size determination 319
    8.1.1 Classical light scattering 319
    8.1.2 Photon correlation spectroscopy (PCS) 320
    8.1.3 Turbidimetry 321
    8.1.4 Stopped-flow spectroscopy 323
    8.1.5 Small-angle neutron and X-ray scattering 323
    8.1.6 Chromatographic methods 324
    8.1.7 Field-flow fractionation 326
    8.1.8 Electron microscopy 328
    8.1.9 Centrifugation 328
    8.1.10 Titration of latices with an aqueous solution of emulsifier 329
    References 330
8.2 Methods for characterizing the nature of polymer particles 331
    8.2.1 Conductometric and potentiometric titrations 332
    8.2.2 Electrophoresis 335
    8.2.3 Other methods 337
    References 338
8.3 Cleaning of polymer latices 339
    8.3.1 Dialysis 339
    8.3.2 Ion exchange 340
    8.3.3 Ultrafiltration 341
    8.3.4 Other methods 341
References 341

9 Future trends in dispersion polymerization 342
References 344

Subject index 345

Preface

Polymer dispersions, also called polymer colloids or latices, play an important role in the production of synthetic elastomers, lacquers, adhesives, additives, and many other products. Development of the industrial production of polymer dispersions, especially dispersions in aqueous media was necessitated not only by ecological problems which arise when systems based on organic solvents are used but also by the fact that dispersion polymerization is readily applicable in industry. Radical polymerization in disperse systems may be used for preparation of polymeric spherical nano- and milliparticles (10^{-9}–10^{-3} m) with a required structure, thus allowing a variety of applications, including the unconventional.

The world production of polymer dispersions as final products and also as intermediates accompanying the production of certain polymers and copolymers represents nowadays several hundred thousands tons per year. Dispersions of polymers and copolymers based on acrylic, vinyl chloride, vinyl acetate, butadiene, and styrene esters serve as examples. Research into the preparation and use of polymer dispersions is still underestimated. This statement mainly concerns the preparation of unconventional polymer dispersions for special applications. The situation is also reflected in published information about polymer dispersions.

The present monograph can fill the gap in the literature dealing with polymer dispersions only partially. The field of polymer dispersions is today so extensive that full treatment of a single aspect (e.g. classical oil-in-water polymer dispersions) would require a scope much greater than we can provide in this book. The same is true for unconventional polymer dispersions, e.g. systems based on inverse micelles and polymer dispersions of the core/shell structure. Taking these facts into account, the authors aimed at providing a brief but, as far as possible, a comprehensive insight into the problems of preparing polymer nano- or milliparticles by radical polymerization of unsaturated monomers in classical and unconventional disperse systems. The necessary selection of the material and the extent of their analysis reflect the personal approach of the authors. In spite of this, the authors believe they have chosen the optimum alternative, which characterizes polymer dispersions in an objective and comprehensive way. It is up to the reader to judge to what extent the authors have succeeded in their aim. The book put forward is a completely revised, amended and extended version of the 1991 Slovak edition.

The presentation and explanation of the topic require the reader to be at least partially acquainted with the basic concepts of macromolecular chemistry and the kinetics of radical polymerization in homogeneous reaction systems. The area of polymer dispersions cuts across several scientific disciplinary borders. Some knowledge of physical chemistry, physics of colloidal systems, and surface science is therefore necessary to understand the material put forward. Those interested in a deeper insight into, and analysis of, the topic described will find representative references to the original books and journal publications.

The authors wish to express their gratitude to journal and monograph publishers for their kind permission to use certain data from tables and figures appearing in their publications.

Bratislava, August 1992 *Authors*

1
Introduction

The demand for materials with improved or novel properties and high functionality requires a search for the development of new materials and methods of their application. Research into polymer dispersions, i.e. of systems containing polymer particles of varying size, chemical and physical structures, may help significantly in satisfying this demand. The classical industrial production of polymer dispersions is an important aspect of radical-polymerized unsaturated monomers. Dispersions of copolymers of butadiene and styrene, poly(butadiene), poly(vinyl acetate), copolymers of vinyl chloride and vinyl acetate, copolymers of acrylic acid esters, poly(vinylidene chloride), copolymers of acrylamide and its derivatives, etc. may serve as examples. Development of the methods for the preparation of polymer dispersions of controlled particle size was stimulated not only by the needs of the coating and lacquer industries but also of other activities which use polymers as basic materials or additives in the production and processing (manufacturing) of textile, natural and synthetic leather, rubber products, etc. The fact that at relatively high concentrations of the polymer component of a dispersion (30–60 mass %), the viscosity of the polymer dispersions remains rather low, is one of the advantages of polymer dispersions. As regards the preparation of a dispersion, the high polymerization rate, the high degree of polymerization of the product, and the fact that the reaction heat of polymerization can be readily dissipated in the continuous phase, cannot be neglected.

The application of emulsion thermoplastic polymers in various branches of national economics, but mainly in the coating and lacquer industry and in the textile and paper industry, was stimulated at the end of the 1950s owing to the development of crosslinkable polymer dispersions [1]. The crosslinking of a component of a dispersion polymer during application was achieved by using functionalized (thermo)reactive monomers, e.g. acrylamide, hydroxyalkyl methacrylate, glycidyl methacrylate, etc. in their copolymerization with styrene, acrylic and methacrylic esters or in the terpolymerization of these monomers. The formation of crosslinks between the originally linear, uncrosslinked macromolecules leads to changes in the physical properties, such as insolubility in solvents, different dynamic mechanical properties as a function of temperature, etc.

The successful preparation of monodisperse poly(styrene) particles aroused interest in microparticle systems [2]. Monodisperse polymer particles are widely used, e.g. as calibration standards, standards for determining the pore size and effectiveness of filters, column packings for chromatographic separation, carriers of biochemicals and as part of diagnostic tests. The success of each of these applications depends on the size and particle size distribution, on their chemical composition, and the degree of porosity. Methods of modification of polymer dispersions based on classical monomers have recently been intensively studied. Efforts are oriented towards the directed control of the mainly physical structure of colloidal polymer particles related to their properties. Special attention has been devoted to polymer particles of heterogeneous core/shell structure aimed at obtaining tough polymer networks used for damping sound and vibration. The preparation of polymer microparticles of diameter 1×10^3–5×10^4 nm has also been a point of interest. Attempts to prepare polymer particles with these dimensions via the mechanical homogenization of monomer droplets in the aqueous phase, followed by suspension monomer polymerization, failed.

An elegant method for the preparation of monodisperse micron polymer particles was developed by Ugelstad *et al.* [2]. The method is based on the gradual swelling of submicron polymer particles prepared via classical emulsion polymerization, i.e. low-molecular mass water-insoluble material and monomer in the presence of an aprotic solvent miscible with water is followed by polymerization of the monomer in swollen polymer particles. The preparation of micron particles in a gravity-free state on an orbital station has been proposed [3]. Another possibility for the preparation of polymer particles of size *ca.* 10^3 nm is offered by an oil-in-oil (O/O) dispersion polymerization. The monomer polymerizes in an organic solvent, which, though miscible with monomer, functions as a precipitant for the polymer being formed. To hinder formation of the coarse polymer dispersion (and finally of a coagulate) the polymer particles being formed have to be stabilized. This is why this type of nonaqueous dispersion polymerization makes use of polymer dispersing agents on block or graft polymers. The composition of these copolymers has to fulfil the condition of compatibility with both the dispersed phase and the polymer particle which is to be stabilized. Such a property of the copolymers will be achieved through suitable composition of the graft branches and the main chain of the graft copolymer or individual blocks of the block copolymer.

Papers dealing with the morphology of polymer particles have stimulated research into the field of the dispersion polymerization and practical applications of polymer dispersions. In the oil-in-water (O/W) dispersion polymerization, i.e. in the classical emulsion polymerization of a mixture of monomers, the particle morphology is influenced by the different solubilities of monomers in the aqueous phase, the copolymerization parameters, and the mode of preparation of the polymer dispersion (batch, semicontinuous or continuous). In batch dispersion polymerization, the more rapidly copolymerizing monomer enriches the polymer product at the beginning of the copolymerization; conversely, the product formed at the end of the copolymerization is depleted of this monomer. Particles are thus formed with morphology similar to core/shell particles, i.e. the compositions of the core and the shell are chemically different. The copolymerization of two monomers with different water solubilities leads to a structure of this type. The more hydrophilic reaction product will mostly form the particle shell, whereas the core will be formed by the more hydrophobic product. A typical shell/core structure will be achieved via monomer polymerization in the presence of a feed of polymer particles, i.e. by adding a separately prepared polymer dispersion to the reaction system undergoing dispersion polymerization. To inhibit or lower the compositional heterogeneity of copolymer particles, we must keep the ratio of polymerizing comonomers in the reaction system approximately constant and to select such a pair of comonomers which at least partially meet the requirement of the same solubility in the aqueous phase. The first requirement may be realized by using a semicontinuous process of preparation of the copolymer dispersion, or better, by gradual addition of the more reactive monomer to the reaction mixture so as to keep the ratio between both comonomers in the reaction feed constant. The second requirement is difficult to fulfil since the overall aim is the preparation of a polymer dispersion of specified chemical and physical properties; this is usually predetermined by the choice of both comonomers.

A special case of dispersion polymerization is the preparation of three-block copolymers of type ABA via sequential radical polymerization, i.e. the polymerization proceeds in two steps. In the first step, the polymer initiator is prepared through reaction between monomer A and a multifunctional initiator. In the second step, monomer B undergoes emulsion polymerization with the polymer initiator [4]. The preparation of two interpenetrating polymer networks can be conducted by using the technique of emulsion polymerization of the monomer in the presence of a polymer dispersion. Interpenetrating polymer networks may be prepared from both compatible and incompatible polymers. The resulting structure of the interpenetrating polymer networks is formed by core/shell polymer particles [5–8]. In addition to utilizing these polymer dispersions, microparticle systems currently find more and more important application in electronics, e.g. in the preparation of conducting coatings and plastisols for displays.

To modify the surface properties of inorganic pigments, methods have been developed for the polymerization of vinyl monomers in an adsorbed double layer (see Chapter 2 and section 6.5) formed from molecules of a surfactant on the surface of pigment particles [9]. Remarkable progress has been achieved in the development of composites based on inorganic and organic compounds, the so-called dispersed composites, where an inorganic core is covered with a layer of organic polymer, e.g. the encapsulation of CdS with poly(styrene) or poly(methyl methacrylate) [10]. New methods of encapsulation of powders of various materials of submicron dimensions via the emulsion polymerization of methyl methacrylate in the absence of emulsifier are described in ref. [11]. Blended polymer dispersions are prepared through simple mixing of dispersions or through precipitation of a polymer component of one dispersion into the polymer particles of another dispersion. This approach was used to prepare composite microparticles (core/shell type) composed of organic (shell, organic polymer) and inorganic (core, SiO_2) components [12]. Other data concerning the preparation and properties of composite microparticle systems based on organic and inorganic polymers by heterocoagulation are reported by Furusawa & Anzai [12].

Much attention has been devoted to study of the effect of aggregates of amphiphilic molecules on the course of chemical and physical processes [13–17]. The majority of studies were carried out using classical micelles. Other organized systems, such as spherically-closed monolayers and vesicles and inverse micelles have recently attracted attention. Aggregates of amphiphilic molecules are excellent tools which enable the separation of an organic phase from an aqueous phase and hence separation of the components of a reaction system if they differ in solubility in the aqueous and organic phases. Poly(styrene) nanoparticles (diameter 20–40 nm) were prepared in this way [18] via the polymerization of styrene in an oil-in-water (O/W) microemulsion, and of poly(acrylamide) and copolymers of acrylamide in water-in-oil (W/O) microemulsion [19–26]. Microparticles of polymer dispersions are, with regard to the possibility of chemical (incorporation of various functional groups to bind a substrate) and physical variability (microparticles of the core/shell type, porous or nonporous structure of microparticles, crosslinking of microparticle macromolecules) as well as large specific surface area, suitable carriers of various substrates immobilized on the surface of, or in the pores of, microparticles [27–29]. Polymer nano- and microparticles have been applied in biology and medicine, being used as carriers of pharmacologically effective compounds, cell membrane probes, and in diagnostic tests.

Their role in masking biologically active molecules against the immunological response of an organism is important. Magnetizable polymer particles are used for the directed transport of a drug to a particular organ and for the *in vitro* separation of cancer and normal cells [30]. An example of the biological application of polymer particles is the immobilization of bioactive molecules by a covalent bond, ionic bond, or adsorption on the polymer particle [31]. The immobilization of enzymes on a polymer carrier enables their separation from a reaction system; it sometimes also increases their effectiveness and storage life. It often enables the use of enzymes under abnormal conditions (high temperature, presence of organic solvent). Polymer particles of narrow size distribution are applied in examination of the fagocytosis of cells [32] and in precipitation (agglutination) tests. It is unnecessary to modify the surface of polymer particles for some applications; in most cases, however, suitable functional groups ($-CONH_2$, $-COOH$, etc.) must be introduced onto the particle surface (e.g. of poly(styrene)) in order to bind antigens (a molecule or part of a molecule able to initiate the formation of an antibody by the immunological system of the organism) or antibodies by a covalent chemical bond on the surface of the particle.

Investigation of polymer carriers of catalytically-active compounds from the point of view of both the preparation of suitable polymer structures and the reactivity of the immobilized substrate is motivated by potential attractive possibilities of industrial applications as well as efforts in filling the gaps in our knowledge about this field. Particularly interesting are microparticles with a chemically-inert core of porous or nonporous structure covered by a thin layer of functional polymer [33]. Polymer dispersions are used either as made or with only slight modification. A polymer dispersion is often precipitated in order to obtain a polymer component of the dispersion for further special processing or application (rubber, emulsion PVC). The number of possibilities offered by dispersion systems to research and industry seems infinite; they provide, however, a suitable initial impression. Other details are available in different sections of this monograph and in the references cited.

The meat of the book is found in the chapter describing the preparation and properties of classical polymer dispersions of the polymerization-unsaturated (usually vinyl) monomer in the oil-in-water emulsion polymerization system. Special cases of polymerization in disperse systems are illustrated by an example of inverse emulsion polymerization, the polymerization of monomers with emulsifying properties, the synthesis of dispersions of controlled structure and polymer composites. The part devoted to disperse polymerization in anhydrous media analyses questions of the formation and stabilization of polymer dispersions in organic (nonaqueous) systems. In later chapters, some problems associated with the preparation of dispersions via suspension radical polymerization are discussed and methods for the characterization of polymer dispersions are reported. The final chapter outlines the present trends and future development in the research and applications of polymer dispersions.

References

1. Warson, H. (1975) *Polym. Prepr.* **16** 280
2. Ugelstad, J., Mork, P. C., Kaggerud, K. H., Ellingsen, T. & Berge, A. (1980) *Adv. Colloid Interface Sci.* **13** 101
3. Lovelace, A. M., Vanderhoff, J. W., Micale, F. J., El-Aasser, M. S. & Kornfeld, D. M. (1981) *U. S. 4247434*
4. Su, J. S. N. & Piirma, I. (1987) *J. Appl. Polym. Sci.* **33** 272
5. Sperling, L. H. (1981) *Interpenetrating polymer networks and related materials.* Plenum Press, New York
6. Sperling, L. H., Chiu, T. W., Granlich, R. G. & Thomas, D. A. (1974) *J. Paint Technol.* **46** 47
7. Hourston, D. J. & Satgurunathan, R. (1984) *J. Appl. Polym. Sci.* **29** 2969
8. Hourston, D. J., Satgurunathan, R. & Varma, H. (1986) *J. Appl. Polym. Sci.* **31** 1955

References

9. Megurot, K., Yabe, T., Ishioka, S., Kato, K. & Esumi, K. (1986) *Bull. Chem. Soc. Jpn.* **59** 3019
10. Haga, Y., Nakajima, M. & Yosomiya, R. (1987) *Angew. Makromol. Chem.* **153** 71
11. Hasegawa, M., Arai, K. & Saito, S. (1987) *J. Polym. Sci., Polym. Chem. Ed.* **25** 3117, 3231
12. Furusawa, K. & Anzai, C. (1987) *Colloid Polym. Sci.* **265** 1
13. Fendler, J. H. (1982) *Membrane mimetic chemistry.* J. Wiley & Sons, New York
14. Bunton, C. A. & Savelli, G. (1986) *Adv. Phys. Org. Chem.* **22** 213
15. Harbour, J. R. & Hair, M. L. (1986) *Adv. Colloid Interface Sci.* **24** 103
16. Harbour, J. R. & Hair, M. L. (1980) *J. Phys. Chem.* **84** 1500
17. Harbour, J. R. & Bolton, J. R. (1978) *Photochem. Photobiol.* **28** 231
18. Atik, S. S. & Thomas, J. K. (1981) *J. Am. Chem. Soc.* **103** 4279
19. Leong, Y. S., Reiss, G. & Candau, F. (1981) *J. Chim. Phys.* **78** 279
20. Leong, Y. S. & Candau, F. (1982) *J. Phys. Chem.* **86** 2269
21. Candau, F., Leong, Y. S., Pouyet, G. & Candau, S. (1984) *J. Colloid Interface Sci.* **101** 167
22. Candau, F., Leong, Y. S. & Fitch, R. M. (1985) *J. Polym. Sci., Polym. Chem. Ed.* **23** 193
23. Gobe, M., Konno, K. & Kitahara, A. (1982) In: *Proceedings of the 35th Symposium on Colloid and Interface Chemistry*, Kirin, Japan, p. 36.
24. Candau, F., Zekhnini, Z., Heatley, F. & Franta, E. (1986) *Colloid Polym. Sci.* **264** 676
25. Candau, F. (1989) In: El-Nokaly, M. (ed.) *Polymer association structures: microemulsions and liquid crystals.* ACS Symposium Series **384**. Washington DC, p. 47
26. Candau, F. (1990) In: Candau, F. & Ottewill, R. H. (eds) *Scientific methods for the study of polymer colloids and their applications.* NATO ASI Series. D. Reidel Publ. Co., Dordrecht, p. 73
27. Kitano, H., Nakamura, K. & Ise, N. (1982) *J. Appl. Biochem.* **4** 34
28. Clark, D. S., Bailey, J. E., Yen, R. & Rembaum, A. (1984) *Enzyme Microb. Technol.* **6** 317
29. Suen, C. H. & Morawetz, H. (1985) *Makromol. Chem.* **186** 255
30. Kálal, J. (1987) *Makromol. Chem., Macromol. Symp.* **12** 259
31. Okubo, M., Kamel, S., Tosaki, Y., Fukunaga, K. & Matsumoto, T. (1987) *Colloid Polym. Sci.* **265** 957
32. Williams, C. A. & Chase, M. W. (eds) (1976) *Methods in immunology and immunochemistry*, vol. V. Academic Press, New York, p. 280
33. Verlaan, P. J., Bootsma, J. P. C. & Challa, G. (1982) *J. Mol. Catal.* **14** 211

2

Methods of preparation and classification of dispersed systems

Two methods are known for the preparation of colloidal dispersions in liquid (sols) or in gas (aerosol, smoke) media: dispersion and condensation methods.

In the dispersion method, the material to be dispersed is subjected to grinding or milling until the desired degree of fineness is achieved. The presence of a stabilizer is necessary to prevent the fine particles from agglomeration. Depending on the character of the dispersed phase, the stabilizer is usually a low-molecular-mass species, either ionic or polar, which is adsorbed on the surface of the particles. During the adsorption of ions, particles become electrically charged and repel each other (electrostatic particle stabilization). Through the adsorption of a polar species or of a polymer which contains hydrophilic and lipophilic moieties, steric or entropic stabilization of the particles is achieved. This lies in the situation that a part of the stabilizer molecule is adsorbed on the particle (the so-called anchoring part) and the rest of the stabilizer molecule (the so-called soluble part) interacts with the disperse phase (water, alkane, etc.). The utilization of such compounds is advantageous in those cases where their molecules fulfil both functions, i.e. electrostatic and steric stabilization of the particles.

In the condensation method, the particles are formed via some chemical or physical process of the ions or molecules in solution or by reactions in the gas phase. Typical examples of a dispersion formed by the condensation method are the gold sol* prepared by the reduction of tetrachloroaurate anion with formaldehyde in aqueous solution, the preparation of a polymer dispersion via emulsion polymerization, and the formation of aerosols (see p.18). The presence of a particle stabilizer is again necessary to prevent the particles from forming agglomerates. It is often advantageous to use both methods for the preparation of a dispersed system. The material which is to be dispersed by milling or grinding is first prepared by the condensation method from low-molecular-mass starting material of a certain degree of dispersion and the product formed is finally subjected to milling or grinding in the presence of a stabilizer until the degree of dispersion desired (i.e. desired particle size) is

---

* The word sol denotes a disperse system composed of solid dispersed particles of a compound in a disperse liquid phase.

reached. This two-step condensation-dispersion procedure provides the necessary control of the preparation of a dispersion with the desired properties. We should say that, with the exception of the preparation of polymer emulsions and photographic dispersions prepared by condensation methods, in technical preparations dispersion methods are mostly used. Here we can mention, for example, the preparation of dispersed dyestuffs, lacquers, pharmaceutical and agricultural emulsions and suspensions.

The dispersed systems may be divided into several types according to the state of the dispersed phase and disperse medium.

The dispersed liquid-in-liquid systems, i.e. emulsions, can be prepared by strong agitation of the two mutually insoluble or only partially-soluble liquids together, usually oil (designation for a water-insoluble liquid) in water (classical, direct emulsions, O/W) or water-in-oil (inverse emulsions, W/O). In the absence of an emulsion stabilizer, on interruption of the stirring, the macrophases gradually separate. Formation of a stable oil-in-water (O/W) or water-in-oil (W/O) emulsion is mainly determined by the character of the stabilizer (emulsifier) but also by the ratio of both macrophases. Emulsions prepared by stirring are usually dimensionally polydisperse, the dimensions of the droplets of the dispersed phase in the continuous phase being rather large, i.e. of the order of 10^4 nm and more. The formation of an emulsion with droplet size below this limit requires high-intensity stirring, i.e. it requires large amounts of energy. This procedure is therefore disadvantageous if a high degree of dispersion of the emulsion particles is required. The use of a mixture of an ionic emulsifier and higher alkanol, in the presence of water and oil phases accompanied with the formation of the O/W emulsion provides a means. The term miniemulsion has been proposed for these emulsions since the size of the dispersed particles of a miniemulsion (100–400 nm) lies between that of a conventional macroemulsion and that of microemulsion [1–3]. Under certain circumstances, spontaneous emulsification may occur with the minimum energy applied to stir the system. A typical example is the formation of W/O microemulsions in mixtures of water, oil, surfactant and a higher aliphatic alcohol or amine. The size of the particles (droplets) of the dispersed phase is of the order of 10 nm and the particles are practically monodisperse. In emulsions the particle size of the dispersed phase is determined by the mechanism of growth of the polymer particles in the emulsion polymerization. If the conditions are created where droplets of the dispersed monomer in the continuous phase are able to compete with micelles effectively by trapping free radicals generated in the aqueous phase, then the droplets can be a site for the polymerization process [4, 5]. Monomer droplets can successfully compete with monomer-containing and polymer/monomer particles by trapping radicals generated in the aqueous phase only when the overall droplet surface is large enough (i.e. their size approaches that of micelles or the size of polymer/monomer particles). Changing the ratio of the aqueous and oil phases (e.g. by adding oil to the system) leads to a structural change of the system and to inversion of the O/W emulsion to a W/O emulsion. Ideas about the mechanism for the inversion of an emulsion system are not unequivocal and the desired result, i.e. the inversion of the emulsion system, cannot always be achieved [6–8].

Inverse micelles, i.e. the W/O system, are able to solubilize hydrophilic molecules in a hydrocarbon solvent, e.g. enzymes, and even plasmids, with dimensions much greater than those of a water droplet of an inverse micelle. The biopolymer-containing inverse micelles can be considered as a new type of microreactor. The ability of inverse micelles containing an enzyme to react with a water-insoluble substrate or to be soluble in the oil disperse phase is remarkable. The reaction between lipogenase and linoleic acid provides an example [9]. There also exist nonaqueous inverse microemulsions prepared by using formamide instead of water [10]. These are significant because many organic compounds cannot be, because of their insolubility, introduced into a water droplet of an inverse micelle. An advantage in some cases is also the relative permittivity of formamide which is lower than that of water. Our main attention will be directed toward dispersions of the liquid-in-liquid type, i.e. to emulsion systems, since they are the basis of preparation of polymer dispersions if they consist of radically-polymerizable monomer.

Coarse dispersions (10^5 nm and more) of a solid in a liquid are called suspensions. Suspensions, or better microsuspensions, are also formed by the polymerization of monomer in an emulsion system. The final product of emulsion polymerization, i.e. a polymer dispersion, is in principle a dispersion of a solid (polymer) in a liquid and thus, because of its character, is a microsuspension (sol) and not an emulsion. The expression 'polymerization in suspension' is therefore incorrect since the word suspension expresses exactly the status of the dispersed system only after polymerization.

Before, as well as during, polymerization we have in fact a liquid-in-liquid dispersion, i.e. an emulsion. The emulsion of a suspension polymerization has a low degree of dispersion since the droplet diameter is of the order of 10^5–10^6 nm or more. The use of the incorrect name, i.e. suspension polymerization is, however, so deep-rooted that there is no reason to adhere rigorously to the correct designation of this method of preparation of a polymer dispersion.

For the sake of completeness, we shall mention another three types of dispersion, currently not very important in the preparation of polymer dispersions.[*] Dispersions of liquid-in-gas are denoted as fogs, dispersions of solid-in-gas as smokes. The common name for both types is aerosol. Gas-in-liquid dispersions are foams. The formation of foams in the preparation of polymer dispersions is usually undesirable because it complicates the process (foaming of the reaction system). There are also solid-in-solid dispersions, e.g. organic composite materials and dispersions of gas-in-solid (colloidal inclusions).

The term polymer dispersion generally refers to the dispersion of a synthetic or natural polymer in the liquid phase, most often in water. This term will designate dispersions of polymers prepared by the radical polymerization of dispersed unsaturated monomers in a liquid disperse phase. We shall not consider polymer dispersions prepared in other ways, e.g. by milling synthetic or natural polymer in the presence of water and a dispersing agent or by the emulsification of a polymer in aqueous solution in the presence of a surfactant followed by evaporation of the organic solvent.

Classical oil-in-water emulsion polymerization, as well as inverse emulsion polymerization, is characterized by the emulsification of an immiscible monomer in the continuous water or oil phase followed by the radical-initiated polymerization. The result is a colloidal sol of polymer particles, i.e. a polymer dispersion. The presence of a surface-active compound, the emulsifier, in the emulsion polymerization of an almost water-insoluble monomer is necessary. The emulsifier forms an emulsion of oil in water and the initiator producing reactive radicals by thermal or redox decomposition must also be present. Water-soluble initiators are used most often although oil-soluble initiators may also be used. The average particle diameter of the polymer dispersion usually ranges between 100 and 300 nm. This contrasts strongly with the size of dispersed (emulsified) monomer droplets,

---

[*] Preparation of poly(acrylamide)-based aerosol was described by E. J. Davis *et al.* (1987) (*J. Colloid Interfcce Sci.* **118** 343).

which is one or two orders of magnitude higher (1000–10 000 nm). This indicates that the mechanism of classical emulsion polymerization cannot be determined by polymerization in monomer droplets and that any proposed mechanism for emulsion polymerization has to take account of the one- order-of-magnitude reduction in polymer particle size compared with the dimensions of the monomer droplets. Fikentscher [11] was the first who, as early as 1938, called attention to this fact.

Inverse emulsion polymerization (water-in-oil) requires emulsification of a water-soluble monomer, usually as an aqueous solution in the continuous oil phase using an emulsifier to form a water-in-oil emulsion. To initiate radical polymerization in inverse emulsion polymerization systems, water-soluble and oil-soluble initiators are used. The polymerization results in water-swollen polymer particles dispersed in the oil phase, of average size between 50 and 300 nm. The original size of the emulsified monomer droplets varied from 50 to 10 000 nm. We see almost no difference between the size of polymer particles prepared by classical emulsion polymerization and that of those prepared by inverse emulsion polymerization.

Another approach to polymer dispersions is the preparation of nonaqueous oil-in-oil dispersions. Monomer dissolved in the organic phase, which must not be a solvent for the polymer being formed, polymerizes under the effect of an oil-soluble initiator in the presence of a polymer dispersing agent (block or graft copolymer). The polymer particles formed are insoluble in the disperse phase and are stabilized by the polymer-dispersing agent. A stable dispersion arises of average particle size 100–10 000 nm.

Microemulsions are colloidal dispersions of varying stability with dimensions of the dispersed phase from 5 to 100 nm. Microemulsions usually contain water, hydrocarbon solvent, emulsifier, and co-emulsifier (usually C_4–C_8 n-alkanols) [6, 12–15]. Microemulsions prepared from water, hydrocarbon, and alcohol are also known [16,17]. Microemulsions of this type are in essence water-in-oil dispersions. Microemulsions not containing water have also been prepared [18]. The presence of a monomer soluble in water is naturally essential for the preparation of a polymer microemulsion, however, the polymer particles are smaller, approximately between 30 and 100 nm. There is some evidence that microemulsions contain domains both rich in the aqueous phase and in the oil phase. These phases are separated from each other by a layer of an amphiphilic compound. The mutual arrangement and the shape of the domains, i.e. their microstructure has been the subject of current research efforts of many groups [19, 20]. Submicron polymer particles prepared in microemulsions show a narrow size distribution and strong stability in the system. Details on the mechanism of their formation in water-in-oil (W/O) microemulsions are not yet known. The mass of the polymer particles is usually higher than that of the monomer dissolved in a water pool of an inverse micelle.* However, this means that the individual water pools of inverse micelles [21] are not isolated from each other but their rapid association and dissociation take place. Thus the transfer occurs of monomer dissolved in a water pool of an inverse micelle to another water pool of another inverse micelle. The resulting polymer particles are monodisperse in spite of this process which should lead logically to an increase in polydispersity of the particles. This contradiction may be explained by the idea of the function of the water pools of inverse micelles as being a

---

* The water pool is the inner part of an inverse micelle. It is by means of the water pools that the interaction between inverse micelles, where mass exchange between individual droplets occurs, takes place. A water droplet without a protective shell of surface-active compound cannot exist as a stable entity in the system.

reservoir of monomer molecules by analogy with a reservoir of monomer in the form of monomer droplets in the aqueous phase for the growth of polymer particles in classical emulsion polymerization. The difference is, however, that the transfer of monomer proceeds by collisions of the water pools and not by the diffusion of monomer through the disperse phase [22]. This picture assumes that particle nucleation (the formation of growth centres in individual water pools) takes place not only in the first phases of the reaction and that in some water pools of inverse micelles, polymerization is not initiated. If coagulation of the polymer microparticles should take place, i.e. if aggregates of several inverse micelles should be formed because of a shift of the balance between association and dissociation in favour of the association of inverse micelles, polydisperse polymer particles would inevitably be formed. A known example is the coagulation of primary polymer particles in classical oil-in-water emulsion polymerization caused by insufficient stability of the polymer particles.

The polymerization of vinyl monomers in the double layer formed by molecules of the surface-active compound and adsorbed on the solid surface offers interesting possibilities [23]. This is, in fact, a polymerization at the liquid/solid interface.* Other examples of the polymerization of monolayer and multilayer structures at gas/liquid and liquid/solid interfaces were reported by Lando *et al.* [24, 25]. Recent years have seen a growing interest in studying phospholipid aggregates in nonpolar media since the aggregates may serve as a model of cell membranes. In aqueous solution, these aggregates (liposomes) become spherically-closed mono- and multilayer (lamellar) formations, in which individual layers are separated by a water layer. The lipophilic part of phospholipid molecules makes up the inside of the layer (lamella). The layer contains two rows of phospholipid molecules with their polar ends turned outwards from the layer and the lipid part inwards to the layer. Various compounds of lipophilic character may be solubilized in the lipid part of the monolayer. Liposomes act as transport agents of pharmacologically-effective preparations into selected cells of an organism [26, 27]. With regard to a better defined structure of the double layer and substantially lower dynamics of exchange of the molecules of the double layer with those of the surface-active compounds in the aqueous phase, double layers provide a better model for the investigation of chemical reactions in heterogeneous systems than classical or inverse micelles [28].

Monolayers and double layers formed by aggregation of amphiphilic compounds with a radically-polymerizable double bond (amphiphilic monomer) can be polymerized [29–31] There are some limitations, however. The exchange of a molecule of amphiphilic monomer, which is a part of an aggregate, with that in solution is a rather fast process (of the order of μs and less [32]) which depends on the type of amphiphilic monomer and on the critical concentration at which the formation takes place of aggregates of the amphiphilic monomer in the continuous liquid phase. Polymerization of the molecules of the amphiphilic compound in the aggregate may be expected only when the time needed for the dissociation of individual molecules of the amphiphile from the aggregate is longer (or at least of the same order) as that needed for the formation of macromolecules from amphiphilic monomer in the aggregate [33]. The question of polymerizability or nonpolymerizability in aggregates of amphiphilic compounds containing a radically-polymerizable double bond has not been

---

*   The authors in [24] use the term admicelle for this system. A more suitable name is probably 'adsorbed double layer' since the word micelle covers spherical, lamellar or even cylindrical ordering of the molecules of a surface-active compound in water on exceeding a certain critical concentration (see section 3.3).

sufficiently answered as yet [34, 35]. It is without doubt that the polymerization of amphiphile monomers may lead to the formation of stabilized, spherically-closed layer formations suitable for longer-term applications.

Other cases of polymerization in organized systems leading to the formation of products of desired properties have also been reported (a particular steric configuration of the structural units of a macromolecule, ordering of the different structural units in the sequence and the length required). The polymerization of monomers on the surfaces of inorganic compounds and complexes is an example (Ziegler–Natta catalysts for the polymerization of 1-alkenes [36]), the Merrifield synthesis [37], the polymerization of monomers being thermotropic liquid crystals of smectic, nematic or cholesteric structure [38], polymerization on matrices [39], etc. These do not lead, however, to the formation of polymer particles and the mechanism for the polymer formation is often nonradical. A proposal has recently been put forward [40] for the nomenclature and characterization of individual types of heterophase polymerization based on the chemical and colloidal criteria of the specified reaction systems.

References

1. Ugelstad, J., El-Aasser, M.S. & Vanderhoff, J.W. (1973) *J. Polym. Sci., Polym. Lett. Ed.* **11** 503
2. Ugelstad, J., Hansen, F.K. & Lange, S. (1974) *Makromol. Chem.* **175** 507
3. Hansen, F.K. & Ugelstad, J. (1979) *J. Polym. Sci., Polym. Chem. Ed.* **17** 3069
4. Choi, Y.T., El-Aasser, M.S., Sudol, E.D. & Vanderhoff, J.W. (1985) *J. Polym. Sci., Polym. Chem. Ed.* **23** 2973
5. Delgado, J., El-Aasser, M.S., & Vanderhoff, J.W. (1986) *J. Polym. Sci., Polym. Chem. Ed.* **24** 861
6. Benett, K.E., Hatfield, J.C., Davis, H.T., Macosco, C.W. & Scriven, L.E. (1982) In: Robb, I.D. (ed.) *Microemulsions*. Plenum Press, New York
7. Talmon, Y. & Prager, S. (1978) *J. Chem. Phys.* **69** 2984
8. Saito, H. & Shinoda, K. (1970) *J. Colloid Interface Sci.* **32** 647
9. Luisi, P. (1985) *Angew. Chem., Int. Ed.* **24** 439
10. Lattes, A., Rico, I., de Savignac, A. & Ahmad-Zadeh Samii, A. (1987) *Tetrahedron* **43** 1725
11. Fikentscher, H. (1938) *Angew. Chem.* **51** 433
12. Friberg, S. & Lapzynska, I. (1975) *Prog. Coll. Polym. Sci.* **56** 16
13. Prince, L.M. (ed.) (1977) *Microemulsions*. Academic Press, New York
14. Tadros, Th. F. (ed.) (1984) *Surfactants*. Academic Press, London
15. Shah, D.O. (ed.) (1985) *Macro- and microemulsions*. ACS Symp. Ser. 272. Am. Chem. Soc., Washington DC
16. Kaiser, B.A. & Holt, S.L. (1982) *Inorg. Chem.* **21** 2323
17. Smith, G.D., Donelan, C.E. & Barlen, R.E. (1977) *J. Colloid Interface Sci.* **60** 488
18. Rico, I. & Lattes, A. (1984) *Nouv. J. Chim.* **8** 429
19. Jahn, W. & Strey, R. (1988) *J. Phys. Chem.* **92** 2294
20. Kahlweit, M. et al. (1987) *J. Colloid Interface Sci.* **188** 436
21. Menger, F.M., Donahue, J.A. & Williams, R.F. (1973) *J. Am. Chem. Soc.* **95** 286
22. Sugimoto, T. (1987) *Adv. Colloid Interface Sci.* **28** 65
23. Wu, J., Harwell, J.H. & O'Rear, E.A. (1987) *J. Phys. Chem.* **91** 623
24. Puterman, M., Fort, T., Jr. & Lando, J.B. (1974) *J. Colloid Interface Sci.* **47** 705
25. Letts, S.A., Fort, T., Jr. & Lando, J.B. (1976) *J. Colloid Interface Sci.* **56** 64
26. Torchilin, V.P., Klibanov, A.L., Ivanov, N.N., Ringsdorf, H.R. & Schlarb, B. (1987) *Makromol. Chem., Rapid Commun.* **8** 457
27. Ostro, M.J. (1987) *Sci. Am.* **256** 102
28. Fendler, J.H. (1980) *Acc. Chem. Res.* **13** 7
29. Fendler, J.H. (1984) In: Mittal, K.L. & Lindman, B. (eds) *Surfactants in solution*, vol. 3. Plenum Press, New York, p.1947.
30. Dorn, K., Klingbiel, R.T., Specht, D.P., Tyminski, P.N., Ringsdorf, H.R. & O'Brien, D.F. (1984) *J. Am. Chem. Soc.* **106** 1627
31. Paleos, C.M. (1985) *Chem. Soc. Rev.* **14** 45
32. Aniansson, E.A.G., Wall, S.N., Almgren, M., Hoffmann, H., Kielmann, I., Ulbright, W., Zana, R., Lang, J. & Tondre, C. (1976) *J. Phys. Chem.* **80** 905
33. Hamid, S. & Sherrington, D. (1986) *J. Chem. Soc., Chem. Commun.* p. 937

References

34. Elias, H.G., Chung, D.C., Donkai, N., Hellman, G.P., Solc, K. & Nagai, K. (1987) *Makromol. Chem.* **188** 537
35. Nagai, K. & Elias, H.G. (1987) *Makromol. Chem.* **188** 1095
36. Paecht-Horowitz, M. (1977) In: Elias, H.G. (ed.) *Polymerization of organized systems.* Gordon and Breach Science Publishers, New York, p. 89
37. Merrifield, R.B. (1965) *Science* **150** 178
38. Krentsel, B.A. (1977) In: Elias, H.G. (ed.) *Polymerization of organized systems.* Gordon and Breach Science Publishers, New York, p. 117
39. Bartoň, J. & Borsig, E. (1988) *Complexes in free-radical polymerization.* Elsevier, Amsterdam, p.193
40. Hunkeler, D., Candau, F., Pichot, C., Hamielec, A.E., Xie, T., Bartoň, J., Vašková, V., Guillot, J., Dimonie, M.V. & Reichert, K. H., *personal communication*

3

Polymerization and copolymerization in aqueous emulsions

3.1 CLASSICAL AND MORE RECENT THEORIES OF EMULSION POLYMERIZATION

Several models have been proposed to explain the time profiles of emulsion polymerization and copolymerization. The models can be classified according to various criteria, e.g. according to the nature of the water solubility of the monomer there are kinetic models of (a) emulsion polymerization of water-insoluble vinyl monomers and (b) polymerization of unsaturated partially water-soluble monomers. Another classification (according to the site of the reaction centre) differentiates four kinetic models:

(i) micellar (polymerization starts in emulsifier micelles saturated by vinyl monomer),
(ii) the model of interfacial polymerization (polymerization starts in an emulsifier layer adsorbed on the surface of polymer particles),
(iii) the model of the start and the course of the emulsion polymerization in the aqueous phase, and
(iv) the model according to which the initiation and growth of the polymer chain take place in monomer droplets.

3.1.1 Harkins' theory

The first hypotheses about the kinetics and mechanism of emulsion polymerization appeared in the early 1940s. The principles of the general theory of the formation and course of emulsion polymerization were presented by Harkins [1–3]. His ideas have been the starting point for current models describing this complex process. Harkins assumes that the polymerization process starts with the decomposition of a water-soluble initiator into primary radicals which are absorbed into monomer-swollen emulsifier micelles. The propagation itself proceeds inside micelles. As reaction proceeds, the micelles become polymer/monomer particles; their number is controlled by the concentration of micellar emulsifier and the formation of new particles ends when they are consumed. The monomer is present in the system as monomer droplets from which it diffuses into the polymer particles and thus maintains dynamic equilibrium until the time at which the monomer droplets disappear. The droplets serve only as a reservoir of monomer. In parallel with the growth of the polymer particles and with the diffusion of monomer from the monomer droplets into the particles, the molecules of free emulsifier diffuse onto the surface of the particles. The role of the emulsifier is to stabilize the polymer particles being formed and growing. After it is consumed, no further particles are formed and their number stabilizes at a constant value. The polymerization then proceeds only in the particles and the emulsifier adsorbed on their surface is in dynamic equilibrium with the emulsifier dissolved in water. On consumption of the monomer droplets, the dynamic equilibrium is broken and the centres of polymerization are fed only with monomer present in the polymer particles. This is the micellar model and is shown in Fig. 1.

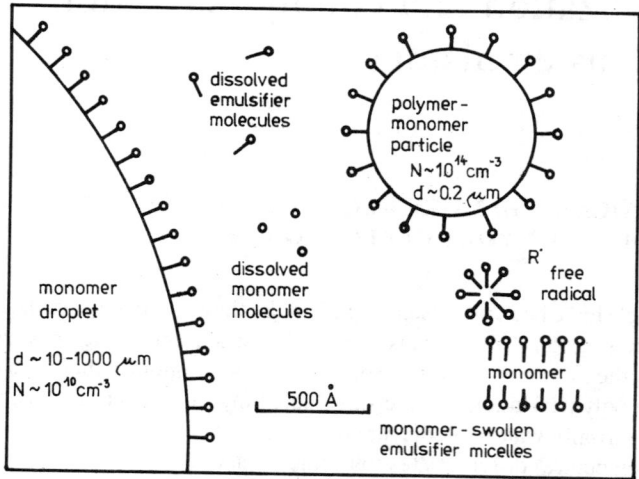

Fig. 1 — Harkins' micellar model of emulsion polymerization. Number of monomer molecules dissolved in water: $\sim 10^{18}$ cm^{-3}, number of monomer-swollen emulsifier micelles: $\sim 10^{18}$ cm^{-3}, number of monomer droplets: $\sim 10^{10}$ cm^{-3}.

To sum up, the essential features of Harkins' model are as follows:

(i) An appreciable amount of the emulsifier is in the form of micelles.
(ii) The principal centres of polymerization are the monomer-swollen emulsifier micelles and the monomer-swollen polymer particles formed.
(iii) The monomer droplets act as a reservoir of monomer. Negligible quantities of polymer are formed in the emulsified monomer droplets.
(iv) The growth of the polymer particles leads to an increase in their surface area and occurs via the consumption of the monomer in the particles. The polymer particles then tend to absorb monomer from the aqueous phase.
(v) The monomer molecules diffuse from the emulsified monomer droplets through the aqueous phase into the polymer particles.
(vi) The continual adsorption of micellar emulsifier on the surface of the growing polymer particles leads to the disappearance of emulsifier micelles, usually at low conversions.
(vii) The continual absorption of monomer into the growing polymer particles leads to the disappearance of the monomer droplets, usually at medium conversions.
(viii) The steady rate of polymerization is proportional to the initial concentration of emulsifier.

(ix) The small number of particles formed in the aqueous phase decreases with increasing emulsifier concentration.

3.1.2 Smith–Ewart's theory

Harkins' ideas about the micellar model of emulsion polymerization had been applied in the kinetic model which was experimentally verified by Smith & Ewart [4, 5]. The model is based on the postulates:

(i) The emulsion polymerization system consists of 4 phases: water, emulsified monomer droplets, monomer-swollen emulsifier micelles, and monomer/polymer particles stabilized by adsorbed emulsifier.
(ii) The decomposition of a water-soluble initiator generates primary radicals, which, on entering the monomer-swollen micelles initiate polymerization, nucleate micelles and form new polymer particles. Monomer/polymer particles are formed only by the entry of radicals into emulsifier micelles.
(iii) The surface area of the micelles under classical conditions of emulsion polymerization is, on average, three orders of magnitude larger than the surface area of the emulsified monomer droplets; the probability of entry of free radicals into the monomer droplets is therefore very small and is neglected in kinetic considerations. Only the effective entry of radicals into micelles is taken into account (the contribution due to polymerization in the monomer droplets is neglected). The oil drops serve only as a reservoir of monomer.
(iv) The number of particles increases until the free emulsifier is consumed, then it stabilizes at a constant value. During polymerization, a dynamic equilibrium is maintained between the emulsifier adsorbed on the surface of the polymer particles and the free emulsifier, monomer in water, in the monomer droplets, and in the monomer/polymer particles. In their theory, Smith and Ewart suggest three possibilities for the occurrence of a propagating radical in a particle: kinetic model 1 assumes the average number of radicals per latex particle ($\bar{n}$) to be less than 0.5; kinetic model 2 describes the case when this number is equal to 0.5, and model 3 is valid when $\bar{n} > 0.5$.
(v) The polymerization proceeds in three stages and the polymer is formed only in the latex particles. In the first stage, the latex particles are generated from the monomer-swollen emulsifier micelles by radicals entering the micelles from the aqueous phase. After depletion of the micellar emulsifier, the formation of new particles finishes. In this stage the polymer particles grow by means of the diffusion of monomer from the monomer droplets. After depletion of the monomer droplets, the rate of polymerization decreases abruptly.

Kinetic model 1 describes a process in which the number of free radicals per particle is much less than unity. Hence there are a large number of particles which have no radicals at all.

This condition will be fulfilled when

$$\rho N \ll k_0 a_p / v, \qquad (1)$$

where ρ is the rate at which radicals enter the micelles or polymer particles contained in unit volume of the aqueous phase, N is the number of particles, a_p is the surface area of a particle, v is the volume of a polymer particle and k_0 is the specific rate constant for the process.

The rate of polymerization per unit volume of aqueous phase can be written:

$$R_p = d[M]/dt = k_p [M] V c_p, \qquad (2)$$

where k_p is the rate constant for chain propagation, [M] is the monomer concentration in the polymer particles, V is the total volume of polymer particles per unit volume of the aqueous phase and c_p is the average concentration of free radicals in the particles.

The radical termination may take place either in the aqueous phase or in the polymer particles.

If termination occurs mainly in the aqueous phase, the rate of polymerization per unit volume of aqueous phase will be

$$R_p = k_p [M] V \alpha_r \left(\frac{R}{2k_t} \right)^{0.5}, \qquad (3)$$

where α_r is the partition coefficient for radicals between the particles and the aqueous phase, R is the rate of generation of radicals per unit volume of the aqueous phase, and k_t is the rate constant for termination.

Another quantity of interest is the average lifetime of a radical in the polymer particles

$$\bar{t} = \frac{\alpha_r V}{(2k_t R)^{0.5}}. \qquad (4)$$

If termination takes place mainly in the polymer particles, and if these particles are so small that termination proceeds very rapidly every time two radicals are in the particle, then the rate of termination is simply twice the rate at which radicals enter the polymer particles. Thus under these conditions the rate of polymerization and the average lifetime of a radical can be written:

$$R_p = k_p [M] \left(\frac{RNv}{2k_0 a_p} \right)^{0.5} = k_p [M] \left(\frac{RV}{2k_0 a_p} \right)^{0.5} \qquad (5)$$

and

$$\bar{t} = \frac{N}{2\rho} = \left(\frac{V}{2k_0 a_p R} \right)^{0.5}. \qquad (6)$$

Thus the overall rate of polymerization is a function of the termination mechanism operating in the system.

The case, $\bar{n} \ll 1$ is applicable for the emulsion polymerization, for example, of monomers with a high chain-transfer constant or for a polymerization conducted in the presence of chain-transfer agents.

Kinetic model 2 derived from the above considerations assumes that:

(i) a maximum of one propagating radical occurs in an active latex particle. On entry of

another radical, immediate chain termination and deactivation of the polymer particle take place. The activity of the particle is regenerated on entry of another radical. The processes of activation and deactivation of the polymer particle repeat periodically until complete consumption of the monomer;
(ii) each particle is active half of the time and passive half of the time during polymerization;
(iii) the average rate of polymerization is equal to one half the polymerization rate of an active monomer/polymer particle;
(iv) in a system with a larger number of particles, half of the polymer particles will be active at a particular time and the other half inactive.

To derive the kinetic model 2, simplifying assumptions are used as follows:

– a steady rate of polymerization is reached after the number of reaction centres has reached its maximum,
– there is no desorption of radicals from the polymer particles,
– the termination of a radical on entry of a second radical happens immediately,
– the absence of coagulation and coalescence of the latex particles,
– at any time, all polymer particles have similar dimensions,
– the radical distribution among the latex particles is of a quasi-steady-state character.

The requirements for this situation can be stated as

$$k_0 (a_p /v) \ll \rho N < k_t /v .\tag{7}$$

The Smith–Ewart kinetic model 2 provides data about the course of emulsion polymerization, i.e. about the kinetic parameters of the reaction system and their dependence on the reaction conditions. The rate of polymerization in monomer/polymer particles may be expressed by the equation

$$R_p = \frac{k_p N}{N_A} [M]\bar{n} ,\tag{8}$$

where $\bar{n}$ is the average number of radicals per particle and N_A is Avogadro's number.

To find out the number of polymer particles per unit volume formed at the time of consumption of free emulsifier, Smith and Ewart proposed the equation

$$N = k(\rho/\mu)^{0.4} (a_s S)^{0.6} ,\tag{9}$$

where μ is the rate of increase in volume of a particle, a_s is the area occupied by one emulsifier molecule on the surface of a micelle or polymer particle, S is the amount of emulsifier adsorbed on the surface of the polymer particles and k is a constant of value between 0.37 and 0.53.

Equation (9) corresponds to the following assumptions:

(i) In the initial steps of polymerization the emulsifier is present in the micelles and only a negligible part is dissolved in water; in kinetic measurements, this amount is neglected. The amount of emulsifier adsorbed on the surface of the monomer droplets is also neglected.

(ii) During polymerization the emulsifier is distributed between the micellar fraction and the fraction adsorbed on the surface of polymer particles, where dynamic equilibrium is maintained until the free micelles are consumed. If we denote the overall amount of emulsifier in the system by the symbol S, the micellar fraction as S_m, and the fraction of emulsifier adsorbed on the surface of the polymer particles as S_p, then, until the micelles are consumed, we always have: $S = S_m + S_p$. Free emulsifier micelles act as a reservoir of emulsifier, from which emulsifier molecules diffuse onto the surface of propagating and formed particles.

(iii) In the stationary state, the mass ratio monomer/polymer is constant and does not change until the complete consumption of monomer droplets.

(iv) The area occupied by an emulsifier molecule is the same for both micelles and polymer particles.

(v) The rate of polymerization inside the polymer particle under steady-state conditions is constant and the same for all particles present.

Combination of Eqs (8) and (9) gives a relation for the rate of polymerization after the completion of nucleation:

$$R_p = \frac{k}{2} k_p [M]_{eq} (\rho/\mu)^{0.4} (a_s S)^{0.6}, \qquad (10)$$

where $[M]_{eq}$ is the equilibrium monomer concentration in monomer/polymer particles. After the consumption of free emulsifier micelles and before the disappearance of monomer droplets, the propagation rate constant, k_p, and the equilibrium monomer concentration, $[M]_{eq}$, are constant, and the rate of polymerization with respect to monomer concentration may be considered as a zero-order reaction.

Under stationary conditions of polymerization, the parameter ρ is proportional to the concentration of initiator. If we assume that the values of the parameters do not vary, we find:

(i) the rate of polymerization is proportional to the 0.4 power of the initiator concentration and to the 0.6 power of the emulsifier concentration if that part of the emulsifier dissolved in water is very small and represents a negligible fraction of the total amount of emulsifier used,

(ii) the rate of polymerization per unit mass of emulsifier is inversely proportional to the 0.4 power of the concentration of micellar emulsifier.

The total number of radicals present in the polymer particles is given by

$$n_T = (N/2)(1 + 1/\beta - 1/3\beta^2 + ...), \qquad (11)$$

where the quantity β is defined as

$$\beta = k_t N/\nu\rho. \qquad (12)$$

The mean lifetime of a free radical under normal polymerization conditions where each free radical produced enters a polymer particle is given by

$$\bar{t} = N/2\rho . \tag{13}$$

Equation (11) shows that the number of growing radicals will be nearly equal to half the number of particles. Since it is possible to have a large number of particles present, a fast rate of polymerization follows. Equation (13) indicates that the average lifetime of a free radical increases with increase in the number of polymer particles; thus it is possible to have high rates of polymerization with simultaneous high molecular masses in emulsion polymerizations.

The third case of the Smith–Ewart model covers the situation when the number of radicals per polymer particle is large compared with unity. This situation holds when

$$S/N \gg k_t v . \tag{14}$$

The radicals thus enter the polymer particles more rapidly than they are deactivated in termination reactions and desorption processes. Under such conditions the polymer particles function as a unique 'microreactor' where the polymerization is controlled by the kinetics of the oil phase (homogeneous) polymerization. So the rate of polymerization, the average lifetime of a radical and the average number of radicals per particle will be, respectively,

$$R_p = k_p[M] \left(\frac{\rho N v}{2 k_t}\right)^{0.5} = k_p[M] \left(\frac{\rho V}{2 k_t}\right)^{0.5} , \tag{15}$$

$$\bar{t} = \left(\frac{N v}{2 k_t \rho}\right)^{0.5} = \left(\frac{V}{2 k_t \rho}\right)^{0.5} \tag{16}$$

and

$$\bar{n} = \left(\frac{\rho V}{2 k_t V}\right)^{0.5} . \tag{17}$$

Regarding the relationships between the three cases discussed, it is evidently possible, under suitable conditions, to go from one case to the next by increasing the rate of radical formation and decreasing the rate of radical deactivation.

If the condition for case 2

$$k_0(a_p/v) \ll \rho/N < k_t/v \tag{18}$$

is not fulfilled, then an increase in the rate of formation of radicals will cause a transition from case 1 to case 3 without realizing case 2.

Smith and Ewart restricted themselves to treating certain cases of emulsion polymerization and did not present general results which could be obtained from their model.

Stockmayer [6] and O'Toole [7] have given a general solution to the Smith–Ewart model involving Bessel functions.

They give the following expressions for the average number of radicals per particle:

if $m \leq 1$, $\bar{n} = \frac{a}{4} \frac{I_0(a)}{I_1(a)}$ (19)

and

if $m \geq 1$, $\bar{n} = -\frac{m-1}{2} + \frac{a}{4} \frac{I_0(a)}{I_1(a)}$, (20)

where

$$a = [(8\rho v)/(N k_t)]^{0.5}$$ (21)

and I_0 and I_1 are the Bessel functions.

Stockmayer [6] prepared curves of $\bar{n}$ and a for various values of m involving the possibility of escape of the radicals from particles which achieve their asymptotic values at $a = 4$ corresponding to values of $\bar{n}$ close to 1. This means that when $\bar{n} > 1$, it is also possible to use the formula

$$\bar{n} = \frac{a}{4} + \frac{1-2m}{8}.$$ (22)

If the rate of escape of radicals is very low then $m = 0$.

The Smith–Ewart theory was later treated by Parts et al. [8], who suggested an expression for the number of particles at any stage of polymerization. Their theory shows that the rate of polymerization passes through a maximum at the point where the emulsifier micelles disappear.

Napper & Parts [9] have treated the Smith–Ewart and Stockmayer theories to obtain the instantaneous rate of polymerization at any stage of reaction.

Ugelstad et al. [10] extended the micellar model in the absorption-desorption process of radicals. They assume that the desorbed radicals contribute to the stationary concentration of radicals in the aqueous phase.

3.1.3 Gardon's theory

Gardon [11–16] modified the micellar model and thus extended its applicability to a larger number of monomers, comonomer pairs and reaction conditions by introducing new parameters and by mathematical specification of the existing equations.

Gardon divided the process of emulsion polymerization into three states. Stage I covers the decomposition of initiator, the start of the propagation of the polymer chain in the micelles, and the nucleation of micelles, and ends when a constant number of polymer particles is reached. This event takes place at 10–20% conversion.

The stationary state, Stage II, attributed by Gardon as interval II, covers conversions from 20 to 60 and/or 70% conversion. The particle number and the mass ratio monomer/polymer in the polymer particles are constant in Stage II.

Stage III is nonstationary and covers high conversions (above 70%).

In the initial stage of polymerization (Stage I) primary radicals formed by decomposition of a water-soluble initiator enter the micelles, initiate the growth of the polymer chain and the nucleation of micelles. As reaction proceeds, the number of polymer particles increases and the concentration of free emulsifier decreases. At the end of Stage I, free emulsifier is consumed and the particle number becomes stabilized.

Sec. 3.1] Classical and more recent theories

In Stage II (if coalescence of the particles does not take place), the particle number is constant and only an increase in particle size is observed. In Stages I and II the reaction system consists of three phases: water, which is the polymerization medium, emulsified monomer droplets, and monomer/polymer particles.

In Stage III the reaction mixture has only two phases: the aqueous phase and the monomer/polymer particles. The particle number is constant just as in Stage II (if coalescence of the particles does not occur).

Gardon based his calculations on the following assumptions:

(i) The rate of decomposition of initiator is very low as compared with the rate of polymerization in the polymer particles. Hence the initiator concentration and the rate of formation of initiating radicals do not vary as the polymerization proceeds.
(ii) The absorption of radicals by the micelles and polymer particles is an irreversible process and the rate of desorption of radicals from polymer particles is zero or negligible.
(iii) The emulsifier critical micellar concentration (CMC) is very low with respect to the total emulsifier concentration and therefore is not taken into account in kinetic considerations.
(iv) Emulsified monomer droplets act as a reservoir of monomer and maintain the aqueous phase during polymerization with the monomer at saturation.
(v) The monomer/polymer mass ratio in polymer particles and the equilibrium monomer concentration are constant at the site of the reaction centre and do not depend either on the conversion or on the size of the particles (Stages I and II).
(vi) Initiation, propagation, and termination proceed in the micelles and in the polymer particles via a radical mechanism.

Gardon has reported that at the beginning of polymerization about 10^{17} micelles are normally present in 1 cm^3 of water; when the polymerization is complete, the same volume contains 10^{12} to 10^{14} polymer particles on average. It follows that only one of every 100 or 1000 micelles absorbs a radical and becomes a polymer particle. Other micelles serve as reservoirs of the emulsifier and, as polymerization proceeds, their concentration decreases to zero (micelle nucleation is complete).

The monomer is in water, in micelles of diameter between 2 and 5 nm, in emulsified drops of diameter from 10^3 to 10^4 nm (10^7 to 10^9 monomer droplets per cm^{-3} water) and in monomer/polymer particles with diameter 50–300 nm. The emulsified monomer droplets have perhaps 100 times larger a diameter than that of the monomer/polymer particles and as much as 1000 times greater than the diameter of the micelles. We know that the rate of absorption of radicals into spherical particles is proportional to their surface area, and most radicals are therefore absorbed by micelles and polymer particles. The amount of radicals absorbed into monomer droplets is very small so that the micellar model neglects their contribution to the rate of polymerization.

Gardon characterizes the influence of the initiator on the polymerization process by a parameter R, which is equal to ρ in the original Smith–Ewart model. R is the number of radicals produced in 1 cm^3 of water per unit time:

$$R = 2 N_A k_d [I], \qquad (23)$$

where k_d is the decomposition rate constant of the initiator I, [I] is its molar concentration, N_A is Avogadro's number, and the coefficient 2 expresses the formation of two radical fragments from one initiator molecule. The parameter R is constant for a particular concentration of initiator and temperature.

In the presence of a micellar emulsifier, the total surface area of the micelles and polymer particles does not change as polymerization proceeds and is assumed to be a constant value although the character of the micelles and of the monomer/polymer particles is different. This surface area is expressed quantitatively by:

$$S = N_A \, a_s \{[S] - \text{CMC}\}, \qquad (24)$$

where a_s is the area occupied by an emulsifier molecule, [S] is the emulsifier concentration, and CMC is the critical micellar concentration of emulsifier.

Another parameter used by Gardon is the rate of increase in volume of a polymer particle K:

$$K = 3/4 \, \pi \, (d_m/d_p) \, (k_p/N_A) \, \Phi_m/(1-\Phi_m), \qquad (25)$$

where d_m and d_p are the densities of monomer and polymer, respectively, k_p is the propagation rate constant, and Φ_m is the volume fraction of monomer in the monomer/polymer particle (the degree of swelling of a polymer particle by monomer). Table 1 gives these values for monomers currently used in emulsion polymerization.

Table 1 – Monomer parameters

Monomer	$d_{m, 30°C}$ [a] / g cm^{-3}	$d_{p, 20-25°C}$ [b] /g cm^{-3}	Φ_m	Ref.	$k_{p, 30-50°C}$ /10^5 cm^3 mol^{-1} s^{-1}	Ref.
MA	0.936	1.22	0.85	[12]	7.2	[23]
EA	0.915	1.12	—	—	9.0	[24]
n-BA	0.889	1.09	0.68	[17]	21.0; 4.5	[23, 25]
MMA	0.929	1.17	0.73	[12]	1.4	[23]
n-BMA	0.889	1.06	0.60	[18]	3.7; 6.0	[23, 25]
St	0.897	1.05	0.61	[19]	1.1	[23]
VAc	0.918	1.18	0.85	[12]	10.0	[23]

[a] Ref. [20, 21].
[b] Ref. [22].

MA — methyl acrylate, EA — ethyl acrylate, n-BA — n-butyl acrylate, MMA — methyl methacrylate, n-BMA — n-butyl methacrylate, St — styrene, VAc — vinyl acetate. d_m — monomer density, d_p — polymer density, k_p — propagation rate constant, Φ_m — volume fraction of monomer in monomer/polymer particles.

The value of the volume fraction of monomer is determined by the swelling of the final latex by monomer or by studying the kinetics of polymerization [12,13,17].

The time dependence of conversion at the beginning of polymerization (Stage I) can be written as

$$P = 0.351 \, (k_p/N_A) \, (d_m/d_p) \, \Phi_m \, R t^2 \, . \tag{26}$$

Thus the symbol P expresses the amount of converted monomer in terms of the volume of polymer per unit volume of water. It follows that for a very short initial period, while the polymer particles become nucleated, the conversion is proportional to the initiator concentration and to t^2 and is independent of the emulsifier concentration. Particle nucleation stops at a time t_{cr} when the conversion corresponds to P_{cr}

$$t_{cr} = 0.365 \, (S/R)^{0.6}/K^{0.4} \, , \tag{27}$$

$$P_{cr} = 0.209 \, S^{1.2} \, (K/R)^{0.2} \, (1 - \Phi_m) \, . \tag{28}$$

The critical conversion, P_{cr}, is reached at a time when the total surface of the polymer particles is equal to the value of the parameter S. The value of P_{cr} is obtainable from the dependence of the rate of polymerization on conversion or from the number of particles. For instance, for the emulsion polymerization of vinyl acetate, a value of P_{cr} ca. 0.08 was obtained at low concentrations and while at higher concentrations, P_{cr}, was ca. 0.25 [26].

The Smith–Ewart Stage I gives:

$$N = 0.208 \, S^{0.6} \, (R/K)^{0.4} \, . \tag{29}$$

The total number of radicals formed in Stage I can be obtained from the following equation:

$$R \, t_{cr} = 0.365 \, S^{0.6} \, (R/K)^{0.4} \, . \tag{30}$$

The initial molecular mass is calculated from the average volume of a particle with one terminated chain in it:

$$M_{initial} = 1.44 \, k_p \, \Phi_m \, d_m \, N/R \, . \tag{31}$$

It was stated that the initial molecular mass should not differ significantly from the molecular mass prevailing in Stage II. During Stage II the conversion proceeds at an approximately constant monomer concentration in the polymer particles and the particle volumes increase with time. In the range of conversions $P_{cr} \leq P \leq P_{2-3}$ the rate of polymerization is predicted to have the following constant value:

$$R_p = dP/dt = \bar{n} \, (k_p/N_A)(d_m/d_p) N \, \Phi_m \, . \tag{32}$$

This rate is called the Smith–Ewart rate and the symbol B is assigned to it. Here $\bar{n}$ is the average number of radicals per particle, N is the final, time-independent value corresponding to the swelling equilibrium. With the restriction $P > P_{cr}$, N is constant. This implies, following the assumptions relating to Stage I, that all radicals originating in the aqueous phase are absorbed into the monomer-swollen polymer particles. If they were not, new particles would be nucleated. As in Stage I, Φ_m is assumed to stay constant. Morton et al. [27] have reported that only a limited amount of monomer can enter a polymer particle from the monomer-saturated aqueous phase. This is equally true for a polymerizing or for a 'dead' polymer particle. Besides, Φ_m is a relatively insensitive function of particle size.

Equation (32) can be written in terms of the parameters K, S and R.

$$B = 0.185 \, (k_p \, \Phi_m \, S \, d_m/d_p \, N_A)^{0.6} \, [R \, (1 - \Phi_m)]^{0.4} \tag{33}$$

or

$$B = 0.435 \, (1 - \Phi_m) \, (KS)^{0.6} \, R^{0.4}. \tag{34}$$

This equation can be expressed in terms of [M], the molar concentration of monomers in the particles at swelling equlibrium: thus B, the number of moles of monomer converted per cm^3 of water per second, is given by:

$$B = \bar{n} \, (k_p/N_A) \, N \, [M]. \tag{35}$$

The time-dependence of conversion is described by

$$P = At^2 + Bt + C. \tag{36}$$

Here B is the Smith–Ewart rate and parameter A is defined in Eq. (37)

$$A = 0.102 \, (k_p^{1.94}/k_t^{0.94}) \, (d_m/d_p \, N_A) \, [\Phi_m^{1.94}/(1 - \Phi_m^{0.94})] \, R. \tag{37}$$

For most cases C is of the order of 10^{-2} cm^3 polymer/cm^3 water. Thus the following equation represents the theory with adequate accuracy:

$$P = At^2 + Bt. \tag{38}$$

The term At^2 is analogous to homogeneous polymerization while the term Bt comes from the micellar model of emulsion polymerization.

In systems with small polymer particles, the term Bt plays a dominant role. The conversion curve in Stage II is linear and the rate of polymerization is high. As the size of the polymer particles increases, the term At^2 becomes comparable with term Bt; this causes curvature in the time-dependence of conversion in Stage II and a retardation of polymerization. The experimental results indicate that the linear regions of the conversion curves occur at low initiation rates with the formation of latices with small particles. The variation of the shape of the conversion curve with the polymer particle size and the rate of initiation was supported by results obtained in the emulsion polymerizations of methyl methacrylate [28], vinyl acetate [29] and vinyl chloride [24].

In Stage II of emulsion polymerization Φ_m is constant, hence the viscosity in the particle is independent of conversion. It follows that throughout Stage II k_t is a conversion-independent constant. Because its value is not the same as in other polymerization processes, k_t is not an independent parameter, it can be determined by analysing emulsion polymerization data.

It is convenient to express termination rates in terms of the average number of radicals per particle, $\bar{n}$

$$\bar{n} = V_s N_A [R^\bullet]. \tag{39}$$

Here V_s is the volume of the monomer-swollen particle and $[R^\bullet]$ is the concentration of radicals.

The equation for emulsion polymerization can be written in the following way:

$$d\bar{n}/dt = R/N - (k_t/V N_A) \bar{n} (\bar{n} - 1), \tag{40}$$

where R/N is the average rate of entry of radicals into a particle. The termination rate, given in the second term above, decreases as the particle volume increases.

The parameters B, N, Φ_m and k_p combined with values of A and R define k_t through the following equation:

$$k_t/k_p = 0.158 (RB/AN)^{1.062} \Phi_m/(1 - \Phi_m). \tag{41}$$

Gardon reported that the termination rate constant for emulsion systems differs considerably from the value of k_t determined for homogeneous polymerizations. The internal viscosity of the polymer particles reduces the translational motion of polymer radicals and also, in parallel, the rate of termination. The values obtained for the ratio k_t/k_p through an analysis of the experimental data of emulsion polymerization varied between 10^2 and 10^4.

The value of the average number of radicals per particle can be calculated from the experimental variable P and the parameters A and B using the following equation:

$$\bar{n} = 0.5 [1 + (4A/B^2)P]^{0.5}. \tag{42}$$

This equation defines limits of the validity of the Smith–Ewart theory of Stage II: $4AP/B^2 \ll 1$. Since A is proportional to R and B is proportional to N, the Smith–Ewart theory of Stage II is valid if the latex particle size is small and the initiation rate is low.

The number-average molecular mass of all terminated chains in a single polymer particle can be calculated from the following equation:

$$\overline{M}_n = (4A N_A d_p/BR) P / \left\{ [1 + (4A/B^2) P]^{0.5} - 1 \right\}. \tag{43}$$

The Smith–Ewart kinetic model 2 is based on an assumption that the radicals in polymer particles are immediately terminated on entry of another radical. Gardon modified this approach and submitted a theory of slow termination. He says that termination of the two radicals proceeds at a particular rate and that radicals in a particle coexist for some time. If the rate of termination is lower than the rate of entry of radicals into a particle, then, not only two but also more radicals may co-exist in a particle and their average number increases with increasing particle volume. Gardon also supposes that the reactivity of the radicals does not change even after staying in the particle for a longer time. He also neglects the desorption of radicals from a particle. The assumption about the co-existence of several radicals in a particle was confirmed by investigating the emulsion polymerization of styrene [30] and chloroprene [31]; the rate of polymerization was enhanced by further addition of initiator. The validity of the modified micellar model was also supported by other results obtained from emulsion polymerizations, mainly of lipophilic monomers [17, 19, 32]. By contrast, Hrabák et al. [33] showed that no rate enhancement of emulsion polymerization of chloroprene with respect to one particle was observed with increasing initiator concentration. Other papers have also appeared [34–36] which contained experimental data that could not be interpreted according to the micellar models given above.

After the monomer droplets disappear at conversion P_{2-3}, most of the unconverted monomer is present in the monomer-swollen polymer particles and a small fraction of monomer may be dissolved in water. P_{2-3} can be calculated:

$$P_{2-3} = (m/w)(1 - \Phi_m)(dw/dp)/[(1 - \Phi_m + (dm/dp)\Phi_m]. \tag{44}$$

In Stage III, conversion proceeds at the expense of the monomer concentration in the polymer particles and the monomer-swollen particle volume may actually shrink due to contraction on polymerization. As the monomer concentration in the particles decreases with increasing conversion in Stage III, the internal viscosity of the particles increases and the value of k_t decreases as predicted by Trommsdorff et al. [37].

Thus the polymerization process seems to be controlled by two opposing processes. With increasing conversion the molecular mass and conversion rate should decrease because the propagation rate is proportional to the (decreasing) monomer concentration in the polymer particles. On the contrary, with increasing conversion, the termination rate decreases and this causes an increase in the rate and in the molecular mass via the Trommsdorff gel effect. The rate of polymerization may continue to increase for a short time after the conversion P_{2-3}, if the gel effect is more important than the effect of the decreasing rate of propagation. At higher conversion the rate of polymerization and the molecular mass decrease with increasing conversion because here the slow propagation can be expected to predominate.

3.1.4 Medvedev's theory

The micellar models (Smith, Ewart, Gardon) which were sufficient to describe the course of emulsion polymerizations of lipophilic monomers are inadequate for monomers with increasing hydrophilic character. As the hydrophilicity of the monomers is increased, remarkable deviations from the models were observed [34–36]. Deviations were also observable in systems where the polymer formed is insoluble in its own monomer.

One of the models presenting a new approach to the interpretation of experimental results is Medvedev's model [38–41]. Its author has proposed that the *surface* of the polymer particles is the site of propagation of a polymer chain. Initiation, propagation, and termination all proceed at the micelle (polymer particles)/water interface. The emulsifier layer adsorbed on the surface of the polymer particles acts as a stabilizer of the latex and, at the same time, it is also the site of initiation and propagation of the polymer chain. The influence of changes in pH and the addition of inorganic emulsifiers on the polymerization supports the idea of polymerization at the interface.

It is assumed that, after adsorption of an initiator molecule on the surface of the micelles or monomer/polymer particles, the mechanism of initiation starts to be affected. Interactions between the molecules of emulsifier and initiator lead to an increase in the concentration of primary radicals.

In the presence of an oil-soluble initiator, the overall polymerization rate, R_p, is proportional to the concentration of emulsifier [S] to the first power:

$$R_p = k_1 [M][S][I_1]^{0.5} \; ; \; \overline{P}_n = k'_1 [M][I_1]^{0.5}, \qquad (45)$$

where k_1 and k'_1 are constants, [M] is the monomer concentration in the effective volume of latex particles, [S] is the emulsifier concentration in the aqueous phase, [I$_1$] is the concentration of initiator in the interphase zone, and $\overline{P}_n$ is the average degree of polymerization. In this case the oil-soluble initiator enters the polymer particles together with the monomer.

In the emulsion polymerizations initiated by a water-soluble initiator, the rate of polymerization is proportional to the 0.5 order with respect to the emulsifier concentration

$$R_p = k_2 [M][S]^{0.5} [I_2]^{0.5} \; ; \; P_n = k'_2 [M][S]^{0.5} [I_2]^{-0.5}, \qquad (46)$$

where k_2 and k'_2 are constants and [I$_2$] is the concentration of initiator in the aqueous phase. In this case the radicals initiating the polymerization are generated in water.

Medvedev found a number of discrepancies with respect to the micellar theory which led to the following conclusions:

(i) The rate of polymerization does not depend on the transfer of emulsifier from micelles to the surface of the monomer-swollen polymer particles.
(ii) Throughout the polymerization the total surface area of the latex particles remains constant.
(iii) The nature of the surface layers influences the stability of the latex particles.
(iv) The adsorbed emulsifier layers influence the rate of the initiation steps.

The rate of polymerization is, according to Eqs (45) and (46), proportional to the 0.5 power of the concentration of both initiator and emulsifier and is directly proportional to the monomer concentration. The reaction order with respect to initiator follows from the surmise about the bimolecular termination of the macroradicals at the interphase and about the greater mobility of these radicals on the surface of the polymer particles.

Medvedev predicts that nucleation proceeds on the surface of emulsifier micelles which are swollen with monomer or are on the surface of the emulsified polymer particles. The propagation of a polymer chain on the surfaces of micelles or polymer particles is initiated by the entry of a radical from the aqueous phase or by a radical formed through the decomposition of initiator adsorbed on the surface of the micelles and monomer/polymer particles.

The adsorbed emulsifier layers thus determine not only the stability of the latex particles but also the rate of radical formation and the site of polymerization. Polymerization proceeds in the narrow zone within the monomer-swollen polymer particles adjoining the interphase. The rate of polymerization is thus determined by the surface area of the latex particles and by the concentration and nature of the emulsifier. It was proposed that the depth of the reaction zone is a function of the diffusion properties of the macromolecules, the radicals and the reaction components. Initiating radicals are formed either in the surface area of the particles or emerge from the aqueous phase into the reaction zone.

Medvedev's model of emulsion polymerization proposes that the growing latex particles consist of a polymer-enriched nucleus surrounded by a monomer-saturated shell, which is the main reaction site of polymerization in the polymer particles. Medvedev considers the polymer particles to be heterogeneous even under equilibrium conditions of monomer swelling. Thus, the diffusion of monomer is not controlled by the polymerization and each chain grows in a zone enriched with monomer. This process is considered as an encapsulation of the polymer-enriched core by a monomer-enriched shell.

This model of interfacial polymerization describes rather well emulsion polymerizations proceeding under extreme conditions, e.g. at high concentrations of initiator and emulsifier, where micellar models usually fail.

3.1.5 Theory of homogeneous nucleation

Roe [42] introduced the first ideas about the theory of homogeneous nucleation. He found during the emulsion polymerization of partially water-soluble vinyl monomers that emulsifier molecules were unnecessary in some systems in view of the initiation of polymerization. The emulsifier only acted as a stabilizer of the polymer particles. Roe assumed that new particles are formed via interaction between oligomeric radicals and emulsifier molecules (or micelles) and that the process takes place until the complete consumption of free emulsifier. The association of the oligomeric radicals and their swelling give primary particles stabilized by the emulsifier present in the system. Unstabilized oligomeric radicals are absorbed into polymer particles, where they initiate or terminate chain propagation. As for the kinetics of emulsion polymerization, Roe accepts the mathematical model 2 developed by Smith and Ewart.

Fitch et al. [43, 44] worked out the first principles of the theory of homogeneous nucleation, which is applicable to emulsion polymerization of hydrophilic monomers. One of the assumptions of this theory is that the initiation of polymerization starts in the aqueous phase through the decomposition of initiator into primary radicals followed by the addition of monomer molecules dissolved in water. The growth of the oligomeric radical proceeds in the aqueous phase up to a particular critical chain length. On exceeding this limited length, the macroradicals precipitate from solution and form a new phase — a water-insoluble polymer in the form of spherical particles (primary particles).

In the initial stages of polymerization (% conv. ~ 0), the rate of particle formation is equal to the rate of radical generation:

$$dN/dt = R_e = fR_d, \qquad (47)$$

where R_e is the effective rate of radical formation, f is the efficiency of initiation, and R_d is the rate of decomposition of initiator in water. If the system contains monomer/polymer particles, Eq. (47) becomes more complicated:

$$dN/dt = R_e - R_c - R_{fl}, \qquad (48)$$

where R_c is the rate of absorption of oligomeric radicals into the polymer particles and R_{fl} is the rate of flocculation of polymer particles.

The increase in number of polymer particles (conversion of the monomer present and the flocculation of particles) is accompanied by reduction in the active surface of these particles and increase in their stability through the migration of hydrophilic groups to the surface. Stable particles arise once having reached the critical value of the potential of the particle surface. If flocculation of the particles does not take place, the equation for the rate of particle formation is simplified:

$$dN/dt = R_e - R_c . \qquad (49)$$

As can be inferred from Eqs (47)–(49), primary particles are formed from the time of attaining equilibrium between the rate of radical formation in the aqueous phase and the rate of their absorption by polymer particles

$$R_e - R_c = 0 . \qquad (50)$$

A relation was derived for the rate of adsorption of oligomeric radicals, R_c, by means of collision theory

$$R_c = \pi r^2 L R_e N , \qquad (51)$$

where r is the radius of the monomer/polymer particle and L is the average distance travelled by the radical from its time of formation to precipitation from solution.

A relation was proposed for the overall number of particles:

$$N = \int_x^{t_{max}} R_e \left[1 - N \pi^{1/3} \left(\frac{3V}{4} \right)^{2/3} L \right] dt , \qquad (52)$$

where V represents the instantaneous volume of the monomer/polymer particle and t_{max} is the time necessary to reach the point where dN/dt is zero.

The model of homogeneous nucleation was formulated on the basis of considerations about the reactions of radicals which may proceed in the aqueous phase. If there is no emulsifier in the system, radicals produced by the decomposition of initiator may:

(i) add to a monomer dissolved in water,
(ii) be absorbed into polymer particles,
(iii) terminate with other radicals in the aqueous phase or
(iv) precipitate from solution as a primary particle if the radical has reached the critical value of the chain length.

The low concentration of oligomeric radicals dissolved in water is the reason for their low tendency to coagulate. Bimolecular termination (combination) leads to the formation of an inactive 'dead' molecule, its size being the sum of the sizes of the mutually-terminated radicals. With increasing length of the polymer chain, the probability of precipitation of the polymer from solution increases. The overall balance of the formation of primary particles should involve the nature of the oligomeric radical (i.e. whether it contains ionic groups) since the termination of ionic group-containing radicals leads to an increase in the critical chain length. Oligomeric radicals precipitating from solution are stabilized by interactions with emulsifier molecules, thus enhancing the stability of the primary polymer particles.

Hansen & Ugelstad [45–47] derived kinetic equations of homogeneous nucleation for the dependence of particle formation on the character of the reaction components and reaction conditions. Radical growth, radical adsorption, and the formation of primary particles can be described in the initial steps of a polymerization without 'seed particles' as

$$dR_i/dt = \rho_i - k_{pi} M_w R_i - k_{twi} R_i R_{tot} , \qquad (53)$$

$$dR_1/dt = k_p M_w R_i - k_p M_w R_1 - k_{c1} N R_1 - k_{twi} R_1 R_i - k_{tw} R_1 R_{tot} , \qquad (54)$$

$$dR_j/dt = k_p M_w R_{j-1} - k_p M_w R_j - k_{cj} N R_j - k_{twi} R_j R_i - k_{tw} R_j R_{tot} , \qquad (55)$$

$$dN_1/dt = k_p M_w R_{jcr-1} . \qquad (56)$$

Here ρ_i ist the rate of radical production from initiator, R_i is the number of radicals formed via decomposition of an initiator, R_{tot} is the total number of radicals propagating in the aqueous phase and is equal to the sum $R_1 + R_2 + ... + R_{jcr-1}$, M_w is the number of moles in the aqueous phase, k_{twi} is the constant for termination of the initiator radical and propagating radical, k_{tw} is the termination constant for the reaction of two propagating radicals, k_{cj} is the constant for radical capture with chain length between 1 and j in a particle, k_p is the propagation rate constant for a radical with the chain length 1 to j, and N_1 is the number of precipitated primary particles. Equation (56) expresses the rate of formation of primary particles which arise through precipitation from a solution of radicals with chain length $j$cr.

For the total number of particles we have in general

$$dN/dt = dN_1/dt - R_{fl} , \qquad (57)$$

where R_{fl} stands for the rate of self-coagulation of primary particles. It is known that primary particles are unstable in an emulsifier-free system and coagulate rapidly with themselves.

Note that in any case the amount of emulsifier is less than the critical micelle concentration, so that nucleation is solely homogeneous.

Addition of Eqs (53)–(55) up to $j = j$cr – 1 gives

$$dR_{tot}/dt = \rho_i - k_p M_w R_{jcr-1} - 2k_{twi} R_i R_{tot} - k_{tw} R_{tot}^2 - N \sum_{j=1}^{jcr-1} k_{cj} R_j . \qquad (58)$$

If a steady state is assumed for all radicals R_j up to R_{jcr-1}, including R_i radicals, and the approximation that the termination with initiator radicals can be neglected, then we obtain

$$dN_1/dt = \rho_i/[(1 + k_cN/k_pM_w) + (k_{tw}R_{tot}/k_pM_w)]^{jcr-1}. \tag{59}$$

Noting that

$$R_{tot} = (\rho_i/k_{tw})^{1/2} \tag{60}$$

and solving Eqs (60) and (59), gives an expression which is easily integrated to give:

$$N_1(t) = (1/k_1)([k_1\rho_i\, t + (k_2 + 1)_{jcr}] - k_2 - 1), \tag{61}$$

where $k_1 = k_c/k_pM_w$ and $k_2 = (k_{tw}\rho_i)^{0.5}/k_pM_w$.

Equation (63) shows that the maximum rate of formation of primary particles is reached in Stage I. As polymerization proceeds, primary particles are formed continuously but more and more slowly.

With the approximation that the absorption of all oligomers is neglected except those of chain length $jcr - 1$ and by including terms for the loss of primary particles by coagulation with themselves and with 'seed' or 'premature' particles, the expression for the rate of nucleation of particles in emulsifier-free systems becomes

$$\frac{dN}{dt} = \rho\frac{k_pM_w}{k_pM_w + k_cN + k_{cs}N_s} - k_fN^2 - k_{fs}N_sN, \tag{62}$$

where k_f and k_{fs} are the coagulation constants for primary-primary and primary-seed particles and N_s is the number of seed particles.

When $N_s = 0$, the expression for N when $k_cN \gg k_pM_w$ is

$$N = \frac{1}{4}\left(\rho_i\frac{3\eta k_pM_wW_{11}}{32\eta D_w\bar{k}T + F}\right). \tag{63}$$

When F and $W_{11} = 1.0$, and with the parameters used below, the approximation $k_cN \gg k_pM_w$ is only valid when $N \gg 10^{14}$.

When added seed particles dominate capture and coagulation, the expression N becomes

$$N = \rho_i\frac{3\eta k_pM_wrW_{1s}}{8\eta D_w\bar{k}TF_s(N_sr_s)^2}, \tag{64}$$

where η is the viscosity of water, W_{11} and W_{1s} are the Fuchs' stability ratios for primary-primary and primary-seed particles, and F and F_s are the absorption efficiencies of oligomer radicals of chain length $jcr - 1$.

The high concentration of seed particles should therefore be inversely proportional to $(N_sr_s)^2$.

The total rate of particle nucleation in the systems with emulsifier is

$$\frac{dN}{dt} = \frac{dN_{pm}}{dt} + \frac{dN_h}{dt}, \tag{65}$$

where dN_{pm}/dt is the rate of micellar nucleation and dN_h/dt is the rate of homogeneous nucleation.

The rate of micellar nucleation is described by

$$\frac{dN_{pm}}{dt} = \rho_i P_m + k_{des} N_1 P'_m, \qquad (66)$$

where P_m is the probability of absorption of initiator radicals (or, more correctly, oligomeric radicals produced from initiator radicals) in micelles, P'_m is the probability of absorption of desorbed monomer radicals, N_{pm} is the total number of primary particles produced from micelles, N_1 is the number of particles with one radical, and k_{des} is the desorption constant expressed by

$$k_{des} = 4\pi D_m r / v_p, \qquad (67)$$

where D_m is the effective diffusivity of monomer radicals out of the particles and v_p is the volume per particle.

The rate of homogeneous nucleation is

$$\frac{dN_h}{dt} = \rho_i P_p + k_{des} N_1 P'_h. \qquad (68)$$

The expression for the number of particles with one radical (N_1) is

$$\frac{dN_1}{dt} = \frac{dN}{dt} + \rho_i P_{p0} + k_{des} N_1 P'_{p0} - \rho_i P_{p1} - k_{des} N_1 P'_{p1} - k_{des} N_1, \qquad (69)$$

where N is the number of particles produced from micelles and homogeneously ($N = N_{pm} + N_h$), P_{p0} and P'_{p0} are the absorption probabilities of initiator and monomer radicals by particles with zero radicals (dead particles), and P_{p1} and P'_{p1} are the same probabilities for particles with one radical. The absorption probabilities for seed particles are P_s and P'_s and the probabilities for homogeneous nucleation are P_h and P'_h.

Oligomeric (initiator) and monomer radicals are distinct, as the latter have no electric charge and may therefore have different probabilities of absorption. We have $P_m + P_{p0} + P_s + P_h = 1.0$ and $P'_m + P'_{p0} + P'_s + P'_h = 1.0$.

The rate of total growth in volume of the particle, as long as there is enough monomer present, is given by

$$\frac{dV}{dt} = \frac{k_p}{N_A} \frac{\Phi_m}{1-\Phi_m} \frac{dm}{dp} \bar{n} N + v_m \frac{dN}{dt}. \qquad (70)$$

Here v_m is the volume of one micelle. The final term accounts for the volume of the new particles nucleated from micelles and formed by the homogeneous mechanism.

The mass of new particles formed (g dm^{-3}) is calculated from V_p and Φ_m:

$$P = V_p (1 - \Phi_m) dp, \qquad (71)$$

where V_p is the total volume of primary particles.

The volume fraction of monomer, Φ_m, may be calculated from the equation

$$\Phi_m = -0.06821\left(\log\frac{r}{\gamma}\right)^2 + 0.3375\left(\log\frac{r}{\gamma}\right) + 0.505,\tag{72}$$

where γ is the interfacial tension and r is the particle radius.

The degree of conversion, ω (weight fraction of polymer), is equal to

$$\omega = \left(1 + \frac{\Phi_m}{1-\Phi_m}\frac{dm}{dp}\right)^{-1}.\tag{73}$$

Barrett [48] reported on the distribution of radicals between aqueous and polymer phases. He considered two limiting cases: the distribution of radicals between aqueous and polymer phases or between the polymer phase and the particle surface. Experimental results indicate that the latter situation is more probable since small particles block fresh nucleation more than large ones. On entry into a particle, a radical need not be deactivated by chemical reaction but may also diffuse out from the particle. The diffusion (values of diffusion constants) of radicals is not markedly influenced by monomer concentration. The role of the distribution of radicals between aqueous and polymer phases (or between water and the polymer particle surface) has been shown by Dankwerts [49] to be dominant. However, Dankwerts has not observed any differences in the distribution of radicals between the aqueous phase and small or large particles. In Stage I, the rate of radical adsorption by particles was proportional to the product Nr^2 or Nr. The conclusion of the theory about radical adsorption by particles (using emulsifier at a concentration above the CMC) is that particle nucleation proceeds in micelles and the maximum rate of adsorption is reached at the end of Stage I. In the initial step of polymerization (Stage I), the average number of radicals in a particle is equal to 1 and decreases as reaction proceeds; in Stage II, its value is 0.5. A higher rate of adsorption is observed in particles which contain a free radical.

Hull & Vanderhoff [50] and Alexander & Napper [51] reported that the absorption of (negatively-charged) radicals decreases because of the electrostatic repulsion between charged radicals and the surface particles. The decrease of radical adsorption is also caused by their reduced solubility in the organic phase represented by polymer particles and monomer-swollen micelles.

In systems without emulsifier or with low emulsifier concentrations (and in some cases also at medium emulsifier concentrations [43]), the formation of particles with a maximum in Stage I is observed. Primary polymer particles exhibit low stability and are subject to coagulation during polymerization. At higher conversions, these particles are trapped by large polymer particles. The mutual coagulation of primary particles gives larger particles of higher surface charge which results from the diffusion of ionic and polar groups onto the particle surface. After reaching their critical size, the particles have a sufficient concentration of dissociated groups able to stabilize them. Owing to coagulation, particles of varying size, surface charge density and stability arise during polymerization. The coagulation of the primary particles is also accompanied by the propagation of monomers in the particles which lowers the surface charge density. If an active particle associates with an inactive, then the propagation is unaffected. On the other hand, an inactive particle can be formed by the association of active particles.

In the presence of an emulsifier (acting as a particle stabilizer), the kinetics of coagulation are similar to those in emulsifier-free systems; it is, however, more strongly influenced by the surface charge density. The rate of coagulation affects the rate of adsorption of emulsifier on the primary particles which in turn is a function of the nature of the polymer. The adsorption of emulsifier on polar polymers is ineffective and slow and the limiting coagulation will therefore be made more effective by using higher emulsifier concentrations [52].

The increasing rate of formation of polymer particles and the attainment of a maximum at the end of the nucleation period have been explained by Napper *et al.* [53–55] by a two-step model of coagulative nucleation. According to these authors, unstable small primary polymer particles are formed in the first step, which mutually coagulate and form true polymer particles in the second step. Primary (precursor) particles are formed by the precipitation of individual or more mutually associated oligomeric radicals from the aqueous phase or by the entry of radicals into micelles in the presence of emulsifier. Primary particles differ from true latex particles in at least two ways:

(i) they are colloidally-unstable and thus coagulate mutually or with true latex particles and
(ii) they polymerize very slowly.

Their low stability follows from their smallness and the strong curvature of their surface. According to DLVO theory [56], the structure of the electrical double layer of these particles does not create the conditions for the formation of the larger surface charge needed to stabilize the colloidal system. The low values of the rate of polymerization follow from the reduced ability of the particles to be swollen by the monomer, which is due to the hydrophilic nature of the primary particles [55], from the small particle size [27] and from the high rate of radical desorption [57]. The growth of primary polymer particles is mainly due to coagulation and the contribution of the polymerization of the monomer to the particle growth increases with enlargement of the particle volume.

The theory of coagulative nucleation combines Müller and Smoluchowski's kinetics of coagulation [58–60] with DLVO theory [56]. The rate of particle formation is given by:

$$dN_c/dt = \sum_{i=1}^{m} v_i \left(\sum_{j=m-1+i}^{m} B_{ij} v_j \right), \qquad (74)$$

where N_c is the number of latex particles, v is the rate of coagulation, B_{ij} is Müller's rate of coagulation between i-fold and j-fold flocculated particle. The model neglects particle growth via polymerization of the monomer in a particle. This assumption is justified only in the case of small particles, very low degrees of swelling, and a very high rate of desorption.

The coagulative model states that in systems with constant initiator concentration, increase in the emulsifier concentration is accompanied by an increase in the number of particles and with a monotonic decrease of the reaction order with respect to emulsifier. The lifetime of the primary particles increases in parallel and their influence on the growth and number of latex particles becomes greater. Collision of the primary particles with the latex are more frequent. A smaller number of latex particles is formed and their limiting constant concentration is reached. The fact that the number of particles reaches a constant value at high emulsifier concentrations was experimentally confirmed elsewhere [42, 61]. The theory of coagulative nucleation differs from that of Fitch only in the rapid drop in the rate of particle nucleation following the consumption of free emulsifier.

The theory of homogeneous nucleation predicts that initiation and propagation in the aqueous phase may be described as conforming with the model of homogeneous radical polymerization; the average degree of polymerization, X_n, can therefore be written:

$$X_n = R_p/R_i = k_p [M] (1/R_i k_t)^{0.5}, \tag{75}$$

provided that $R_i = R_e$ and that the terminated radicals do not precipitate from solution. The values calculated for styrene, vinyl acetate and methyl methacrylate are 7, 1320, and 82, respectively [62]. The critical length of the oligomer chain was calculated from Eq. (75) and from the literature data for k_p and k_t reported earlier [23], from the values of R_i [63–65], and from experimental values of the water solubility of individual monomers [63]. Homopolymerization of styrene in water gives rise to a surface-active oligomer with the same characteristics as molecules of sodium dodecyl sulfate or Aerosol MA (dihexylester of sodium sulfosuccinic acid).

3.1.6 Polymerization in emulsified monomer droplets

The micellar and coagulative theories discussed above assume the active participation of emulsifier micelles and monomer/polymer particles in the process of polymerization; the monomer droplets were considered to be an inactive component, acting only as a reservoir of monomer. The activity of particles was associated with the efficiency of radical capture, which is proportional to their surface area. The monomer droplets are relatively large and unable therefore to compete with the homogeneous and micellar nucleation mechanisms. The amount of emulsifier adsorbed on the droplets is also small, so that nearly all emulsifier added is present in solution in the aqueous phase or as micelles. The number of new particles formed is much higher than the number of droplets, and as nucleation proceeds the droplets are depleted of monomer, which is transferred to the polymer particles. Large particles which are often detected on electron micrographs, originate from monomer droplets, having been initiated by the absorption of radicals from the aqueous phase.

By lowering the average size of the monomer droplets from the usual micron size to submicron, the probability of radical capture, as well as the concentration of reaction sites in emulsifier monomer particles, increase. The situation described can arise from using special emulsifier systems and higher rates of stirring during the emulsification of a monomer in the aqueous phase. Mixtures of anionic emulsifier with higher alcohols [66] or alkanes [67] exemplify such emulsifier systems; the higher alcohols or alkanes function as co-emulsifiers.

The emulsification of vinyl monomers by conventional emulsifiers leads to the formation of macroemulsions even at higher emulsifier concentrations and under constant stirring. A macroemulsion is formed as a result of the effective coalescence of monomer particles and of diffusion of the monomer from smaller droplets into larger ones. The latter process can be reduced by adding a slightly water-soluble lipophilic substance which inhibits escape of the monomer from the monomer particles into water.

The formation of a crystalline complex between the emulsifier and coemulsifier positively influences the formation of smaller monomer particles. The formation of such complexes has been confirmed by electron diffraction and electron microscopy of mixed systems of hexadecyltrimethylammonium bromide and cetyl alcohol and of lauryl sulfate and lauryl alcohol or cetyl alcohol [68–70]. Such mixed emulsifier systems encourage the formation of a finely-dispersed monomer system with a particle size close to those of micelles and polymer particles.

Ugelstad et al. [66, 69–73] have described methods by which the monomer could be emulsified in much smaller droplets than those utilized in conventional emulsion polymerizations. These methods made use of mixed emulsifier systems of ionic emulsifiers and long-chain alkanols. It was shown that with oil-soluble initiators these systems produce latices with particles similar in size to the monomer droplet emulsions, although smaller particles, depending on the emulsifier concentration in the aqueous phase, are also found. With a water-soluble initiator the number of small particles was considerably higher, although large particles similar in size to the monomer droplets were also formed, which indicated initiation to occur in the droplets.

Hansen & Ugelstad [73] have shown that the number of new particles formed depends not only on the efficiency of homogeneous nucleation, but also on the amount of emulsifier present to stabilize the new particles. During normal emulsion polymerizations below the CMC, the probability of homogeneous nucleation (P_h) is rapidly reduced from 1 to of about 10^{-3} or lower. If the probability of homogeneous nucleation (P_h) in the presence of droplets is of the same order of magnitude as this low value, the reduction in P_h during most of the nucleation time will be relatively low and the number of new particles will not be much lower than it would have been in an ordinary emulsion polymerization with the same amount of emulsifier. It must also be remembered that the emulsifier adsorbed on the droplets may be desorbed and readsorbed on the new particles; the amount of emulsifier available for the stabilization of new particles will therefore be greater than the free amount.

As the droplet number is increased, P_h will decrease and so will the number of new particles. The conditions for nucleation in the aqueous phase will become less favourable. The simultaneous increase in the droplet number and decrease in P_h will therefore favour droplet initiation, as also will the fact that the emulsifier adsorption increases with increasing droplet number. If the free emulsifier concentration is higher than the CMC, it would be expected that the number of new particles would be still less affected by the monomer droplets because the competition between micellar and droplet initiation would be even more in favour of micellar initiation.

In droplet-initiated emulsion polymerization, the particle growth kinetics are different from normal emulsion polymerization. In the emulsion polymerization of styrene, Ugelstad et al. [71] have found that the total polymerization rate decreases as initiation in the aqueous phase decreases because the total number of droplets plus particles decreases. The lowest rate was achieved when very little new nucleation took place and the rate conformed to the Smith–Ewart case 3 kinetics with $\bar{n} \gg 1$. In conventional emulsion polymerization in which $\bar{n}$ is constant, the rate is essentially constant in Stage II because the monomer concentration in the particles is constant. In droplet-initiated polymerization, however, the monomer concentration is already decreasing at the start of reaction. The rate therefore decreases continuously when $\bar{n}$ is constant. Another interesting feature of droplet polymerization is the marked change in particle size distribution, which is nearly monodisperse.

In references [74–77] are recorded the formation of monomer particles with diameters ca. 200 nm using a mixed emulsifier system at concentrations of 1–3 mass % per monomer. The results of the emulsion polymerizations of vinyl monomers using such mixed emulsifier systems may be taken as evidence of the high activity of monomer particles in nucleation. The emulsified monomer droplets of this size compete with micelles for the adsorption of radicals and in the formation of the reaction centres of propagation of the polymer chain.

A contribution from the polymerization of styrene monomer droplets to the overall reaction rate has been demonstrated by Chamberlain et al. [78]. They used lauryl sulfate and lauryl alcohol as an emulsifying system, and nucleated monomer particles of styrene with a diameter of 200 nm were reported to be the main source of nucleation of the polymer particles.

Choi et al.[79] state that, under particular conditions, the monomer particles do not act as a reservoir for monomer but also take part in the polymerization process. The high nucleation ability of styrene monomer droplets emulsified by sodium lauryl sulfate and cetyl alcohol was achieved by using peroxodisulfate or 2,2'-azobis(2-methyl)butyronitrile as initiators. Analysis of the kinetic data of emulsion polymerization with emulsified monomer particles taking an active part in the polymerization has shown:

(i) The duration of particle nucleation is rather long. This is usually explained by the reduced adsorption of radicals initiated by monomer particles. The fraction of polymer particles formed on an entry of a radical into monomer particles is determined by the initiator concentration which then strongly affects the number of particles.

(ii) Emulsion polymerization with the involvement of monomer particles is not characterized by the constant rate at medium conversions, as found in conventional emulsion polymerization. After consumption of free monomer droplets, the polymerization rate decreases because of the decrease in monomer concentration in the latex particles.

In spite of unfavourable statistical factors, a small fraction of the monomer particles may be nucleated. Durbin et al.[80] confirmed the participation of a small fraction of monomer particles in nucleation. Analysis of the final latex particles has shown that this system also contains micron particles. The latex contained a large fraction of submicron polymer particles with a narrow distribution, and a small fraction of large particles of wide distribution. The large particles result from the nucleation of monomer particles and from monomer propagation, and do not greatly influence the properties of the resulting latex.

References

1. Harkins, W. D. (1945) *J. Chem. Phys.* **13** 381
2. Harkins, W. D. (1946) *J. Chem. Phys.* **14** 47
3. Harkins, W. D. (1947) *J. Am. Chem. Soc.* **69** 1428
4. Smith, W. V. & Ewart, R. H. (1948) *J. Chem. Phys.* **16** 592
5. Smith, W. V. & Ewart, R. H. (1948) *J. Am. Chem. Soc.* **70** 3695
6. Stockmayer, W. H. (1957) *J. Polym. Sci.* **24** 314
7. O'Toole, J. T. (1965) *J. Appl. Polym. Sci.* **9** 1291
8. Parts, A. G., Moore, D. E. & Watterson, J. G. (1965) *Makromol. Chem.* **89** 156
9. Napper, D. H. & Parts, A. G. (1962) *J. Polym. Sci.* **61** 113
10. Ugelstad, J., Mork, P. C. & Aasen, J. O. (1967) *J. Polym. Sci., A-1* **5** 2281
11. Gardon, J. L. (1968) *J. Polym. Sci., A-1* **6** 623
12. Gardon, J. L. (1968) *J. Polym. Sci., A-1* **6** 643
13. Gardon, J. L. (1968) *J. Polym. Sci., A-1* **6** 665
14. Gardon, J. L. (1968) *J. Polym. Sci., A-1* **6** 687
15. Gardon, J. L. (1968) *J. Polym. Sci., A-1* **6** 2853
16. Gardon, J. L. (1968) *J. Polym. Sci., A-1* **6** 2859
17. Capek, I., Bartoň, J. & Orolínová, E. (1984) *Chem. Zvesti* **38** 803
18. Capek, I., Bartoň, J. & Kárpátyová, A. (1987) *Makromol. Chem.* **188** 703
19. Goldwasser, J. M. & Rudin, A. (1982) *J. Polym. Sci., Polym. Chem. Ed.* **20** 1993
20. Zonis, S. A. (ed.) (1963) *Spravochnik khimika II*. Gos. Nauchno-tekhn. Izd. Khim. Lit., Leningrad
21. Gladyshev, B. B. & Gibov, K. M. (1968) *Polymerization at high conversion and methods of its investigation.* Nauka, Alma-Ata (in Russian)
22. Lewis, O. G. (1968) *Physical constants of linear homopolymers.* Springer-Verlag, Berlin, Heidelberg
23. Brandrup, J. & Immergut, E. H. (1975) *Polymer handbook.* 2nd ed. J. Wiley–Interscience, New York, II-45
24. Soh, S. K. & Sundberg, D. C. (1982) *J. Polym. Sci., Polym. Chem. Ed.* **20** 1345
25. Maxwell, I. A., Napper, D. H. & Gilbert, R. G. (1987) *J. Chem. Soc., Faraday Trans. 1* **83** 1449
26. French, D. M. (1958) *J. Polym. Sci.* **32** 396
27. Morton, M., Kaiserman, S. & Altier, M. W. (1954) *J. Colloid Sci.* **9** 300

28. Zimmt, W. S. (1963) *J. Appl. Polym. Sci., A-1* **1** 877
29. Napper, D. H. & Parts, A. G. (1962) *J. Polym. Sci.* **61** 113
30. Gerrens, H. & Kohnlein, E. (1960) *Z. Elektrochem.* **64** 1199
31. Morton, M., Cala, J. A. & Altier, M. W. (1956) *J. Polym. Sci.* **19** 547
32. Sathpathy, U. S., Paul, T. K., Banerjee, M. & Konar, R. S. (1981) *J. Macromol. Sci., Chem.* **15** 1495
33. Hrabák, F., Bezděk, M., Hynková, V. & Peltbauer, J. (1967) *J. Polym. Sci.* **16,** 1345
34. Gerrens, H. (1963) *Ber. Bunsenges. Phys. Chem.* **67** 741
35. Pavlyuchenko, V. H. & Ivanchev, S. S. (1983) *Acta Polym.* **34** 521
36. Chatterjee, S. P., Banerjee, M. & Konar, R. S. (1978) *J. Polym. Sci., Polym. Chem. Ed.* **16** 1517
37. Trommsdorff, E., Kohle, H. & Legally, P. (1947) *Makromol. Chem.* **1** 169
38. Medvedev, S. S. (1957) In: *Proceedings of the International Symposium on Macromolecular Chemistry,* Prague. Pergamon, London, p. 174.
39. Medvedev, S. S. (1957) *Collect. Czech. Chem. Commun.* **22** 160
40. Medvedev, S. S. (1968) In: *Kinetics and mechanism of formation of macromolecules.* Nauka, Moscow, p. 5 (in Russian)
41. Zuikov, A. V., Gritsova, I. A. & Medvedev, S. S. (1968) *Vysokomol. Soedin., B* **10** 591
42. Roe, C. P. (1968) *Ind. Eng. Chem.* **60** 20
43. Fitch, R. M. & Tsai, C. H. (1971) In: Fitch, R. M. (ed.) *Polymer colloids.* Plenum Press, New York, pp. 73, 103
44. Fitch, R.M. & Kamath, Y. K. (1976) *J. Colloid Interface Sci.* **54** 6
45. Hansen, F. K. & Ugelstad, J. (1978) *J. Polym. Sci., Polym. Chem. Ed.* **16** 1953
46. Hansen, F. K. & Ugelstad, J. (1979) *J. Polym. Sci., Polym. Chem. Ed.* **17** 3033, 3047
47. Hansen, F. K. & Ugelstad, J. (1982) In: Piirma, I. (ed.) *Emulsion polymerization.* Academic Press, New York, p. 51.
48. Barrett, K. E. J. (ed.) *Dispersion polymerization in organic media.* J. Wiley & Sons, New York
49. Dankwerts, P. V. (1951) *Trans. Faraday Soc.* **47** 1014
50. Hull van den, H. J. & Vanderhoff, J. W. (1970) *Br. Polym. J.* **2** 121
51. Alexander, A. E. & Napper, D. H. (1971) *Prog. Polym. Sci.* **3** 145
52. Eliseeva, V. I. (1972) *Acta Chim.* **71** 465
53. Feeney, P. J., Napper, D. H. & Gilbert, R. G. (1987) *Macromolecules* **20** 2922
54. Lichti, G., Gilbert, R. G. & Napper, D. H. (1983) *J. Polym. Sci., Polym. Chem. Ed.* **21** 269
55. Feeney, P. J., Napper, D. H. & Gilbert, R. G. (1984) *Macromolecules* **17** 2520
56. Krumrine, P. R. & Vanderhoff, J. W. (1980) In: Fitch, R. M. (ed.) *Polymer colloids,* vol. 2. Plenum Press, New York, p. 289
57. Hawkett, B. S., Napper, D. H. & Gilbert, R. G. (1980) *J. Chem. Soc., Faraday Trans. 1* **76** 1323
58. Smoluchowski von, M. (1916) *Phys. Z.* **17** 557
59. Overbeek, J. Th. G. (1960) In: Kruyt, H. R. (ed.) *Colloid science.* Elsevier, Amsterdam
60. Müller, H. (1928) *Kolloid.-Beih.* **26** 257
61. Sutterlin, N. (1980) In: Fitch, R. H. (ed.) *Polymer colloids,* vol. 2. Plenum Press, New York, p. 583
62. Vanderhoff, J. W. (1985) *J. Polym. Sci., Polym. Symp. Ser.* **72** 161
63. Kolthoff, I. M. & Miller, I. K. (1951) *J. Am. Chem. Soc.* **73** 3055
64. Liu, L. & Krieger, I. M. (1981) *J. Polym. Sci., Polym. Chem. Ed.* **19** 3013
65. Zukoski, C. F. & Saville, D. A. (1985) *J. Colloid Interface Sci.* **104** 583
66. Ugelstad, J., El-Aasser, M. S. & Vanderhoff, J. W. (1974) *Polym. Lett.* **11** 505
67. Ugelstad, J. (1978) *Makromol. Chem.* **179** 815
68. Grimm, W. L. (1982) *M. S. Thesis.* Lehigh University
69. Ugelstad, J., El-Aasser, M.S. & Vanderhoff, J. W. (1973) *J. Polym. Sci., Polym. Lett. Ed.* **11** 503
70. Hansen, F. K., Baumann Ofstad, E. & Ugelstad, J. (1978) In: Smith, A. L. (ed.) *Theory and practice of emulsion technology.* Academic Press, New York, p. 13
71. Ugelstad, J., Hansen, F. K. & Lange, S. (1974) *Makromol. Chem.* **175** 507
72. Azad, A. R. M., Ugelstad, J., Fitch, R. M. & Hansen, F. K. (1976) *Am. Chem. Soc., Symp. Ser.* **24** 1
73. Hansen, F. K. & Ugelstad, J. (1979) *J. Polym. Sci., Polym. Chem. Ed.* **17** 3069
74. Chou, Y. J., El-Aasser, M. S. & Vanderhoff, J. W. (1980) *J. Dispersion Sci., Technol.* **1** 129
75. Chou, Y. J., El-Aasser, M. S. & Vanderhoff, J. W. (1980) *Polym. Colloids* **2** 599
76. El-Aasser, M. S., Lack, C. D., Choi, Y. T., Min, T. I., Vanderhoff, J. W. & Fowkes, F. M. (1984) *Colloids Surfaces* **12** 79
77. Grimm, W. L., Min, T. I., El-Aasser, M. S. & Vanderhoff, J. W. (1983) *J. Colloid Interface Sci.* **94** 531
78. Chamberlain, B. J., Napper, D. H. & Gilbert, R. G. (1982) *J. Chem. Soc., Faraday Trans. 1* **78** 591
79. Choi, Y. T., El-Aasser, M. S., Sudol, E. D. & Vanderhoff, J. W. (1985) *J. Polym. Sci., Polym. Chem. Ed.* **23** 2973
80. Durbin, P. D., El-Aasser, M. S., Poehlein, G. W. & Vanderhoff, J. W. (1979) *J. Appl. Polym. Sci.* **24** 703

3.2 CONTEMPORARY IDEAS ABOUT KINETIC PROCESSES AND MECHANISM OF EMULSION POLYMERIZATION

3.2.1 Some microscopic kinetic processes in emulsion polymerization

Many mechanistic aspects of emulsion polymerization have not so far been explained despite the large number of kinetic studies published. These studies have presented several alternative models for emulsion polymerization, applicable to different types of monomer.

The multistep polymerization mechanism for emulsion polymerization currently accepted emphasizes initiation, propagation, termination, chain transfer, particle nucleation, and the growth of polymer particles. Radicals are produced in the aqueous phase and the formation of polymer proceeds exclusively in monomer-swollen polymer particles. The complex character of emulsion polymerization follows from the heterogeneity of the polymerization medium, from the presence of several phases, and from variation of the kinetic parameters of polymerization with adsorption, diffusion, and addition processes. In kinetic considerations, in addition to the effect of the aqueous and organic phases on radical activity, the influence of the interphase also must be taken into account. The presence of an interphase enables consideration of the new microscopic radical processes during emulsion polymerization. Napper & Gilbert [1] introduced to these processes the entry of radicals generated in the aqueous phase into the polymer particles (characterized by the rate constant ρ), the desorption (exit) of radicals from monomer/polymer particles into the aqueous phase (described by coefficient k), the re-entry of radicals (characterized by parameter α, where $-1 \leq \alpha \leq +1$), and the bimolecular deactivation of radicals (determined by the constant c). Figure 2 is a schematic illustration of this model.

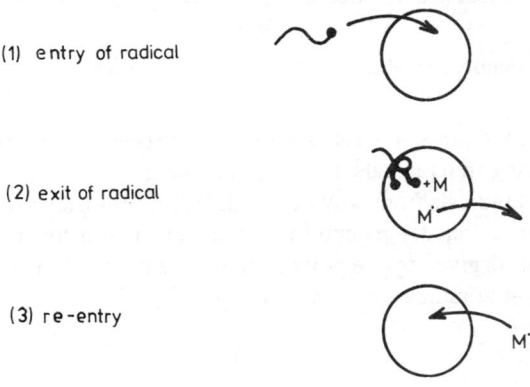

Fig. 2 — Model of entry and exit of radicals from particles.

The radicals are assumed to originate in the aqueous phase via the decomposition of initiator (I) and, before entering micelles or a monomer/polymer particle, add to several monomer molecules (M):

$$I \xrightarrow[M]{k_I} R^\bullet. \qquad (1)$$

The process depicted by Eq. (1) thus incorporates (i) decomposition of the initiator, (ii) propagation of the free radical species produced by initiator decomposition, and (iii) any mutual termination events that occur before this degree of polymerization is reached [2]. If process (i) is one of the rate-determining steps, we have $k_I \propto [I]$.

The entry of a radical into a monomer/polymer particle is characterized by the rate constant k_e:

$$R^\bullet + P \xrightarrow{k_e} \text{entry}, \qquad (2)$$

where P denotes a latex particle of molar concentration $N' = N/N_A$; the contribution of (2) to ρ, the entry rate coefficient per particle being $k_e[R^\bullet]$, N_A is Avogadro's constant (mol^{-1}), and N is the number of particles per unit volume. The propagation in the aqueous phase also enables the involvement of deactivation processes which is why the efficiency of radical adsorption by the polymer particles is less than the 100 % usually assumed [3]. The entry of radicals into particles may proceed not only via the adsorption of water-soluble radicals by the particles but also through adsorption of the primary polymer radicals formed by the precipitation of oligomeric radicals from the aqueous phase.

k_e can vary between 5×10^5 and 5×10^6 $\text{dm}^3 \text{ mol}^{-1} \text{ s}^{-1}$ which is several orders of magnitude lower than for the diffusion-controlled entry of radicals [4]. This difference follows from the presence of an energy barrier for radical entry into the particles and from the contribution of radical deactivation characterized by the process:

$$2R^\bullet \xrightarrow{k_t} \text{inactive products.} \qquad (3)$$

The rate coefficient for this process, k_t, is expected to be similar to that for the bimolecular termination of growing free radicals in an organic phase.

The desorption of radicals from polymer particles is a multistep process. The radical exit may take place after a transfer reaction to monomer or to a transfer agent. The rate of desorption of radicals is given by the product of the rate constant for desorption k_{des} and the average concentration of radicals in a particle $\bar{n}$ ($k_{des}\bar{n}$):

$$P \xrightarrow{k_{des}\bar{n}} E^\bullet, \qquad (4)$$

where $E^\bullet$ is the exiting free radical. The radical $E^\bullet$ differs chemically from the oligomeric radical $R^\bullet$, which in most cases in electrically charged. By contrast, radical $E^\bullet$ is electrically neutral and is partially deactivated by a cross-termination reaction, Eq. (5), with radicals $R^\bullet$ generated by decomposition of the initiator I (peroxodisulfate) and by propagation with monomer dissolved in water:

$$E^\bullet + R^\bullet \xrightarrow{k_{ht}} \text{inert product}. \qquad (5)$$

Reaction (5) is important because of the different character of radicals E• and R• and the high concentration of R•. Homotermination of radicals E• is neglected because of their low concentration in water.

Once a free radical has dissolved in the aqueous phase, it may move into the nonaqueous phase. Any exited free radical may, in principle, undergo subsequent re-entry into a polymer particle. Re-entry probably requires the addition of one or more monomer molecules to the exited free radicals since the latter moved out of the particles in response to a chemical potential gradient. Re-entry without addition would thus require movement against a chemical potential gradient. Note that the re-entering radicals may be chemically different from the entering radicals derived from decomposition of initiator since it will lack the associated fragments of decomposition [1].

Re-entry of E• into a latex particle, as suggested by Ugelstad & Hansen [5], can be written:

$$E^{\bullet} + P \xrightarrow{k_r} \text{entry.} \tag{6}$$

The contribution of (6) to the overall entry rate per particle is k_r [E•], and thus:

$$\rho = k_e [R^{\bullet}] + k_r [E^{\bullet}]. \tag{7}$$

The overall entry rate can be written as:

$$\rho = \rho_A + \alpha k \bar{n}, \tag{8}$$

where ρ_A is the entry rate coefficient in the absence of the exit process and α is a dimensionless 'rate parameter' expressing the relative importance of cross-termination, Eq. (5) and re-entry, Eq. (6); $\alpha = +1$ if re-entry is complete, and $\alpha = -1$ if cross-termination is complete. ρ_A is independent of k, $\bar{n}$, and x (the fractional conversion of monomer to polymer), and depends only on [I] and N.

The rate coefficient ρ_A for some unsaturated monomers may be factorized into (i) a contribution ρ_0 from any background free-radical generating process [6, 7] and (ii) a contribution ρ_I which depends directly on the initiator concentration

$$\rho_A = \rho_0 + \rho_I. \tag{9}$$

Note that it is inadmissable to assume completely efficient capture of aqueous-phase free radicals by the polymer particles. Thus, for example, it has been found for peroxodisulfate-initiated systems that $\rho_I \propto [I]^{0.6}$, suggesting that there is extensive self-termination of the free radicals formed in the aqueous phase.

If the rate coefficient k_e is defined so that the rate for Eq. (2) is $k_e[R^{\bullet}]$ (N_c/N_A), one could identify ρ_I as

$$\rho_I = k_e [R^{\bullet}]. \tag{10}$$

References [1, 6] have shown that for a number of different monomers the dependence of ρ_I on the initiator concentration exhibits exponents of less than unity. For example, for certain seeded emulsion polymerizations using peroxodisulfate anions as the initiator,

$$\rho_I / s^{-1} = 0.16 \, ([I]/\text{mol dm}^{-3})^{0.58}, \tag{11}$$

where [I] is the initiator concentration. An exponent of unity would be expected if the efficiency of radical capture were 100% at all initiator concentrations.

The foregoing results were also discussed in terms of the relatively low solubility of styrene in the aqueous phase and the different concentrations of free radicals established in the aqueous phase in response to differences in the initiator concentration. At very low initiator concentrations, the time required for the primary free radicals to add on monomers for entry is less than the time required for bimolecular termination events. Radical capture efficiencies near to 100% are then achieved. At higher initiator concentrations, however, the time required to generate free oligomeric radicals for entry is long compared with the time scale for bimolecular termination events, the latter being reduced significantly by the higher free radical concentration. Radical capture efficiencies significantly less than 100% are then observed.

With acrylate and methacrylate monomers, similar factors appear likely to govern the entry rate coefficient. However, the exponents for the dependence of ρ_I on [I] are greater for these monomers [8–10], implying somewhat higher radical capture efficiencies than for styrene, e.g. chain transfer [6, 11]. In addition to the high water solubility of monomer, chain transfer can also influence the radical capture efficiency [6, 11].

It was found [12–14] that the entry rate coefficient is one of the main factors controlling monodispersity after long times. Monodispersity is favoured by a high radical entry rate which is promoted, in the absence of secondary nucleation, by a reduced number of latex particles. The rate of entry of free radicals into the latex particles was found to be a function of the properties of the emulsifier. One such property is the adsorption isotherm of the emulsifier onto the polymer particle surface, which affects particle growth indirectly by changing the particle number mainly during nucleation and hence the entry rate during growth.

All rate-determining parameters including the microscopic rate coefficients can in fact be determined by at least three different methods: (i) from the approach to the steady state in chemically-initiated systems [6], (ii) from relaxation studies of γ-initiated systems [7], and (iii) from the time evolution of the particle size distributions [13] and/or their combinations [8]. These methods have been successfully applied to a number of monomers [8, 10].

It is accepted that the exit (or desorption) of free radicals from latex particles proceeds by a stepwise mechanism involving at least three discrete processes. Thus, the exit of radicals from particles does not occur by the sudden dissolution of propagating polymer chains in the aqueous phase. The generally accepted mechanism for desorption postulates that a free radical of low degree of polymerization diffuses out of the particle. The quantitative description of this process has been reviewed by Nomura *et al.* [15–17].

Nomura's model features the following steps:

(i) the transfer of free radical activity from the growing polymer chain to a monomer molecule or a chain transfer agent,
(ii) diffusion of the resulting low molecular mass radical to the surface of the particle, and then, if propagation beyond a certain degree of polymerization has occurred,
(iii) diffusion of the free radicals away from the particle through the aqueous phase.

Thus, this theory is based on the transfer-diffusion mechanism. When allowance is made for the possibility that transfer to monomer may be the sole rate-determining step, the theory predicts that

$$k_{des} = \min\left\{k_{tr}C_M, (3D_w z/r_s^2)[(k_f/k_p)/(q + 6D_w/D_p)]\right\}, \tag{12}$$

where D_w, D_p are the diffusion coefficients of the exiting species in water and the latex particle, respectively, z (mean) is the degree of polymerization of the exiting species, r_s is the swollen radius of the latex particles, k_f is the free radical transfer rate coefficient and q is the partition coefficient of the exiting free radicals between the organic and aqueous phases. The latter quantity can be approximated by $q \simeq C_M/C_w$, where C_M and C_w are the concentrations of monomer in the latex particles and aqueous phase, respectively. Note that 'min' in Eq. (12) refers to the smaller of the two values given in brackets.

In principle, any or all of the three sequential steps involved in the exit may be rate limiting, whereupon Eq. (12) may adopt a simpler form. If step (i), the transfer to monomer (with rate coefficient k_{fm}), is rate determining, then the exit rate constant is given simply by

$$k_{des} = k_{fm} C_M \tag{13}$$

in the absence of other chain transfer. If diffusion of the low molecular mass radical through the monomer-swollen polymer particle is also rate determining, then

$$k_{des} = 1.5 z D_p (k_{fm}/k_p)/r_s^2. \tag{14}$$

Finally, if diffusion of the exited free radicals away from the particle is rate determining, k_{des} is given by

$$k_{des} = 3 z D_w (k_{fm}/k_p)(C_w/C_M)/r_s^2. \tag{15}$$

Diffusion in the aqueous phase might be expected to be the rate-determining step for monomers, such as styrene, butyl methacrylate, divinylbenzene, which are only sparingly soluble in water.

Equation (15) implies that the exit rate coefficient may, in principle, be altered in several different ways, e.g., (i) by varying the radius of the latex particles, (ii) by changing the solubility of the monomer in the continuous phase, and (iii) by altering the temperature. Further, the value of k_{des} should be increased by the presence of chain transfer agents whose solubility in the aqueous phase is comparable to or greater than that of the monomer, provided, of course, that they are incorporated into the latex particles. Finally, it is expected that the transition from one rate-determining mechanism to another takes place. The validity of each of the steps of Nomura's model has been tested experimentally for seeded emulsion polymerization of styrene [7, 18–20]. The absolute values measured for k_{des} for styrene at several different temperatures were found to be predicted well by Eq. (15) [21]. Experimentally [7] it was found from relaxation studies of particles with $r_s = 79$ nm that

$$k_{des}/s^{-1} = 10^{2.3 \pm 0.5} \exp[-(32 \pm 10 \text{ kJ mol}^{-1})/RT], \tag{16}$$

whereas Eq. (15) predicts under these conditions

$$k_{des}/s^{-1} = 10^{2.8} \exp[-(34 \text{ kJ mol}^{-1})/RT] . \qquad (17)$$

This good agreement between theory and experiment suggests that Nomura's model incorporates correctly most of the physicochemical processes that determine the exit rate from polymer particles.

Equation (15) predicts that the coefficient of the exit rate should be inversely related to the square of the radius of the swollen latex particles. Experimentally [6] it was found from seeded styrene emulsion polymerizations for particles with swollen radii in the range 44–79 nm that

$$k_{des}/s^{-1} = 1.7/(r_s/\text{nm})^{1.7 \pm 0.3} . \qquad (18)$$

Napper & Gilbert [1] also investigated the effect of the solubility of the exiting free radicals on k_{des}. The value of k_{des} was found to increase with increasing C_w (by the addition of methanol) until a plateau was reached. This suggests that the addition of methanol increases the exit rate constant by increasing the solubility of the exiting styrene free radicals in the continuous phase and increasing the driving force for diffusion away from the latex particles.

Additions of a chain transfer agent, carbon tetrabromide, or hydrocarbon diluents, such as ethylbenzene and toluene, to a seeded styrene emulsion polymerization were found to increase the exit rate coefficient dramatically without changing the nature of the rate-determining step [11, 22].

Nomura et al. [23, 24] also applied the model of radical desorption from particles to the copolymerization system. They calculated several relations for the mean rate coefficient for radical desorption. The authors found that desorption of radicals does not affect the composition of the copolymer.

The average number of radicals per particle in the copolymerization was defined by [23, 24]

$$\bar{n}_t = 0.5 (-C + \sqrt{C^2 + 2C}) = \bar{n}_a + \bar{n}_b , \qquad (19)$$

where $C = R / \bar{k}_{des} N_T$, R is the rate of radical production, $\bar{k}_{des}$ is the average rate coefficient for radical desorption, N_T is the total number of particles (particles/cm^3 water), and $\bar{n}_a$ and $\bar{n}_b$ are the average numbers of a-radicals and b-radicals per particle.

Additionally, strict application of the 'zero-one' assumption led Nomura et al. [15–17] to consider that instantaneous termination occurs in the polymer particles when a radical enters a particle already containing one radical. However, as shown in [25], the probability of desorption of a monomeric radical from a particle containing two radicals is higher than the probability of termination and therefore a significant fraction of the reabsorbed monomeric radicals will redesorb.

Asua et al. [25] presented a model which takes into account possible reactions of the desorbed radical in the aqueous phase as well as the competition between desorption and termination in polymer particles containing more than one radical. This model overcomes some important limitations of the previous models.

Contemporary ideas on kinetics and mechanism

The desorption rate coefficient, k_{des}, derived by Asua et al. [25] is given by:

$$k_{des} = k_{fm} [M]_p \frac{K_{0j}}{\beta K_{0j} + k_p [M]_p}, \qquad (20)$$

where k_{fm} is the monomer chain transfer constant, β is the probability that the desorbed single-unit monomeric radical reacts in the aqueous phase either by propagation or termination, and K_{0j} is the rate of diffusion of a radical chain of length j out of the particle.

The probability β is given by

$$\beta = \frac{k_p[M]_w + k_{tw}[R^{\bullet}]_w}{k_p[M]_w + k_{tw}[R^{\bullet}]_w + k_a N_T \Phi_w^w / N_A}, \qquad (21)$$

where $[M]_w$ is the monomer concentration in the aqueous phase, k_{tw} is the termination constant in the aqueous phase, $[R^{\bullet}]_w$ is the concentration of free radicals in the aqueous phase, k_a is the absorption rate coefficient and Φ_w^w is the volume fraction of water in the aqueous phase.

The value of β controls the lifetime of a single-unit monomeric radical in the aqueous phase and, together with the probability of desorption, the number of subsequent desorption-reabsorption steps that one single-unit radical may undergo. If $\beta = 0$, desorbed monomeric radicals do not react in the aqueous phase and thus reabsorb into the particles. If $\beta = 1$, none of these will reabsorb as single-unit radicals and therefore, no desorption can take place other than that from chain transfer to monomer.

Assuming a diffusion mechanism and no additional resistance the interphase, Nomura et al. [15–17] derived the following equation for K_{0j}

$$K_{0j} = \frac{12 D_{wj} / m_{dj} d_p^2}{1 + 2 D_{wj}/m_{dj} D_{pj}}, \qquad (22)$$

where D_{wj} and D_{pj} are the diffusion coefficients of a radical of length j in the aqueous phase and the polymer particle, respectively, m_{dj} is the partition coefficient of a radical of length j between the particles and the aqueous phase and d_p is the particle diameter. Since K_{0j} decreases sharply with radical chain length, it is assumed that only single-unit radicals can desorb from the polymer particles.

Two limiting cases arise from Eq. (20):

(i) when reactions in the aqueous phase are negligible (i.e., for sparingly water-soluble monomers) all monomeric radicals are reabsorbed into the particles, $\beta \to 0$ and

$$k_{des} = k_{fm} [M]_p \frac{K_{0j}}{k_p [M]_p}, \qquad (23)$$

(ii) when reactions in the aqueous phase are important (i.e., for highly water-soluble monomers), $\beta \to 1$ and

$$k_{des} = k_{fm}[M]_p \frac{K_{0j}}{K_{0j} + k_p[M]_p}. \qquad (24)$$

However, when the mass transfer coefficient has a value comparable to the propagation term and $K_{0j} \gg k_p[M]_p$, Eq. (20) tends to

$$k_{des} = k_{fm}[M]_p/\beta, \qquad (25)$$

whereas Nomura's equation tends to

$$k_{N, des} = k_{fm}[M]_p / \bar{n}. \qquad (26)$$

The difference between Asua's and Nomura's approaches is illustrated by considering the effect of the solubility of monomeric radicals in the aqueous phase on the desorption coefficients reported by Adams et al. [26]. These authors studied the emulsion polymerization of styrene in the presence of varying amounts of methanol. The methanol served to increase the solubility of both monomer and single-unit monomeric radicals in the aqueous phase. Adams et al. [26] found that k_{des} increased significantly on addition of up to 20 % by volume methanol in water; and approached a limiting maximum value as the methanol concentration was increased further. The authors attributed the plateau value of k_{des} to the fact that desorption becomes controlled by the monomer transfer reaction.

According to Ashua's model [25], the effect of methanol concentration on the emulsion polymerization is interpreted as follows: when there is no methanol in the system, the concentration of styrene in the aqueous phase is low ($[M]_w = 3 \times 10^{-6}$ mol cm^{-3}) and both β and K_{0j} have small values (β = 0.45, K_{0j} = 280). There is little difference between the predictions of Asua's and Nomura's equations in this region. The increase in the amount of methanol in the system increases both the value of the mass transfer coefficient K_{0j} and the probability of reaction of the desorbed single-unit radical in the aqueous phase. Throughout the experimental region, the transfer rate ($k_{fm}[M]_p$) is found to be much smaller than the mass transfer rate (K_{0j}). Consequently, the plateau region is not a result of the increase of the mass transfer rate with respect to the monomer transfer reaction rate. Instead, the plateau was suggested to be the result of the competition between the mass transfer rate and the propagation rate in the particles in addition to the increase in the probability of reaction in the aqueous phase.

3.2.2 Growth and deactivation of polymer particles

The rate of growth of the polymer chain in polymer particles is a function of the concentration of monomers and radicals in the particles and of the propagation rate constant. The average degree of polymerization in these particles reaches 10^4. Some tens of addition steps in the aqueous phase during the generation of the oligomeric radical or during production of the primary polymer particle cannot markedly influence the average polymer chain length and therefore do not need to be considered in kinetic analyses.

The primary particles are extremely small particles (radius ≤ 3 nm) and their properties differ from those of 'true' latex particles. They are only swollen by monomer to a limited extent so that propagation processes occur at a reduced rate inside them. These primary particles may undergo both homo- and heterocoagulation to form higher primary and 'true' polymer particles. Ultimately, coagulation and growth by propagation of these particles leads to mature (true) latex particles that are fully swollen with monomer.

It is assumed that in the steady state, the monomer concentration does not change in the polymer particles. Morton et al. [27] proposed the following semiempirical relation for calculating the equilibrium monomer concentration:

$$2\Phi_m \sigma / RTr = -(\ln(1 - \Phi_p) + (1 - 1/P)\Phi_p + \chi \Phi_p^2), \qquad (27)$$

where Φ_m and Φ_p are the volume fractions of monomer and polymer, respectively, in the monomer/polymer particle, σ is the interfacial tension, T is the absolute temperature, P is the average degree of polymerization, χ is the interaction parameter, and r denotes the radius of the polymer particles. The validity of the equation was confirmed by experimental results obtained from an investigation of the variation of the equilibrium monomer concentration in the polymer particles with the polarity of the monomer, published by Eliseeva et al. [28, 29].

The propagation of nonpolar monomers in emulsion polymerization is characterized by lower propagation rate constants and therefore the equilibrium monomer concentration in the particles of such systems is worth considering [30, 31]. On the other hand, emulsion polymerizations of polar monomers, e.g. acrylates, show high values of the propagation rate constant and fluctuation of the monomer in the particles. It is the diffusion of monomer from the monomer phase into the polymer particles that controls the rate of polymerization in this case.

A nonequilibrium state, or unbalanced monomer distribution in a polymer particle, was observed in the emulsion polymerizations and copolymerizations of acrylates and vinyl acetate [32–36] and in systems where the polymer being formed does not dissolve in its own monomer (vinyl chloride polymerization) [37–39].

There are some speculations that the transfer of monomer from monomer droplets to polymer particles does not have to proceed exclusively through the aqueous phase, but may take place by direct contact of the polymer particles with the monomer droplets. This mechanism was applied to emulsion copolymerizations of monomers with different water solubility whereby the copolymers produced are similar in composition to those formed in homogeneous polymerizations. The copolymerization parameters are also similar: e.g. the pairs isoprene/acrylonitrile [40] or butyl acrylate/vinyl acetate [41, 42].

Emulsion copolymerizations of monomers of different polarity and water solubility lead mostly to the production of a copolymer of composition different from that of the homopolymer formed in the homogeneous medium [32–34]. These deviations are due to the different rates of diffusions of monomers into the polymer particles and to the different compositions of the monomer blend around the reaction centre and of the monomer feed.

Reference [40] describes the first application of emulsion polymerization to determine the temperature dependence of the propagation rate constant of chloroprene:

$$k_p = 2.9 \times 10^9 \, e^{(-9.7 \, kJ \, mol^{-1}/RT)} \, dm^3 \, mol^{-1} \, s^{-1}. \qquad (28)$$

Several additional polyempirical relations were proposed for the calculation of k_p from experimental data. The relationship for calculating k_p for the emulsion polymerization of styrene for the temperature range from 45 to 65 °C provides an example [7]

$$k_p = 10^{7.1 \pm 0.5} \, e^{(-29 \pm 3 \, kJ \, mol^{-1}/RT)} \, dm^3 \, mol^{-1} \, s^{-1}. \qquad (29)$$

This equation gives for styrene emulsion polymerization $k_p = 257$ dm^{3}mol^{-1}s^{-1} at 50°C. This value agrees well with k_p for the homopolymerization of styrene reported elsewhere [45]. Similar values were obtained for the emulsion polymerization of styrene for Stages II [6] and III [46]. However, according to Soh [47], in Stage III, k_p decreases in comparison with Stages I and II and the value of k_p for styrene is estimated to be 156 dm^3 mol^{-1} s^{-1}. Investigations of propagation reactions taking place in the polymer particles in the stationary state led to a proposal of a constant value for the propagation rate coefficient, k_p; its decrease was predicted only at very high conversions [43, 44].

Recent results from an investigation of the relation between k_p and conversion for the emulsion polymerization of methyl methacrylate by ESR showed that the value of k_p ($\sim 1 \times 10^3$ dm^3 mol^{-1}s^{-1}) does not vary up to high conversion (80%) [48] in agreement with refs [43, 44]. The value is only a little higher than those values published for the block polymerization of methyl methacrylate [49]. The decrease of k_p at conversions exceeding 80% has been explained by applying the conditions for the glassy state of the system where the propagation-determining step is diffusion of monomer [51].

Soh & Sundberg [51] have adopted the Rabinovitch model [52] for diffusion-controlled processes to account for diffusion-controlled reactions. They justified their application of this model, which is strictly applicable only to small molecules of similar size, on the basis that the propagating ends of the macroradicals possess a mobility comparable to that of the monomer molecules. According to Soh and Sundberg, the observed propagation rate coefficient k_p at high conversions can be written in terms of two limiting rate coefficients:

$$1/k_p = 1/k_p^0 + 1/k_{pvf}. \tag{30}$$

Here k_p^0 corresponds to the rate coefficient in the absence of diffusion control and k_{pvf} is its value when there is absolute diffusion control. The latter rate coefficient is given by the Smoluchowski-type expression [53]:

$$k_{pvf} = (N_A/10^3)\, 4\pi R_{AB}\, D_{MO}\, e^{(1/v_{fm} - 1/v_f)}, \tag{31}$$

where R_{AB} is the collision radius for the monomer and the growing free radical, D_{MO} is the self-diffusion coefficient for the pure monomer, and v_{fm} and v_f are the fractional free volumes of the pure monomer and swollen latex particles, respectively.

The diffusive term of k_p can also be given by the following expression [53–55]:

$$k_{pd} = 4\pi(D_M + D_R)(R_M + R_R), \tag{32}$$

where D_M and D_R are the diffusion coefficients for monomer and free radicals and R_M and R_R the radii of interaction of monomer and free radicals, respectively. Both R_M and R_R may be approximated by $\sigma/2$, where σ is the Lennard–Jones diameter of the monomer [54].

The diffusion of the macroradical was found to be sensitive both the conversion and the size of the free radical [55]. In the 'rigid chain limit', where translational motion of the polymer chains is assumed to be completely frozen, the reactive chain end can grow and simultaneously move by means of propagation, which increases the chain length. This was termed 'reaction diffusion' [54], and the diffusion rate coefficient for this process was calculated from the expression

$$D_R = k_p C_m a^2/6, \qquad (33)$$

where C_m is the concentration of monomer in the particles, and a is the root-mean-square end-to-end distance of polymer chain per square root of the number of monomer units.

When the monomer concentration in the latex particles fell below that corresponding to the glass transition point at the polymerization temperature, the value of k_p was found to decrease with increasing mass fraction of polymer in the particles, w_p [10].

The results presented in [48] suggest that k_p decreases approximately exponentially with the mass fraction of polymer in the latex particles at high mass fractions of polymer. The authors obtained the following expression for k_p (for $w_p \geq 0.84$) for the emulsion polymerization of methyl methacrylate

$$k_p = k_p^0 \, e^{[-29.8 (w_p - 0.84)]}, \qquad (34)$$

where k_p^0 corresponds to the rate coefficient for $0.33 \leq w_p \leq 0.84$, i.e., $k_p = k_p^0$. The observed decrease in k_p at high conversions was associated with diffusion control of the propagation reaction in the glassy state. The rate-determining step under these conditions was expected to be the diffusion of monomer to the sites of the propagating free radicals. The monomer concentration at the onset of the observed reduction in k_p was $ca.$ 1.8 mol dm^{-3} corresponding to a mean distance of separation of the monomer molecules of about 1 nm. The consumption of monomer in the propagation step thus requires monomer molecules to diffuse significant distances through a medium composed largely of polymer particles.

In reference [48] there is applied the free volume theory [50] to follow the dependence of k_p on w_p. The authors reported that near the glass transition point the free volume theory predicts values of k_p that are in fair agreement with the experimental results. For large values of w_p beyond the glass transition point the free volume theory predicts, however, an extremely rapid decrease in k_p with w_p. The experimentally determined decrease in k_p is much less dramatic, resulting in the theoretical values for k_p being several orders of magnitude less than the observed values if $w_p > 0.9$.

The peculiar character of bimolecular termination reactions in polymer particles follows from the high microviscosity of the core of the polymer particle and from the low mobility of propagating radicals. The termination rate constants are on average 2–4 orders of magnitude lower than for polymerizations taking place in homogeneous media [1, 45, 49–59]. The lowering of the rate of termination is accompanied by an increase in the number of radicals in a polymer particle. Friis & Hamielec [60] assume that the polymerization in polymer particles proceeds under the conditions of the gel effect and the measure of the significance of this effect is proportional to the particle size and thus also to the concentration of radicals.

It is generally accepted that in free-radical polymerization, bimolecular termination can be regarded as a consecutive three-step reaction [61]. Firstly, the two radical coils must gain proximity to each other by translational diffusion. Then the free-radical chain ends must make contact: this is achieved through conformational reorientations of the polymer molecules, the so-called process of segmental diffusion. Finally, the barrier to chemical reaction must be overcome. The activation energy of this final step is accepted to be small. When this is coupled with the fact that macroradicals are large and for which diffusive processes are slow, it follows that termination is exclusively diffusion-controlled. At low conversions (bulk or solution polymerization), the translational mobility of the monomer-lapping polymer coils is sufficiently rapid for the subsequent free-radical encounter process to be rate determining. At high conversions (emulsion polymerization), the chains of the polymer matrix will be so entangled that centre-of-mass diffusion will essentially finish. It has therefore been proposed 'that the active chain ends wander by the residual termination' (see later).

It is assumed that instantaneous termination of radicals occurs in small particles and that the average number of radicals per particle is 0.5. During polymerizations in particles larger than 100 nm, termination is retarded by the possible growth of an oligomeric radical after its entry into the particle. This growth reduces the mobility and ability to terminate immediately [59, 62]. The concentration of radicals in small particles is also effectively controlled by desorption processes.

The termination rate constant varies significantly in the presence of the polymer matrix that forms the latex particle in comparison with the value of k_p [63]. At low conversions (Stage I) where particle nucleation and stabilization of the formation and number of monomer-polymer particles take place, termination reactions are mainly affected by the translational diffusion of radicals. In Stage II characterized by the constant ratio of polymer and monomer in latex particles (for most polymers the mass fraction of the polymer w_p varies between 0.3 and 0.5), the radical termination is governed by the rate of segmental diffusion of the radical ends of the polymer chain. This mode of termination follows from the high molecular masses of the polymer being formed and from the high viscosity of the latex particle core.

The bimolecular termination process within the latex particles is characterized by the pseudo-first-order coefficient $C = k_t/N_A V_s$, where N_A is Avogadro's constant and V_s is the swollen volume of the latex particles [1].

The second-order termination rate coefficient is expected to be a strong function of w_p. This is because the bimolecular termination process becomes progressively more diffusion-controlled as the mass fraction of polymer increases. A quantitative theory for the dependence of k_t on w_p has been given by Hamielec et al. [43, 44, 64].

As the fraction of polymer in the polymer particle increases ($w_p \geq 0.6$), the viscosity of the system and also the molecular mass increase. With increasing molecular mass the degree of entanglement of the macromolecules also increases and influences the mobility of the macroradicals in the particles and the mechanism of their termination. The mechanism of termination at high viscosities of the reaction mixture has also been studied by several authors [50, 65–71], who assume the mobility of entangled macroradicals in particles to be limited and the macroradicals not to be subject to termination by translation diffusion but only as a result of the movement of the radical centre caused by propagation of both radicals: this type termination is called residual termination.

One formulation of residual termination has been given by Gardon [71]. He derived the expression

$$(k_t/k_p)_{min} = 0.185\Phi_m(1 - \Phi_m). \tag{35}$$

Equation (35) gives a hypothetical limiting ratio for k_t and k_p for which all free radicals would survive in emulsion polymerization. Equation (35) seems inadequate in that it contains no terms dependent on the nature of the reaction system. Recognizing this, Gardon developed a latice model (denoted by subscript L) for termination

$$(k_t/k_p)_L = 2\mu (d_m/d_p) \Phi_m , \tag{36}$$

where μ is the latice parameter, i.e. the number of sites surrounding a given latice site. Gardon suggests $\mu = 8$ for a closely packed, hexagonal latice.

Later, Soh & Sundberg [51, 65–67] proposed their theory of excess chain-end mobility, which yields the expression:

$$k_t(\text{residual}) = f_t \pi\sigma^2 a N_A k_p C_M j_c^{0.5} , \tag{37}$$

where σ is the radius of termination given by:

$$\sigma = \frac{1}{r_c} \ln\left[\frac{r_c^3}{N_A[R^\bullet]\pi^{1.5}}\right]^{0.5} . \tag{38}$$

Here $[R^\bullet] = \bar{n}/N_A V_s$, $r^2 = 3/(2 j_c a^2)$, $j_c = P_{cr}/2 w_p$, where P_{cr} is the critical degree of polymerization for an entanglement of pure polymer, f_t is a so-called 'efficiency parameter', and r_c is a coefficient equal to 1 for disproportionation and to 2 for recombination.

For the chain-entanglement component of k_t, Soh and Sundberg give

$$k_t (\text{ent.}) = z k_t^* \exp\left(\frac{1}{v_{fm}} - \frac{1}{v_f}\right), \tag{39}$$

where

$$v_f = v_{fp} (1 - \Phi_m) + v_{fm} \Phi_m ,$$

$$\Phi_m = (1 - w_p)/(1 - \varepsilon w_p),$$

$$z = 1 - e^{-\beta} + \beta \int_1^\infty Y^{-2.4} e^{-\beta} dY . \tag{40}$$

Here $P_c = P_{cr}/(1 - \Phi_m)$, z and β are dimensionless parameters, $\beta = P_c/i$, $i = M_g/M_0$, k_t^* is the termination rate coefficient for a monomeric free radical, v_f is the fractional free volume, v_{fm} is the free volume of monomer, v_{fp} is the free volume of polymer, ε is the volume shrinkage factor, equal to $d_p/d_m - 1$, Y is the reduced degree of polymerization and M_0 and M_g the molecular masses of monomer and of growing chain, respectively.

The overall value of the termination rate coefficient is then given by

$$k_t = k_t(\text{res.}) + k_t(\text{ent.}) . \tag{41}$$

Soh and Sundberg's model was applied to the emulsion polymerization of methyl methacrylate by Ballard et al. [72].

One finds that $k_t(\text{ent.})$ or $k_t(\text{total})$ for the system with a chain transfer agent (CTA) is significantly greater than that calculated in the absence of a CTA in accord with experimental observation (Table 2). The *calculated* value for k_t with a CTA is considerably greater than the *observed* value [72]. The difference in k_t values was ascribed to the uncertainties in various parameters in Eqs. (38)–(41) and to the sensitivity of k_t(ent.) to many of these parameters. However, these differences are less important than the values obtained for k_t elsewhere [10, 54].

The results obtained by Ballard *et al.* [72] have been interpreted semiquantitatively (based on the model of Soh and Sundberg) in terms of two mechanisms: translational diffusion determined by chain entanglements (dominant for growing chains of lower molecular mass) and residual excess chain-end mobility (dominant for growing chains of higher molecular mass).

More recently, Russell *et al.* [54] have proposed their model for the residual termination. They derived the expression

$$k_t(\text{res.}) = 8\pi k_p C_M a^2 r_a / 3, \tag{42}$$

where r_a is the radius of interaction for termination. There are two limiting cases for r_a.

Table 2 — Experimental and calculated values of k_t (dm^3 mol^{-1} s^{-1}) for methyl meth-acrylate

Symbol	No CTA	2% CBr$_4$	Ref.
w_p	0.33	0.33	
k_t(ent.)	0.5×10^5	9.8×10^5	[72]
k_t(res.)	0.3×10^5	0.3×10^5	[72]
k_t(total)	0.8×10^5	1.0×10^6	[72]
k_t(exp.)	0.3×10^5	1.5×10^5	[72]
	0.6×10^4	—	[10]
	1.0×10^6	—	[54]
k_t(exp.)^a	4.0×10^7	—	[54]

$^a w_p = 0$.

The first of these is that of the rigid chain. The resulting expression is the minimum value for k_t (res.) and is

$$k_t(\text{res. min}) = 4\pi k_p C_M a^2 \sigma / 3, \tag{43}$$

where σ is the Lennard–Jones diameter of the monomer ($r_a = \sigma$).

On the other hand, the maximum possible value for the residual termination coefficient resulted in the case where the chain end is totally flexible. Here, r_a was estimated to be

$$r_a \simeq \langle r_a^2 \rangle^{0.5} = a j_c^{0.5}, \tag{44}$$

where j_c is normally of the order of 100 monomer units [73].

The resulting expression for k_t(res. max) is

$$k_t \text{ (res. max)} = 8\pi k_p C_M a^3 j_c^{0.5}/3 . \qquad (45)$$

Equations (43) and (45) represent the two limits between which the residual termination coefficient must be encompassed.

The model developed by Russell et al. [54] gives readily evaluated expressions for the upper and lower limits of k_t (Table 3). The model suggests that at high conversion, the rate coefficient for termination is governed by that for propagation. Comparison of theoretical predictions with experimental indicates that residual termination predominates throughout much of the conversion of monomer to polymer. It is also expected that this model should be applicable to crosslinked systems for wide ranges of conversion. Moreover, it was suggested that the model of Russell et al. has some advantages with regard to all the above theories:

(i) no calibration with other values of k_t is required and
(ii) no adjustable parameters are introduced.

The model of residual termination of macroradicals in a particle was verified by Maxwell et al. [8] for the emulsion polymerization of butyl acrylate ($w_p = 0.58$). The value of k_t(res.) obtained through application of Soh–Sundberg's model is very high ($7 \times 10^3 \text{dm}^3\text{mol}^{-1}\text{s}^{-1}$) and residual termination strongly affects the mechanism of termination.

Table 3 — Experimental and calculated values of k_t ($\text{dm}^3 \text{mol}^{-1} \text{s}^{-1}$) for butyl acrylate (BA), butyl methacrylate (BMA) and methyl methacrylate (MMA) [54]

Symbol	BA	BMA	MMA
w_p	0.57	0.47	0.55
k_t (exp.)	8×10^2	7×10^3	1×10^{4} [a]
k_t (res. min)	9×10^2	1.5×10^3	2×10^3
k_t (res. max)	3×10^4	5×10^4	5×10^4
k_t (exp.) / k_p	1.7	12	—
k_t (res. min) / k_p	2.3	2.5	—
(k_t / k_p) L	6.4	8.2	—

[a] Ref. [10].

Capek [74] discussed the contribution of residual termination in the steady state. He obtained low values of k_t (res.) for the emulsion polymerization of butyl acrylate carried out at a high temperature (70°C) and at high initiator concentration ($\sim 10^{-2}$ mol dm^{-3}), i.e. residual termination does not play any dominant role in termination. The high temperature, and the concentrations of monomer and initiator in the system rather caused the termination of radicals by diffusion processes. The temperature enhancement lowers the viscosity of the system and the gel effect, as well as the contribution of the particle entanglement, and, by contrast, it increases fraction of termination by diffusion processes. The high concentrations of primary or oligomeric radicals in the polymer particles that follows from the high concentration of the initiator used favours the termination of macroradicals by monomeric or oligomeric radicals.

3.2.3 Mathematical modelling of emulsion polymerization

Since the pioneering work of Smith & Ewart [75], there has been a considerable effort to develop mathematical models to achieve a fundamental understanding of emulsion homopolymerization. The critical point in the development of these mathematical models is the elucidation of the mechanisms involved in emulsion polymerization and the estimation of the corresponding parameters. Significant contributions to this field have been added by the models of investigators such as Gilbert & Napper [21], Min & Ray [76], Hansen & Ugelstad [77], Hamielec & MacGregor [78], Feeney et al. [79, 80], Rawlings & Ray [81], and more recently by Asua et al. [82]. Mathematical models for emulsion homopolymerization have been extended to copolymerization by Haskel & Settlage [83], Broadhead et al. [84], Nomura & Fujita [85], Dougherty [86, 87], Storti et al. [88, 89], Delgado et al. [90] and Richards et al. [91].

The physical picture of emulsion polymerization or copolymerization is complex due to the presence of multiple phases, multiple monomers, radical species and other ingredients, the role of complex particle formation mechanisms and the probable involvement of several different types of reactor. The mechanistic events involved in an emulsion polymerization reaction are as follows:

(i) generation of free radicals by decomposition of the initiator in the aqueous phase (for water-soluble initiators;
(ii) propagation of free radicals in the aqueous phase;
(iii) termination of free radicals in the aqueous phase;
(iv) entry of free radicals into the latex particles;
(v) desorption of the free radicals from the polymer particles;
(vi) propagation within the polymer particles;
(vii) termination within the particles.

In these events, free radicals with different characteristics are involved, namely, radicals of different chain length and different chemical composition, according to whether they arose from the initiator or from a desorbed monomeric radical. In principle, the rate of a particular event is different for each type of oligomer. Were a distinction to be made between all these oligomers, it would result in a model that contains so many parameters that they could not be estimated from experimental data. Therefore, some kinds of averaged parameters have to be used.

Min & Ray [76] and Hamielec & MacGregor [78] have developed detailed, computer-based models that treat the polymerization kinetics as well as the reactor dynamics. Such models indicate how the process dynamics influence polymer properties.

Considerable effort has been made to develop strategies for kinetic investigation of the emulsion polymerization process by the group at the University of Sydney with Napper, Gilbert et al. [6, 8, 10, 13, 14, 20, 21, 46, 48, 72, 92]. Two main approaches have been developed by this group. The first is based on the analysis of the time dependence of conversion during Stage II or III of a seeded emulsion polymerization using both chemical and γ-radiolysis initiation (see sections 3.2.1 and 3.2.2). The second approach is based on analysis of the evolution of the particle size distribution during Stage II of a seeded emulsion polymerization.

In order to analyse experimental data from the emulsion polymerization experiments, the Sydney group uses the material balance of monomer coupled with the population balances for particles containing i radicals:

$$\frac{dx}{dt} = \frac{k_p[M]_p \bar{n} N_T}{M_0 N_A}, \qquad (46)$$

$$\frac{dN_i}{dt} = \rho(N_{i-1} - N_i) + k_{des}[(i+1)N_{i+1} - iN_i] + c[(i+2)(i+1)N_{i+2} - i(i-1)N_i], \qquad (47)$$

where x is the conversion, t is time, M_0 is the initial number of moles of monomer per volume of water, N_i is the number of particles containing i radicals per volume of water and k_p, $[M]_p$, k_{des}, c, $\bar{n}$ and k_t have been defined in previous sections.

The Sydney group [8, 20, 21] proposed different methods for data analysis for a system in which the average number of radicals per particle is less than 0.5, a zero–one system, and systems where $\bar{n}$ is greater than 0.5. In their work, attention was focused on systems with experimenal conditions which result in $\bar{n} < 0.5$. In general, such experimental conditions may be reached by lowering the initiator concentration, increasing the number of polymer particles and/or decreasing the particles diameter. For a zero–one system, Eqs (46) and (47) reduce to

$$\frac{dx}{dt} = \frac{k_p[M]_p N_1}{M_0 N_A}, \qquad (48)$$

$$\frac{dN_1}{dt} = \rho(N_T - 2N_1) - k_{des}N_1. \qquad (49)$$

Equations (48) and (49) may be integrated analytically, under given initial conditions, to yield a solution for $x(t)$ by assuming that the rate parameters are constant with time and that the monomer concentration is known, being constant in Stage II or changing in a predictable way in Stage III.

The solution for $x(t)$ becomes

$$x = at + b, \qquad (50)$$

where a and b, the slope and intercept of the straight line portion of an x versus t curve, are functions of the parameters ρ_A, α, and k_{des}. Thus, for an individual experiment, ρ_A and k_{des} can be obtained if α is known.

Whang et al. [20] used two criteria to estimate a single value of α:

(i) the exit rate coefficient obtained from the slope and intercept analysis should be independent of initiator concentration and particle number and
(ii) the value of k_{des} should agree with the value deduced from relaxation studies where α was assumed to equal +1.

Lichti *et al.* [14] used the time evolution of the particle size distribution (PSD) to estimate the rate coefficients for entry, exit, and propagation of free radicals in seeded and *ab initio* styrene emulsion polymerizations. The authors found the PSD to be insensitive to the value of the radical desorption rate constant for two out of the three cases studied.

The model proposed by Lichti *et al.* has been used by Chen & Wu [93] to determine kinetic parameters from particle size distribution. Surprisingly, they found the desorption rate constant to be proportional to the particle surface area, i.e. the rate of desorption increases with increasing particle size. In addition, in order to obtain the basic equations in their approach, these authors considered both ρ and k_{des} to be independent of the volume of the particle but then used the resulting equations to determine the dependence of ρ and k_{des} on the particle size.

Nomura [94] has proposed an approach to determine kinetic parameters by using the steady-state portion of the conversion versus time curves of seeded polymerizations. This approach involves an assumption of the extent of termination in the aqueous phase.

A new method for the estimation of kinetic parameters in emulsion polymerization systems has been presented by Asua *et al.* [82]. This method is based on studies of the evolution of monomer conversion in chemically-initiated seeded emulsion polymerization systems. In this study, homopolymerization under zero–one conditions is considered. The method is based on a fundamental model that includes the free radical balance in the aqueous phase and fundamental parameters such as the entry (k_e) and exit rate (k_{des}) coefficients, the termination rate constant (k_{tw}) in the aqueous phase, and the rate coefficient for initiator decomposition (k_I). By using average parameters, the material balance for free radicals in the aqueous phase was written as follows:

$$\frac{d[R]_w}{dt} = \frac{2k_I[I]}{\Phi_w} + \frac{k_{des}\bar{n}N_T}{N_A} - \frac{k_e[R]_w N_T}{N_A} - \frac{2k_{tw}[R]_w^2}{\Phi_w}, \tag{51}$$

where Φ_w is the volume fraction of water in the aqueous phase and the left-hand side represents the accumulation of free radicals in the aqueous phase.

The first term of the right-hand side accounts for the generation of radicals through initiator decomposition, the second for the desorption of radicals from the latex particles, the third for the entry of radicals into the polymer particles, and the fourth for the consumption of radicals by bimolecular termination. Note that Eq. (51) does not make any distinction between the different types of radical.

For most systems, the contribution of the aqueous phase polymerization to the overall conversion is negligible and, hence, the monomer material balance for a zero–one system is given by Eq. (48). Here, the population balance for N_1 is as follows

$$\frac{dN_1}{dt} = k_e[R]_w(N_t - N_1) - k_e[R]_w N_1 - k_{des}N_1. \tag{52}$$

For a chemically-initiated system, the initial conditions for Eqs (48, 51, 52) are

$$t = 0, x = 0, [R]_w = 0, N_1 = 0. \tag{53}$$

Equations (48, 51–53) are a system of initial-value stiff differential equations containing four unknown parameters k_e, k_{des}, k_I and k_{tw}.

Equations (48, 51, 52) have been written as follows

$$\frac{dS}{dt} = F(t, S, Y, K),\tag{54}$$

where S is the vector of the state variables, Y is vector of the independent variables and K is the vector of adjustable parameters, where

$$S = \{x, N_1, [R]_w\},\tag{55}$$

$$K = \{k_e, k_{des}, k_I, k_{tw}\}.\tag{56}$$

These parameters, together with the propagation rate constant, are estimated by using all the available experimental data at the same time. Therefore, the method does not depend on any literature value for a parameter. It has been found that accurate values of the parameters are obtained by this approach if a sufficient number of experiments with a minimum range of variation are available. This minimum ranges between 9 and 12 and the use of three different values of the number of particles is advised. An important conclusion of this study is that the entry rate coefficient and the termination rate constant in the aqueous phase are correlated, with the result that these parameters cannot be determined unambiguously. It has been reported that the values of the estimated parameters are unaffected by experimental noise due to random errors. However, when the experiments contain systematic errors, the estimated parameters are greatly affected.

Dougherty [86, 87] developed the SCOPE (simulation and control of polymer emulsions) computer-based dynamic process model. The model describes typical industrial batch copolymerizations. The user supplies the SCOPE computer programme with the emulsion recipe, such process conditions as flow rates and temperatures, and kinetic parameters, such as rate constants and reactivity ratios. The model then calculates the species concentrations, particle size, and molecular mass as a function of time.

The species considered in the model include the following: water (or other diluent), one or more monomers, (co)polymer, emulsifier, promotor, activator, and/or chain transfer agent. Some (or all) of these species may be present in the starting reaction mixture, or some (or all) may be added to the reactor as a feed stream of monomer emulsion, in which case the monomers polymerize as they are fed in.

The model includes material and energy balance equations as well as the associated proportional integral derivative control equations. Mathematical expressions relate the reaction kinetics and process variables to conversion, species concentrations, molecular mass, particle size and copolymer composition. This model gives good agreement with several experiments reported by Nomura et al. [23]. However, it slightly underestimates the effects of particle concentration on conversion rate at fixed monomer concentrations.

Storti et al. [89] generalized the Smith–Ewart model for multimonomer emulsion polymerization. The authors developed a procedure for reducing the large number of population balances to a smaller one, where each particle population is characterized only by the overall number of radicals of any type contained inside each particle. An approximation procedure is thus proposed for reducing the original multimonomer system to that typical of homopolymerization processes. The reliability of such a 'pseudo-homopolymerization approach' has been tested by comparison with experimental data for binary and ternary systems. The results obtained indicate that the behaviour of the ternary system can be predicted from the binary polymerization parameters involved.

Delgado et al. [90] developed a model that describes monomer transport between monomer droplets, the aqueous phase, and polymer particles during emulsion polymerization. The model was used for investigating the role of co-emulsifier (hexadecane) in the miniemulsion copolymerization of vinyl acetate (VAc) and butyl acrylate (BA), as well as the effect of different components and process variables on the rate of copolymerization, the monomer distribution between phases, and the composition of the copolymer.

The polymerization may take place in any (or several) of the following phases: the aqueous, monomer-swollen polymer particles and monomer droplets.

The material balance for the monomer present in the monomer/polymer particles was given by

$$dN_{i,p}/dt = K_{i,a-p} A_p (C_{ei,p} - C_{i,p}) - \sum_{j=0}^{n} k_{p_{ij}} C_{i,p} N_p / N_A , \quad (57)$$

where $K_{i,a-p}$ is the mass transfer coefficient for i between the aqueous phase (a) and the monomer-swollen polymer particles (p), A_p is the transfer area of phase p, $C_{ei,p}$ is the equilibrium concentration of component i in phase p, $C_{i,p}$ is the concentration of component i in phase p, $k_{p_{ij}}$ is the propagation rate constant between monomer i and radical j, N_p the total number of particles and $\bar{n}_j$ the average number of radicals j per particle.

The material balance for the monomer in the aqueous phase is given by

$$dN_{i,a}/dt = K_{i,d-a} A_d (C_{i,d} - C_{ei,d}) - K_{i,a-p} A_p (C_{ei,p} - C_{i,p}) - \sum_{j=1}^{n} K_{p_{ij}} C_{i,a}[R^\bullet]_{j,a} V_a, \quad (58)$$

where the subscript a represents the aqueous phase and d the monomer droplets, $[R^\bullet]_{j,a}$ is the concentration of radicals j in the aqueous phase and V_a is the volume of the aqueous phase. The term $K_{i,d-a} A_d (C_{i,d} - C_{ei,d})$ which expresses the flow of monomer out of the droplets is equivalent to the expression $-K_{i,d-a} A_d (C_{ei,a} - C_{i,a})$.

Two processes had to be taken into account in the material balance for the monomer droplets. The first is the monomer transported via flow from the oil phase through the continuous phase to the polymer particles. The second is the change in the status of the phase when (i) a monomer droplet becomes a polymer particle owing to the entry of a radical, and therefore the monomer originally belonging to a monomer droplet belongs now to a polymer particle and when (ii) droplet-particle coalescence occurs. These two processes are expressed mathematically by

$$dN_{i,d}/dt = K_{i,d-a} A_d (C_{ei,d} - C_{i,d}) - \frac{N_{i,d}}{N_d} R_{d-p} , \quad (59)$$

where R_{d-p} is the rate of change of monomer droplets to polymer particles, given by the expression

$$R_{d-p} = \rho_D N_d + k_1 N_p N_d . \quad (60)$$

Here ρ_D is the rate of radical entry per droplet, k_1 is the coagulation rate constant, N_d is the total number of monomer droplets and $N_{i,d}$ is the number of moles of component i in the monomer droplets.

The change in the number of droplets (N_d) as the polymerization proceeds will be given by the disappearance of droplets due to particle nucleation and the disappearance of droplets due to coagulation with themselves and the polymer particles already formed:

$$-dN_d/dt = \rho_D N_d + k_2 N_d^2 + k_1 N_p N_d, \tag{61}$$

where k_2 is the coagulation rate constant.

The equilibrium concentrations of different species in each phase were determined from the equilibrium thermodynamics of swelling [27, 90, 95, 96]. In general, the following expression was developed for the equilibrium concentrations:

$$C_{ei,q} = \Phi_{i,q}/\overline{V}_i, \tag{62}$$

where $\Phi_{i,q}$ is the volume fraction of component i in phase q and $\overline{V}_i$ is the molar volume of i.

For a miniemulsion copolymerization of VAc and BA, if the mass transfer coefficient was greater than 10^{-4} cm s^{-1}, then the polymer particles were found to be under equilibrium-swelling conditions. For values of the mass transfer coefficient decreasing from 10^{-4} to 10^{-8} cm s^{-1}, the polymerization process becomes progressively transport-controlled.

The model predicted the disappearance of the monomer droplets for the conventional emulsion copolymerization process, as found experimentally. It also showed that when hexadecane is present in the monomer droplets, they will not disappear as a result of transfer of monomer from the oil phase through the continuous phase to the particles. Moreover, it shows that hexadecane in fact plays two opposing roles. Its presence in the monomer droplets increases their capability to retain the amount of monomers originally present in them on the entry of radicals and the formation of polymer. On the other hand, it reduces the equilibrium concentration of monomer in the polymer particles. The model was found to confirm the experimental and theoretical findings of Poehlein et al. [97] that the dispersion of monomers in small droplets increases the swellability of polymer latexes.

Richards et al. [91] presented a physical mechanism and mathematical structure of a comprehensive model for dynamic emulsion polymerization and/or copolymerization. This model combines the theory of coagulative nucleation of homogeneously nucleated precursors with detailed material and energy balances to calculate the time evolution of the concentration, size, and colloidal characteristics of the latex particles, the monomer conversion, the copolymer composition and the molecular mass.

The equilibrium distribution of the two monomers between the latex particles, the aqueous phase, and the comonomer droplets was described using empirical partition coefficients. The volume of the separate monomer phase V_m is calculated from the total volume of the emulsion V_e and the volumes of the aqueous and polymer phases V_w and V_p as follows:

$$V_m = V_e - V_p - V_w. \tag{63}$$

This model distinguishes between two types of free radical species in the particle, namely, radical chains ending with A (concentration C_{ap}) and radical chains ending with B (concentration C_{bp}). The following equations were derived for C_{ap} and C_{bp} in terms of the average number of radicals per particle, $\bar{n}$:

$$C_{ap} = \frac{\bar{n} N_e V_e}{V_p N_A}\left(\frac{1}{1+\gamma_p}\right), \qquad (64)$$

where

$$\gamma_p = \frac{C_{bp}}{C_{ap}}. \qquad (65)$$

The steady-state Smith–Ewart [75] equation for a comonomer system is as follows:

$$0 = \rho(N_{n-1} - N_n) + k_{des}[(n+1)N_{n+1} - nN_n] + c[(n+2)(n+1)N_{n+2} - n(n-1)N_n]. \qquad (66)$$

The kinetic constants ρ, k_{des}, and c that describe the entry of radicals into particles, radical desorption, and bimolecular termination are related here to the corresponding constants for copolymerization.

The average number of radicals was calculated by using the expression for $\bar{n}$ suggested by Ugelstad et al. [98].

The following differential equation was proposed to calculate the time evolution of the emulsion concentration of colloidally-stable latex particles of any size, N_e

$$V_e \frac{dN_e}{dt} = -N_e \frac{dV_e}{dt} - N_e Q_e + G_{ce} V_e, \qquad (67)$$

where Q_e is the volumetric reactor outflow and G_{ce} is rate of generation of the particles.

The average particle volume v_p and the average radius of the monomer swollen and unswollen particles (r_p and r_q, respectively) were calculated from the following expressions:

$$v_p = V_p/N_e V_e, \qquad (68)$$

$$r_p = (3v_p/4\pi)^{1/3}, \qquad (69)$$

$$r_q = r_p \Phi_q^{1/3}, \qquad (70)$$

where Φ_q is the volume fraction of the latex particle that consists of dead polymer.

The following steady-state radical balance was used to calculate the concentration of copolymer oligomer radicals in the aqueous phase, C_{aw} and C_{bw}

$$2R_w = R_{ew} + R_{tw}. \qquad (71)$$

This balance equates the radical generation rate $2R_w$ with the sum of the rate of radical entry into the particles and precursors R_{ew}, and the rate of radical termination R_{tw}.

The authors [91] developed the following equation for the rate of formation of primary precursors (particles) G_{he}:

$$G_{he} = \frac{2R_w N_A (V_w/V_e)}{(1 + R_{ew}/R_{pw})^{j_{cr}-1}}, \quad (72)$$

where R_{pw} is the rate of propagation in the aqueous phase and j_{cr} is the critical degree of polymerization of oligomer radicals at which the growing chains become large enough to form an insoluble precursor particle of radius r_1 and volume v_1. The primary particles (precursors) grow by propagation and coagulation until they reach a size that makes them colloidally stable. The generation rate G_{ce} of latex particles was developed as

$$G_{ce} = \frac{1}{2} \sum_{i=1}^{m-1} v_i \left(\sum_{j=m-1}^{m-1} B_{ij} v_j \right) \frac{K_{vm-1} v_{m-1}}{v_1} - 2 B_{m,m} N_e^2, \quad (73)$$

where B_{ij} is the coagulation coefficient and m is the critical number of primary precursors in a given precursor particle in which the stable latex particle takes place. Equation (73) contains terms corresponding to the coagulation of precursor particles of volumes v_i and v_j, a term corresponding to the growth in volume of precursor particles by propagation and a term that represents the possibility of the death of the latex particles by flocculation.

The coagulation coefficients are calculated by a lengthy procedure which is based on the DLVO theory of colloidal stability [99].

This model (i) successfully predicts the variation over 4 orders of magnitude of the particle number as the emulsifier concentration ranges from zero to a high value and the S-shaped curve observed experimentally for this dependence; (ii) predicts the dependence of the particle number on ionic strength and initiator concentration; (iii) gives quite acceptable accord with batch and continuous emulsion polymerization conversion, particle size, and molecular mass data; and (iv) copolymerization data are also successfully modelled using the Gear algorithm [100].

3.2.4 Site of initiation-partitioned polymerization approach

Recent study of the emulsion polymerization of butyl methacrylate initiated by 2,2'-azoisobutyronitrile (AIBN) has shown that in this system part of the AIBN is transferred from the oil to the aqueous phase [101]. This fact can be used for initiating the polymerization of the system consisting of the aqueous phase containing a water-soluble monomer, and the oil phase formed by an organic solvent containing the initiator. If transfer of the monomer to the oil phase is improbable, then the initiation may occur only in the aqueous phase, namely by radicals formed via thermal decomposition of the transferred initiator in the aqueous phase. Such a reaction system is a simple model of partitioned polymerization.

In this example, it is the initiator that passes from the oil phase to the aqueous phase to initiate polymerization. The oil phase containing the initiator acts as a reservoir for the initiator, just as monomer droplets dispersed in the aqueous phase serve as a reservoir of monomer during emulsion polymerization. Another example of partitioned polymerization is classical emulsion polymerization. Here the theory [75] involves radical formation by the thermal or catalytic decomposition of the initiator in the aqueous phase, and the initiation of polymerization by radicals which pass from the aqueous phase to the oil phase formed by emulsifier micelles containing monomer or, in the more advanced stages of polymerization, by polymer/monomer particles.

The principle of the initial distribution of monomer and initiator between various phases can be applied usefully in reaction systems containing components which affect the transfer of monomer or initiator from one phase to the other or which suppress or eliminate, for example, the initiation reaction. The emulsion copolymerization system acrylonitrile/butyl acrylate provides an example. In the presence of butyl acrylate the solubility of acrylonitrile in the aqueous phase decreases [102]. Suppression or elimination of radical polymerization in a particular phase may readily be achieved on adding a suitable inhibitor of radical reactions. Of the possible combinations of the three-component partitioned polymerization system, monomer (1)/inhibitor (2)/initiator (3), with respect to the initial deposition of the reaction components in the aqueous phase/oil phase, the systems of interest are: 1/2,$\overleftarrow{3}$ (the arrow denotes the component passing from one phase to the other; in this case transfer of the initiator to the aqueous phase), monomer being in the aqueous phase, inhibitor and initiator in the oil phase, and the system 3/$\overleftarrow{1}$,2. Systems 2,$\overrightarrow{3}$/1 and $\overrightarrow{1}$,2/3 require transfer of the initiator from the aqueous to the oil phase and the transfer of monomer from the aqueous to the oil phase. To carry out polymerization in the system 1/2,$\overleftarrow{3}$, the initiator (3) should pass from the oil to the aqueous phase containing monomer (1). The transfer of monomer to the oil phase, where both initiator and inhibitor are located, does not lead to polymerization as long as the oil phase contains inhibitor.

A necessary condition for successful application of the principle of partitioned polymerization of the systems monomer/inhibitor/initiator is that the inhibitor should not pass from one phase to another and that the monomer and/or initiator should be capable of transfer. This is, of course, an idealized case since a particular fraction of a compound always passes from one phase to the other. However, by suitable arrangement of the concentrations of inhibitor and initiator, the effect of the transferred inhibitor can be eliminated.

The system 3/$\overleftarrow{1}$,2 was studied as another example of partitioned polymerization [103]. Ammonium peroxodisulfate (APS) located in the aqueous phase was used as initiator. The oil phase consisted of the vinyl monomers: butyl methacrylate, methyl methacrylate, styrene or acrylonitrile, and mixtures of butyl methacrylate/acrylonitrile and styrene/acrylonitrile monomers. The stable radical 2,2,6,6-tetramethyl-4-octadecanoyloxypiperidine-1-N-oxyl (STMPO) was used as inhibitor in all systems studied.

The shape of the conversion curve of the partitioned emulsion polymerization of methyl methacrylate is characterized by a short rise in the conversion to about 5–8% and by a relatively long period of constant polymerization rate at conversions between 10 and 80%. The rate of polymerization gradually decreases over the conversion region above 80%. Neither the shape of the conversion curve nor its parameters change in the presence of STMPO inhibitor used at concentrations up to 5×10^{-4} mol dm^{-3} monomer/oil phase, i.e., until the equivalent ratio is obtained with respect to the concentration of APS (5×10^{-4} mol dm^{-3} water). No inhibition period is observed.

The polymerization of styrene in the presence and/or absence of STMPO proceeds from 5% conversion at a constant rate up to 80–90% conversion (Fig. 3).

From an analysis of the kinetics of emulsion polymerization of butyl methacrylate initiated by AIBN, the initiation mechanism and the site of initiation were proposed. On the decomposition of initiator transferred from the oil phase to the aqueous phase, the radicals formed react with the monomer molecules in the aqueous phase giving rise to oligomeric radicals. The growth of the oligomeric radicals continues in the aqueous phase until the solubility limit is reached whereupon the radicals precipitate and aggregate with the formation (nucleation) of oligomer (polymer) particles. Such a mechanism for the initiation of emulsion polymerization with the so-called oil-soluble initiator (AIBN) contradicts the mechanism of initiation of emulsion polymerization based on the idea of the escape (desorption) of one of the pair of radicals from a monomer-containing micelle or from a polymer/monomer particle [104]. However, on the basis of the results obtained it was possible to exclude the formation of polymer/monomer particles by gradual change of the micelles of the monomer-containing emulsifier, i.e. the initiation of polymerization by primary and/or oligomeric radicals entering micelles from the aqueous phase.

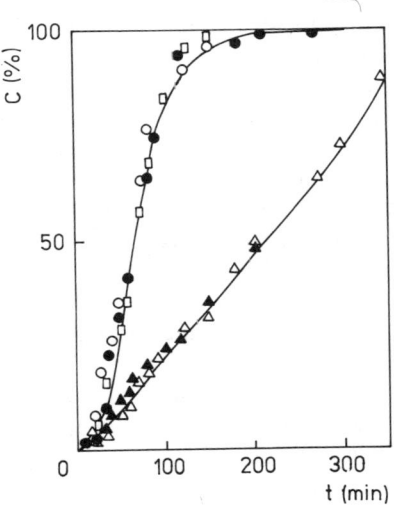

Fig. 3 — Plot of the conversion (C) of the partitioned emulsion polymerization of methyl methacrylate and/or styrene versus reaction time at 60 °C. System: water (80 cm^3)/methyl methacrylate (styrene) (40 cm^3), sodium dodecylsulfate, SDS = 1×10^{-2} mol dm^{-3} (water); [(NH$_4$)$_2$S$_2$O$_8$] = 5×10^{-4} mol dm^{-3} (water). Concentration of 2,2,6,6-tetramethyl-4-octadecanoyloxypiperidine-1-N-oxyl radical, STMPO); (□): [STMPO] = 0; (○): [STMPO] = 1×10^{-4} mol dm^{-3} of monomer (oil) phase; (●): [STMPO] = 5×10^{-4} mol dm^{-3} of monomer (oil) phase (for emulsion polymerization of methyl methacrylate). Concentration STMPO; (△): [STMPO] = 0; (▲): [STMPO] = 1×10^{-4} mol dm^{-3} of monomer (oil) phase (for emulsion polymerization of styrene). (Reprinted from ref. [103] with permission of Hüthig and Wepf Verlag, Basel.)

The results from study of the partitioned emulsion polymerization systems with the inhibitor in the monomer (oil) phase define more clearly the conclusions presented earlier and extend the validity of the mechanism of nucleation of polar (partially water-soluble) monomers to the emulsion polymerization systems of classical nonpolar monomers [105, 106]. The high concentration of inhibitor in the oil phase in the initial stage of polymerization eliminates the possibility of initiation in monomer droplets and in micelles swollen by monomer. The polymerization can, therefore, be initiated only in the aqueous phase where a part of monomer is dissolved (methyl methacrylate, 0.15 mol dm^{-3} of water). After reaching a certain molecular mass, oligomeric radicals formed via polymerization become water-insoluble and form aggregates or particles, in which the propagation proceeds. The monomer needed for propagation in the particles is provided by monomer droplets and by monomer-swollen micelles. Termination proceeds in these polymer/monomer particles by primary or oligomeric radicals entering the particle from the aqueous phase or by STMPO radicals. The latter possibility is probable, especially in the later stages of polymerization, when we might expect that the loss of monomer in the form of droplets will lead to a greater transfer of STMPO to the oil phase formed mostly by the polymer/monomer particles and to termination reactions with growing radicals. The formation of a stable emulsion showing no tendency towards phase separation also probably contributes to the transfer of STMPO from monomer droplets to a site where their termination with primary or growing radicals can take place. The transfer of STMPO through the aqueous phase might proceed not only via the diffusion mechanism but also on particle collision [104].

A polymerization system where monomer and initiator are located in a common phase is exemplified by the emulsion polymerization of methyl methacrylate and/or butyl methacrylate initiated by 2,2′-azobisisobutyronitrile [107].

In the presence of STMPO inhibitor, a relatively short inhibition period is observed, which is, for the partitioned emulsion polymerization of methyl methacrylate, ca. 10 min and for BMA ca. 25 min. The inhibition period of the system with BMA is 2.5 times longer than that with MMA, although the concentration of STMPO was five times lower as compared with the system containing MMA. At a concentration of STMPO of 10^{-3} mol dm^{-3} in the oil (monomer) phase, no polymerization occurred after ca. 280 min with BMA. Similarly, if a lower concentration of AIBN was used (5×10^{-4} mol dm^{-3} in the oil (monomer) phase at an STMPO concentration of 10^{-4} mol dm^{-3} in the oil (monomer) phase), no polymer was formed after 250 min.

These results confirm the assumption of a finite solubility of STMPO in the aqueous phase containing a surface-active compound [103]. In a study of partitioned polymerization using transfer of initiator [103] using a much higher monomer concentration in the phase where polymerization took place (e.g. acrylamide with a concentration of 0.5 mol dm^{-3} in the phase), an inhibition period of only 4 min was observed. As for the emulsion polymerization of methyl methacrylate using the same concentrations of AIBN in the oil phase (10^{-2} mol dm^{-3}) and the fivefold higher STMPO concentration in the oil phase (5×10^{-4} mol dm^{-3}), the inhibition period was ca. 10 min. On the other hand, for BMA, the inhibition period is already 25 min, despite a fivefold lower STMPO concentration. Of course, the solubility of BMA in the aqueous phase is only a fraction of the solubility of MMA, i.e. 0.0035 mol dm^{-3} in water [101].

The different inhibition periods may be interpreted by considering the highly differing monomer concentration in the aqueous phase. This interpretation is based on the assumption that the concentration of the transferred AIBN from the oil to the aqueous phase is the same for both MMA and BMA systems. Similarly, one can assume an equal (small) concentration of transferred STMPO in the aqueous phase of both MMA and BMA systems. The rate of reaction between the primary radical and monomer and of the primary radical with STMPO is determined by the concentrations of the reacting species and the values of the rate constants for the respective reactions. Although such a comparison clearly favours termination, a comparison of the concentrations of the monomers and of STMPO in the aqueous phase (the concentration of STMPO in the aqueous phase is of the order 10^{-7} to 10^{-6} mol dm^{-3}) shows that both reactions may proceed in the system, the monomer solubility in the aqueous phase exerting a decisive influence on their relation. Due to the higher concentration of MMA in the aqueous phase as compared with that of BMA, the addition of primary radicals with respect to their termination with STMPO is higher in the system with MMA than that with BMA. Evidence is provided by the values of the inhibition periods.

The emulsion polymerization of 2-ethylhexyl acrylate, despite a very low monomer concentration in the aqueous phase and a rather high concentration of the inhibitor in the oil phase (1×10^{-4} mol dm^{-3}), starts without any significant inhibition period if ammoniun peroxodisulfate (5×10^{-4} mol dm^{-3} of water) is used as the water-soluble initiator. This is in line with the foregoing discussion.

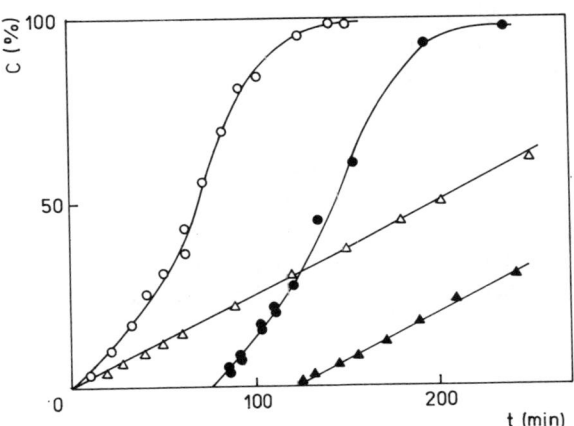

Fig. 4 — Plot of the conversion (C) of the partitioned emulsion polymerization of methyl methacrylate and/or styrene versus reaction time at 60 °C. System: water (80 cm^3), methyl methacrylate (40 cm^3) and/or styrene (40 cm^3), sodium dodecylsulfate, SDS = 1×10^{-2} mol dm^{-3} (water); [(NH$_4$)$_2$S$_2$O$_8$] = 5×10^{-4} mol dm^{-3} (water). Concentration of potassium nitrosodisulfonate, FS; (○): [FS] = 0; (●): [FS] = 1×10^{-4} mol dm^{-3} (water) (MMA system); (△): [FS] = 0; (▲): [FS] = 1×10^{-4} mol dm^{-3} (water) (styrene system). (Reprinted from ref. [109] with permission by Hüthig and Wepf Verlag, Basel.)

Relatively short inhibition periods in the partitioned emulsion polymerization of MMA and BMA are also caused by the fact that STMPO transferred to the aqueous phase is captured by the oligomer/monomer and polymer/monomer particles being formed. This lowers the concentration of STMPO in the aqueous phase and the termination of STMPO with primary or oligomeric radicals is suppressed. STMPO thus participates in the termination reactions of propagating radicals in polymer/monomer particles along with primary and oligomeric radicals. The termination of STMPO in the polymer/monomer particle is probably the reaction mainly responsible for the decay of STMPO radicals in an emulsion polymerization system.

Identical courses of the polymerization of the monomers studied, either in the presence or absence of the inhibitor in the oil phase, lead to the conclusion as to the correctness of the supposed mechanism for initiation and nucleation of particles and on the site of initiation of emulsion polymerization both for the oil-soluble initiator AIBN and for the classical water-soluble initiators like APS.

If, instead of the oil-soluble inhibitor STMPO, partially water-soluble inhibitors such as 4-hydroxy-2,2,6,6-tetramethylpiperidinyl-1-oxy radical and 2,2-diphenyl-1-picrylhydrazyl radical [108] or the completely water-soluble potassium nitrosodisulfonate (KSO$_3$)$_2$NO$^\bullet$ (Fremy's salt) were used, remarkable periods of inhibition were observed [109] as is clearly seen from the example given in Fig. 4. The rate of polymerization reaches, after the period of inhibition, the value observed for emulsion polymerization system without Fremy's salt. If, however, an oil-soluble initiator like dibenzoyl peroxide is used together with Fremy's salt in the emulsion polymerization of methyl methacrylate and/or styrene, the polymerization starts without any period of inhibition. The inefficiency of the inhibitor is due to the initiation of polymerization by oil-soluble initiator in the oil (monomer) phase. The rate of polymerization is low at low conversions of methyl methacrylate and/or styrene (up to 10% conversion) and resembles the rate of homogeneous free-radical polymerization in the bulk monomer (in monomer droplets). The nucleation of polymer particles in the aqueous phase is inhibited by the presence of Fremy's salt. The further course of polymerization system is characterized by the formation of a coagulate of highly swollen (by unreacted monomer) sticky nature. The polymerization system is macroscopically heterogeneous and, on standing, quickly separates into two phases.

Confirmation of the universality of the mechanism for oligonucleation of the particles of emulsion polymerization, regardless of the type of initiator and the type of monomer used (as long as its solubility in the aqueous phase does not reach the solubility limit of acrylonitrile), is an important contribution of partitioned emulsion polymerization to our understanding of the mechanism for initiation and particle nucleation of emulsion polymerization.

References

1. Napper, D. H. & Gilbert, R. G. (1987) *Makromol. Chem., Macromol. Symp.* **10/11** 503
2. Priest, W.J. (1952) *Phys. Chem.* **56** 1077
3. Vanderhoff, J. W. (1969) In: Ham, G. E. (ed.) *Vinyl polymerization*, vol. 2. M. Dekker, New York
4. Penboss, I. A., Gilbert, R. G. & Napper, D. H. (1986) *J. Chem. Soc., Faraday Trans. 1* **82** 2247
5. Ugelstad, J. & Hansen, F. K. (1976) *Rubber Chem. Technol.* **49** 536
6. Hawkett, B. S., Napper, D. H. & Gilbert, R. G. (1980) *J. Chem. Soc., Faraday Trans. 1* **76** 1323
7. Landsdowne, S. W., Gilbert, R. G., Napper, D. H. & Sangster, D. F. (1980) *J. Chem. Soc., Faraday Trans. 1* **76** 1344
8. Maxwell, I. A., Napper, D. H. & Gilbert, R. G. (1987) *J. Chem. Soc., Faraday Trans. 1* **83** 1449
9. Halman, L. F., Napper, D. H. & Gilbert, R. G. (1984) *J. Chem. Soc., Faraday Trans. 1* **80** 2851
10. Ballard, M. J., Napper, D. H. & Gilbert, R. G. (1984) *J. Polym. Sci., Polym. Chem. Ed.* **22** 3225
11. Lichti, G., Sangster, D. F., Whang, B. C. Y., Napper, D. H. & Gilbert, R. G. (1982) *J. Chem. Soc., Faraday Trans. 1* **78** 2129
12. Feeney, P. J., Napper, D. H. & Gilbert, R. G. (1987) *J. Colloid Interface Sci.* **118** 493
13. Ballard, M. J., Napper, D. H. & Gilbert, R. G. (1981) *J. Polym. Sci., Polym. Chem. Ed.* **19** 939
14. Lichti, G., Hawkett, B. S., Gilbert, R. G. & Napper, D. H. (1981) *J. Polym. Sci., Polym. Chem. Ed.* **19** 925
15. Nomura, M. (1982) In: Piirma, I. (ed.) *Emulsion polymerization*. Academic Press, New York, Chapter 5
16. Nomura, M. & Harada, M. (1981) *J. Appl. Polym. Sci.* **26** 17
17. Nomura, M., Harada, M., Nakagawara, K., Eguchi, W. & Nagata, S. (1971) *J. Chem. Eng. Jpn.* **4** 160
18. Whang, B. C. Y., Lichti, G., Gilbert, R. G., Napper, D. H. & Sangster, D. F. (1980) *J. Polym. Sci., Polym. Lett. Ed.* **18** 711
19. Lichti, G., Gilbert, R. G. & Napper, D. H. (1983) *J. Polym. Sci., Polym. Chem. Ed.* **21** 269
20. Whang, B. C. Y., Napper, D. H., Ballard, M. J., Gilbert, R. G. & Lichti, G. (1982) *J. Chem. Soc., Faraday Trans. 1* **78** 1117
21. Gilbert, R. G. & Napper, D. H. (1983) *J. Macromol. Sci., Rev. Macromol. Chem. Phys.* **C23** 127
22. Lichti, G., Sangster, D. F., Whang, B. C. Y., Napper D. H. & Gilbert, R. G. (1984) *J. Chem. Soc., Faraday Trans. 1* **80** 2911
23. Nomura, M., Yamamoto, K., Horie, I., Fujita, K. & Harada, M. (1982) *J. Appl. Polym. Sci.* **27** 2483
24. Nomura, M., Kubo, M. & Fujita, K. (1983) *J. Appl. Polym. Sci.* **28** 2767
25. Asua, J. M., Sudol, E. D. & El-Aasser, M. S. (1989) *J. Polym. Sci., Polym. Chem. Ed.* **27** 3903
26. Adams, M. E., Napper, D. H., Gilbert, R. G. & Sangster, D. F. (1986) *J. Chem. Soc., Faraday Trans. 1* **82** 1979
27. Morton, M., Kaiserman, S. & Altier, M. W. (1954) *J. Colloid Sci.* **9** 300
28. Eliseeva, V. I., Ivanchev, S. S., Kukhanov, S. I. & Lebedev, A. V. (1976) *Emulsion polymerization and its application in industry*. Khimiya, Moscow (in Russian)
29. Eliseeva, V. I. (1971) *Formation of polymer films*. Khimiya, Moscow, p. 8 (in Russian)
30. Gritskova, I. A., Sedakova, K. N., Muradyan, D. S. & Pravednikov, A. N. (1978) *Dokl. Akad. Nauk SSSR* **238** 607
31. Shvetsov, O. K., Zhukova, T. D. & Ustavshchikov, B. F. (1981) *Vysokomol. Soedin., B* **23** 36
32. Eliseeva, V. I., Karapetyan, N. G. & Botnyakov, I. S. (1965) *Vysokomol. Soedin., A* **7** 497
33. Avetisyan, I. S., Eliseeva, V. I. & Lapionov, O. T. (1967) *Vysokomol. Soedin., A* **9** 570
34. Brooks, B. W. (1971) *Br. Polym. J.* **3** 2617

References

35. Grancio, M. & Williams, D. (1970) *J. Polym. Sci., A-1* **8** 2617
36. Shchepetilnikov, B. V., Eliseeva, V. I. & Zuykov, A. V. (1978) *Vysokomol. Soedin., A* **20** 2097
37. Okubo, M., Katsuta, Y. & Matsumoto, T. (1980) *J. Polym. Sci., Polym. Lett. Ed.* **18** 481
38. Okubo, M., Katsuta, Y. & Matsumoto, T. (1982) *J. Polym. Sci., Polym. Lett. Ed.* **20** 45
39. Shchevchuk, L. M., Batueva, L. I., Kuvarina, N. M., Duiko, N. V. & Kulikova, A. E. (1981) *Vysokomol. Soedin., A* **23** 913
40. Svetsov, O. K., Zhukova, T. D., Ustavshchikov, B. F., Kanebskii, I. M. & Petukhov, N. P. (1981) *Acta Polym.* **32** 403
41. Makawinata, T., El-Aasser, M. S., Vanderhoff, J. W. & Pichot, C. (1981) *Acta Polym.* **32** 583
42. Nikolaev, A. F., Vishnakova, L. P., Gromova, O. N., Grigorieva, M. M. & Kleshcheva, M. S. (1969) *Vysokomol. Soedin., A* **11** 2418
43. Marten, F. L. & Hamielec, A. E. (1978) *ACS Symp. Ser.* **104** 104
44. Harris, B., Hamielec, A. E. & Marten, F. L. (1980) *ACS Symp. Ser.* **165** 315
45. Brandrup, J. & Immergut, E. H. (1981) *Polymer handbook*, vol. 2. Wiley–Interscience, New York
46. Hawkett, B. S., Gilbert, R. G. & Napper, D. H. (1981) *J. Chem. Soc., Faraday Trans. 1* **77** 2395
47. Soh, S. K. (1980) *J. Appl. Polym. Sci.* **25** 2993
48. Ballard, M. J., Gilbert, R. G., Napper, D. H., Pomery, P. J., O'Sullivan, P. W. & O'Donnel, J. H. (1986) *Macromolecules* **19** 1303
49. Mahabadi, H. K. & O'Driscoll, K. F. (1977) *J. Macromol. Sci., Chem.* **11** 967
50. Friis, N. & Hamielec, A. E. (1976) *ACS Symp. Ser.* **24** 82
51. Soh, S. K. & Sundberg, D. C. (1982) *J. Polym. Sci., Polym. Chem. Ed.* **20** 1331
52. Rabinowitch, E. (1937) *Trans. Faraday Soc.* **33** 1225
53. Smoluchowski, M. (1919) *Z. Phys. Chem.* **92** 129
54. Russell, G. T., Napper, D. H. & Gilbert, R. G. (1988) *Macromolecules* **21** 2133
55. Mills, M. F., Gilbert, R. G. & Napper, D. H. (1990) *Macromolecules* **23** 4247
56. Gardon, J. L. (1968) *J. Polym. Sci., A-1* **6** 687
57. Van der Hoff, B. M. E. (1960) *J. Polym. Sci.* **44** 241
58. Ivanchev, S. S., Pavlyuchenko, V. N. & Rozhkova, D. A. (1976) *Vysokomol. Soedin., A* **18** 2725
59. Friis, N. & Hamielec, A. E. (1973) *J. Polym. Sci., Polym. Chem. Ed.* **11** 3321
60. Friis, N. & Hamielec, A. E. (1974) *J. Polym. Sci., Polym Chem. Ed.* **12** 251
61. Benson, S. W. & North, A. M. (1962) *J. Am. Chem. Soc.* **84** 935
62. Van der Hoff, B. M. E. (1958) *J. Polym. Sci.* **33** 487
63. Friis, N. & Nyhagen, L. (1973) *J. Appl. Polym. Sci.* **17** 2311
64. Friis, N. & Hamielec, A. E. (1976) *ACS Symp. Ser.* **24** 82
65. Soh, S. K. & Sundberg, D. C. (1982) *J. Polym. Sci., Polym. Chem. Ed.* **20** 1299
66. Soh, S. K. & Sundberg, D. C. (1982) *J. Polym. Sci., Polym. Chem. Ed.* **20** 1315
67. Soh, S. K. & Sundberg, D. C. (1982) *J. Polym. Sci., Polym. Chem. Ed.* **20** 1345
68. Tulig, T. J. & Tirrel, M. (1981) *Macromolecules* **14** 1501
69. De Gennes, P. G. (1971) *J. Chem. Phys.* **55** 572
70. Stickler, M. (1983) *Macromol. Chem.* **184** 2563
71. Gardon, J. L. (1968) *J. Polym. Sci., A-1* **6** 2851
72. Ballard, M. J., Napper, D. H., Gilbert, R. G. & Sangster, D. F. (1986) *J. Polym. Sci., Polym. Chem. Ed.* **24** 1027
73. Ferry, J. D. (1970) *Viscoelastic properties of polymers*. Wiley–Interscience, New York
74. Capek, I. (1989) *Chem. Papers* **43** 527
75. Smith, W. V. & Ewart, R. H. (1948) *J. Chem. Phys.* **16** 592
76. Min, K. W. & Ray, W. H. (1974) *J. Macromol. Sci., Rev. Macromol. Chem.* **C11** 177
77. Hansen, F. K. & Ugelstad, J. (1978) *J. Polymer. Sci., Polym. Chem. Ed.* **16** 1953
78. Hamielec, A. E. & MacGregor, J. F. (1983) In: Reichert, K. K. & Geisler, E. (eds) *Polymer reaction engineering.* C. Hanser, München, p. 21
79. Feeney, P. J., Napper, D. H. & Gilbert, R. G. (1984) *Macromolecules* **17** 2520
80. Feeney, P. J., Napper, D. H. & Gilbert, R. G. (1987) *Macromolecules* **20** 2922
81. Rawlings, J. B. & Ray, W. H. (1988) *Polym. Eng. Sci.* **28** 237
82. Asua, J. M., Adams, M. E. & Sudol, E. D. (1990) *J. Appl. Polym. Sci.* **39** 1183
83. Haskell, V. C. & Settlage, P. H. (1971) In: Fitch, R. M. (ed.) *Polymer colloids*, vol. I. Plenum Publ. Corp., New York, p. 583
84. Broadhead, T. O., Hamielec, A. E. & MacGregor, J. F. (1985) *Makromol. Chem., Suppl.* **10/11** 105
85. Nomura, M. & Fujita, K. (1985) *Makromol. Chem., Suppl.* **10/11** 25
86. Dougherty, E. P. (1986) *J. Appl. Polym. Sci.* **32** 3051
87. Dougherty, E. P. (1986) *J. Appl. Polym. Sci.* **32** 3079
88. Storti, G., Vitalini, L., Albano, M., Carra, S. & Morbidelli, M. (1987) *Int. Symposium of Free-Radical Polymerization, Kinetics and Mechanism*, S. Margherita Ligure (Italy), p. 214
89. Storti, G., Carra, S., Morbidelli, M. & Vita, G. (1989) *J. Appl. Polym. Sci.* **37** 2443

90. Delgado, J., El-Aasser, M. S., Silebi, C. A., Vanderhoff, J. W. & Guillot, J. (1988) *J. Polym. Sci., Polym. Phys.* **26** 1495
91. Richards, J. R., Congalidis, J. P. & Gilbert, R. G. (1989) *J. Appl. Polym. Sci.* **37** 2727
92. Maxwell, I. A., Sudol, E. D., Napper, D. H. & Gilbert, R. G. (1988) *J. Chem. Soc., Faraday Trans. 1* **84** 3107
93. Chen, S. A. & Wu, K. W. (1988) *Polymer* **29** 545
94. Nomura, M. (1985) *Makromol. Chem., Suppl.* **10 /11** 25
95. Flory, P. J. (1953) *Principles of polymer chemistry.* Cornell University Press, Ithaca
96. Ugelstad, J., Mork, P. C., Futakamba, H. R. M., Soleimany, E., Nordhuus, I., Nustad, K., Schmid, R., Berge, A., Ellingsen, T. & Aune, O. (1983) In: Poehlein, G. W., Ottewill, R. H. & Goodwin, J. W. (eds) *Science and technology of polymer colloids*, vol. 1. NATO ASI Ser. M. Nijhoff Publ., Dordrecht, p. 51
97. Jansson, L. H., Wellons, M. C. & Poehlein, G. W. (1983) *J. Polym. Sci., Polym. Lett. Ed.* **21** 937
98. Ugelstad, J., Mork, P. C. & Aassen, J. O. (1967) *J. Polym. Sci., A-1* **5** 2281
99. Overbeek, J. Th. G. (1952) In: Kruyt, H. R. (ed.) *Colloid science*. Elsevier, Amsterdam
100. Numerical Algorithms Group (1984) *Library manual: Routines CO5NBF, DO2EBF, EO4ABF*, vol. 1
101. Bartoň, J. & Kárpátyová, A. (1987) *Makromol. Chem.* **188** 693
102. Capek, I., Bartoň, J. & Orolínová, E. (1985) *Acta Polym.* **36** 187
103. Bartoň, J., Juraničová, V. & Hloušková, Z. (1988) *Makromol. Chem.* **189** 501
104. Pavlyuchenko, V. N. & Ivanchev, S. S. (1983) *Acta Polym.* **34** 521
105. Roe, C. P. (1968) *Ind. Eng. Chem.* **60** 20
106. Fitch, R. M. & Tsai, C. H. (1970) *J. Polym. Sci., Polym. Lett. Ed.* **8** 703
107. Bartoň, J. & Juraničová, V. (1989) *Makromol. Chem.* **190** 763
108. Bartoň, J. & Juraničová, V. (1989) *Makromol. Chem.* **190** 769
109. Bartoň, J. & Juraničová, V. (1991) *Makromol. Chem., Rapid Commun.* **12** 6693.3

3.3 CHARACTERIZATION OF MICELLAR SYSTEMS AND PRINCIPAL COMPONENTS OF EMULSION POLYMERIZATION SYSTEMS

3.3.1 Micellar systems: oil/water

3.3.1.1 Micelles and their formation

An emulsifier dissolves in water and forms true solutions at high dilution. Dilute solutions show the features of single-phase homogeneous systems. As the emulsifier concentration in water increases, the emulsifier molecules associate spontaneously to form larger, usually regular aggregates called micelles.

An emulsifier molecule consists of a hydrophilic and a lipophilic part. When an emulsifier molecule dissolves in water, its hydrophilic part is hydrated by water molecules.

Its lipophilic part is attracted to others by cohesive forces, and tries to escape from water. Its solubility is given by the value of the critical micellar concentration (CMC), above which, on adding further emulsifier, only self-association of its molecules is observed.

According to the nature of the hydrophilic part of emulsifier, we distinguish anionic, cationic, nonionic, and amphoteric emulsifiers. Anionic emulsifiers make up the largest group, being based mainly on sulfate and sulfonate groups as scrutiny of their industrial applications has shown (Fig. 5). Table 4 records some familiar types of emulsifier.

The solubility of emulsifiers in the aqueous phase depends on the length of the lipophilic alkyl chain, the degree of alkyl branching, the presence of unsaturated bonds and the nature of the hydrophilic group, including the nature of the ion. The emulsifier molecules dissolve in water because water has an open structure due to three-dimensional hydrogen-bonding. This type of bonding enables the existence of clusters of water molecules containing cavities of specific sizes, which can accommodate nonpolar chains [1–3]. At a given temperature,

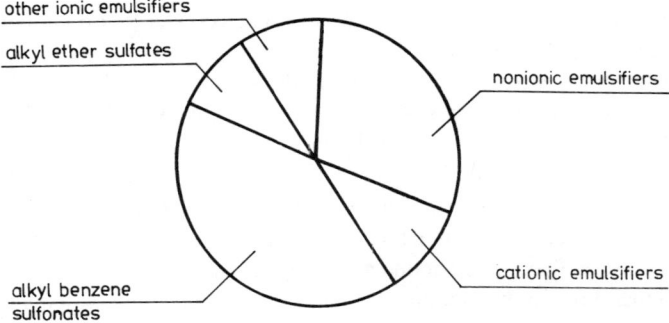

Fig. 5 — Use of synthetic emulsifiers in Europe in 1990 [1].

Table 4 — Some types of commonly-used emulsifier

Cationic emulsifiers

$CH_3-(CH_2)_n-N^+$ $(CH_3)_3 X^-$ $CH_3-(CH_2)_n-N^+\langle\rangle X^-$

$X^- = F^-, Cl^-, Br^-, I^-, NO_3^-$

Anionic emulsifiers

$CH_3-(CH_2)_n-OSO_3^- M^+$

$CH_3-(CH_2)_n-OPO_3^- M^+$ $M^+ = Li^+, Na^+, K^+, Ca^{2+}, Mg^{2+}$

Nonionic emulsifiers

$CH_3-(CH_2)_n-\langle\rangle-O-(CH_2-CH_2-O)_m-H$

$CH_3-(CH_2)_n-N\langle\rangle N-R$

Zwitterionic (amphoteric) emulsifier

$$CH_3-(CH_2)_n-\underset{\underset{CH_3}{|}}{\overset{\overset{CH_3}{|}}{N^+}}-CH_2-SO_3^-$$

only a certain amount of emulsifier penetrates into the cavities, and the rest forms associates. In other words, further addition of emulsifier provides a driving force to minimize the contact of the emulsifier hydrocarbon chains with water. Therefore, according to Langmuir's principle of differential solubility, the hydrocarbon chains cluster to form a 'core', while the polar groups interact with water [4].

McBain & Martin [5] were the first to picture the structure of micelles. They proposed that a micelle is an aggregate composed of two layers of the emulsifier; its lipophilic alkyl parts are oriented against each other and the hydrophilic parts are directed outwards into the aqueous phase (lamellar micelles). Others have shown later [6–8] that the shape of a micelle is usually spherical and that there are also other aggregates (e.g. cylindrical, rod-like, irregular). Currently we consider a micelle to be an aggregate of dynamic structure. In this cluster, the emulsifier molecules are bound by forces which allow contraction and expansion of the micelle depending on the medium. An emulsifier molecule is assumed to remain in a micelle for about 10^{-3} to 10^{-8} s, i.e. as a function of the length of the lipophilic chain [6].

Each micelle contains a specific number of emulsifier molecules, indicated by an aggregation number which determines the size and the shape of the micellar associates. The formation of spherical micelles prevails in dilute solutions; Fig.6 illustrates their shapes. The micellar core is formed by the alkyl chains of emulsifier molecules; hydrophilic groups hydrated by water molecules form a Stern layer [7]. The region directed from the Stern layer towards the aqueous phase is called the Gouy–Chapman layer and is characterized by a high concentration of counter-ions of the polar heads and separates the lipophilic interior from the aqueous phase [6]. This model visualizes a micelle as an oil drop with an ionic or polar shell [7].

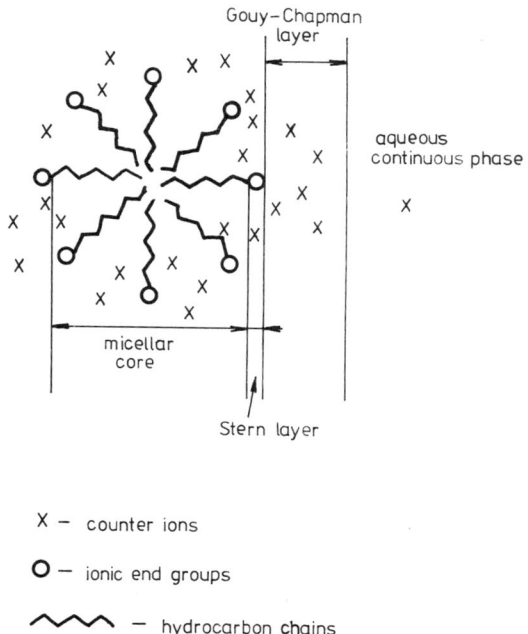

Fig. 6 — Two-dimensional presentation of a model of a spherical ionic micelle.

The investigation of micellar systems by sedimentation, diffusion, and light-scattering methods supports the formation of spherical-ellipsoidal micelles formed by a particular number of emulsifier molecules [9, 10].

Data about the structure of micelles are obtained from examination of the water penetrating into the micelle nuclei and from measurements of the internal viscosity of micelles, their aggregation number, the CMC, and the effect of temperature and of various additives on the composition and character of the surface layer of the micelle. Opinions about the structure of the micelle surface layers are in general agreement; differences occur, however, in opinions about the structure of the core (e.g. whether it contains water or not, etc.). The differences do not follow from any shortage of experimental data but from their interpretation.

The authors of ref. [11] used results from small-angle neutron scattering by emulsifier micelles to suggest a model of the core without water molecules, where the chains of the emulsifier molecules are distributed statistically.

The investigation of micelles by ^{1}H NMR and ^{13}C NMR spectroscopy [12, 13] confirmed the presence of water in the micelle cores. The results have also been supported by fluorescence measurements [14] which examined the operation and effectiveness of hydrogen bonds in micellar aggregates. Reference [15] depicts a micelle as an aggregate with a porous structure.

In addition to these ideas, there have also appeared compromise views of the partial penetration of water molecules into the micelle core. According to Corkill et al. [16] and Kurz [17], the methylene groups adjacent to the ionic heads maintain an aqueous atmosphere. Kresheck [18] proposed a model in which the first four to six carbons next to an ionic group are in a hydrated form.

The values of the internal viscosity of micelles are mostly obtained through spectral methods. The interpretations of such spectral data are different, which leads to different values of the internal density of micelles as has been indicated in several papers [19–29].

By measuring depolarization of the fluorescence of 2-methylanthracene located in the core, Shinitzky et al. [19] estimated the viscosity of the core to be 17 to 50 cP; these values are higher than those of water (1.0 cP), dodecane (1.3 cP), or 1-octanol (8.9 cP). The higher viscosity of the core was ascribed to the reduced fluidity of the lipophilic components of the emulsifier molecules in the core. The same method of determination, but a different interpretation of the results, led Ohnishi et al. [20] to estimate the viscosity of the core as being only 17 cP. Menger and Jerkunica [21] reported a value of 83 cP for the viscosity of the micelle core which is comparable to that of the monomer component.

There are also some speculations that the inside of a micelle is formed by a solid phase. Pownall & Smith [22] came to this conclusion on the basis of measurements of the fluorescence spectrum of monomeric pyrene and its excimer, which were present in the micelle cores of hexadecyltrimethylammonium bromide, and estimated the viscosity of the core to be 151 cP. Dorrance & Hunter [23] have also made fluorescence measurements which led them to the conclusion that the interior of the micelle is extremely viscous, comparable with a solid phase.

Investigation of the formation of micelles as a function of the number of emulsifier molecules forming an associate provides partial data on the mechanism of intramicellar interaction but mainly on the size of micelles. The number of emulsifier molecules is apparently not always constant for a particular type of micelle, and can vary as a function of the reaction conditions, e.g. dodecyltrimethylammonium bromide reaches aggregation numbers of 45, 61, 73, and 84 [24]. The balance of the micelles in such systems is described by an average aggregation number [25]. It is generally assumed that the number of molecules in a micelle differs from the average aggregation number by ±10%, and consequently statements about the monodispersity of micelles follow from the lower sensitivity and precision of experimental methods and most probably arise from inadequate evaluation of the experimental data [26].

The aggregation number of micelles formed by ionic emulsifiers increases with the chain length [27].

As the hydrophilic moiety of a nonionic emulsifier containing the same lipophilic part is increased, micelles of lower aggregation number are formed [28]. As a polyoxyethylene chain lengthens, the affinity of the emulsifier to water increases and its micellar affinity decreases. The reverse effect of ethylene oxide groups on micellar affinity is observed for ionic emulsifiers (alkyl ether sulfates) which is explained by the hydration of the ethylene oxide groups present [29].

One of the most important characteristics of micellar systems is the critical micellar concentration (CMC) which determines the solubility of the emulsifier and the beginning of spontaneous association of emulsifier molecules. It decreases with the increasing lipophilic chain length of emulsifiers with the same hydrophilic group (Table 5). The dependence of the CMC on the character of the emulsifier is described by the semiempirical equation:

$$\log(\mathrm{CMC}) = A - Bn, \qquad (1)$$

where A and B are constants and n is the number of carbon atoms in the lipophilic moiety of the emulsifier [30].

Table 5 — Dependence of critical molar concentration (CMC) of carboxyl-type emulsifier on length of lipophilic alkyl group

Emulsifier	Temperature / °C	CMC / mol dm^{-3}	Ref.
n-C$_7$H$_{15}$CO$_2$K	26	0.035^a	[30]
n-C$_9$H$_{19}$CO$_2$K	26	0.095	[30]
n-C$_{11}$H$_{23}$CO$_2$K	26	0.024	[30]
n-C$_{13}$H$_{27}$CO$_2$K	26	0.006	[30]
n-C$_{12}$H$_{25}$(CH$_3$)$_3$ NOH	28	0.031	[31]
n-C$_{14}$H$_{29}$(CH$_3$)$_3$ NOH	28	0.008	[31]
n-C$_{12}$H$_{25}$(CH$_3$)$_2$ (C$_2$H$_5$) NBr	25	0.014	[32]
n-C$_{12}$H$_{25}$(CH$_3$)$_2$ (C$_3$H$_7$) NBr	25	0.012	[32]

^{a} Unpurified emulsifier.

As the fraction of hydrophilic ethylene oxide groups in a molecule of a nonionic emulsifier increases, the value of the CMC and the amount of surface area of the emulsifier located on the micelle surface increases, and, on the other hand, the tendency to association of the emulsifier effectively decreases.

Increase in branching of the lipophilic chain leads to higher CMCs and causes a decrease in the ability of the emulsifier to associate and form micelles [33, 34]. Table 6 contains the CMCs of some common emulsifiers.

Table 6 — Critical micellar concentrations (CMC) of some emulsifiers at 25°C

Emulsifier	Formula	CMC /mol dm^{-3}	Ref.
Dodecylammonium chloride (DAC)	$CH_3(CH_2)_{11}NH_3^+ Cl^-$	1.2×10^{-2}	[35]
Sodium dodecyl sulfate (SDS)	$CH_3(CH_2)_{11}SO_4^- Na^+$	8.1×10^{-3} 8.2×10^{-3}	[35] [36]
Cetyl pyridinium chloride (CPC)	$CH_3(CH_2)_{15}N^+(CH)_5Cl^-$	9.0×10^{-4}	[35]
Sodium 1,2-bis-(2-ethylhexyloxy-carbonyl)-ethanesulfonate (AOT)		6.8×10^{-4}	[36]

The CMC is determined by measuring the surface tension [37] and conductivity [35] of aqueous solutions of the emulsifier as a function of its concentration. The method for surface tension measurement has been used for all types of emulsifier. The plot of the dependence of the surface tension on emulsifier concentration consists of two linear parts. At lower emulsifier concentrations, the increase of concentration is accompanied by a linear decrease in surface tension and, on reaching the CMC, the surface tension becomes stabilized and does not change even on adding further emulsifier. In contrast to the surface tension, the specific conductivity increases linearly with emulsifier concentration and the CMC corresponds to a break in the linear dependence (at concentrations higher than the CMC, the slope of the straight line is less). The so-called equivalent conductivity (the ratio of specific conductivity and equivalent concentration) exhibits a different dependence. A rapid decrease in this quantity occurs around the CMC. On the other hand, in high-frequency conductometry, a rapid rise in conductivity is observed near the CMC. Conductometry is, however, only applicable to ionic emulsifiers since the magnitude of the CMC is indicated by a sudden decrease in the conductivity of the solution above a particular emulsifier concentration.

To determine the CMC, spectral methods based on the interaction between emulsifier and dye molecules are often used. When the emulsifier used is at a concentration below the CMC, the spectra of the solutions do not change. Changes are observed on reaching the CMC when solubilization of the dyes takes place in the micelles of emulsifier [38, 39].

The water-solubility of an emulsifier is also a function of temperature, as evident from the temperature dependence of the solubility of sodium dodecyl sulfate (Fig. 7). This relationship is also characteristic for other types of emulsifier and increases linearly with increasing temperature to a particular critical value called the Krafft point. The Krafft point may be defined as the temperature above which the solubility of an emulsifier increases steeply. At this temperature, the solubility of the emulsifier becomes equal to its CMC, and therefore emulsifier micelles only exist at temperatures above the Krafft point. This is a triple point at which the emulsifier coexists in the monomeric and micellar forms and in the hydrated solid state. Below the Krafft point, the emulsifier dissolves in a molecularly-dispersed manner until the saturation concentration is reached. At higher concentrations, the hydrated solid is in equilibrium with individual molecules. Above the Krafft point, the hydrated solid is in equilibrium with both micelles and individual molecules [40, 41].

Thus, the physical meaning of the solubility curve of an emulsifier is different from that

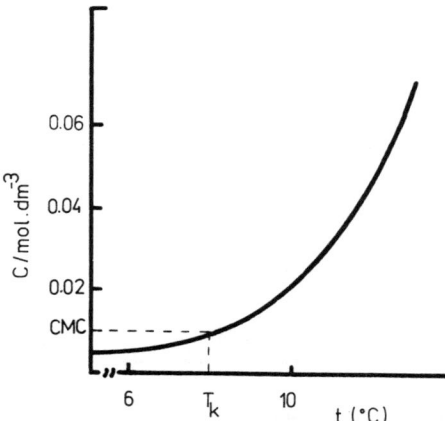

Fig. 7 — Temperature-dependence of the solubility of sodium dodecyl sulfate [41].

of ordinary substances (Fig. 7). Above the CMC the thermodynamic functions, e.g. the partial molar free energy, the activity and the enthalpy, remain more or less constant. For this reason, micelle formation can be considered as the formation of a new phase. The Krafft point depends therefore on a complicated three-phase equilibrium.

As the emulsifier molecule becomes more lipophilic, the value of the Krafft point is shifted to higher temperatures. A similar trend has also been observed by increasing the portion of the branched chains in micellar aggregates. The Krafft point is also affected by the charge of the metal cation of the emulsifier. Sodium tetradecyl sulfate is characterized by a Krafft point of 21°C, and, for the corresponding calcium salt this value is 67°C [42]. Table 7 shows the dependence of the increase of the Krafft point on the increase of the charge of the cation in salts of n-dodecyl ether sulfate.

Table 7 — Krafft points for various derivates of n-dodecyl ether sulfates [1, 43] (Reprinted from ref.[1] with permission of American Chemical Society, Washington)

Emulsifier anion	Metal cation			
	Na^+	Ca^{2+}	Sr^{2+}	Ba^{2+}
$C_{12}H_{25}OSO_3^-$	9	50	64	105
$C_{12}H_{25}OCH_2CH_2OSO_3^-$	5	15	32	62
$C_{12}H_{25}(OCH_2CH_2)_2OSO_3^-$	−1	0	—	35
$C_{12}H_{25}(OCH_2CH_2)_3OSO_3^-$	0	0	0	12

These conclusions are not unambiguous. Reference [44] found that, contrary to the statements given above, the Krafft point *decreases* with increasing length of the hydrocarbon chain, and the atomic mass and charge of the metal ion.

Although micellar colloids are in a dynamic association-dissociation equilibrium, the theoretical approach to micelles depends on whether they are regarded as a chemical species or a separate phase. The mass action model, which has been adopted ever since the discovery of micelles, is associated with the former viewpoint [45], while the phase separation model is used to describe the latter approach [46]. The mass action model must include every association constant over all stepwise associations from monomer to micelle. Consequently, this model has the disadvantage that either monodispersity of the micellar aggregation number is assumed, in spite of the fact that polydispersity exists, or numerical values of each association constant must be assumed [45]. On the other hand, the phase separation model is based on the assumption that the emulsifier activity and/or surface tension of emulsifier solutions remain constant above the CMC [47, 48]. The appearance of many different theories of micelle formation is a result of the differing properties of micelles [45, 46]. That is, a micelle may not have such a large aggregation number that it can be regarded as a phase in the usual sense, but it has properties similar to those of a phase. At the same time there are too many aggregated monomer molecules in each micelle for it to be regarded as a chemical species, even a bulky chemical species.

The aggregation of an anionic emulsifier, ME, can be written as [49]:

$$m_1 M + n_1 E \underset{}{\overset{K_1}{\rightleftarrows}} M_1,$$

$$m_2 M + n_2 E \underset{}{\overset{K_2}{\rightleftarrows}} M_2,$$

$$\vdots$$

$$m_i M + n_i E \underset{}{\overset{K_i}{\rightleftarrows}} M_i, \tag{2}$$

where ME is an emulsifier molecule, M is the counter-ion with charge z_K and E emulsifier ions with charge z_a.

From Eq. (2) the micellization constant is written as

$$K_i = [M_i] / \left([M]^{m_i} [E]^{n_i} \right). \tag{3}$$

It is known to be impossible to determine every micellization constant of Eq. (2) experimentally. Therefore, the monodispersity of the micelles is assumed to treat these species theoretically. Hence the following relation is tacitly assumed for the micellization constants:

$$K_n = \sum n_i K_i [M]^{m_i} [E]^{n_i} / (n [M]^m [E]^n). \tag{4}$$

The micelle concentration is given by

$$[M_n] = K_n [E]^n [M]^m. \tag{5}$$

Equation (5) has three parameters to be determined, K_n, n, and m, which are the most important factors for the mass action model of micelle formation.

When the temperature of a solution of a nonionic emulsifier is increased, the well-known cloud-point phenomenon is observed [50, 51], i.e. upon heating, an aqueous solution of nonionic emulsifier becomes cloudy (reversible) at the temperature of incipient phase separation (cloud point).

The increase in temperature causes partial dehydration and finally results in separation of the emulsifier-rich phase. The cloud point is characteristic for each nonionic emulsifier and varies regularly with the length of the polyoxyethylene (POE) chain. The cloud point increases fairly rapidly with the number of ethylene oxide units [51], but is markedly affected by some additives [52]. The sign (increase or decrease) and magnitude of the effect depend on the species and amount of the additive. The addition of electrolytes to aqueous solutions of nonionic emulsifiers is known to modify their cloud points [53, 54]. Two effects can be observed depending on the nature of the salt: the cloud point decreases due to the dehydration of the emulsifier by the salt (salting-out) or it increases, reflecting the enhanced solubility of the emulsifier in the aqueous phase (salting-in) (see also section 3.3.1.2).

According to Schott *et al.* [55], the shifts in the cloud point can be considered as the sum of the contributions of the individual ions forming the electrolyte. The additivity law can be written as:

$$\Delta T_{p,salt} = \Delta T_{p,anion} + \Delta T_{p,cation} . \tag{6}$$

Much attention has been paid for decades to the particle solute–solvent interaction due to the lipophilicity of the hydrocarbon chains. Emulsifiers have been classified according to the HLB value (hydrophilic-lipophilic balance), which constitutes an empirical numerical method of correlating their emulsifying and solubilizing properties. Griffin [56, 57] and Davies & Rideal [58] have shown that it is possible to evaluate the HLB of a compound from its chemical formula from the equation

$$HLB = \sum A - mB + 7 , \tag{7}$$

where m corresponds to the number of lipophilic groups and A and B are two parameters, which are respectively, characteristic of the hydrophilic and lipophilic groups. The HLB value can also be calculated from the well-known Griffin equation [57]:

$$HLB = 20 M_H / (M_L + M_H), \tag{8}$$

where M_H and M_L are the masses of the hydrophilic and lipophilic moieties, respectively.
For emulsifiers whose sole hydrophilic moiety is polyoxyethylene,

$$HLB_G = E/5 , \tag{9}$$

where E is the mass percentage of ethylene oxide of the emulsifier molecule [56, 57]. The subscript G refers to Griffin. For example, for polyoxyethylated dodecanols or octoxynols (TRITON X), Eq.(9) becomes

$$HLB_G = \frac{881 p}{44.05 p + A} , \tag{10}$$

where p is the number of ethylene oxide repeat units per molecule and A is the molecular mass of the lipophilic moiety, namely 206.3 for octylphenol and 186.3 for dodecanol.

Davies & Rideal [58] treated the HLB as an additive and constitutive property of the emulsifier, assigning the following numerical values to functional groups of nonionic emulsifiers: 0.33 for $-CH_2CH_2O-$, 1.9 for $-OH$, and 0.475 for $-CH-$, $-CH_2-$, $-CH_3$ and $=CH-$ groups of the hydrocarbon moiety.

The equation for Davies' HLB was

$$HLB_D = \sum \text{hydrophilic group values} - 0.475 \begin{pmatrix} \text{number of carbon atom} \\ \text{in hydrocarbon moiety} \end{pmatrix} + 7 \qquad (11)$$

which for nonionic polyoxyethylated emulsifiers becomes

$$HLB_D = 0.33p + B . \qquad (12)$$

For octoxynols, $B = 2.25$ and for polyoxyethylated dodecanols, $B = 3.2$.

3.3.1.2 Influence of additives on the formation and nature of micelles

In the preceding part of this chapter we have mentioned the effect of the character of the cation on the properties of the micellar system. The addition of a simple electrolyte, such as NaCl, markedly affects the CMC of an ionic emulsifier. Table 8 highlights some examples of this effect. Fluctuations in CMC values caused by the addition of electrolyte result from changes in the mutual repulsion of polar groups; this also leads to a more compact arrangement of the molecules in a micelle and/or their greater number in an aggregate. The decrease in mutual repulsion leads to closer packing of emulsifier molecules in a micelle and hence to an increase in aggregation number (Table 9), and the size of the micelles [63, 64]. In other words, if the electrical surface potential is reduced, more polar heads, and hence more emulsifier molecules, can make up a given micelle.

The decrease in CMC of an emulsifier on adding electrolyte can be explained by the Frank–Evans 'iceberg' theory [2]; the micelle formation is an entropy-directed process and is influenced by changes in the water structure surrounding the emulsifier ions. A disturbance in the regular structure of water positively influences micelle formation as a result of the lowering of the water solubility of the emulsifier molecules [65].

An increase in the ionic strength of an aqueous solution increases the tendency to associate molecules of ionic emulsifiers [66, 67]. At certain NaCl concentrations, not only the size but also the shape of the micelles changes, e.g. spherical micelles of sodium dodecyl sulfate become rod-like on adding more NaCl [68].

We can observe and evaluate the interactions between emulsifier and electrolyte and their influence on micelle formation by spectral measurements (through changes in the absorbance of a dye added to the system). The spectrum of Bromopyrogallol in an aqueous solution of cetylpyridinium bromide (pH 4, ionic strength 0.2 M) shows an absorption maximum at 580 nm when cetylpyridinium bromide micelles are present. The addition of NaCl releases the dye from the micelles and the absorption spectrum changes to resemble that of the dye in the absence of emulsifier micelles and raises the value of the pK_a of the dye-emulsifier associate [69, 70].

Table 8 — Effect of adding of electrolyte on CMC of sodium dodecyl sulfate (SDS) [59] and cetyl pyridinium bromide (CPB) [60]

	CMC^a /10^{-4} mol dm^{-3}
Aqueous solution of emulsifier :	
CPB	68.6
Electrolyte added:	
KCl, 0.01 M	37.6
KCl, 0.1 M	42.0
KCl, 1 M	71.4
HCl, 0.2 M	35.0
HCl, 1 M	49.6
NaCl, 0.1 M	34.0
NaCl, 1 M	49.6
NaCl, 1.5 M	45.6
Aqueous solution of emulsifier:	
SDS	82
Electrolyte added:	
NaCl, 0.06 M	20
CaCl$_2$, 0.006 M	70
CaCl$_2$, 0.012 M	60

^{a} Obtained from surface tension measurements at 20°C.

Table 9 — Aggregation number of micelles of some emulsifiers at 25°C [37, 61, 62]

Emulsifier	Without additive	Aggregation number with NaCl			
		0.1 M	0.3 M	0.5 M	0.8 M
$C_{12}H_{25}OSO_3Na$	60	91	110	212	270
$C_{12}H_{25}OCH_2CH_2OSO_3Na$	72	96	—	—	—
$C_{14}H_{29}OSO_3Na$	80	—	—	—	—

The light scattering from aqueous solutions of sodium dodecyl sulfate in the presence of sodium halide (concentration ~ 0.8 M) indicates a dependence of the equilibrium between the formation of spherical and rod-like micelles on the emulsifier concentration [71]. At low emulsifier concentrations, the equilibrium is shifted towards the formation of spherical micelles while at higher emulsifier concentrations, rod-like associates are formed.

In contrast to ionic emulsifiers, the associates and CMC of nonionic emulsifiers are only slightly influenced by addition of electrolyte [72]. For most nonionic emulsifiers, the formation of spherical micelles is observed both in the presence and absence of electrolyte. A few papers have also appeared pointing to the production of larger rod-like aggregates

[73, 74]. Association into such entities was observed for nonionic emulsifiers, where the lipophilic moiety dominated over the hydrophilic, e.g. in the case of hexaoxyethylene dodecyl ether [73] and heptaoxyethylene cetyl ether [74]. The formation of large rod-like micelles has also been observed on using dimethylolyl aminoxide [75] in the presence of NaCl. The average molecular mass of the associates reached a value as high as 10^5; their average size varied around 60 Å [76].

In the absence of electrolyte, the length of the rod-like emulsifier micelle (dimethylolyl aminoxide) varies around 10^3 Å; on adding NaCl solution with a concentration of 0.01 M, the length becomes 4×10^3 Å. The radius of gyration of the micelles changes from 200 Å (in the absence of electrolyte) to 300 Å on adding electrolyte [77, 78].

Gamboa & Sepulveda [79] found that with increasing electrolyte concentration (NaCl, NaBr, NaNO$_3$) the viscosity of aqueous solution of anionic (sodium dodecyl sulfate) and cationic (cetyl trimethylammonium bromide) emulsifier increased. The high viscosities observed in these systems are interpreted in terms of a micellar sphere-to-rod transition which occurs over a certain range of concentrations of either emulsifier or added salt.

Although the origin of the effects is still a matter of discussion, a generally-accepted interpretation is based on a modification of the water structure around the ions, which affects the water–emulsifier interactions [41, 80]. Some ions (Na^+, K^+) are structure-makers, so that water molecules surrounding them are more organized than in pure water. This result in a lower hydration of the emulsifier polar head since the work required to introduce an emulsifier molecule in a cavity near the ion is increased because of the ordering of the water molecules [79].

By contrast, other ions are effective in destroying the water structure. This promotes solubilization of the polar heads of the emulsifier since the work required to form a water cavity close to the polar head is decreased; consequently, the emulsifier is made more soluble in water due to this salting-in effect. This behaviour is particularly observed for large, polarizable anions such as thiocyanate or iodide [81].

In addition, the growth of micelles above a given salt concentration can be the result of two independent factors: the small repulsive energy due to the small degree of ionization and the small attractive energy contribution corresponding to a large Hamaker constant [82].

Hayashi & Ikeda [68] suggest that micelle formation in concentrated salt solutions can occur in two steps according to the processes

$$m D \rightleftarrows D_m , \qquad (13)$$

$$n D_m \rightleftarrows D_{mn} , \qquad (14)$$

where D is the monomer, D_m is the primary or spherical micelle, and D_{mn} the rod-like micelle. The first step would correspond to the aggregation of monomers to give spherical micelles and the second step to the transition of sphere-to-rod micelles.

The structure and shape of micelles are affected not only by electrolytes like NaCl but also by interactions with organic components, mostly alcohols. These enter micelles without causing any substantial change of their volume; they alter, however, their structure (mixed micelles appear). The hydroxyl groups of the alcohol are bound through hydrogen bonds to the polar groups of the emulsifier and reduce thus repulsive forces between the polar groups

[83]. Like electrolytes, alcohols lower the CMCs of the emulsifiers. The extent of the influence of the alcohol depends on its character; more precisely, as the chain length increases, the force of lipophilic interaction, which positively influences micelle formation, increases. 1-Propanol thus lowers the CMC of n-dodecyltrimethylammonium bromide and dodecyl sulfate more effectively than ethanol [84].

The addition of an organic molecule generally reduces the CMCs of emulsifiers. The degree of reduction is proportional to the length of the alkyl group. The CMC of n-dodecyltrimethylammonium bromide will decrease on adding a compound of type $CH_3-(CH_2)_n-COOH$ with increase of the number of $-CH_2-$ groups, from 0 to 5–40 times. At the same time the aggregation number decreases [85].

3.3.1.3 Microemulsions and miniemulsions

The term microemulsion was first used by Hoar & Schulman [86] thus characterizing a transparent disperse water system containing a large amount of emulsifier. The system exhibits excellent thermodynamic stability and low turbidity.

The spontaneous formation and thermodynamic stability of microemulsions require a low interfacial tension between water and oil [87], which is obtained by the presence of a large amount of one or more emulsifiers. The distribution of the emulsifier between the droplets, the continuous medium and the interface separating them, is largely determined by their solubilities in the fluid media and their adsorbabilities [88].

The properties of a microemulsion are given by the nature of the specific interactions of the components of the emulsifier system on the surface of the micelles or at the interface. By effective applying these interactions, a stable emulsion occurs even on less vigorous stirring. By contrast, even vigorous stirring does not lead to a stable microemulsion [89].

Microemulsions are often droplet-type dispersions, either of oil-in-water (O/W) or of water-in-oil (W/O). However, the droplet sizes are very small, usually 10 nm, i.e. ca. 100 times smaller than the sizes of typical emulsion droplets. They usually behave like Newtonian fluids; their viscosity is comparable to that of water, even at high droplet concentration, probably because of reversible droplet coalescence. Indeed the microstructure evolves constantly due to exchanges of the constituents. This important feature strongly affects the dynamic properties of a microemulsion, which sometimes behaves almost as a molecular mixture of its constituents [90].

The existence of ordered structures, spherical droplets and interfacial properties has been the basis for the interpretation of a large number of different theoretical and experimental studies of microemulsions [91–106].

The formation and properties of a microemulsion depend on the ability of the emulsifier to coil the lipophilic and hydrophilic sequences; this also determines the extent and direction of the curvature at the interface. In order to characterize an emulsifier with respect to the formation of a microemulsion, Mitchell & Ninham [98] introduced an expression $v/(a_0 l_c)$ — the 'packing ratio' where v is the partial volume of emulsifier, a_0 is the surface area of the active group, and l_c is the maximum chain length. The magnitude of the packing ratio of the emulsifier shows whether production of a microemulsion is possible and also provides information about its type. If, with increase of temperature the value of $v/(a_0 l_c)$ increases, but does not reach the value of 1, the solubilization of hydrocarbons increases and a normal type of microemulsion is formed (O/W). If, however, $v/(a_0 l_c)$ reaches the value of 1 or higher, then a W/O microemulsion results.

Mitchell & Ninham [98] also reported that the emulsifier film possesses elasticity. It is usually bent due to the spontaneous curvature C_0 of the film. An emulsifier film adsorbed at the O/W interface is a two-dimensional system. The emulsifier film bends spontaneously towards the medium in which the emulsifier is more soluble. If the polar part of the emulsifier is more bulky than its lipophilic part, the interface will curve spontaneously towards water and O/W structures will be favoured. The spontaneous curvature C_0 is consequently positive. If the polar part is less bulky than the lipophilic part, the interface will curve in the opposite direction and W/O structures will be favoured; the spontaneous curvature is then negative.

Some space-filling models have been developed to describe the structures of microemulsions. Talmon & Prager [91] proposed a model using polyhedra. The polyhedra are filled with either oil or water and the volume of the emulsifier film is neglected. If the volume fraction of oil is large, the water polyhedra are isolated: this represents the W/O structure. If the volume fraction of water is large, the oil polyhedra are isolated: an O/W structure is favoured. A simpler space-filling model (with cubes) was proposed by de Gennes & Taupin [92]. This model predicts that two-phase systems with symmetrical O/W and W/O microemulsion can coexist; this is never observed in practice, because of the preferred requirements of curvature.

Droplet structures are very common in microemulsion systems [99]. Since the interfacial film is mainly incompressible, the area per emulsifier molecule is close to or equal to the saturated state Σ. The droplet radius can be calculated easily by assuming the emulsifier molecules to sit at the droplets' surface. One obtains

$$r = 3\varphi_o/C_E \tag{15}$$

and

$$r = 3\varphi_w/C_E \tag{16}$$

for, respectively, O/W and W/O droplets. C_E is the number of emulsifier molecules per unit volume, φ_o is the volume fraction of oil and φ_w that of water.

Osmotic virial coefficients can give some information about the interaction between droplets and the microemulsion structure [99]. This approach predicts that when φ is small ($\varphi \lesssim 0.1$), the measured diffusion coefficient D varies with droplet volume fraction φ as

$$D = D_0(1 + \alpha\varphi), \tag{17}$$

$$D_0 = kT/6\pi\eta R_H, \tag{18}$$

where α is the virial coefficient, η is the oil viscosity, and R_H the droplets' hydrodynamic radius. For hard spheres $\alpha \simeq 1.5$. If a supplementary attractive potential is present, $\alpha > 1.5$.

When the interaction potential between droplets is sufficiently attractive, an appreciable number of higher order aggregates can be found. In O/W and W/O microemulsions, the number of clusters increases when the attractive interactions increase, as well as the lifetime of the clusters.

When the volume fraction φ of droplets is sufficiently large, an infinite cluster appears:

this corresponds to the percolation threshold: $\varphi = \varphi_p$. In W/O microemulsions where the potential is attractive, a large increase in electrical conductivity is observed around φ_p [100]. Space-filling models neglect the role of the emulsifier and predict the occurrence of bicontinuous structures above $\varphi_p \sim 0.16$.

In the so-called 'pseudophase' model the components of the microemulsion are considered to be distributed in equilibrium among three different types of domain in the solution: (i) an aqueous domain containing some emulsifier and coemulsifier, (ii) a hydrocarbon domain containing all of the nonpolar hydrocarbon, some emulsifier, and water plus coemulsifier in certain proportions and (iii) a membrane domain that separates the aqueous and hydrocarbon domains and contains most of the emulsifier and some coemulsifier [97].

The Winsor bicontinuous microemulsions have been the most widely studied [95, 96]. Experimental evidence for the existence of bicontinuous structures was found, especially in cases where the microemulsions are in equilibrium with both excess oil and water. The microemulsions are in an inversion region, and the interfacial film between the oil and water microdomains has a spontaneous radius of curvature close to zero. The structure is therefore lamellar-like.

Microemulsions can coexist with many different types of other phase: oil, water, liquid crystals or other microemulsions [99].

Microemulsion properties are extremely varied, and the enormous diversity of their practical applications is a consequence. One of their disadvantages is the large amount of emulsifier required to stabilize them because of the small droplet size. Although many microemulsion properties are well-understood, much remains to be explained, i.e. the inversion mechanisms, structures other than droplets, the role of thermal fluctuation, and, finally, the dynamic properties.

Of other theories concerning the formation and stability of microemulsions, we can mention the interfacial theory also called the theory of mixed films [89, 101, 102], and the solubilization [103, 104], and thermodynamic [105, 106] theories.

The formation of a microemulsion is known to be manipulated with any of a multitude of variables: electrolyte concentration, temperature, emulsifier structure, and coemulsifier concentration. The system variables which govern the microemulsion environment may be modelled in terms of HLB [107]. At high HLB, the emulsifier prefers an aqueous environment and therefore, one can observe microemulsions in brine. On the other hand, a low HLB favours microemulsion formation in oil. At intermediate values, a microemulsion which can be said neither to be oil-continuous nor water-continuous may be in equilibrium with excess brine and oil phases. Bicontinuous structures have been proposed to describe these intermediate-phase microemulsions [108]. By changing the environment of the emulsifier, for example, by varying the coemulsifier concentration, the system can be transformed from a microemulsion which is water-continuous, and which can exist in equilibrium with an excess of oil phase, to systems in which the microemulsion exists in equilibrium with the aqueous phase.

The nature and structure of microemulsion also depend on the values of the diffusion coefficients of individual components of the emulsion system (emulsifier, coemulsifier, oil and water) [109]. The measurements of FT-NMR spectra of these components gave rather high values for the diffusion coefficients which were comparable with those of the corresponding pure liquids [110]. This fact eliminates a model of microemulsion formation through the aggregation of emulsifier and coemulsifier into larger ordered aggregates; these are probably irregular clusters of emulsifier and coemulsifier molecules.

The high diffusion coefficients of the components of microemulsion systems obtained by NMR have also been described elsewhere [111], varying around 10^{-9} m^2 s^{-1}. These values are approximately two orders of magnitude higher than predicted for discontinuous media. Based on them, microemulsions can be regarded as smaller aggregates of emulsifier molecules with two bicontinuous phases (water and oil), with an easily deformable, flexible interphase. In the presence of an alkanol $C_nH_{2n+1}OH$, $n < 6$ no separation of the lipophilic and hydrophilic regions was observed and the structure may be pictured as a bicontinuous model. Higher alcohols ($n > 6$) favour the formation of defined cores and hydrophilic zones of microemulsion aggregates.

Ionic emulsifiers generally require use of a coemulsifier to form microemulsions and emphasis has been laid on medium-chain-length alcohols [112] or occasionally tertiary amines [113]. Microemulsions typically require 6–10% by mass of emulsifier and 8–14% coemulsifier [114]. Friberg and Buraczewska [115] have also observed that, as the hydrocarbon level exceeds 50% of the components other than water, the ability to solubilize water decreases sharply. This limits one's ability to dilute the microemulsion system with the hydrocarbon or to add large amounts of water.

Mixtures of aqueous solutions of anionic emulsifier (e.g. sodium lauryl sulfate) with alcohol (coemulsifier) with $n > 4$ [116] give stable microemulsions of excellent stability even after adding additives including water. The systems show the characteristic S-shaped conductivity curve, typical for stable microemulsions. When using alcohols with $4 > n > 2$, the increase of water concentration was accompanied by a conductivity increase up to a particular water concentration.

According to Lang *et al.* [117] a coemulsifier (alcohol with shorter alkyl chain) considerably shortens the lifetime of an emulsifier molecule in a micelle. Its presence causes a disorganization of the inner structure of the micelle, and increases the flexibility of the interphase and the whole micelle. Microemulsions formed by an anionic emulsifier and alcohol exhibit high structural flexibility and a dynamic character which enables the rapid decomposition and re-formation of the associates.

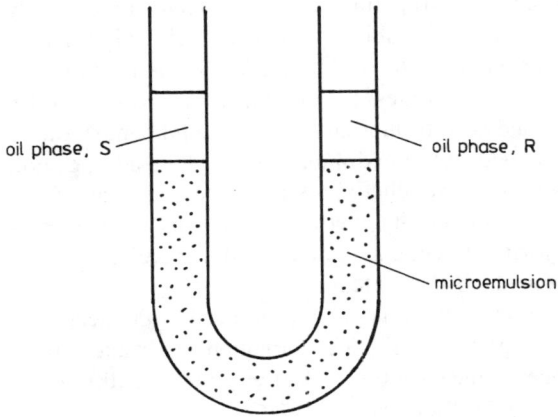

Fig. 8 — Scheme of a reactor used for investigating the transfer of compounds of lipophilic character via an aqueous medium.

As regards their dynamic nature, microemulsions are used for the transfer of lipophilic compounds through aqueous media. The mechanism of transfer of these compounds (e.g. arenes) was discussed by Xenakis & Tondre [118, 119] (see Fig. 8). The lipophilic-solubilized compound dissolved in oil S passes to an oil phase R via diffusion in microemulsion particles. The transfer is possible via the rapid solubilization and desolubilization of the lipophilic substance in the aqueous phase.

Greatly enhanced water-solubilizing ability is observed at high hydrocarbon levels when quaternary ammonium salts are used in place of the commoner anionic emulsifiers [120]. These systems still require the usual amounts of emulsifier and coemulsifier.

Of the emulsifier systems (sodium dodecyl sulfate, sodium laurate, and tetradecyltrimethylammonium bromide) described in [121], hexylamine was found to give excellent water-solubilizing capacity at high hydrocarbon levels with extremely low emulsifier concentrations and very low levels of coemulsifier; in general, hexylamine is far superior to either pentylamine, pentanol or hexanol. This may be due to the good solubility of water in the coemulsifier coupled with sparing solubility of the coemulsifier in water.

In contrast to microemulsions, miniemulsions are milk-like, opaque and of lower viscosity; their particle diameters vary between 100 and 200 nm. They are obtained by the emulsification of an oil phase in an aqueous solution of emulsifier or of an emulsifying blend. In many cases, they are products of the emulsion polymerization and copolymerization of vinyl monomers.

The stability of miniemulsions is a function of temperature, the stirring mechanism, and the nature of the emulsifier system. Ugelstad *et al.* [122] reported that the order of mixing the components of disperse system influences the properties of the emulsion. An emulsion of higher stability is prepared by adding an oil phase to an aqueous solution of emulsifier and coemulsifier (fatty alcohol). If, however, the alcohol is first dissolved in the oil phase, and only then the mixture is emulsified in an aqueous solution of anionic emulsifier, no stable emulsion is formed but phase separation is observable. If additional emulsifier is added to such a mixture, the emulsification will lead to stabilization.

Conductometric and centrifugation methods were used for investigating the stability of microemulsions of a cationic emulsifier and cetyl alcohol coemulsifier [123, 124]. Variation of the components and analysis of the corresponding emulsions by conductometric titrations and centrifugation led to the conclusion that only a particular type of conductometric curve is attributable to a stable miniemulsion. The conductometric curve of a stable emulsion is characterized by a rapid decrease in conductivity on adding a small amount of oil and reaches a minimum at a particular oil concentration. After reaching the CMC and with further increase of the oil concentration, a rapid increase in conductivity is observed and a maximum is reached; on adding another dose of oil phase beyond this maximum, the conductivity slightly decreases and finally reaches a plateau. Fluctuations in the conductivity were explained by a change in the association of molecules which form a micelle associated with the 'release' or 'tying-up' of the most conducting components of the system.

The conductometric method has also been used with success to test the stability of miniemulsions by others [125–127]. The formation of a stable emulsion was accompanied by negligible changes in the solution conductivity even for different conditions of preparation and larger amounts of additives.

The kinetic studies of Grimm *et al.* [128] have shown that the composition of the emulsifying system remarkably affects the formation and stability of a miniemulsion. Stable emulsions have only been prepared at certain ratios of ionic emulsifier and coemulsifier. In systems containing sodium lauryl sulfate or hexadecyltrimethylammonium bromide combined with cetyl or lauryl alcohols, stable emulsions were formed only at ratios: of emulsifier to coemulsifier of 1:1 or 1:3. It has also been shown that the use of an emulsifier and coemulsifier with the same alkyl chain lengths leads to emulsions with higher stability. The stability of an emulsion was positively influenced by the duration of stirring, reduction of the rate of addition of the oil component, and the use of temperatures higher than the melting point of the fatty alcohol.

3.3.1.4 Solubilization

An important feature of emulsifiers and emulsifying systems is that they increase the water solubility of lipophilic compounds. Lipophilic benzene dissolves in water in the absence of emulsifier to less than 0.1 mol dm^{-3}. On adding 0.25 mol dm^{-3} of sodium dodecyl sulfate or hexadecyltrimethylammonium bromide, the water-solubility of benzene increases tenfold, and at higher emulsifier concentration, the water solubility of benzene may be increased up to 0.9 mol dm^{-3} [129].

The term 'solubilization ratio' was introduced to classify emulsifiers according to their solubilizing efficiency. The ratio expresses the maximum number of molecules of lipophilic compound solubilized by a given number of emulsifier molecules that form a micelle [130]. The solubilization ratio can be determined by various methods, i.e. turbidimetry, light (neutron) scattering, spectrophotometry and extraction methods.

Turbidity is apparent after adding the lipophilic compoud, due to the change in particle size of the dispersions [130, 131].

Spectrophotometry can be used to characterize the solubilization capacity of emulsifiers or micelles which are spectrally inactive (i.e. do not absorb light at certain wavelengths) and, on the other hand, for spectrally active lipophilic compounds in either the visible or UV regions [132]. This method can also be used to investigate the solubilization of several substances if their absorption bands do not overlap.

Liquid extraction combined with mass spectrometry is used for evaluating blends of lipophilic compounds solubilized in micelles [133]. A solubilizer or a corresponding blend is extracted by a lipophilic solvent, e.g. CCl4, and the solution obtained is analysed mass spectrometrically.

The stripping method is mainly used for determining the solubility of gases in aqueous micellar solutions [134]. Data on the solubility of a gas introduced into a dispersion are obtained by analysis of the solution and carrier gas by gas chromatography. An advantage of this method is the possibility of using small amounts of sample, the simple and rapid procedure, and high reproducibility.

The use of the same method by different authors for the determination of solubilization parameters can give different results according to the various interpretations of the experimental data. By applying UV spectroscopy to determine the solubility of benzene in micellar systems, Rehfeld [135] came to the conclusion that benzene is localized in micellar cores (cationic cetyltrimethylammonium bromide (CTAB) and anionic sodium dodecyl sulfate (SDS) were used as emulsifiers). However, spectral determinations of the solubility of benzene and its alkyl derivatives in micellar solutions of CTAB and SDS led Mukerjee & Cardinal [136] to the conclusion that benzene is located in the micelle shell. The authors regard benzene as a polar component; and, by contrast, alkyl benzene derivatives are considered as nonpolar and are situated in the micelle cores.

Using ^{1}H NMR, Eriksson & Gilbert [137] have found the benzene to be located only at the polar interface. Fendler & Patterson [138] studied the reactivity of hydrated electrons generated by pulse radiolysis in micellar systems, and proposed that benzene is located in the interfacial region in CTAB micelles, whereas in SDS micelles it is located in the interior.

The water solubility of lipophilic compounds was found to be a function of the molar volume of the solubilized compound. Table 10 contains some values supporting this picture. Some correlation was observed between the values of the surface tension and the solubility of lipophilic substances. The addition of cationic or anionic emulsifiers remarkably increases their solubility, as seen from Table 10. The solubilizing ability of an emulsifier as characterized by the molar solubilization ratio depends on the volume of the solubilized compound [130, 134, 139] and is also somewhat affected by its polarity [140].

Table 10 — Properties of solubilizates and molar solubilization ratios S_R of lipophilic molecules at 25°C [134, 141, 142]

Solubilizate	Water solubility /mol dm^{-3}	Molecular volume /Å^3	Interfacial tension against water /mN m^{-1}	S_R 0.1 M CPC^a	S_R 0.1 M SDS
n-Decane	6.0×10^{-8}	323	52.0	0.32	0.20
n-Hexane	7.0×10^{-4}	217	50.70	0.77	0.39
Cyclohexane	8.9×10^{-4}	179	50.20	—	—
Cyclohexene	2.6×10^{-3}	167	44.20	1.72	1.13
Toluene	7.2×10^{-3}	176	36.10	2.22	1.36
CCl$_4$	7.54×10^{-3}	161	45.00	1.47	0.69
Benzene	2.3×10^{-2}	146	33.93	2.99	1.68

^{a} Cetylpyridinium chloride.

Hoskins & King [143] studied the effect of co-emulsifier (n-pentanol) on the mechanism of solubilization of ethane in dodecyl sulfate micelles. The authors have found that, when using lower concentrations of coemulsifier, its adsorption takes place only on the surface of the micelles. At higher concentrations of n-pentanol, the penetration of coemulsifier molecules into micelle cores was observed which changed the interior of the micelles. Through swelling, the volume of the micelle core increases, and, at the same time, conditions for the entry of larger amounts of an organic compound into the micelles are improved.

Small-angle neutron scattering [144] assigned another role to the coemulsifier (n-pentanol) during its interaction with micelles of sodium octanoate. An increase in the concentration of coemulsifier brings about a parallel increase in water solubility. The penetration of n-pentanol molecules into the micelles favours a release of the molecules of micellar emulsifier which diffuses into the aqueous phase. Even though the surface and interior of the micelles change as a consequence of the adsorption of n-pentanol molecules, no change of aggregation number is observed.

Cationic emulsifiers display a high solubilizing ability for aromatic hydrocarbons [145]; their high activity results from a strong interaction between the cationic emulsifier and lipophilic compounds and increases with the length of the alkyl chains [146]. Anionic emulsifiers also solubilize aromatic compounds effectively [147], but nonionic emulsifiers are rather ineffective [148].

The increase in the solubilization of lipophilic compounds in water in the presence of micelles is explained by the dynamic nature of the formation and decay of micelles [149]. The micelles decompose into monomer emulsifier and, conversely, they aggregate to create organized structures. In the presence of oil drops, the emulsifier molecules adsorb onto the surface of the drops and form micelles with oil-phase swollen cores. Mutual interactions of the emulsifier molecules as well as their interactions with the oil and/or the aqueous phases obviously determine the mechanism for micelle formation and the degree of solubilization.

The increase in the solubilization of organic compounds in water is accounted for by the adsorption-desorption mechanism [149, 150]. The lipophilic substance is subjected to adsorption and desorption processes which are influenced by the nature of both the micelle surface and the solubilized compound. The micellar core is considered to be a nonpolar phase and the surface or interface to be polar. The solubilization of polar compounds is therefore controlled by the nature of the micelle surface; these compounds are localized in an interphase, whereas lipophilic compounds are located only in the micellar cores.

The solubilization of mixtures of oils by micellar systems is of interest because of the possible existence of differential solubilization processes in certain circumstances. When the oil phase is contacted with the aqueous micellar solution, the oil components become solubilized in the micelles. Let us consider an oil phase of two components. As the oil phase composition is varied from pure component 1 to pure component 2, the amounts of 1 and 2 solubilized must vary between the amounts found for the respective single-component oil systems. Since the rate of solubilization in these systems is proportional to the amounts solubilized [149], this quantity likewise varies between its values for the two single-component oil systems. The functions describing both these quantities are continuous.

The rate of solubilization is defined as the rate of transport of oil from the bulk phase across unit interfacial area [149, 151]:

$$\text{rate} = -(1/A)(dV/dt). \tag{19}$$

If the molecular volumes are $\overline{V}_1$ and $\overline{V}_2$, respectively, the solubilization rate can be written as

$$\text{rate} = \text{const}(b_1\overline{V}_1 + b_2\overline{V}_2), \tag{20}$$

where the constant term depends on the properties of the micellar system and the emulsifier micelle solubilizes b_1 molecules of 1 and b_2 molecules of 2.

As in the case of micellization, an approach to the transfer process can be made in terms of the following models [152–156]:

(i) pseudo-phase transfer model. Here the solubilizate is partitioned between the aqueous and micellar phases,
(ii) mass-action model. In this model the distribution of the solubilizate can be regarded as a binding equilibrium of the solubilizate to the micelles.

The pseudo-phase model is useful for a calorimetric approach. With this method it is possible to obtain at the same time the thermodynamic functions (ΔG, ΔH, and ΔS) for the transfer process of the solubilizate from water to micelles.

If the solubilizate behaves ideally both in the aqueous and micellar phases, then the partition constant, on the molarity scale, is

$$K_r^c = C_M^e / C_W^e, \qquad (21)$$

where C_M^e and C_W^e are the molar concentrations of solubilizate at equilibrium in the micellar aqueous phases, respectively.

The solubilizate concentration at equilibrium in the two phases is given by Eqs (22) and (23) [152]

$$C_M^e = \frac{Q}{\Delta H_t^0 V ([E] - CMC) \overline{V}_E}, \qquad (22)$$

$$C_W^e = \frac{[C]V_c - Q/\Delta H_t^0}{V\{1 - ([E] - CMC)\overline{V}_E\}}, \qquad (23)$$

where Q is the experimental enthalpy of transfer, ΔH_t^0 is the molar enthalpy of transfer, $\overline{V}_E$ is the partial molar volume of the solubilizate, V and $[E]$ the volume and concentration of the emulsifier after the mixing process and V_c and C the volume and concentration of the solubilizate.

The thermodynamic functions can be calculated from the following equation [155, 156]:

$$\frac{[C]V_c}{Q} = \frac{1}{\Delta H_t^0 K_r^c} \left\{ \frac{10^3}{\left(\frac{[E]V_E}{V} - CMC\right)\overline{V}_E} - 1 \right\}. \qquad (24)$$

From the intercept and the slope of the straight line, as predicted by Eq. [24], it is possible to calculate K_r^c and ΔH_t^0 at the same time and then ΔG_t^0 and ΔS_t^0.

The scheme of the stepwise mass-action model is the following [155]:

$$E_n + A \underset{k_i}{\overset{k_b}{\rightleftarrows}} E_n A,$$

$$E_n A + A \underset{2k_i}{\overset{k_b}{\rightleftarrows}} E_n A_2,$$

$$\vdots$$

$$E_n A_{m-1} + A \underset{mk_i}{\overset{k_b}{\rightleftarrows}} E_n A_m, \qquad (25)$$

where E_n indicates the micelle, A the solubilizate, and m the number of solubilizate molecules per micelle.

The following equilibrium constants correspond to these equations:

$$K_b = K_1 = \frac{k_b}{k_i} = \frac{[E_nA]}{[E_n][A]},$$

$$\vdots$$

$$K_m = \frac{K_b}{m} = \frac{[S_nA_m]}{[S_n][A]^m}. \tag{26}$$

The multistep equilibrium model is too complex if one assumes that the equilibrium constant of each step is independent of all the others. Consequently, it is necessary to assume some correlation between the sequential equilibrium constants or to make some approximation.

To correlate the fraction of solubilizate in the micellar phase (N_b) to the distribution constant and to the emulsifier concentration, the following equation was developed [155]

$$N_b = \frac{K([E] - CMC)}{1 + K([E] - CMC)}, \tag{27}$$

where K is the ratio between the binding constant (K_b) and the aggregation number of the micelles (n).

The following equation (strictly valid only as $m \to 0$) correlating the unmicellized emulsifier concentration ([m]) to the solute and to the emulsifier concentrations [154]:

$$\frac{[m]}{CMC} = 1 - \frac{m}{1+\gamma}\left(\frac{2.3K_s + (1+\beta)K}{1 + K([E] - CMC)}\right), \tag{28}$$

where K_s is Setchenov's constant, and γ and β are the degree of dissociation of 1:1 ionic emulsifier and micelles, respectively.

3.3.1.5 Reactions in micellar systems

An interface formed by the micellar surface of emulsifier molecules and the aqueous phase creates suitable conditions for the course of various catalysed reactions. The properties of this micromedium are a complex function of the character of the interface (i.e. lipophilicity, hydrophilicity, polarity, charge, water activity, concentration, and the type of functional group and the segmental and rotational mobilities of the groups located at the micelle surface) and determine the special features of catalysed reactions.

The structure of the micelle surface is very complicated; even in the simplest case the surface consists of strongly anchored electrically-charged ions, mobile ions (counter-ions), and functional groups solvated by water molecules. The question of attraction and localization of organic molecules in a micellar medium has not yet been clarified; lipophilic molecules and hydrophilic ions are assumed to associate with fragments of the micelle surface and are transported into the surface or inner spheres of the micelles [6]. These associations lead to an increase in the concentration of additives and their lifetimes in micelles.

The character of micellar surfaces is significantly affected by the polarity of the groups or molecules present. The value of the interfacial dielectric constant of commonly used emulsifiers was estimated to be 35 (a value lower than the dielectric constant of water and almost the same as for ethanol) [157–159]. By using ionic emulsifiers, the molar concentration of charged polar groups located on the surface of the micelles varies between 3 and 5 mol dm^{-3}. Approximately 80% of the charge neutralizes the counter-ions that form the Stern layer; the residual mobile ions form the diffusion Gouy–Chapman layer.

While investigating the mechanism for catalysed reactions, the question of the activity of water bonded on the surface (hydrated form) has arisen. Results from some studies [160, 161] have shown, however, that water localized at the interface has the same activity as that occurring in solution.

The micelles of ionic emulsifiers show the properties characteristic of enzymes (i.e. the ability to catalyse specifically only a particular reaction) in a given direction. Micelles of ionic and nonionic emulsifiers are generally capable of speeding up some reactions, which however, do not proceed via any mechanism characteristic of enzyme catalysis [162].

The catalytic effect of emulsifier micelles on the course of reaction, which is exhibited as an increase in the reaction rate, is attributed to a decrease in the standard free energy for the transition state of reaction. In reactions proceeding under homogeneous conditions, the free energy difference between the reactants and transition state is discussed in terms of the Brönsted–Bjerrum equation [159]

$$\Delta G^{\neq} = RT \ln \frac{f^{\neq}}{f_A f_B} . \tag{29}$$

According to Eq. (29) reactions taking place at the interface are influenced by the concentration and type of substrate. These reactions may also be influenced by the interaction of substrate molecules with the surface segments of micelles. The binding of substrate molecules causes an increase in the local concentration of the reaction components; it also leads to orientation of the molecules of the reaction component which is associated with the shielding of certain functional groups.

An important contributor to catalysis in micellar or enzymatic systems for second-order or higher-order reactions derives from the decrease in entropy of the reactants by virtue of their binding to the catalyst surface.

The activity coefficient of a molecule in the micellar phase may not be informative in attempting to account for an increase in reaction rate. The fact that an organic substrate associates with micelles with an equilibrium constant greater than unity requires that its activity coefficient decreases on going from the aqueous phase to the micellar. However, an overall decrease in the activity coefficient of the reactant may be accompanied by an increase in the activity coefficient at the site of chemical reaction. This consideration is not evident on the basis of the Brönsted–Bjerrum equation.

An understanding of catalysis for organic reactions in the presence of micelles requires that one should separate the two factors indicated above: the entropic contribution reflecting concentration effects and the effects of the relative activity coefficients at the site of reaction for substrates and transition states.

In unimolecular reactions, the entropic contribution is unimportant. Rate changes reflect changes in the nature of the medium in which the reaction occurs. An example of micellar catalysis for a unimolecular reaction is the decarboxylation of 5-nitrobenzisoxazole-3-carboxylate [163]:

$$\underset{NO_2}{\text{[5-nitrobenzisoxazole-3-carboxylate structure]}} \longrightarrow \underset{NO_2}{\text{[2-cyano-4-nitrophenolate structure]}} + CO_2 \qquad (30)$$

5-nitrobenzisoxazole-3-carboxylate

This reaction is catalysed by cationic, anionic and zwitterionic emulsifiers. The most probable explanation for the fact that micelles are effective catalysts for this reaction is destabilization of the substrate in the less polar environment provided by the micellar surface [164]. The rate of decarboxylation increases slightly on adding salt and is also positively influenced by the choice of emulsifiers with higher molecular mass [165].

An example of micellar catalysis which derives from medium effects as opposed to entropic ones, is provided by the unimolecular hydrolysis of phosphate esters [166]. In this case, cationic micelles are good catalysts for hydrolysis. For example, the rate of hydrolysis of 2,5-dinitrophenyl phosphate dianion is approximately 25 times faster in the presence of an optimal concentration of hexadecyltrimethylammonium bromide than in water.

The examples of micellar catalysis for unimolecular reactions indicate that the utilization of binding forces between substrate and micelle can bring reactive functionalities of substrate into an environment in which their reactivity is increased.

The addition of hydroxyl ion to a stable dye of the triphenylmethyl cation variety is a bimolecular reaction catalysed by micellar systems [167, 168]. The rate enhancement of the addition results from the increase in local concentration of hydroxyl ions on the micelle surface and of the destabilization of the dye which takes place via their interactions with emulsifier molecules [169]. A similar effect has also been observed during the addition of cyanide ion to a pyridine ring [170]:

$$\text{[N-alkyl-3-carbamoylpyridinium ion]} + CN^- \longrightarrow \text{[CN adduct structure]} \qquad (31)$$

N-alkyl-3-carbamoylpyridinium ion

An investigation of the catalysis of the addition of cyanide ion to substituted pyridinium ions shows that not only the reaction rate constant increases but also a shift occurs of the equilibrium towards the resulting product [171]. As the length of the chain was increased from 8 to 16 carbon atoms, the reaction rate increased 64 times and the equilibrium constant 35.5 times (see Table 11). Table 11 shows that the effect of the lipophilic character of the substrate and emulsifier is prominent. Micellar catalysis strongly resembles that of enzymes.

Emulsion systems, mainly those composed of anionic emulsifiers, are effective catalysts of the hydrolysis of esters. This property follows from the ability of micelles to attack directly the substrate, destabilize it and thus facilitate an attack on the ester group by a nucleophile [172, 173].

Table 11 — Rate and association constants [a] for addition of cyanide ions to a series of N-substituted 3-carbamoylpyridinium ions in the presence of n-alkyltrimethylammonium bromide in water at 25°C (Reprinted from ref. [171] with permission of Pergamon Press, Oxford)

Substrate	Decyl	Dodecyl	Tetradecyl	Hexadecyl
Octyl				0.21; 135
Decyl			1.1; 530	1.35; 710
Dodecyl		2.5; 1100		
Tetradecyl	0.28; 330		6.6; 3600	10.4; 4500
Hexadecyl		6.4; 4500		13.3; 4800

[a] Emulsifier concentration: 0.02 mol dm^{-3}. The first figure is the second-order rate constant (dm^3 mol^{-1} s^{-1}), the second figure is the association constant (mol^{-1}).

Arai et al. [174] have reported an effective rate enhancement of the hydrolysis of carbohydrates in water solubilized in an organic solvent with dodecylbenzenesulfonic acid (DBSH), which acts both as emulsifier and as catalyst. They found that these reactions have disadvantages: for example, recovery of the catalyst and of the products is difficult. In order to solve these problems, Arai et al. [175] have attempted to fix the micelle structure. The authors carried out the polymerization of styrene containing a small amount of divinylbenzene, solubilized in a normal micelle of sodium dodecyl benzenesulfonate (DBSNa) in water. The crosslinked polystyrene beads obtained were found to contain a significant amount of unextractable DBSNa and that the DBSH-containing beads (converted from DBSNa) showed higher catalytic activity than a free acid-form (commercial sulfonic acid-type cation exchanger for the hydrolysis of acetates), especially for the more lipophilic esters.

Arai et al. [176] tested the catalytic activity of DBSH and the crosslinked poly(methacrylic acid) gels. They reported that the crosslinked product containing DBSH is a better catalyst in the triphase catalysis system. This system has the advantage of easy separation of the catalyst, reactant and product in the hydrolysis of more lipophilic substrates.

The first kinetic model describing the catalysis of unimolecular reactions by micelles was postulated by Menger & Portnoy [177, 178]:

$$D_n + S \xrightleftharpoons{K} D_n^{\cdot} \cdot S \qquad (32)$$

$$k_w \downarrow \qquad \qquad k_m \downarrow$$

$$\text{products} \qquad \qquad \text{products}$$

Employing certain simplifying assumptions, this kinetic scheme provides the following rate law:

$$k = \frac{k_w + k_m K C_m}{1 + K C_m}, \qquad (33)$$

where K is the equilibrium constant for the association of substrate S with micelles of the emulsifier D_n, and C_m is the concentration of micelles given by expression $(C_D - \text{CMC})/N$, where C_D is the overall concentration of emulsifier used and N is the micellar aggregation number; k_w and k_m are the rate constants for formation of the product from unassociated and associated reaction components, respectively. The model predicts a sigmoidal increase in the rate constant with increasing emulsifier concentration. Such behaviour is typical for unimolecular reactions. In bimolecular reactions, the maximum rate is reached at a particular critical emulsifier concentration; above this value, there is usually found only a decrease.

Martinek et al. [179, 180] treated the reaction of two uncharged organic molecules. The equation which they derived is:

$$k_{app} = \frac{k_m P^n C_D V + k_w (1 - C_D V)}{1 + C_D V (P - 1)^n}, \qquad (34)$$

where V is the molar volume of emulsifier and P is the partition coefficient of the substrate between the two phases. This relationship is also applicable to reactions between ions and organic molecules.

Another approach with a considerably wider area of application and with a good estimate of the mechanism for reactions catalysed by micelles is the Romsted theory [181]. This theory enables one to obtain values of the equilibrium constant for an associate micelle-substrate and to follow its change in the reaction system on condition that no substantial changes occur in the structure of the micelles. The Romsted theory takes into account saturation of the Stern layer by counter-ions; this makes the theory different from those discussed above. In other words, an ion is considered to be bound to the micellar surface and not to be free in solution. This assumption is taken into account in the kinetic relations by introducing an equilibrium constant K which expresses the exchange of counter-ions on the micellar surface

$$I_m + X_w \xrightleftharpoons{K} K_w + X_m, \qquad (35)$$

in which I is taken to be a reactive and X an unreactive counter-ion. The rate constant k for a bimolecular reaction in the micellar phase can be obtained from:

$$k = \frac{k_m \beta S K_a (C_D - CMC)}{[K_a(C_D - CMC)+1][I_t + X_t K]} + \frac{K_w}{[K_a(C_D - CMC) + 1]}, \quad (36)$$

where β is the degree of binding of the counter-ions to the Stern layer, S is the molar density of the micellar phase and K_a is the equilibrium constant for incorporation of the organic substrate into the Stern layer. In the case of first order reactions, Eq. (36) simply reduces to the equation of Menger and Portnoy (Eq. (32)). The utility of Eq. (36) lies in the fact that it accounts quantitatively for the basic features of reaction kinetics in micellar systems. A plot of the rate constant versus emulsifier concentration described by Eq. (36) displays a maximum reached at a particular emulsifier concentration. Its applicability has been supported experimentally (bimolecular reactions catalysed by the micellar phase). It has been observed in a number of cases that increasing lipophilicity of substrate and emulsifier results in larger maximal rates which are obtained at lower emulsifier concentrations. Computer-generated plots based on Eq. (36) reproduce this behaviour very nicely [181].

A particularly pleasing success of Eq. (36) is that it accords with the important observation of Bunton & Wolfe [182] that second-order-rate constants for the specific acid-catalysed hydrolysis of p-nitrobenzaldehyde diethyl acetal in the presence of sodium dodecyl sulfate decrease with increasing acid concentration. In addition there are many examples of the inhibition of reaction in micellar systems through increasing concentrations of unreactive counter-ions. This phenomenon finds a ready explanation in terms of Eq. (36).

3.3.2 Initiating systems

The initiation of polymerization in emulsion systems is a complex process incorporating the decomposition of initiator in the aqueous phase, the formation of oligomeric radicals, the entry of radicals into micelles and monomer/polymer particles, formation of the reaction centres in the aqueous phase and in polymer particles and the escape of radicals associated with their re-entry into particles.

The initiators used in emulsion polymerization systems can be divided into two groups: dissociation and redox. Dissociation initiators are compounds which generally decompose at higher temperatures ($\geq 40°C$) into radical fragments mostly of ionic nature. Redox initiators are two- or multicomponent systems with components showing different oxidation and reduction properties. Oxidation–reduction interactions of the components of the initiating system lead even at slightly higher temperatures to the formation of free radicals mostly of neutral character.

3.3.2.1 Dissociative initiators

3.3.2.1.1 Peroxodisulfates
Peroxodisulfates are water-soluble compounds used for initiating the polymerization of both hydrophilic and lipophilic monomers in homogeneous or heterogeneous aqueous disperse media. Potassium peroxodisulfate and ammonium peroxodisulfate are the best known representatives of the group.

Kolthoff *et al.* [183, 184] discussed the kinetics of the decomposition of peroxodisulfate in water in detail and proposed a model which is still valid

$$S_2O_8^{2-} \underset{k_{-1}}{\overset{k_1}{\rightleftarrows}} 2SO_4^{\bullet-}, \tag{37}$$

$$SO_4^{\bullet-} + H_2O \overset{k_a}{\rightarrow} HSO_4^- + OH^{\bullet}, \tag{38}$$

$$2OH^{\bullet} \overset{k_{OH}}{\rightarrow} H_2O + \tfrac{1}{2}O_2. \tag{39}$$

The decomposition of peroxodisulfate in water is a first-order reaction with respect to its concentration and is a function of pH and of the ionic strength of the medium.

The effect of pH on peroxodisulfate decomposition can be expressed as

$$k_d = k_1 + k_2 [H^+], \tag{40}$$

where k_d is the overall decomposition rate constant, k_1 is the contribution from the noncatalysed decomposition and k_2 from the catalysed decomposition. An activation energy of 108.8 kJ mol^{-1} was obtained for the catalysed decomposition of peroxide and 140.7 kJ mol^{-1} for the noncatalysed one. The rate constant for decomposition in acid medium rapidly decreases with increasing ionic strength but, in alkaline medium, the decomposition rate constant is unaffected by varying ionic strength.

By studying the effect of temperature, pH, ionic strength and concentration of the reaction components, Hakoila [185] came to a similar conclusion as other investigators, i.e. that the decomposition rate of peroxodisulfate is strongly affected by the pH of the medium.

The mechanism for decomposition of peroxodisulfate according to Eqs (37)–(39) in neutral or slightly alkaline media has been experimentally supported by House [186] and Banerjee & Konar [187]; however, they neglect the deactivation of ion radicals by recombination

$$2SO_4^{\bullet-} \overset{k_{-1}}{\rightarrow} S_2O_8^{2-}, \tag{41}$$

both inside and outside the cage (with regard to their low concentration caused by the high effectiveness of reaction (38)).

The values of the rate constants k_a and k_{-1} reported in other papers [189, 190] dealing with the mechanism for peroxodisulfate decomposition varied within $10^3 \leq k_a \leq 10^4$ s^{-1} [188, 189] and $3.7 \leq k_{-1} \leq 8.8 \times 10^8$ dm^3 mol^{-1} s^{-1} [189, 190]. The values agree with the idea of a limited termination of $SO_4^{\bullet-}$ by recombination.

Prediction of the low effectiveness of the deactivation of $SO_4^{\bullet-}$ radicals by recombination has also been confirmed by other authors [187, 191]. If the recombination of $SO_4^{\bullet-}$ radicals proceeded in the stationary state, we would have

$$d[SO_4^{\bullet-}]/dt = 2k_1[S_2O_8^{2-}] - 2k_{-1}[SO_4^{\bullet-}] - k_a[SO_4^{\bullet-}][H_2O] = 0, \tag{42}$$

or

$$[SO_4^{\bullet-}] = (k_a/4k_{-1}) \pm (k_a^2 + 16k_1k_{-1}[S_2O_8^{2-}])^{0.5}/4k_{-1}. \tag{43}$$

Solution of the above equations using the values of k_a and k_{-1} from other papers [188–190], gave values for $[SO_4^{\bullet -}]_{ss} \leq 0$ which are unacceptable in terms of the above-mentioned mechanism.

In radical polymerizations of vinyl monomers in aqueous media, the decomposition of initiator is somewhat influenced by the monomer present, which, by capturing $SO_4^{\bullet -}$ radicals, lowers the probability of other routes of radical deactivation

$$SO_4^{\bullet -} + M \xrightarrow{k_3} {}^-O_3S-O-M^\bullet (M_1^\bullet), \tag{44}$$

$$SO_4^{\bullet -} + M \xrightarrow{k_4} HSO_4^- + R^\bullet, \tag{45}$$

where $R^\bullet$ is the water-soluble monomer radical.

The mechanism for deactivation of $SO_4^{\bullet -}$ radicals described by Eq. (44) has also been confirmed by the results from a study of the polymerization of acrylonitrile in water [192, 193]. However, the transfer of radical to monomer (Eq. (45)) has not been confirmed. The rate of thermal decomposition of peroxodisulfate in aqueous solution in the presence of acrylonitrile monomer (M) and of nitrogen was written as

$$-d(S_2O_8^{2-})/dt \propto [S_2O_8^{2-}]^{1.35}[M]^{1.26} \tag{46}$$

in the concentration ranges of peroxodisulfate $(1.8 \text{ to } 18.0) \times 10^{-3}$ mol dm^{-3}, and of monomer (M), 0.3 to 1.2 mol dm^{-3}. The secondary or induced decompositions of peroxodisulfate were suggested to be due to the following elementary reactions:

$$S_2O_8^{2-} + M \xrightarrow{k_5} {}^\bullet M-O-SO_3^- + SO_4^{\bullet -} \tag{47}$$

and

$$S_2O_8^{2-} + M_j^\bullet \xrightarrow{k_6} M_j-O-SO_3^- + SO_4^{\bullet -}, \tag{48}$$

where $M_j^\bullet$ radicals ($j = 1$ to 10) are water-soluble oligomeric or polymeric free radicals, k_5 and k_6 at 50°C have been estimated as 1.7×10^{-5} and 5.1×10^3 dm^3 mol^{-1} s^{-1}, respectively.

Maruthamuthu [194] investigated the decomposition of peroxodisulfate in the presence of methyl methacrylate and obtained a rather high rate constant for the reaction between monomer and $SO_4^{\bullet -}$ radicals ($k_3 = 1.0 \times 10^9$ dm^3 mol^{-1} s^{-1} at 25°C).

Adhikari et al. [195] studied the dependence of the rate constant for thermal decomposition on the concentration of ethyl acrylate dissolved in water and on the concentration of peroxodisulfate, and proposed

$$d[S_2O_8^{2-}]/dt \sim [S_2O_8^{2-}]^{1.0}[M]^{0.92}. \tag{49}$$

This relation shows that the rate of decomposition increases in proportion to the concentrations of ethyl acrylate (ranging between 4.62×10^{-2} and 23.1×10^{-2} mol dm^{-3}) and of peroxodisulfate. The reaction order with respect to monomer (0.92) is direct evidence of the induced decomposition of peroxodisulfate by monomer and therefore the rate of decomposition can be expressed by

$$-d[S_2O_8^{2-}]/dt = k_1[S_2O_8^{2-}][M]^{0.92} + k_5[M]^1[S_2O_8^{2-}]^1. \qquad (50)$$

The increase in conversion of ethyl acrylate was, however, accompanied by a decrease in the rate of decomposition of peroxodisulfate. The decrease was ascribed to a smaller contribution from the catalysed decomposition of peroxodisulfate as a result of the decrease of monomer concentration. On changing monomer to polymer, the fraction of the decomposition of peroxodisulfate activated by monomer decreased. The value for the rate constant for the catalysed decomposition of peroxodisulfate was in this case estimated to be 5.4×10^{-5} $dm^3\ mol^{-1}\ s^{-1}$ (at 50°C).

The accelerating effect of monomer on the rate of decomposition of peroxodisulfate has also been observed in the emulsion polymerization of allyl acetate [196] and for a series of other vinyl monomers [197].

Dunn [198] has proposed another, partially different mechanism for the monomer-catalysed decomposition of peroxodisulfate. In the first step, monomer radicals are first formed

$$SO_4^{\bullet-} + RH(\text{monomer}) \rightarrow HSO_4^- + R^{\bullet}, \qquad (51)$$

which induce the decomposition of peroxodisulfate in the following step

$$R^{\bullet} + S_2O_8^{2-} \xrightarrow{k'} R-O-SO_3^- + SO_4^{\bullet-}. \qquad (52)$$

The author predicts the formation of radical $R^{\bullet}$ also to proceed in polymer particles, through chain transfer to monomer (vinyl acetate)

$$M_j^{\bullet} + RH \rightarrow M_jH + R^{\bullet}, \qquad (53)$$

which desorbs from the particles into the aqueous phase and takes part in reaction (52). The course of these reactions should lead to a decrease in the pH value of the aqueous medium.

Kuanhsiung [199] followed the catalytic effect of vinyl acetate on the decomposition of peroxodisulfate and came to the conclusion that activation of peroxodisulfate obeys Eq. (52) not only in the aqueous phase but also in the interphase. He reported that the high value of k' ($2 \times 10^7\ dm^3\ mol^{-1}\ s^{-1}$) results from both contributions and reflects the active participation of vinyl acetate in the decomposition of peroxodisulfate.

The emulsifier is another component of the polymerization system that affects the decomposition of peroxodisulfate. Studying the effect of emulsifier on the decomposition of peroxodisulfate, the question arises as to which form of the emulsifier (micellar or nonmicellar) contributes to the process more significantly. Jurzhenko et al. [200, 201] ascribe the rate enhancement of the decomposition to micellar emulsifier. They assume that the catalysed decomposition proceeds only on the micelle surface and that nonmicellar emulsifier does not influence the decomposition of peroxodisulfate. Ryabova et al. [202, 203] studied the effect of anionic emulsifier (potassium dodecyl sulfate) below and above the CMC and, in contrast to the preceding investigation, observed an increase in the decomposition rate of peroxodisulfate in both concentration ranges of emulsifier. The complexity of this mechanism follows from the fact that the activation effect of emulsifier was eliminated in the presence of aerial oxygen. If, however, the system contained styrene, the rate enhancement of the decomposition of peroxodisulfate was not influenced by oxygen (the rates relate to the decomposition rate of peroxodisulfate in pure water). At lower concentrations of styrene (0.0014–0.0045 mol dm^{-3}) the rate constant obtained for peroxodisulfate decomposition was $k = 3.2 \times 10^{-5}$ dm^3 mol^{-1} s^{-1} and was comparable with that in pure water; at higher concentrations of monomer (dissolved in water) the rate constant increased to 5.5×10^{-5} dm^3 mol^{-1} s^{-1}. On adding emulsifier, the value increased about 1.5 times.

Brooks & Makanjuola [204] have reported that the decomposition of potassium peroxodisulfate is greatly accelerated in the presence of sodium dodecyl sulfate (SDS) and/or polymer particles. The decomposition rate constant was found to increase ca. five times in the presence of poly(styrene) particles compared to that obtained in pure water. The surface of the polymer particles was occupied up to 75% by SDS molecules.

Okubo & Mori [205] have investigated the decomposition rate of potassium peroxodisulfate in various aqueous solutions, pure water, SDS solution, emulsifier-free poly(styrene) emulsion, SDS solution containing acrylamide monomer and an SDS-containing emulsion polymerization system for styrene. 'Free' SDS molecules in the molecular dispersed state were found to increase markedly the decomposition rate, whereas those which formed micelles and became adsorbed onto the polystyrene particles did not increase it. The acceleration in the former case disappeared in the presence of a small amount of monomer. The authors assumed that the acceleration of the peroxodisulfate decomposition in the presence of SDS proceeds by the following steps:

$$CH_3CH_2(CH_2)_{10}OSO_3^- + SO_4^{\bullet -} \rightarrow HSO_4^- + CH_3\overset{\bullet}{C}H(CH_2)_{10}OSO_3^-, \qquad (54)$$

$$CH_3CH_2(CH_2)_{10}OSO_3^- + S_2O_8^{2-} \rightarrow SO_4^{\bullet -} + SO_4^{2-} + [X], \qquad (55)$$

where [X] is a reactive intermediate.

Vinogradov et al. [206] examined the effect of the type of cation of anion-active emulsifier on the decomposition of peroxodisulfate. They used sodium, potassium, and ammonium salts of oleic acid and observed an effect of the emulsifier in all cases but did not observe any differences between the individual types of salts.

Crematy [207] reported the effect of cationic emulsifier on the thermal decomposition of peroxodisulfate in the aqueous phase. According to the author, cationic emulsifier radicals were produced by hydroxyl radicals and chain termination occurred due to reactions between the emulsifier radical and $SO_4^{\bullet -}$. He found that the decomposition of peroxodisulfate was accelerated in the presence of cationic emulsifier above the CMC; the decomposition rate was found to be independent of the emulsifier concentration below the CMC.

Antonova et al. [208] and Capek et al. [209] investigated the emulsion polymerization of vinyl monomers initiated by peroxodisulfate in the presence of cationic emulsifiers. They found that the interaction between peroxodisulfate and cationic emulsifier led to the formation of a water-insoluble product. The rate of polymerization was found to decrease with increasing concentration of the peroxodisulfate/emulsifier product.

The mechanism for induced peroxodisulfate decomposition caused by emulsifier molecules is not yet known exactly. Mutual interactions between the initiator and emulsifier molecules are assumed to lead to the formation of associates with a lower activation energy for peroxodisulfate decomposition [210].

3.3.2.1.2 Other dissociative initiators
Oil-soluble initiators, such as benzoyl peroxide and 2,2′-azobisisobutyronitrile can also be used for initiating the emulsion polymerization of vinyl monomers. However, the majority of the initiator, which is in the monomer particles, is inactive for several reasons: (i) the cage effect plays an efficient role in the deactivation of radicals and (ii) the diffusion of these radicals and initiator molecules into the aqueous phase is very limited because of their low water solubility. The fraction of initiator dissolved in water is active and mostly determines the rate of initiation of emulsion polymerization [211].

The influence of benzoyl peroxide on the character of disperse polymerization in water has been discussed elsewhere [212]. Using an initiator concentration lower than 0.2 mass % (with respect to monomer), the course of polymerization resembles that of classical emulsion polymerization. The rates of growth of polymer particles observed were the same as in systems containing a water-soluble initiator. The increase in emulsifier concentration positively affected the water-solubility of benzoyl peroxide and the contribution from emulsion polymerization.

Aromatic diazoamines (I) and diazoethers (II) are representatives of initiators of higher water solubility which are applicable in emulsion polymerizations

$$Ar-N=N-NH-Ar', \qquad\qquad Ar-N=N-S-Ar'. \qquad(56)$$

$$(\,I\,) \qquad\qquad\qquad (\,II\,)$$

These decompose like AIBN [213–216].

Nuyken *et al.* [213] have tested the effect of some aromatic diazoamines and diazoethers on the emulsion polymerization of styrene and butadiene. The effect of the diazoamines increased as the substituent became more electron-donating and their use led to the formation of poly(styrene) with an RMM (relative molecular mass) lower than when initiated by peroxodisulfates. The decrease in RMM is attributed to the active participation of $ArS^{\bullet}$ and $ArNH^{\bullet}$ radicals in transfer reactions. Only $Ar^{\bullet}$ radicals function in the initiation step.

The group of azo initiators is typified by 2,2′-azobisisobutyronitrile(AIBN)-based high-molecular initiators; these are prepared via a reaction between AIBN and multifunctional alcohols, such as tetraethylene glycol or 1,6-hexanediol. The reaction is catalysed by protons [214, 215]. DSC measurements confirmed that high-molecular initiators decompose more rapidly than AIBN monomer, an effect associated with the lower activation energy for decomposition of the polyinitiator (*ca.* 125 kJ mol^{-1}) compared with that for AIBN (~ 130 kJ mol^{-1}) [216]. The reaction of these polyinitiators with acrylamide produces special polyinitiators with surface-active properties. Their radical products adsorb strongly to the surface of the polymer particles where they mainly initiate chain growth and contribute to the stability of latex particles [217].

By modifying AIBN, we can prepare water-soluble initiators with surface-active properties, such as 4,4′-azobis-4-cyanovaleric acid and its salts, and alkyl derivatives [218].

The polyinitiators have wide applicability and are used for preparing special polymer dispersions with a reduced emulsifier content or without using emulsifier [219].

Peroxides and hydroperoxides represent another group of initiators used for initiating polymerization in disperse aqueous media. Hydrogen peroxide is a typical water-soluble peroxide which produces initiating hydroxyl radicals via thermal decomposition. To lower the relatively high O–O bond dissociation energy, it is mostly used in combination with a reducing agent.

Hydroperoxides are soluble in both organic and aqueous phases and, in emulsion polymerization systems, they are therefore sited both in monomer particles and in the aqueous phase. The decomposition of hydroperoxides generates hydroxyl (hydrophilic) and alkoxyl (lipophilic) radicals

$$ROOH \rightarrow RO^\bullet + {}^\bullet OH \tag{57}$$

which diffuse and concentrate into their preferred respective phases. The emulsion polymerization of styrene initiated by cumyl hydroperoxide where $OH^\bullet$ radicals diffuse into water and cumyloxy radicals into monomer particles, provides an example [220]. The increased rate of decomposition of cumyl hydroperoxide is attributed to the active role of the interphase where the catalysed decomposition of the hydroperoxide takes place.

The increased rate of the decomposition of hydroperoxides in the presence of emulsifier micelles has also been observed elsewhere [221, 222]. The micellar catalysis reflects the accumulation of hydroperoxide molecules in the interphase, where the interaction of some emulsifier segments with the peroxide leads to a reduction in the activation energy for peroxide decomposition. The increase in initiator concentration at the particle surface is particularly marked if an initiator is used with functional groups able to interact positively with the surface layer of the polymer particle.

An increased concentration of peroxide groups on the surface of or within the polymer particles is also achieved via the copolymerization of vinyl monomers with unsaturated peroxides. These peroxides decompose much more rapidly: the methacrylic ester of α-oxyethyl-*tert*-butyl peroxide forming a copolymer with butyl acrylate or with styrene in disperse media decomposed much more rapidly than under comparable conditions in a homogeneous phase (in chlorobenzene, $k_d = 6 \times 10^{-8}\ s^{-1}$) [223]. The decomposition of peroxides and hydroperoxides is catalysed most efficiently by cationic emulsifiers [224, 225]. The associated mechanism resembles that of peroxide decomposition induced by tertiary amines [226].

3.3.2.2 Redox initiating systems

3.3.2.2.1 Hydrogen peroxide/metal salt systems

Haber and Weiss [227] observed the rate enhancement of the decomposition of hydrogen peroxide in the presence of iron(II) salts. The authors explained this phenomenon by the following chain reaction

$$H_2O_2 + Fe^{2+} \rightarrow HO^- + HO^\bullet + Fe^{3+}, \tag{58}$$

$$HO^\bullet + H_2O_2 \rightarrow H_2O + HO_2^\bullet, \tag{59}$$

$$HO_2^{\bullet} + H_2O_2 \rightarrow O_2 + H_2O + HO^{\bullet}, \tag{60}$$

$$HO^{\bullet} + Fe^{2+} \rightarrow Fe^{3+} + HO^-. \tag{61}$$

Thus the decomposition of H_2O_2 is catalysed by Fe^{2+} ions and, as for every chain reaction, it consists of initiation, propagation, and termination steps. Reaction (58) is the initiation, Eqs (59) and (60) describe the propagation step, and reaction (61) characterizes the deactivation of the chain reaction. The species initiating chain propagation is $HO^{\bullet}$. $HO^{\bullet}$ is very reactive and, besides adding to double bonds, it is able to abstract hydrogen from alkanes and to produce radicals capable of initiating polymerization. For instance, in its interaction with methanol, $HO^{\bullet}$ is able to abstract not only from C–H but also from O–H [228].

Marvel et al. [229] examined not only the effect of Fe^{2+} but also that of Fe^{3+} salts on the decomposition of H_2O_2; they used the system hydrogen peroxide/Fe^{3+} salt for initiating the emulsion polymerization of styrene and butadiene. At higher monomer concentrations, the deactivation of $HO^{\bullet}$ (Eq. (61)) was strongly suppressed and initiating radicals mainly added to the double bonds of the monomer. Rate enhancements of both the peroxide decomposition and the initiation have also been observed on adding an Fe^{3+} salt. These results led to an extension of the Haber–Weiss model by other steps involving the active role of Fe^{3+} ions in a cyclic reaction mechanism

$$HO^- + Fe^{3+} \rightarrow HO^{\bullet} + Fe^{2+}, \tag{62}$$

$$HO^{\bullet} + Fe^{3+} \rightarrow H^+ + 1/2\, O_2 + Fe^{2+}. \tag{63}$$

The Fe^{3+} ions generated by Eq. (58) enter the cycle through Eqs (62) and (63).

The decomposition of H_2O_2 is also effectively catalysed by Ti^{3+} salts [230]. In the first step, the decomposition of hydrogen peroxide into hydroxyl radicals (capable of initiating polymerization in water) is catalysed, and, in the second step, particularly at high concentrations of the Ti(III) salt, efficient deactivation of $OH^{\bullet}$ radicals takes place.

Iron(III) salts (e.g. nitrate) were successfully used in the catalytic decomposition of H_2O_2 and were applied to the preparation of the emulsion copolymer of styrene and butadiene [231, 232].

3.3.2.2.2 Peroxodisulfate redox systems
A mixture of potassium peroxodisulfate and potassium bisulfite is the classical representative of this type of initiator system. Frank & Haber [233] proposed a mechanism for the decomposition of peroxodisulfate catalysed by bisulfite

$$S_2O_8^{2-} + HSO_3^- \rightarrow SO_4^{2-} + SO_4^{\bullet -} + {}^{\bullet}SO_3H \tag{64}$$

on the basis of their analysis of the results from acrylonitrile polymerization in water. The authors assume both sulfonate and sulfate radicals to be initiating radicals. They also report that the concentration of radicals formed via Eq. (64) and initiating the polymer chain propagation, is reduced by the competing reactions

$$SO_4^{\bullet -} + HSO_3^- \rightarrow SO_4^{2-} + {}^\bullet SO_3H, \tag{65}$$

$$2\,{}^\bullet SO_3H \rightarrow H_2S_2O_6. \tag{66}$$

The mechanism for the induced decomposition of peroxodisulfate via Eq. (64) is supported by results obtained during a study of the incorporation of sulfate ($-OSO_3H$) and sulfonate ($-SO_3H$) groups into the polymer [234]. The content of terminal sulfate and sulfonate groups was found to depend both on the concentration of the components of the redox system and on the monomer concentration. This dependence can be expressed by the semiempirical relation

$$\frac{[\sim SO_3]}{[\sim SO_4]} = 1 + 0.27 \frac{[NaHSO_3]^{1.5}}{[M][K_2S_2O_8]}. \tag{67}$$

The presence of sulfate and sulfonate groups bound to the polymer was supported by analysis of the poly(acrylonitrile) and poly(vinyl acetate) obtained via emulsion polymerization initiated by the peroxodisulfate/bisulfite system [235]. With increasing bisulfite concentration, the content of sulfonate groups in a polymer increased, which was explained by radical transfer to a bisulfite molecule.

The decomposition of the peroxodisulfite/bisulfite system is accelerated by transition metal salts which initiate the decay both of peroxodisulfate and bisulfite [236]. Salts of Fe(II) and Fe(III) are very efficient co-reducing agents [237]. This general mechanism for the radical decomposition of the peroxodisulfate redox system and participation of radicals in the polymerization follows from analysis of the polymer formed in the polymer particles. The results obtained from the emulsion polymerization of styrene initiated by the redox system peroxodisulfate/bisulfite/Fe^{3+} have also indicated the accelerating effect of Fe^{3+} salts on the decomposition rate of the peroxodisulfate/bisulfite pair and on the polymerization rate [238, 239]. Analysis of the poly(styrene) demonstrated the presence of sulfonate and sulfate groups.

Berry & Peterson [240] studied the mechanism for the incorporation of sulfate and sulfonate groups into the polymer using the radioactively labelled components of the initiating system (peroxodisulfate/bisulfite/iron(II) sulfate). Their results confirmed earlier conclusions about the mechanism given above and the participation of both sulfate and sulfonate radicals in the initiation of emulsion polymerization. These authors proposed a mechanism for the participation of Fe^{2+} ions in the decomposition of peroxodisulfate and bisulfite:

$$S_2O_8^{2-} + Fe^{2+} \rightarrow SO_4^{2-} + SO_4^{\bullet -} + Fe^{3+}, \tag{68}$$

$$HSO_3^- + Fe^{3+} \rightarrow {}^\bullet SO_3H + Fe^{2+}. \tag{69}$$

The results obtained from a study of the influence of hydrolysis on the properties of polymers and polymer dispersions have shown that during, and also after, polymerization, hydrolysis of the sulfate groups takes place which considerably alters their presence in the polymer chains [241]. Hydrolysis lowers the concentration of sulfate groups on the particle surface and therefore conclusions derived from the analysis of polymers have to be corrected by the extent of hydrolysis, particularly of sulfate groups.

The redox system peroxodisulfate/iron(II) salt was successfully used to initiate the dispersion polymerizations of acrylonitrile, methyl methacrylate, and tetrafluoroethylene [240, 242, 243]. The system yielded initiating radicals according to Eq. (68). The rate of polymerization increased even after adding a small amount of ferrous sulfate. By contrast, addition of iron(II) sulfate to $K_2S_2O_8$ led to a decrease in the RMM of polymer [244]. The decay of peroxodisulfate was catalysed by the addition of copper salts, which caused an increase in rate of acrylamide polymerization [245].

The combination of peroxodisulfate and silver(I) salts gives an especially efficient redox system [246]. It brings about a rate enhancement of the polymerization accompanied by a decrease in the RMM of the polymer.

A remarkable increase in the decomposition rate of peroxodisulfate was observed by Bawn & Margerison [247] on adding both silver(I) and copper(II) salts. They evaluated the mechanism of decomposition of peroxodisulfate by the decrease in concentration of 1,1-diphenyl-2-picrylhydrazyl (DPPH) as a function of the concentration of silver ions and of peroxodisulfate. This led to a semiempirical relation

$$\frac{d[DPPH]}{dt} = (k_1 + k_2 [Ag^+]) [K_2S_2O_8], \qquad (70)$$

where k_1 and k_2 are the rate constants for the noncatalysed and catalysed decompositions of peroxodisulfate, respectively and the respective associated activation energies are 118.9 kJ mol^{-1} and 83.3 kJ mol^{-1}, respectively.

In emulsion polymerizations of styrene, addition of a Ti^{3+} salt caused a lowering of the rate of the polymerization and of the RMM [248]. The decreases are caused by deactivation of a particular fraction of initiating radicals by Ti^{3+} ions. The expected oxidation-reduction reaction

$$Ti^{3+} + S_2O_8^{2-} \rightarrow Ti^{4+} + SO_4^{\bullet -} + SO_4^{2-} \qquad (71)$$

does not take place, which is why the number of radicals in the system does not increase. In other words, Ti^{3+} reacts with both initiating and propagating radicals and therefore both the rate and the RMM decrease.

Reductions of the RMMs of the polymers have also been observed in the dispersion polymerizations of styrene and acrylonitrile. These have been explained in terms of chain transfer from propagating radicals to Fe^{2+} or Cu^{2+} ions [244, 245].

The rate of decomposition of peroxodisulfates is also increased on addition of quaternary ammonium and phosphonium salts [249, 250]. The system peroxodisulfate/quaternary ammonium salt was used successfully for initiating the emulsion polymerization of butyl methacrylate in water or in a disperse medium, water/acetone [250]. The efficiency of the quaternary ammonium salt in functioning as an initiator and a surface-active compound increased with the increasing length of the alkyl group (octyl, cetyl, ...) and decreased in the order Cl > Br > I. The activation energy for polymerization of 334.9 kJ mol^{-1} obtained using pyridinium cetyl ammonium chloride is higher than that using trimethyl cetyl ammonium chloride (251.2 kJ mol^{-1}). These differences were explained by the different complex-forming abilities of the ammonium salts with peroxodisulfate and the different rates of formation of initiating radicals. Complex formation between these components is mainly affected by the character of the halide. Initiating radicals are formed by electron transfer from nitrogen to peroxodisulfate in agreement with the mechanism for redox reactions. The efficiency of the redox system is influnced by the associated cation and decreases in the order Na$^+$> K$^+$> NH$_4^+$.

The decomposition of peroxodisulfate is accelerated by interaction with the emulsifier which enhances the rate of polymerization [251].

Capek et al. [252] have reported that the rate of peroxodisulfate decomposition is accelerated by the presence of an anionic emulsifier (Dowfax 2A1). With increase of the emulsifier concentration from 0.99×10^{-2} to 4.94×10^{-2} mol dm^{-3}, the rate constant for the decomposition of peroxodisulfate, k_d, increased from 2.5×10^{-5} to 6.5×10^{-5} s^{-1}; in the absence of emulsifier, the value of k_d was 0.6×10^{-5} s^{-1}.

Table 12 — Rate for the decomposition of K$_2$S$_2$O$_8$ (k) in the presence of unsaturated polyester (UP) with different numbers of acidic groups (Reprinted from ref. [253] with permission of Hüthig and Wepf Verlag, Basel)

UP	UP /10^{-3} mol dm^{-3}	k_d /10^{-5} s^{-1}
UP2 (acid number: 1.57 × 10^{-3} mol g^{-1})	0	2.70
	0.3	4.91
	0.6	5.31
	2.5	6.15
	4.2	6.82
	5.8	7.23
	10.4	7.42
	15.6	7.63
UP3 (acid number : 3.92 × 10^{-3} mol g^{-1})	15.6	13.60

Temperature: 70°C, K$_2$S$_2$O$_8$ = 7.5×10^{-3} mol dm^{-3}, NaHCO$_3$ = 15×10^{-3} mol dm^{-3}, CMC (UP2) = 0.79×10^{-3} mol dm^{-3}.

In the presence of unsaturated polyesters (prepared via condensation of 1,6-hexanediol with phthalic anhydride — UP2 or with maleic anhydride — UP3), Funke et al. [253] studied the catalysed decomposition of peroxodisulfate in water. Below the CMC of UP2, the rate of peroxodisulfate decomposition increased more rapidly than above the CMC. Table 12 gives the rate constants for the decomposition of K$_2$S$_2$O$_8$ in the presence of UP2 and UP3. The enhanced effect of UP2 molecules at concentrations below the CMC was explained by the interaction of peroxodisulfate molecules with free molecules of the unsaturated polyester. The decrease in the activating effect of UP2 at higher concentrations (above the CMC) is a result of the shielding of the active carboxyl groups in the micelles. Under these conditions only some surface-localized groups can interact with the peroxodisulfate. The addition of monomer (styrene) substantially reduces the activity of UP2 with respect to the decomposition of peroxodisulfate. By changing the ratio UP2/ styrene from 0.6 to 2.0, k_d increased from 5.3×10^{-5} to 5.9×10^{-5} s^{-1}. These facts show that association of the unsaturated polyesters leads to a reduction in their ability to catalyse the decomposition of peroxodisulfate.

3.3.2.2.3 Other redox systems
The redox system chlorate/bisulfite [254] proved to be very efficient for the emulsion polymerization of vinyl chloride. The use of radioactive components of the initiating system enabled the proposal of mechanisms for the redox decomposition and for the initiation of polymerization. The interaction of a transient intermediate formed between chlorate and bisulfite, $[O_3SOClO_2]^{3-}$, with vinyl chloride monomer leads to the formation of radicals initiating the emulsion polymerization. Chlorine atoms generated during reaction are also partly involved in the initiation.

A mixture of bisulfite and Fe^{3+} salt was shown to be an efficient initiating system of the emulsion polymerization of tetrafluoroethylene [255]. The initiating radicals are in this case $^\bullet SO_3H$ radicals produced via interaction between bisulfite and iron(III) salt. In this polymerization, sulfonate groups were efficiently incorporated into the polymer and positively influenced the emulsifying properties of the polymer formed and the stability of the polymer particles.

The system of bisulfite and Fe^{3+} salt was successfully used for initiation of the emulsion polymerization of styrene [238, 239]. Results from the analysis of poly(styrene) led, however, to the conclusion that the sulfonate groups built into the polymer also originate from another source, i.e. from the emulsifier, and result from chain transfer to the emulsifier. The emulsifier radicals are capable of re-initiating chain propagation and thus enrich the polymer with a structural unit of the emulsifier.

Bisulfite is also able to initiate the polymerization of vinyl monomers in aqueous media in the presence of some metal oxides. Iron(III) oxide is an efficient additive [256, 257]. Poly(methyl methacrylate) with high RMM was formed in the presence of copper(II) oxide [258]. The incorporation of sulfonate groups into the polymer chain increases the surfactant properties of the macroradicals and positively affects the formation of polymer particles and their stability. Mukherjee *et al.* [259] and Moustafa & Abd-el-Hakim [260] successfully used copper and titanium oxides as co-initiating additives in polymerization of methyl methacrylate. With increasing amounts of catalytic additive, an increase in the rate of the polymerization and of the RMM and a decrease in the activation energy for polymerization were observed. The activation energy for the polymerization at temperatures from 40 to 50°C using CuO was 3.3 kJ mol^{-1} and by using MnO_2, 18 kJ mol^{-1}.

The formation of active radicals was observed in the interaction between sulfurous acid and iron(III) oxide. This system was used to initiate methyl methacrylate polymerization in water [256].

In contrast to the systems discussed above where radicals are formed through redox processes, the formation of radicals in the system of sodium bisulfite plus sodium or calcium carbonate has not yet been clarified. These redox systems were successfully used to initiate methyl methacrylate polymerization in water [261, 262].

In determining the efficiency of the redox system peroxodisulfate/iron(II) sulfate, Kolthoff *et al.* [263] found that addition of halide ion caused a remarkable decrease of the radical concentration. The decrease was explained by the reactions

$$SO_4^{\bullet -} + Br^- \rightarrow SO_4^{2-} + Br^{\bullet}, \qquad (72)$$

$$Fe^{2+} + Br^{\bullet} \rightarrow Fe^{3+} + Br^-. \qquad (73)$$

A similar reduction in the concentration of $SO_4^{\bullet -}$ radicals has been observed after adding Cl^- ion to the redox system Ti^{3+}/peroxodisulfate.

The combination of Ti^{3+} and NH_2OH was shown to be a highly efficient system for initiating methyl methacrylate polymerization [264].

Peroxodisulfate/diamine was found to be [265, 266] an excellent water-soluble redox initiator for the polymerization of water-soluble and water-insoluble monomers such as acrylamide and methyl methacrylate. It was found that diamines with two amino groups separated by an ethylene unit are better promoters than those with two amino groups linked by other α,ω-alkylene units such as methylene or propylene. The promoting effects of diamines on vinyl polymerization are in the order: tertiary diamine > secondary diamine > primary diamine. The free radicals from the diamine species, as well as the sulfate-free radicals, are responsible for the initiation of polymerization.

Fernández & Guzmán [267] have used the redox system ceric ammonium nitrate/maltose to initiate the aqueous phase polymerization of methyl methacrylate. Ceric salts are highly reactive in aqueous media. A maltose-derived radical is formed via electron transfer to cerium(IV). The experimental results suggest the following mechanism of primary radical formation:

$$Ce(IV) + R \underset{\rightarrow}{\overset{K}{\leftarrow}} complex, \qquad (74)$$

$$complex \overset{k_d}{\rightarrow} R^{\bullet} + Ce(III) + H^+. \qquad (75)$$

3.3.2.3 Photoinitiating systems

Ultraviolet and γ-radiation have rarely been used to initiate the emulsion polymerization of vinyl monomers although this method enables the preparation of special polymer dispersions. The reason lies in the high demands of experimental technique and on the purity of the reagents used. We should also note that the limited penetration of UV radiation throughout the disperse medium does not create the stationary conditions ideal for studying kinetic processes in polymerization systems.

On absorbing UV light, the compounds enter an excited state; some decompose into radical fragments able to initiate polymerization of vinyl monomers [268]. Donor/acceptor complexes are also in some cases transferred to the (unstable) excited state after absorption of UV radiation and their decomposition leads to free radicals capable of initiating the growth of a polymer chain [269].

The excited donor/acceptor complex of ferrocene and CCl_4 was used to initiate the emulsion polymerization of methyl methacrylate [270]. The formation of initiating $^{\bullet}CCl_3$ radicals occurred through a cyclic mechanism. Through the interaction between ferricinium cation (FcH$^+$) and ascorbic acid (AA), ferrocene (FcH) was regenerated and re-entered the initiating cycle. Considerably higher rates of polymerization than those in the cationic emulsifier-containing system were observed in the presence of anionic emulsifier (sodium dodecyl sulfate). The difference in the rates was explained by the differing abilities of FcH$^+$ to diffuse through the interphase, containing both anionic and cationic emulsifier, into the water.

Excitation of the derivatives of saccharose by UV radiation and their decay led to the production of radical fragments able to initiate polymer chain propagation in the emulsion polymerization of methyl methacrylate [271] and other vinyl monomers [272, 273]. As the saccharose concentration increases, the rate of polymerization is enhanced and is proportional to the 0.45 power of the concentration of saccharide. The rate of polymerization also increases with the 0.6 power of the incident light intensity. The mechanism of initiation by excited saccharose derivatives is not yet established.

The relative efficiencies of water- and oil-soluble photoinitiators in photopolymerization reactions carried out in direct micellar solution has recently become the subject of interesting questions. The dynamics of the excited states of such photoinitiators has been studied in organized media [274].

Carbonyl compounds are efficient sensitizers of the polymerization of vinyl monomers in both homogeneous and heterogeneous systems. After UV light excitation, they decompose into radicals capable of initiating polymerization [275, 276]. Fouassier & Lougnot [277] have proposed a mechanism for the photoinitiation of the emulsion polymerization of acrylate monomers in the presence of benzophenone (BP).

Under UV light excitation, BP is promoted to its first excited singlet state (^{1}BP*) then via fast intersystem crossing it converts into its triplet state (^{3}BP*). This transient state can undergo hydrogen abstraction with hydrogen donors such as an emulsifier (SDS, [277]) or another hydrogen donor [278].

These reactions yield reactive species which attack monomer molecules and initiate polymerization. Moreover, the transient triplet state can be quenched by the monomer or emulsifier by an energy transfer process which does not undergo any chain initiation. This process is thus regarded as an unproductive pathway; the longer the lifetime of the triplet state, the more efficient is the quenching process. This process has been confirmed from phosphorescence quenching of ^{3}BP* by acrylate monomers [279].

Reference [277] reports that the polymerization rate is inversely proportional to the lifetime of the excited triplet benzophenone.

Takeishi et al. [280] indicated the possibility of the use of excited emulsifier molecules as a source of initiating radicals. The authors used the emulsifier N-laurylpyridinium azide for selective photoinitiation of the emulsion polymerization of styrene. This emulsifier has been shown to be inefficient for vinyl acetate, methyl vinyl ketone, acrylonitrile, acrylamide, methyl methacrylate and N-vinylpyrrolidone. The catalytic effect of the excited N-laurylpyridinium salt can be eliminated by adding DPPH and pyridine.

Radical photopolymerization reactions of methyl methacrylate in anionic (sodium dodecyl sulfate) and cationic (cetyl trimethylammonium chloride) micelles were investigated in the presence of ionic (benzophenone) initiators under steady-state illumination by Fouassier & Lougnot [281]. It is assumed that the charge/transfer complex undergoes proton transfer to yield a triplet radical pair between a ketyl radical and an amine-derived radical. This pair converts into a singlet radical pair, the separation of which leads to free radicals. In micelles formed with anionic and cationic emulsifiers, triplet state benzophenone appears as the most efficient initiator: the associated ketyl radicals are weakly reactive, the electron-transfer process with amine is the most efficient, and monomer quenching is less pronounced. An important chemical factor governing reaction efficiency seems to be the reactivity of the ketyl species which act as terminating agents of the growing polymer chains. In the polymerization of MMA in SDS micelles $BP^{\bullet-}$, the ketyl radical and ground-state benzophenone behave as weakly efficient initiators in comparison with triplet benzophenone.

Benzophenone and its derivatives were reported to initiate the emulsion polymerization of butyl acrylate under UV light [282]. The RMMs of the polymer were found to be higher by a factor of 5 when a water-soluble (BP) photoinitiator is used. The general behaviour of this type of photopolymerization is very similar to the classical, thermally-initiated emulsion polymerization.

A photoinitiator with the structure of phenylacetophenone was used in the emulsion polymerization of quaternary ammonium amphiphilic (meth)acrylates [283]. This study shows that high yields of oligomer were obtained from aqueous solutions of monomer where the concentration was above the CMC. For aqueous solutions below the CMC, the conversion to oligomer was virtually zero.

The emulsion polymerization of styrene and methyl methacrylate was photoinduced by dibenzyl ketone [284]. The presence of additives, such as diamagnetic ions, was shown to enhance the size of the micelles, which in turn resulted in a reduced efficiency of free-radical escape from the micelles. The polymerization of MMA was found to be unaffected by the addition of these ions.

Arai [285] investigated the polymerization of unsaturated fatty acids initiated by UV light. The fatty acid acts as a sensitizer. For each fatty acid, the amount of polymer obtained was very small at very low concentrations of the fatty acid, but increased rapidly with increasing monomer concentration at concentrations higher than the first CMC and then moderately at concentrations higher than the second CMC.

A photoinitiator should have many characteristics: excellent reactivity, high molar absorption coefficients, a highly suitable spectral absorption and water-soluble and nonyellowing photolysis products. Water-soluble photoinitiators have received considerable attention over the last few years and very efficient molecules have been investigated such as benzil, benzophenone, thioxanthone, hydroxyalkyl ketone and others [286].

γ-Radiation (characterized by its ability to penetrate disperse systems) has been applied to initiate the emulsion polymerization and copolymerization of n-butyl acrylate, 2-hydroxyethyl methacrylate, and acrylic acid. Latices prepared in this way [287–289] showed higher viscosity, greater density of the surface particle charge, and greater stability compared to those prepared via chemical initiation. The high-energy radiation produces $HO^{\bullet}$ in the aqueous phase, which initiates the growth of polymer chains.

References

1. Schwunger, M.J. (1984) *J. Am. Chem. Soc., ACS Symp. Ser.* **253** 3
2. Frank, H.S. & Evans, M.W. (1945) *J. Phys. Chem.* **13** 507
3. Kavanau, J.L. (1964) *Water and solute–water interactions*. Holden-Day, New York
4. Tanford, C. (1980) *The hydrophobic effect*. 2nd ed. Wiley–Interscience, New York
5. McBain, J.W. & Martin, H.E. (1914) *J. Chem. Soc.* **105** 957
6. Fendler, J.H. & Fendler, E.J. (1975) *Catalysis in micellar and macromolecular systems*. Academic Press, New York
7. Hartley, G.S. (1948) *Q. Rev., Chem. Soc.* **2** 152
8. Menger, F.M. (1977) In: Tamelen van, E.E. (ed.) *Bioorganic chemistry*, vol. III, *Macro- and multimolecular systems*. Academic Press, New York, p. 137
9. Backus, J.K. & Scheraga, H.A. (1951) *J. Colloid Sci.* **6** 508

References

10. Reiss-Husson, F. & Luzzati, V. (1964) *J. Phys. Chem.* **68** 3504
11. Dill, K.A., Koppel, D.E., Cantor, R.S., Dill, J.D., Bendedouch, D. & Chen, S.H. (1984) *Nature* **309** 42
12. Muller, N. & Birkahn, R.H. (1967) *J. Phys. Chem.* **71** 957
13. Menger, F.M., Jerkunica, J.M. & Johnson, J.C. (1978) *J. Am. Chem. Soc.* **100** 4676
14. Menger, F.M., Yoshinaga, H., Venkatusubban, K.S. & Das, A.R. (1981) *J. Org. Chem.* **46** 415
15. Menger, F.M. & Doll, D.W. (1984) *J. Am. Chem. Soc.* **106** 1109
16. Corkill, J.M., Goodman, J.F. & Walker, T. (1967) *Trans. Faraday Soc.* **63** 768
17. Kurz, J.L. (1962) *J. Phys. Chem.* **66** 2239
18. Kresheck, G.C. (1975) In: Franks, F. (ed.) *Water, a comprehensive treatise*, vol. 4. Plenum Press, New York, p. 95
19. Shinitzky, M., Dianoux, A.C., Citler, C. & Weber, G. (1971) *Biochemistry* **10** 2106
20. Ohnishi, S., Cyr, T.J.R. & Fukushina, H. (1970) *Bull. Chem. Soc. Jpn.* **43** 673
21. Menger, F.M. & Jerkunica, J.M. (1978) *J. Am. Chem. Soc.* **100** 688
22. Pownall, H.J. & Smith, L.C. (1973) *J. Am. Chem. Soc.* **95** 3136
23. Dorrance, R.C. & Hunter, T.F. (1972) *J.Chem. Soc., Faraday Trans. 1* **68** 1312
24. Menger, F.M. (1979) *Acc. Chem. Res.* **12** 111
25. Vold, M.J. (1950) *J. Colloid Sci.* **5** 506
26. Aniansson, E.A.G. & Wall, S.N. (1974) *J. Phys. Chem.* **78** 1024
27. Tanford, C., Nozaki, Y. & Rohde, M.F. (1977) *J. Phys. Chem.* **81** 1555
28. Tokiwa, F. (1968) *J. Phys. Chem.* **72** 1214
29. Herrmann, K.W. (1968) *J. Phys. Chem.* **66** 295
30. Harkins, W.D. (1947) *J. Am. Chem. Soc.* **69** 1428
31. Hashimoto, S., Thomas, J.K., Evans, D.F., Mukherjee, S. & Ninham, B.W. (1983) *J. Colloid Interface Sci.* **95** 594
32. Malliaris, A. & Paleos, C.M. (1984) *J. Colloid Interface Sci.* **101** 364
33. Asinger, F., Berger, W., Fanghaenel, E. & Müller, K.R. (1965) *J. Prakt. Chem.* **27** 82
34. Schwunger, M.J. (1970) *Chem. Eng. Technol.* **42** 433
35. Limbele, W.B. & Zana, R. (1986) *Colloids Surfaces* **21** 483
36. Lee, C.H. & Mallinson, R.G. (1990) *J. Appl. Polym. Sci.* **39** 2205
37. Kaler, E.W., Puig, J.E. & Miller, W.G. (1984) *J. Phys. Chem.* **88** 2887
38. Kapoor, R.C. & Mishra, V.N. (1976) *Indian Chem. Soc.* **53** 965
39. Humphry-Baker, R., Grätzel, M. & Steiger, R. (1980) *J. Am. Chem. Soc.* **102** 847
40. Lange, H. & Schwunger, M.J. (1968) *Z. Polym.* **223** 145
41. Shinoda, K. & Becher, P. (1978) In: Shinoda, K. (ed.) *Principles of solution and solubility*. M. Dekker, New York, p. 159
42. Schwunger, M.J. (1969) *Z. Polym.* **233** 979
43. Shinoda, K. (1980) *Pure Appl. Chem.* **52** 1195
44. Hato, M., Tahara, M. & Suda, Y. (1979) *J. Colloid Interface Sci.* **72** 458
45. Mukerjee, P. (1972) *J. Phys. Chem.* **76** 565
46. Moroi, Y., Nishikido, N., Uehara, H. & Matuura, R. (1975) *J. Colloid Interface Sci.* **50** 254
47. Kale, K.M., Cussler, E.L. & Evans, D.F. (1980) *J. Phys. Chem.* **84** 593
48. Elworthy, H. & Mysels, K.J. (1966) *J. Colloid Interface Sci.* **21** 331
49. Moroi, Y. (1988) *J. Colloid Interface Sci.* **122** 308
50. Marszall, L. (1977) *J. Colloid Interface Sci.* **59** 376
51. Schott, H. (1969) *J. Pharm. Sci.* **58** 1443
52. Nakagawa, T. (1967) In: Schick, M.J. (ed.) *Nonionic surfactants*. M. Dekker, New York, Chapter 17
53. Schott, H. (1973) *J. Colloid Interface Sci.* **43** 150
54. Marszall, L. (1981) *Tenside Deterg.* **18** 25
55. Schott, H., Royce, A.E. & Han, S.K. (1984) *J. Colloid Interface Sci.* **98** 196
56. Griffin, W.C. (1949) *J. Soc. Cosmet. Chem.* **1** 311
57. Griffin, W.C. (1954) *J. Soc. Cosmet. Chem.* **5** 249
58. Davies, J.T. & Rideal, E.K. (1961) *Interfacial phenomena*. Academic Press, New York, p. 371
59. Binana-Limbele, W. & Zana, R. (1986) *Colloids Surfaces* **21** 483
60. Diaz Garcia, E.M. & Sanz-Medel, A. (1986) *Talanta* **33** 255
61. Chen, J.M., Su, T.M. & Mou, C.Y. (1986) *J. Phys. Chem.* **90** 2418
62. Kratochvil, J.P. (1980) *J. Colloid Interface Sci.* **75** 271
63. Larsen, J.W. & Magid, L.J. (1974) *J. Am. Chem. Soc.* **96** 5774
64. Tanford, C. (1974) *J. Phys. Chem.* **78** 2469
65. Steigman, J. & Shane, N. (1965) *J. Phys. Chem.* **69** 968
66. Ekwall, P., Mandell, L. & Solyom, P. (1971) *J. Colloid Interface Sci.* **35** 519
67. Kalyanasundaram, K., Grätzel, M. & Thomas, J.K. (1975) *J. Am. Chem. Soc.* **97** 3915
68. Hayashi, S. & Ikeda, S. (1980) *J. Phys. Chem.* **84** 744
69. Minch, M.J., Giacco, M. & Wolff, R. (1975) *J. Am. Chem. Soc.* **97** 3766

70. Rosendorfová, J. & Čermáková, L. (1980) *Talanta* **27** 705
71. Ikeda, S., Hayashi, S. & Imae, T. (1981) *J. Phys. Chem.* **85** 106
72. Ray, A. & Nemethy, G. (1971) *J. Am. Chem. Soc.* **93** 6787
73. Ottewill, R.H., Storer, C.C. & Walker, T. (1967) *Trans. Faraday Soc.* **63** 2796
74. Attwood, D. (1968) *J. Phys. Chem.* **72** 339
75. Imae, T. & Ikeda, S. (1984) *Colloid Polym. Sci.* **262** 497
76. Imae, T., Kamiya, R. & Ikeda, S. (1984) *J. Colloid Interface Sci.* **99** 300
77. Zernike, F. & Prins, J.A. (1927) *Z. Phys.* **41** 184
78. Hansen, J.P. & McDonald, I.R. (1979) *Theory of simple liquids*. Academic Press, New York, London, p. 109
79. Gamboa, C. & Sepulveda, L. (1986) *J. Colloid Interface Sci.* **113** 566
80. Beunen, J.A. & Ruckenstein, E. (1982) *Adv. Colloid Interface Sci.* **16** 201
81. Doren, A. & Goldbarb, J. (1970) *J. Colloid Interface Sci.* **32** 67
82. Dorshow, R., Bunton, C.A. & Nicoli, D.F. (1983) *J. Phys. Chem.* **87** 1409
83. Zana, R., Yiv, S., Strazielle, C. & Lianos, P. (1981) *J. Colloid Interface Sci.* **80** 208
84. Emerson, M.F. & Holtzer, A. (1967) *J. Phys. Chem.* **71** 3320
85. Anacker, E.W. & Underwood, A.L. (1981) *J. Phys. Chem.* **85** 2463
86. Hoar, T.P. & Schulman, J.H. (1943) *Nature* **152** 102
87. Cazabat, A.M. & Langevin, D. (1981) *J. Chem. Phys.* **74** 3148
88. Biais, J., Barthe, M., Clin, B. & Lalanne, P. (1984) *J. Colloid Interface Sci.* **102** 361
89. Prince, L.M. (1977) *Microemulsions, theory and practice*. Academic Press, New York
90. Stilbs, P., Rapacki, K. & Lindman, B. (1983) *J. Colloid Interface Sci.* **95** 585
91. Talmon, Y. & Prager, S. (1978) *J. Chem. Phys.* **69** 2984
92. Gennes de, P.G. & Taupin, C. (1982) *J. Phys. Chem.* **86** 2294
93. Widom, B.J. (1986) *J. Chem. Phys.* **84** 6943
94. de Geyer, A. & Tabony, J. (1985) *Chem. Phys. Lett*. **113** 83
95. de Geyer, A. & Tabony, J. (1986) *Chem. Phys. Lett.* **124** 357
96. Cazabat, A.M., Langevin, D. & Pouchelon, A. (1986) *J. Colloid Interface Sci.* **73** 1
97. Gestblom, B. & Sjöblom, J. (1988) *Langmuir* **4** 360
98. Mitchell, D.J. & Ninham, B.W. (1981) *J. Chem. Soc., Faraday Trans. 2* **77** 601
99. Langevin, D. (1988) *Acc. Chem. Res.* **21** 255
100. Lagües, M., Ober, R. & Taupin, C. (1978) *J. Phys, Lett.* **39** L-487
101. Schulman, J. H., Stoeckenhuis, W. & Prince, L.M. (1959) *J. Phys. Chem.* **63** 1677
102. Prince, L.M. (1967) *J. Colloid Interface Sci.* **23** 165
103. Shinoda, K. & Friberg, S. (1975) *Adv. Colloid Interface Sci.* **4** 281
104. Ahmed, S.I., Shinoda, K. & Friberg, S. (1974) *J. Colloid Interface Sci.* **47** 32
105. Ruckenstein, E. & Chi, J.C. (1975) *J. Chem. Soc., Faraday Trans. 2* **71** 1690
106. Overbeek, J. Th.G. (1978) *Faraday Discuss. Chem. Soc.* **65** 7
107. Shinoda, K., Hanrin, M., Kunieda, H. & Saito, H. (1971) *Colloids Surfaces* **2** 301
108. Scriven, L.E. (1977) *Micellization, solubilization and microemulsions*. Plenum Press, New York
109. Lindman, B., Stilbs, P. & Moseley, M.E. (1981) *J. Colloid Interface Sci.* **83** 569
110. Stilbs, P. & Stilbs, J. (1982) *J. Colloid Interface Sci.* **87** 385
111. Lindman, B., Stilbs, P. & Moseley, M.E. (1981) *J. Colloid Interface Sci.* **83** 569
112. Warnheim, T., Sjöblom, E., Henriksson, U. & Stilbs, P. (1984) *J. Phys. Chem.* **88** 5420
113. Desnoyers, J.E., Quirion, F., Hetu, D. & Perron, G. (1983) *Can. J. Chem. Eng.* **61** 672
114. Friberg, S.E. & Venable, R.L. (1983) In: Becher, P. (ed.) *Encyclopedia of emulsion technology*, vol. 1. M. Dekker, New York, p. 287
115. Friberg, S.E. & Buraczewska, I. (1979) *Prog. Colloid Polymer Sci.* **63** 1
116. Peyrelasse, J., Boned, C., Heil, J. & Clausse, M. (1982) *J. Phys., Solid State Phys.* **15** 7099
117. Lang, J., Djavanbakht, A. & Zana, R. (1982) In: Rubb, I.D. (ed.) *Microemulsions*. Plenum Press, New York, p. 233
118. Xenakis, A. & Tondre, C. (1983) *J. Phys. Chem.* **87** 4737
119. Tondre, C. & Xenakis, A. (1982) *Colloid Polymer Sci.* **260** 232
120. Venable, R.L. (1985) *J. Am. Oil Chem. Soc.* **62** 128
121. Venable, R.L., Elders, K.L. & Fang, J. (1986) *J. Colloid Interface Sci.* **109** 330
122. Ugelstad, J., El-Aasser, M.S. & Vanderhoff, J.W. (1973) *J. Polym. Sci., Polym. Lett.* **11** 503
123. Chou, Y.J., El-Aasser, M.S. & Vanderhoff, J.W. (1980) In: Fitch, R.M. (ed.) *Polymer colloids*, vol. 2. Plenum Press, New York, p. 599
124. Chou, Y.J., El-Aasser, M.S. & Vanderhoff, J.W. (1980) *J. Dispersion Technol.* **1** 129
125. Roehl, E.L. (1972) *Soap Parfum. Cosmet.* **45** 343
126. Brandau, R. & Bold, K.W. (1979) *Fette, Seife, Anstrichm.* **81** 366
127. Garti, N. & Magdasi, S. (1981) *Colloids Surfaces* **3** 221
128. Grimm, W.L., Min, T.I., El-Aasser, M.S. & Vanderhoff, J.W. (1983) *J. Colloid Interface Sci.* **94** 531
129. Bansal, K.M., Patterson, L.K., Fendler, E.J. & Fendler, J.H. (1971) *Int. J. Radiat. Phys. Chem.* **3** 321

References

130. Klevans, H.B. (1950) *Chem. Rev.* **47** 1
131. McBain, M.E. & Hutchinson, E. (1955) *Solubilization and related phenomena*. Academic Press, New York
132. Almgren, M., Greiser, F. & Thomas, J.K. (1979) *J. Am. Chem. Soc.* **101** 279
133. Thomas, D.C. & Christian, S.D. (1981) *J. Colloid Interface Sci.* **82** 430
134. Chaiko, M.A., Nagarajan, R. & Ruckenstein, E. (1984) *J. Colloid Interface Sci.* **99** 168
135. Rehfeld, S.J. (1971) *J. Phys. Chem.* **75** 3905
136. Mukerjee, P. & Cardinal, J.R. (1978) *J. Phys. Chem.* **82** 1620
137. Eriksson, J.C. & Gilbert, G. (1966) *Acta Chem. Scand.* **20** 2019
138. Fendler, J.H. & Patterson, L.K. (1971) *J. Phys. Chem.* **75** 3907
139. Mukerjee, P. (1979) In: Mittal, K.L. (ed.) *Solution chemistry of surfactants*, vol. 1. Plenum Press, New York
140. Nagarajan, R. & Ruckenstein, E. (1983) In: Mittal, K.L. (ed.) *Surfactants in solution — theoretical and applied aspects*, vol. 2. Plenum Press, New York
141. Nagarajan, R., Chaiko, M.A. & Ruckenstein, E. (1984) *J. Phys. Chem.* **88** 2916
142. Good, R.J. & Elbing, E. (1970) *Ind. Eng. Chem.* **62** 54
143. Hoskins, J.C. & King, A.D. (1981) *J. Colloid Interface Sci.* **82** 260
144. Hayter, B., Hayoun, M. & Zemb, T. (1984) *Colloid Polym. Sci.* **262** 798
145. Lianos, P., Viriot, M.L. & Zana, R. (1984) *J. Phys. Chem.* **88** 1098
146. Hoffmann, H. (1978) *Ber. Bunsenges. Phys. Chem.* **82** 988
147. Lianos, P. & Zana, R. (1980) *J. Phys. Chem.* **84** 3339
148. Lianos, P., Lang, J. Strazielle, C. & Zana, R. (1982) *J. Phys. Chem.* **86** 1019
149. Carroll, B.J. (1981) *J. Colloid Interface Sci.* **79** 126
150. Shaeieitz, J.A., Chan, A.F.C., Cussler, E.L. & Evans, D.F. (1981) *J. Colloid Interface Sci.* **84** 47
151. Carroll, B.J. (1986) *J. Chem. Soc., Faraday Trans. 1* **82** 3205
152. Desnoyers, J.E., Caron, G., De Lisi, R., Roberts, D., Roux, A. & Perron, G. (1983) *J. Phys. Chem.* **87** 1397
153. Desnoyers, J.E., Hetu, D. & Perron, G. (1983) *J. Solution Chem.* **12** 427
154. De Lisi, R., Turco Liveri, V., Castagnolo, M. & Inglese, A. (1986) *J. Solution Chem.* **15** 23
155. De Lisi, R. & Turco Liveri, V. (1983) *Gazz. Chim. Ital.* **113** 371
156. De Lisi, R., Milioto, S. & Turco Liveri, V. (1987) *J. Colloid Interface Sci.* **117** 64
157. Mukerjee, P. & Ray, A. (1966) *J. Phys. Chem.* **70** 2144
158. Turner, D.C. & Brand, L. (1968) *Biochemistry* **7** 3381
159. Cordes, E.H. & Gitler, C. (1973) *Prog. Bioorg. Chem.* **2** 1
160. Bunton, C.A. & Huang, S.K. (1972) *J. Org. Chem.* **37** 1790
161. Robb, I.D. & Smith, R. (1974) *J. Chem. Soc., Faraday Trans. 1* **70** 287
162. Jencks, W.P. (1975) *Adv. Enzymol.* **43** 219
163. Bunton, C.A., Minota, M.J., Hidalgo, J. & Sepulveda, L. (1973) *J. Am. Chem. Soc.* **95** 3262
164. Kemp, D.S. & Paul, K. (1970) *J. Am. Chem. Soc.* **92** 2553
165. Kunitake, T., Shinkai, S. & Hirotsu, S. (1977) *J. Org. Chem.* **42** 306
166. Bunton, C.A., Fendler, E.J., Sepulveda, L. & Yang, K.H. (1968) *J. Am. Chem. Soc.* **90** 5512
167. Duynstee, E.F.J. & Grunwald, E. (1959) *J. Am. Chem. Soc.* **81** 4540
168. Duynstee, E.F.J., & Grunwald, E. (1959) *J. Am. Chem. Soc.* **81** 4542
169. Albrizzio, J., Archila, A., Rodulfo, T. & Cordes, E.H. (1972) *J. Org. Chem.* **37** 871
170. Baumrucker, J., Calzadilla, M., Centeno, M., Lehrmann, G., Urdaneta, M., Lindquist, P., Dunham, P., Price, M., Sears, B. & Cordes, E.H. (1972) *J. Am. Chem. Soc.* **94** 8164
171. Cordes, E.H. (1978) *Pure Appl. Chem.* **50** 617
172. Kunitake, T., Okahata, Y. & Sakamoto, T. (1976) *J. Am. Chem. Soc.* **98** 7800
173. Okahata, Y., Ando, R. & Kunitake, T. (1977) *J. Am. Chem. Soc.* **99** 3067
174. Arai, K. & Ogiwara, Y. (1983) *J. Appl. Polym. Sci.* **28** 3309
175. Arai, K., Sugita, J. & Ogiwara, Y. (1986) *Makromol. Chem., Rapid Commun.* **7** 427
176. Arai, K., Maseki, Y. & Ogiwara, Y. (1986) *Makromol. Chem., Rapid Commun.* **7** 655
177. Menger, F.M. & Portnoy, C.A. (1967) *J. Am. Chem. Soc.* **89** 4968
178. Menger, F.M. & Portnoy, C.A. (1968) *J. Am. Chem. Soc.* **90** 1875
179. Martinek, K., Osipov, A.P., Yatsimirskii, A.K., Dadali, V.A. & Berezin, I.V. (1975) *Tetrahedron Lett.* p. 1279
180. Martinek, K., Yatsimirskii, A.K., Osipov, A.P. & Berezin, I.V. (1973) *Tetrahedron* **29** 963
181. Romsted, L.S. (1977) In: Mittal, K.L. (ed.) *Micellization, solubilization and microemulsions*, vol. 1. Plenum Press, New York, p. 509
182. Bunton, C.A. & Wolfe, B. (1973) *J. Am. Chem. Soc.* **95** 3742
183. Kolthoff, I.M., Guss, L.S., May, D.R. & Medalia, A.I. (1946) *J. Am. Chem. Soc.* **1** 340
184. Kolthoff, I.M. & Miller, I.K. (1951) *J. Am. Chem. Soc.* **73** 3055
185. Hakoila, E. (1963) *Ann. Univ. Turk., A* **66** 7
186. House, D.A. (1962) *Chem. Rev.* **62** 185
187. Banerjee, M. & Konar, R.S. (1974) *J. Indian Chem. Soc.* **51** 722
188. Hayon, E., Treinin, A. & Wolf, J. (1972) *J. Am. Chem. Soc.* **94** 47

189. Pennington, D.E, & Haim, A. (1968) *J. Am. Chem. Soc.* **90** 3700
190. Dogliotti, L. & Hayon, E. (1967) *J. Phys. Chem.* **71** 2511
191. Bovey, F.A., Kolthoff, I.M., Medalia, A.I. & Meehan, E.J. (1955) *Emulsion polymerization*. Interscience, New York
192. Sarkar, S., Adhikari, M.S., Banerjee, M. & Konar, R.S. (1988) *J. Appl. Polym. Sci.* **36** 1865
193. Sarkar, S., Dhikari, M.S., Banerjee, M. & Konar, R.S. (1988) *J. Appl. Polym. Sci.* **35** 1441
194. Maruthamuthu, P. (1980) *Makromol. Chem., Rapid Commun.* **1** 23
195. Adhikari, M.S., Sarkar, S., Banerjee, M. & Konar, R.S. (1987) *J. Appl. Polym. Sci.* **34** 109
196. Bartlett, P.D. & Nozaki, K. (1948) *J. Polym. Sci.* **3** 216
197. Morris, C.E.M. & Parts, A.G. (1968) *Makromol. Chem.* **119** 212
198. Dunn, A.S. (1982) In: Piirma, I. (ed.) *Emulsion polymerization*. Academic Press, New York, p. 22.
199. Kuanhsiung, C. (1980) *PhD Thesis*. Case Western University, Ann Arbor
200. Yurzhenko, A.I., Brazhnikova, O.P. & Likholet, N.M. (1955) *Ukr. Khim. Zh.* **21** 586
201. Ivanchev, S.S. & Yurzhenko, A.I. (1958) *Izv. Vyssh. Ucheb. Zaved., Khim. Khim. Tekhnol.* **4** 13
202. Ryabova, M.S., Sautin, S.N. & Smirnov, N.I. (1977) *Zh. Prikl. Khim.* **L** 1719
203. Ryabova, M.S., Sautin, S.N. & Smirnov, N.I. (1978) *Zh. Prikl. Khim.* **LI** 2056
204. Brooks, B.W. & Makanjuola, B.O. (1981) *Makromol. Chem., Rapid Commun.* **2** 69
205. Okubo, M. & Mori, T. (1990) *Makromol. Chem., Macromol. Symp.* **31** 143
206. Vinogradov, P.A., Obintsova, P.P. & Shitova, A.A. (1962) *Vysokomol. Soedin., A* **4** 98
207. Crematy, E.P. (1971) *Makromol Chem.* **143** 125
208. Antonova, L.F., Lepyanin, G.V., Zaitsev, E.E. & Rafikov, S.R. (1978) *Vysokomol. Soedin., A* **20** 687
209. Capek, I., Mlynárová, M. & Bartoň, J. (1988) *Makromol. Chem.* **189** 341
210. Kolthoff, F., Meehan, E. & Carr, E. (1953) *J. Am. Chem. Soc.* **75** 1439
211. Bartoň, J. & Kárpátyová, A. (1987) *Makromol. Chem.* **188** 693
212. Vanderhoff, J.W. & Bradford, E.B. (1956) *Tappi* **39** 650
213. Nuyken, O., Dorn, M. & Kerber, R. (1979) *Makromol. Chem.* **180** 1651
214. Walz, R., Bomer, B. & Heitz, W. (1977) *Makromol. Chem.* **178** 2527
215. Dicke, H.R. & Heitz, W. (1981) *Makromol. Chem., Rapid Commun.* **2** 83
216. Bartoň, J. & Borsig, E. (1984) *Complexes in free-radical polymerization*. Veda, Bratislava, p. 124 (in Slovak)
217. Dicke, R.H. & Heitz, W. (1982) *Colloid Polym. Sci.* **260** 3
218. Blackley, D.C. (1975) *Emulsion polymerization*. Applied Science Publ., London, p. 203
219. Thelisse, J.A. & Quiby, C.A. (1959) *Fr* **123** 3582
220. Van der Hoff, B.M.E. (1960) *J. Polym. Sci.* **48** 175
221. Ivanchev, S.S., Solomko, N.I., Konovalenko, V.V. & Yurzhenko, V.A. (1970) *Dokl. Akad. Nauk SSSR* **191** 593
222. Ivanchev, S.S. & Pavlyuchenko, V.N. (1981) *Acta Polym.* **32** 407
223. Ivanchev, S.S. & Pavlyuchenko, V.N. (1979) *Plaste Kautsch.* **26** 314
224. Ivanchev, S.S., Pavlyuchenko, V.N. & Rozhkova, D.A. (1974) *Vysokomol. Soedin., A* **16** 893
225. Rozhkova, D.A., Pavlyuchenko, V.N. & Ivanchev, S.S. (1975) *Kinet. Katal.* **16** 814
226. Bartoň, J. & Borsig, E. (1984) *Complexes in free-radical polymerization*. Veda, Bratislava, p. 91 (in Slovak)
227. Haber, F. & Weiss, J. (1932) *Naturwissenschaften* **20** 948
228. Kolthoff, I.M. & Medalia, A.I. (1949) *J. Am. Chem. Soc.* **71** 3777
229. Marvel, C.S., Deanin, R., Clous, C.J., Wyld, M.B. & Ceitz, R.L. (1948) *J. Polym. Sci.* **3** 350
230. Dixon, W.T. & Norman, R.O.C. (1962) *Nature* **196** 891
231. Alince, B. & Robertson, A.A. (1979) *J. Appl. Polym. Sci.* **23** 539
232. Alince, B., Robertson, A.A. & Inoue, M. (1978) *J. Colloid Interface Sci.* **65** 98
233. Frank, J. & Haber, F. (1931) *Naturwissenschaften* **19** 450
234. Tsuda, Y. (1961) *J. Appl. Polym. Sci.* **5** 104
235. Peebles, L.H. (1973) *J. Appl. Polym. Sci.* **17** 113
236. Morgan, L.B. (1946) *Trans. Faraday Soc.* **42** 169
237. Banthia, A.K., Mandal, B.M. & Palit, S.R. (1977) *J. Polym. Chem.* **15** 945
238. McCarvill, W.T. & Fitch, R.M. (1978) *J. Colloid Interface Sci.* **64** 404
239. Fitch, R.M. & McCarvill, W.T. (1978) *J. Colloid Interface Sci.* **66** 20
240. Berry, K.L. & Peterson, J.H. (1951) *J. Am. Chem. Soc.* **73** 5195
241. McCarvill, W.T. & Fitch, R.M. (1978) *J. Colloid Interface Sci.* **67** 204
242. Orr, R.J. & Williams, L.J. (1955) *J. Am. Chem. Soc.* **77** 3715
243. Fordham, J.W.L. & Williams, L.J. (1951) *J. Am. Chem. Soc.* **73** 4855
244. Bataille, P., Van, B.T. & Pham, Q.B. (1982) *J. Polym. Sci., Polym. Chem. Ed.* **20** 811
245. Singh, U.C., Manickam, S.P. & Kenkataras, K. (1979) *Makromol. Chem.* **180** 589
246. Manickam, S.P. & Subbaratman, N.R. (1981) *J. Macromol. Sci., Chem.* **15** 1511
247. Bawn, C.E.H. & Margerison, D. (1955) *Trans. Faraday Soc.* **51** 925
248. Bataille, P. & Gonzales, A. (1984) *J. Polym. Sci., Polym. Chem. Ed.* **22** 1409
249. Rasmussen, J.K. & Smith, H.K. (1981) *Makromol. Chem.* **182** 701

250. Simionescu, C., Mihailescu, C. & Bulacovski, V. (1987) *Acta Polym.* **38** 502
251. Alexander, A.E. & Napper, D.H. (1970) *Prog. Polym. Sci.* **3** 145
252. Capek, I., Bartoň, J. & Kárpátyová, A. (1987) *Makromol. Chem.* **188** 703
253. Bauer, H., Ortelt, M., Joos, B. & Funke, W. (1988) *Makromol. Chem.* **189** 409
254. Firsching, F.H. & Rosen, I. (1959) *J. Polym. Sci.* **36** 305
255. Duban, R.C. (1956) *130th Meeting of American Chemical Society*, September. Abstracts of papers
256. Moustafa, A.B. (1974) *Angew. Makromol. Chem.* **39** 1
257. Moustafa, A.B. & Abd-el-Hakim, A.A. (1976) *J. Polym. Sci., Chem.* **14** 433
258. Hamshima, M. & Yamajaki, S. (1964) *Hokoku* **59** 180
259. Mukherjee, A.R., Ghosh, P., Chadha, S.C. & Palit, S.R. (1967) *Makromol. Chem.* **80** 208
260. Moustafa, A.B. & Abd-el-Hakim, A.A. (1977) *J. Appl. Polym. Sci.* **21** 905
261. Moustafa, A.B. & Diab, M.A. (1975) *Angew. Makromol. Chem.* **45** 41
262. Moustafa, A.B. & Diab, M.A. (1975) *J. Appl. Polym. Sci.* **19** 1585
263. Kolthoff, I.M., Medalia, A.I. & Raaen, H.P. (1951) *J. Am. Chem. Soc.* **73** 1733
264. Rubio, S., Serre, B., Sledz, J., Schue, F. & Chapelet-Letourneux, G. (1980) *J. Macromol. Sci., Rev. Macromol. Chem.* **C19** 175
265. Feng, X.D., Guo, X.Q. & Qiu, K.Y. (1988) *Makromol. Chem.* **189** 77
266. Guo, X.Q., Qiu, K.Y. & Feng, X.D. (1990) *Makromol. Chem.* **191** 577
267. Fernández, M.D. & Guzmán, G.M. (1990) *Br. Polym. J.* **22** 1
268. Bartoň, J., Capek, I. & Hrdlovič, P. (1975) *J. Polym. Sci., Polym. Chem. Ed.* **13** 2671
269. Bartoň, J., Capek, I., Arnold, M. & Rätzsch, M. (1980) *Makromol. Chem.* **181** 241
270. Tsunooka, M. & Tanaka, M. (1978) *J. Polym. Sci., Polym. Lett. Ed.* **16** 119
271. Merlin, A. & Fouassier, J.P. (1981) *J. Polym. Sci., Polym. Chem. Ed.* **19** 2357
272. Kubota, H. & Ogiwara, Y. (1976) *J. Appl. Polym. Sci.* **20** 1405
273. Hérold, R. & Fouassier, J.P. (1980) *Angew. Makromol. Chem.* **86** 123
274. Fouassier, J.P. (1983) *J. Chim. Phys.* **80** 339
275. Kuhlmann, R. & Schnabel, W. (1977) *Polymer* **18** 1163
276. Block, H., Ledwith, A. & Taylor, A.R. (1971) *Polymer* **12** 271
277. Fouassier, J.P. & Lougnot, D.J. (1983) *Polym. Photochem.* **3** 79
278. Bartoň, J., Capek, I., Šušoliak, O. & Juraničová, V. (1978) *Makromol. Chem.* **179** 2937
279. Kuhlmann, R. & Schnabel, W. (1976) *Polymer* **17** 419
280. Takeishi, M., Yoshida, H., Niino, S. & Hayama S. (1978) *Makromol. Chem.* **179** 1387
281. Fouassier, J.P. & Lougnot, D.J. (1986) *J. Appl. Polym. Sci.* **32** 6209
282. Bonamy, A., Fouassier, J.P., Lougnot, D.J. & Green, P.N. (1982) *J. Polym. Sci., Polym. Lett. Ed.* **20** 315
283. Hamid, S.M. & Sherrington, D.C. (1987) *Polymer* **28** 332
284. Turro, N.J. & Arora, K.S. (1986) *Macromolecules* **19** 42
285. Arai, K. (1990) *Makromol. Chem., Macromol. Symp.* **31** 227
286. Fouassier, J.P. (1990) *J. Photochem. Photobiol.,* **51** 67
287. Makuuchi, K., Takagi, T. & Araki, K. (1978) *J. Japan Soc. Colour Mater.* **51** 214
288. Egusa, S. & Makuuchi, K. (1981) *J. Colloid Interface Sci.* **79** 350
289. Egusa, S. & Makuuchi, K. (1981) *Radiat. Phys. Chem.* **18** 633

3.4 KINETICS AND MECHANISM OF THE RADICAL POLYMERIZATION OF UNSATURATED MONOMERS IN MICELLAR SYSTEMS

3.4.1 Styrene and its derivatives

Chatterjee *et al.* [1] verified the validity of the micellar model in the emulsion polymerization of styrene carried out at low monomer concentrations (less than mass % of styrene with respect to aqueous phase). The dependences of the rates of polymerization on the concentration of peroxodisulfate and sodium dodecyl sulfate led to values of the reaction orders only slightly differing from the micellar theory (reaction order with respect to initiator was 0.35 and with respect to emulsifier 0.67). The number of particles and the RMM increased linearly up to the limiting value of conversion of 30–40%. As the conversion increased further, the number of particles did not vary but the RMM decreased slightly. The shift of the nucleation stage to higher conversions was explained by the formation of new particles brought about by interactions between oligomeric radicals and free emulsifier molecules.

The authors' conclusion is based on their results of injecting additional amounts of peroxodisulfate at various stages of polymerization. At low conversions, such an addition led to a rate enhancement of the polymerization which caused an increase in the particle number; on the other hand, at higher conversions, addition of further initiator did not influence the polymerization rate.

The polymerization of styrene at monomer concentrations lower than 5 vol.% gave values of the reaction order in peroxodisulfate concentration at conversions below 10% of about 0.5 to 0.55 [2]. At increasing conversions, the reaction order decreased: at 10% conversion it fell to 0.34 and, at 30 %, to 0.26. These results led the authors to conclude that the polymerization of styrene, a sparingly water-soluble monomer, also proceeds to some extent in the aqueous phase. Oligomeric radicals, generated in the propagation, enter emulsifier micelles and initiate polymerization and the formation of new primary particles. The entry of the oligomeric radicals is adversely affected by the electrostatic repulsive forces between the surface of the micelles (negatively charged) and a negatively charged initiator fragment anchored on the other end of the oligomeric radical. The oligomeric radical enters the particles in a conformation whereby the radical moiety is directed towards the particle and its negatively-charged end away from the particle. The decrease in RMM above 30% conversion was accounted for by a decrease in monomer concentration around the reaction centre. After consumption of free monomer, the concentration of radicals in particles increases as a result of the increase in viscosity, but the reduction in monomer concentration is dominant.

Independently of this study, Alexander & Gilbert [3] proposed the same mechanism for initiation of the emulsion polymerization of styrene with the active participation of oligomeric radicals in the formation of new primary polymer particles. The low concentration of styrene and the lipophilic character of the polymer being formed enable the formation of oligomeric radicals with only a very short chain with a degree of polymerization of about 3 to 5. However, such a chain length is sufficient for the separation of a radical from the negatively charged moiety of the sulfate anion. This substantially reduces the barrier between the charged particle and the primary radical with the same charge formed by the decomposition of peroxodisulfate. This barrier allows only about one out of every 10^3 primary radicals present in the aqueous phase to enter a micelle or polymer particle.

Piirma et al. [4] obtained by studying the emulsion batch polymerization of styrene at monomer concentrations around 30–40 vol. %, reaction orders with respect to initiator and emulsifier (sodium dodecyl sulfate and potassium oleate) close to 0.4 and 0.6. At low emulsifier concentrations (around the CMC), the shapes of the conversion curves were typical for emulsion polymerization, i.e. sigmoidal. At higher emulsifier concentrations, the maximum rate of polymerization was reached immediately after the start of polymerization and the maximum RMM was already achieved at ca. 10% conversion and remained almost the same up to 40% conversion. Above the limiting value of conversion, the RMM decreased slightly. The polymer formed showed high RMM ($\overline{M}_n \sim 3 \times 10^6$) with a wide distribution ($M_w/M_n \sim 3$).

The course of the emulsion polymerization is considerably influenced by the nature and concentration of the emulsifier. Al-Shabib & Dunn [5] studied the effect of the alkyl chain length of anionic alkyl sulfates on the rate of polymerization (see Table 13). This (in Stage II) increased by as much as 50% in a series of emulsifiers with alkyl chain from C_{10} to C_{18}. The increase was explained by the increase of the fraction of micellar emulsifier and, in parallel, by the formation of a larger number of monomer/polymer particles in the system.

The emulsion polymerization of styrene proceeded according to the Smith–Ewart model. Good agreement with the model has also been observed in polymerizations under extreme conditions (e.g. by using high initiator concentrations).

Table 13 — Kinetic parameters for batch emulsion polymerization of styrene as a function of length of alkyl group (R) of emulsifier (sodium alkyl sulfate, SAS) (Reprinted from ref. [5] with permission of John Wiley & Sons Inc., New York)

R	10^2 CMC	10^2 SAS[a]	Stage II R_p /10^{-3} mol dm^{-3} s^{-1}	B^b /10^{-4} s^{-1} exp.	B^b /10^{-4} s^{-1} theor.	d /nm t.[c]	d /nm r.[d]	N /10^{15} cm^{-3} exp.	N /10^{15} cm^{-3} theor.
C_{10}	3.4	2.6	4.3	1.6	1.6	54	61	2.3	2.2
C_{12}	0.9	5.1	4.9	1.9	2.4	53	61	3.0	3.3
C_{14}	0.24	5.8	5.0	1.9	2.6	52	60	3.2	3.5
C_{16}	0.062	5.9	5.4	2.1	2.7	51	57	3.6	3.7
C_{18}	0.016	6.0	5.7	2.2	2.7	48	56	4.0	3.7

[a] Micellar emulsifier concentration.
[b] Smith–Ewart rate.
[c] Polymer particle diameter.
[d] Number of polymer particles per 1 cm^3 water.
exp. — experimental value, theor. — theoretical value, t. — value obtained from turbidimetry, r. — value obtained from light scattering.

Van der Hoff [6] has also concluded that only micellar emulsifier takes active part in the polymerization process, through the formation of the monomer/polymer particles and their stabilization. As the emulsifier became more lipophilic, its water solubility and CMC decreased and conversely, the fraction of micellar emulsifier taking part in the formation of the polymer particles increased.

Others [7, 8] who have studied the effect of concentration and the type of emulsifier, initiator and reaction conditions on batch polymerization, have supported (with some modification) the applicability of the micellar model.

In the emulsion polymerizations of styrene, the mechanism of polymerization is also influenced by the thermal initiation, which is more prevalent at higher temperatures. The thermal initiation of styrene is known to follow from the formation of the initiating radical fragments formed by the interaction of the Diels–Alder unstable adduct of two styrene molecules with a third molecule [9, 10].

The initiation of the emulsion polymerization of styrene by radicals originating from the thermal decomposition of Diels–Alder adducts, e.g. at 60°C, leads to the formation of a rather small fraction of the polymer; in the presence of sodium dodecyl sulfate, about 5% conversion is achieved after 10 hours [11]. The thermal polymerization is sensitized by oxygen. The presence of traces of oxygen causes an increase in radical concentration as a consequence of its interaction with the Diels–Alder adducts [12].

Lichti et al. [13] have also considered the contribution of thermal polymerization in kinetic analyses of the emulsion polymerization of styrene. Kinetic processes taking place during the thermally initiated emulsion polymerization of styrene in the presence of anionic sodium dodecyl sulfate were studied in detail by Kast & Funke [14]. These authors observed that the rate of thermally initiated polymerization increases linearly with increasing emulsifier concentration up to a limiting concentration (0.2 mol dm^{-3}) above which it does not change. At emulsifier concentration lower than 0.2 mol dm^{-3}, the reaction order with respect to emulsifier is 0.7 (i.e. close to that for classical emulsion polymerization). The dependence of the number of particles on the emulsifier concentration gives the reaction order with respect to emulsifier of 1.4. A change is also observed in the duration of the nucleation period, which shifts for this system up to 50% conversion. The system is characterized by the formation of polymer particles with a wide size distribution. If the emulsifier used is at 0.1 mol dm^{-3} concentration, a latex is formed with an average particle size D = 80 nm containing particles with diameters from 20 nm to 200 nm. The high value of the reaction order with respect to emulsifier (1.4) is also a result of the participation of transfer reactions to emulsifier and of the active role of the emulsifier radicals formed in the polymerization. The polymerization mainly took place in emulsifier micelles which were swollen with styrene; emulsified monomer droplets were considered only as a source of monomer.

However, the emulsified monomer droplets took an active role in the polymerization process in the thermal emulsion polymerization of divinylbenzene [15]. The emulsion polymerization of this tetrafunctional monomer substantially differed not only from the classical but also from the thermal polymerization of styrene. Polymerization occurred in the monomer particles at a rate comparable with that in the polymer particles. Chain propagation is initiated on entry of a radical into a micelle, and a polymer particle is formed. Along with an increase of the amount of polymer in a micelle (or in a polymer particle) the network density increases and its ability to swell decreases. The diffusion of monomer into the particles, and hence the monomer concentration around the reaction centre, are consequently reduced; this leads to a decrease in the polymerization rate. The formation of a network in a particle causes not only a decrease in the polymerization rate but leads to the formation of small polymer particles (with diameters between 10 and 30 nm). The polymerization in monomer particles (with reference to their initial size) gives much larger particles with diameters from 500 to 10 000 nm. The polymer particles grow mainly by the coalescence of the particles initiated by the reactive vinyl groups on their surface; their growth is brought about by the long reaction times of the polymerization.

In spite of the very considerable interest in polymer latices with cationic groups because of their industrial application, few data are available about the kinetics of their preparation. Sakota & Okaya [16] prepared a cationic polystyrene latex using a water-soluble initiator containing a cationic group. Through the decomposition of initiator, radical fragments were formed with a cationic moiety; these fragments initiated chain propagation and, at the same time, were localized on the surface of the latex particle during its formation, thereby stabilizing it. Stable cationic latices were also prepared via copolymerization of styrene with a small amount of cationic monomer, which was thus built into the polymer and diffused to the surface of the polymer particle being formed. Liu & Krieger [17] used similar procedures for the preparation of monodisperse polystyrene latices with a positively-charged polymer particle surface.

Ohtsuka *et al.* [18] have also described the preparation of a cationic poly(styrene) latex. They used a cationic emulsifier as stabilizer and simultaneously, as a component of the redox initiating system (which contained cerium(IV) salts). Ohtsuka *et al.* [19] have also prepared a cationic copolymer latex by the emulsion copolymerization of styrene (St) with 4-vinylpyridine (VP) in the presence of a nonionic emulsifier. Emulsion copolymerizations of St with VP were carried out at different monomer ratios at pH 2 or 11. A bimodal distribution of the particle diameter was obtained in the polymerization at low emulsifier concentrations and with high VP fractions, f_{VP}, in the monomer feed. The number of larger particles decreased but their size increased with increasing f_{VP} above 0.15. The authors suggested that the formation of large particles is favoured by the slightly preferred polymerization of VP and the interaction between VP and emulsifier at pH 11. Polymerization under acidic conditions is interpreted in terms of the high solubility and amphiphilic character of the growing radicals. Here, the rate of polymerization decreased with increasing f_{VP}. The amount of polyVP on the surface of the latex particles prepared under basic conditions was less than 2% of that of total polyVP. By contrast, the amount of polyVP on the surface of latex particles prepared under acid conditions was much greater than that of polyVP prepared at pH 11. It increased with f_{VP} up to 0.2 and then levelled off.

In the presence of a cationic emulsifier, emulsion polymerizations tend to exclude the use of peroxodisulfates as initiators since their mutual interactions may give inactive products adversely affecting the initiation step [20].

Emulsion polymerizations of styrene in the presence of a nonionic emulsifier are interesting as regards the preparation of polymer dispersions which are stable towards rather drastic changes in pH and in the ionic strength of the reaction medium. This property is conferred by the nature of the nonionic emulsifier and the steric mechanism which is used for stabilizing latex particles.

The kinetic features of the batch emulsion polymerization of styrene in the presence of a nonionic emulsifier (Emulphogene BC-840) were described by Piirma & Chang [21]. The conversion curves of these systems consist of two linear parts. The first part of the curve (i.e. at low and medium conversions, maximum of 40%) is characterized by a low rate of polymerization which is comparable with that of homogeneous polymerization. Above 40% conversion, a sudden rate enhancement occurred, which was attributed to the secondary nucleation of the particles and the gel effect. The dependence of the polymerization rate on the emulsifier concentration yielded a very high value of the reaction order with respect to emulsifier concentration, namely 2.66. This figure was ascribed to the remarkable solubility of emulsifier in the monomer phase (*ca.* 40%) and to the nonstationary micellar concentration of emulsifier during polymerization. With a decrease in the number of free monomer droplets, the emulsifier micellar concentration increases, which positively influences the formation of a new generation of particles and their stability. The formation of polymer particles at higher conversions was confirmed by the particle size distribution. As the conversion increased, the particle size also increased and, simultaneously, at 30–40% conversion, new small particles were formed. At 11% conversion, the particle size diameter reached 260 nm, at 21% it was 330 nm, at 53% as much 360 nm, and at 93%, 400 nm. The secondary fraction of the particles reached, at 53% conversion, a particle diameter of 110 nm and at 93%, 130 nm. An analysis of the RMMs of the polymer led to the conclusion that polymer with higher RMM was formed in the small particles generated by secondary nucleation, in agreement with the principle of emulsion polymerization: the RMM of the polymer is proportional to the reciprocal of the particle size [22].

We have also encountered cases where the use of a nonionic emulsifier is the only way to prepare a polymer dispersion with a higher fraction of polymer. Emulsion polymerization of *tert*-butylstyrene, a water-insoluble monomer, provides an example [23, 24]. No formation of polymer was observed when peroxodisulfate was used as initiator in the presence of anionic emulsifier. This phenomenon was explained by the high hydrophilicity of $SO_4^{\bullet-}$ radicals and the strong electrostatic barrier existing between the negatively-charged radicals and the micellar surface. The presence of trace amounts of *tert*-butylstyrene in water does not enable the formation of oligomeric radicals which would be more lipophilic than the primary radicals. No energy barrier appeared between the radicals and micelles when using a nonionic emulsifier and the polymerization of styrene was quite rapid. The use of the redox system, hydrogen peroxide and L-ascorbic acid led to production of neutral radicals able to enter the micelles of nonionic emulsifier with a positive effect on the polymerization rate.

Similar results have also been obtained by other authors [25] who have reported that *p-tert*-butylstyrene does not polymerize in emulsions using peroxodisulfate as initiator with an anionic emulsifier unless its solubility in the aqueous phase is increased by the addition of a small amount of acetone or methanol, or if a more water-soluble comonomer, even styrene, is present.

Contrary to earlier reports [23–25], *p-tert*-butylstyrene which is less soluble in water than styrene (see later, Table 16) has been found to polymerize in emulsions using peroxodisulfate as initiator [26]. In contrast to styrene, *p-tert*-butylstyrene did not undergo an emulsifier-free emulsion polymerization: however, latex particles were formed immediately on the injection of emulsifier. This observation provides positive evidence that latex particles are nucleated from *p-tert*-butylstyrene monomer solubilized in emulsifier micelles. The similarity of the rates of the emulsion polymerizations of styrene and *p-tert*-butylstyrene in the presence of emulsifier (SDS) micelles indicates the dominance of micellar nucleation of latex particles in the cases of these monomers.

The rate of the emulsion polymerization of styrene can be increased not only by the proper choice of nonionic emulsifier and initiator but also by adding a small amount of anionic emulsifier. Addition of a little anionic emulsifier (sodium dodecyl sulfate) to a system with a nonionic emulsifier (polyoxyethylene type) led to a considerable increase in the rate of polymerization which was attributed to a decrease in the particle size and an increase in their concentration. These procedures produce a very stable polymer dispersion [27, 28]. The values of the kinetic parameters and of the reaction orders with respect to concentrations of the reaction components agreed with the theoretical values according to the Smith–Ewart model. If the proportion of nonionic emulsifier in a mixture of nonionic emulsifier (Triton X-100: octylphenoxypolyethyleneoxyethanol or Emulphogene BC-840: tridecyloxypolyethyleneoxyethanol) with anionic emulsifier (sodium dodecyl sulfate — SDS) increases, a decrease in both the polymerization rate and the stability of the poly(styrene) latex are observed simultaneously [4]. The lower reproducibility of results in the system with Triton X-100 (TR) is due to the ready oxidizability of the nonionic emulsifier by aerial oxygen which leads to the formation of polymerization-initiating peroxide intermediates. The transfer to emulsifier caused the formation of low-RMM polymer ($\overline{M}_n \sim 2 \times 10^5$). In the absence of nonionic emulsifier, a polymer with high RMM ($>10^6$) was formed. The polymerizations were characterized by the linearity of their conversion curves at low and medium conversions. In the system with TR, a deviation from linearity (rate decrease) was observed above 50% conversion and, in the presence of Emulphogene BC-840 (BC) even above 30% conversion. In the system with BC, formation of a polymer with high RMM, $\overline{M}_n \sim 10^6$ was observed. In both systems, a wide RMM distribution of the polymer produced ($\overline{M}_w/\overline{M}_n \sim 2.5$) was observed.

An evaluation of alkylated diphenyl ether disulfonate emulsifiers in styrene/butadiene emulsion copolymerization is summarized in [29]. The monoalkylated and dialkylated diphenyl ether disulfonate emulsifiers (Dowfax and XD type) gave stable highly-solid latices of average particle size between 160 and 200 nm, except the monoalkylated C_6-linear, dialkylated C_{12}-branched, and dialkylated C_{16}-linear substituted emulsifiers, which favoured the formation of particles with much larger sizes. The average particle size was determined principally by the type of emulsifier. The monoalkylated C_{12}-linear substituted emulsifier was the most efficient emulsifier in that only 0.02% was required to give an average particle size of 200 nm, whereas 0.4% was required for Dowfax 2A1 (C_{12}-branched) and more than 1.2% for XD-8292 (C_6-linear). These differences in average particle size correlate well with the surface properties and adsorption characteristics of the emulsifier. When the chain length of the alkyl substituent of the emulsifier was increased, the average particle size decreased owing to the increase in the effectiveness of the emulsifier and the stronger adsorption at the interface. The CMCs of the emulsifier decreased with increasing alkyl chain length for both mono- and dialkylated emulsifiers and were lower for branched substituents than for linear substituents; all showed a CMC equal to, or less than, that for sodium dodecyl sulfate.

The molecular area of the emulsifier increased with increasing chain length of the mono- and dialkylated substituent, except for the dialkylated C_{16}-linear substituted emulsifier, for which it was lower; that for the sodium dodecyl sulfate was lower than all except the dialkylated C_{16}-linear substituted emulsifier, for which it was about the same. The free energies for adsorption and micellization became increasingly more negative with increasing chain length of the linear substituent of both mono- and dialkylated emulsifiers, except for the dialkylated C_{16}-linear substituted emulsifier; the values for sodium dodecyl sulfate were more negative than those for the alkylated diphenyl ether disulfonate emulsifiers.

Application of a water-soluble polyester emulsifier in the emulsion polymerization of styrene has been given by Chen & Liu [30] and McCartney & Piirma [31]. Polyester emulsifiers prepared by the condensation of different dicarboxylic acids or anhydrides and polyoxyethylene are used in a variety of industrial applications. Chen & Liu [30] synthesized a novel series of water-soluble polyester emulsifiers by a two-step condensation of dimethyl 5-sulfoisophthalate, sodium salt, polyoxyethylene and phthalic anhydride. Six polyesters were prepared with polyoxyethylene ranging in RMM from 100 to 1000. These emulsifiers showed excellent emulsifier activity and were investigated for industrial application. Although polyester emulsifiers are of great importance in dispersion-related industries, the mechanism by which these emulsifiers stabilize dispersions and the kinetics of emulsion polymerization (of styrene) using polyesters as emulsifiers has been investigated only very recently by McCartney & Piirma [31]. The structure of the water-soluble polyester emulsifier used is shown below:

$$H\text{-}[(OCH_2CH_2)_n\text{-}OOC\text{-}C_6H_3(SO_3Na)\text{-}COO\text{-}(CH_2CH_2O)_n\text{-}C_6H_4\text{-}CO)_m]OH \qquad (1)$$

The emulsifier exhibited a CMC of 0.25% (w/v) and a surface tension lowering of 2.3×10^{-2} N m^{-1}. The use of this emulsifier for the emulsion polymerization of styrene produced some unusual results. While the rate of polymerization was independent of emulsifier concentration, the number of particles increased as $N \propto [E]^{0.91}$. Under the reaction conditions, relatively small latex particles were observed, i.e., from 50 to 80 nm in diameter. The RMM of the poly(styrene) was not a function of emulsifier. In addition, the RMMs were low for emulsion polymerization, i.e., ranging from $M_n = (3 \text{ to } 10) \times 10^5$ for different monomer concentrations. The low RMMs of polystyrene were attributed to chain transfer occurring during polymerization. It was suggested that the emulsifier radicals generated through the chain transfer retarded or even inhibited polymerization.

Poly(styrene) latices play a particular role in the application of styrene-based polymer dispersions; they are prepared by the batch polymerization of styrene using an amphoteric emulsifier [32]. They are applied as biodegradative, antibacterial, and antistatic additives.

Kato et al. [33] investigated the kinetics and mechanism of the emulsion polymerization of styrene in the presence of an amphoteric emulsifier (N,N-dimethyl-N-dodecyl betaine, DDB). Depending on the pH of the reaction medium, the amphoteric emulsifier can be in its cationic, anionic or nonionic form. The most suitable has been found to be the nonionic form at pH ~ 7, which enabled the preparation of a stable latex without coagulation. By contrast, at low and high pH, the formation of a coagulate was observed. The rate of polymerization was increased in proportion to the emulsifier concentration; the reaction order with respect to emulsifier was 0.6 (at pH ~ 7). The polymerizations led to the formation of a polymer latex with rather small polymer particles and with a narrow size distribution, the RMM distribution being wide. With increase of the DDB concentration from 1 to 12 p.p.h. (referring to 100 mass portions of styrene), the particle diameter fell from 92 to 47 nm, the particle size distribution D_w/D_n from 1.09 to 1.05, the average RMM increased from 2.0×10^6 to 6×10^6 and the RMM distribution $\overline{M_w}/\overline{M_n}$ from 8 to 11.

Other details concerning the emulsion polymerization of styrene in the presence of the amphoteric emulsifier DDB have been put forward by the same authors [34]. As the pH increased, a rate decrease was observed both at low pH (to pH ~ 4) and at high pH (above pH ~ 10) (see Table 14). In the isoelectric region of the emulsifier (from pH 5.5 to 10.5) the rate of polymerization was constant, R_p (% conv. min^{-1}) ~ 3.0. At pH 2 the polymerization rate was maximal, and the minimum polymerization rate was observed at pH 12, R_p ~ 2.7% conv. min^{-1}. The average RMM varied in parallel. The dependence of the particle size showed the opposite trend. In the pH region investigated, the RMM M_n fell from 3.2×10^6 to 2.5×10^6 and the average particle size increased from $D = 42$ to $D = 66$ nm. In the isoelectric region of the emulsifier, the RMM and the particle size were constant.

Table 14 — Influence of pH on kinetic parameters of batch polymerization of styrene at 60°C

pH	$N_{mic}{}^a$ /10^{20} dm^{-3}	$R_p{}^b$ /% conv. min^{-1}	N^c /10^{17} dm^{-3}	$d_n{}^d$ /nm	$M_w{}^e$ /10^6 g mol^{-1}
2	7.7	4.5	4.7	42	3.2
3	6.0	3.4	4.3	51	3.0
4	5.0	3.0	4.2	57	2.9
7	4.8	3.0	4.2	59	2.9
10	4.8	3.0	3.2	60	2.8
11	4.9	2.8	3.2	64	2.6
12	5.0	2.7	—	66	2.5

^{a} Number of emulsifier (DDB) micelles.
^{b} Rate of polymerization in Stage II.
^{c} Number of latex particles per dm^3 water.
^{d} Particle diameter.
^{e} Mass average RMM.

In the presence of N-dodecyl-N,N-dimethyllysine (DDL) with an isoelectric pH ranging from pH 4 to pH 9, the rate of polymerization of styrene varied according to the Eq.[35, 36]

$$R_p \sim [\text{DDL}]^{0.78}. \tag{2}$$

The increase in reaction order was attributed to the nonstationary state conditions applying in Stage II (i.e. the fluctuation of styrene concentration in the polymer particles). The reaction order in emulsifier concentration is lower (0.62) and depends on the particle number. On increasing the concentration of DDL from 5 to 14 p.p.h. (related to styrene), the number-average particle diameter, D_n, decreased from 90 to 47 nm, the polydispersity of the particles D_w/D_n from 1.21 to 1.07 and, conversely, the RMM M_n increased from 1.4×10^6 to 2.0×10^6 and the poly(styrene) polydispersity decreased from 9 to 5.8. Others [30] have also reported the dominant influence of pH on styrene polymerization. On increasing pH (from 2 to 12), the rate and number of polymer particles increased. The rate enhancement was *ca.* 3-fold, the particle number increased 4 times and the average particle size decreased from 75 to 40 nm. As the pH increased, the RMM of the polymer and the CMC of the emulsifier increased. In contrast, the aggregation number of the micelles decreased from 80 (at pH 2) to 20 (at pH 12).

Investigations of the kinetics of batch emulsion polymerization are often adversely influenced by particle nucleation and the wide size distribution of polymer particles.

New particle nucleation reduces the precision and reliability of the kinetic parameters established for the stationary region (Stage II). If, by using a seed latex with defined narrow particle distribution, Stage I will be eliminated (i.e. the stage in which the particles are formed), the polymerization takes place under stationary conditions and the parameters obtained are absolute in nature [37].

Seeded emulsion polymerizations were used by Hawkett *et al.* [38] to explain the mechanisms of both entry and exit of radicals from particles. They found that the value of the average number of radicals (0.5) proposed by the micellar model is not maintained in the emulsion polymerization of styrene and is a function of the initial concentration of peroxodisulfate. With increasing concentration of initiator [I] from *ca.* 10^{-5} to *ca.* 10^{-1} mol dm^{-3}, the average number of radicals in a particle ($\bar{n}$) increased from 0.25 to 0.5. The value of $\bar{n} = 0.5$ was reached at an initiator concentration of 10^{-2} mol dm^{-3}. The escape of radicals from the particle, characterized by the rate constant k_{des}, is inversely proportional to the size of the polymer; at 50°C, its value was estimated to be 1.1×10^{-3} s^{-1}. The concentration of radicals in a particle is a function of the effectiveness of radical trapping by the particle and is characterized by the coefficient of entry ρ. This depends on the initiator concentration

$$\rho \sim [\text{I}]^{0.6}. \tag{3}$$

Napper *et al.* [13, 39] have found that the rate of styrene polymerization decreases on adding transfer agents (CCl$_4$, CBr$_4$) more than would correspond to the simple dilution of monomer. Analysis of the results led to the conclusion that the reduction in rate is caused by increase of the exit of radicals from the particle. The rate of desorption of radicals was proportional to the chain transfer to monomer and increased in the order: styrene (1.4×10^{-2} dm^3 mol^{-1} s^{-1}) < CCl$_4$ (2.3×10^0 dm^3 mol^{-1} s^{-1}) < CBr$_4$ (9.6×10^4 dm^3 mol^{-1} s^{-1}); the values in brackets are transfer constants. The opposite trend was observed in the efficiency of radical exit and is governed by the lipophilic nature of the radical. The exit of radicals from the particle proceeds by the transfer-diffusion mechanism and is determined by the nature of the radicals formed by the transfer. Addition of CBr$_4$ to a polymerization system with a high initiator concentration (above 5×10^{-3} mol dm^{-3}) leads only to a slight decrease in rate, the decrease being proportional to the CBr_4 concentration. At lower initiator concentrations (i.e. below 5×10^{-3} mol dm^{-3}) addition of CBr_4 reduces the rate of polymerization considerably. The dependence of the rate of polymerization on the concentration of additive is described by a curve with a minimum at a certain additive concentration. The increase in rate enhancement of the polymerization with CBr_4 concentration above a certain limiting concentration was ascribed to the high lipophilicity of the radicals (formed via transfer) which positively affected their re-entry into particles.

A decrease in rate of seed emulsion polymerization brought about by transfer was observed on adding small amounts of organic solvents, e.g. n-hexane, cyclohexane, benzene, ethylbenzene and toluene [40]. Radicals formed via chain transfer to hydrocarbons (D) showed lower reactivity, a greater ability of desorption and a lower capability of re-entering the particles. The chain transfer increased in the order: benzene (0.03) < cyclohexane (0.05) < toluene (0.25) < styrene (1.0) < ethylbenzene (1.5); the values in brackets are transfer rate constants of reaction, $k_{tr} /10^{-2}$ dm^3 mol^{-1} s^{-1}. The constant for radical exit $k_{des} /10^{-3}$ s^{-1} increased in the order: 1.8 (styrene, benzene) < 3.6 (cyclohexane) < 6.6 (ethylbenzene) < 6.7 (toluene). A most remarkable decrease in the number of radicals per particle caused by the exit of radicals and in parallel, a decrease in polymerization rate, was observed in the presence of toluene or ethylbenzene. The rate of polymerization is also influenced by the reactivity of diluent radicals formed via transfer, their reactivity being lower when compared with that of styrene radicals in addition reactions with styrene. The measure of radical reactivity was in this case determined by the ratio of the propagation rate constants k_{pD}/k_p and increased in the order: cyclohexane (0.03) < toluene (0.05) < ethylbenzene (0.25) < styrene (1.0). The radicals formed via transfer to solvent are less reactive in propagation with styrene than styrene radical and therefore any increase in their proportion is accompanied by a reduction in the polymerization rate. The decrease is particularly marked at higher conversions after consumption of the free monomer droplets where the proportion of unreactive radicals in the aqueous phase increases.

In the presence of a very reactive transfer agent such as *tert*-butylmercaptan (TBM) using an anionic emulsifier, the rate of styrene polymerization was enhanced with increase in TBM [41]. As TBM concentration increased (from 0 to 5×10^{-3} mol dm^{-3}) at emulsifier concentrations below the CMC of the emulsifier, the polymerization rate increased from 1.0×10^{-6} to 1.0×10^{-5} mol dm^{-3} s^{-1} as a consequence of the increase in the average number of radicals per particle from 0.1 to 1.0 and the particle number from 2.2×10^{16} to 3.5×10^{16} dm^{-3}. The larger number of particles and radicals in the particle and the higher polymerization rate were interpreted in terms of a rate enhancement of the initiation process owing to the induced decomposition of peroxodisulfate (via interaction with TBM) [42]. The authors predicted that processes of radical generation, the entry of radicals into a particle, and particle formation, prevail over the processes of transfer of radicals and their escape from the particle; this is why an increase, and not a decrease, of the polymerization rate is observed. At emulsifier concentrations above the CMC, a decrease in the rate of polymerization occurred by effective chain transfer and radical desorption. The increase in particle number reduces the monomer concentration in the particles and causes a decrease in the polymerization rate.

Whang et al. [43] have found that the concentration of radicals in the particles is not only a function of the concentration of peroxodisulfate, but also depends on the concentration of seed poly(styrene) particles. With increasing peroxodisulfate concentration, e.g. from 1.2×10^{-5} mol dm^{-3} to 1.3×10^{-2} mol dm^{-3}, the reaction order with respect to the particle number N increased from 0.2 to 0.8 (the value obtained from the dependence of the polymerization rate on the concentration of seed particles). An analysis of the experimental results has shown that an increase in initiator concentration leads to an increase in the average number of radicals per particle and of the exponent with respect to the particle number. At low initiator concentrations, the radical capture efficiency of a particle is high; the rate of entry of radicals given by the relation $N\rho$ is constant and the average number of radicals (low values) is proportional to the reciprocal of the particle number. The polymerization rate is in these systems barely sensitive to the change in particle number and the reaction order with respect to N is very low. The values of the reaction order with respect to the particle number lower than 1.0 are at high initiator concentrations caused by lower radical capture efficiency of a particle. The increase in concentration of polymer particles is accompanied by a decrease in the rate of entry of radicals into particles. The radicals desorbed into the aqueous phase are effectively deactivated by cross-termination reactions with primary radicals.

A study of the seed emulsion polymerization of styrene should give information about the mechanism of entry of radicals into particles [44]. The value of the ratio of the rate constants for the entry of radicals into particles and bimolecular termination $k_e^2 / 2k_t$ is considered to be a measure of radical entry into particles. The ratio was shown to be almost constant for different initiators (peroxodisulfate, Fe(II) / H_2O_2 and 2,2'-azobis(2-amidino-propane) hydrochloride) varying only between 1.4×10^4 and 2.3×10^4 dm^3 mol^{-1} s^{-1}. The charge of the primary radical (negative for peroxodisulfate, positive for azo derivative and neutral for the redox system) therefore does not influence the entry of initiating radicals into a particle since oligomeric radicals enter particles and their charge plays only a secondary role. The entry of radicals is influenced by the lipophilic nature of the oligomer. The separation of the charge from the radical centre by two to ten styrene units strongly suppresses the effect of the charge, and, conversely, it promotes the effect of the lipophilic chain on radical entry into a particle. The low effectiveness of radical adsorption by a particle also indicates the dependence of the rate on the initiator concentration $[I]^{0.5-0.9}$. This dependence is controlled by the low concentration of radicals in a particle (average number of radicals in a particle $\bar{n}$ was much below 0.5).

The seed emulsion polymerization of styrene enables us to follow the effects of pH, the ionic strength of solution, and the monomer/polymer ratio as well as other factors on the reaction mechanism. It was found that the ionic strength does not influence the course of polymerization; e. g. by changing the concentrations of Na_2CO_3 and $NaHCO_3$ over the range from 1.3×10^{-2} to 14.8×10^{-2} mol dm^{-3}, no substantial change in the rate of polymerization occurred (only 3.8%). The use of higher electrolyte concentrations led to a lowering of the rate of polymerization which was attributed to the effective coagulation of the particles [45]. Similarly Kamath [46] observed the reduction in the rate of polymerization after exceeding the limiting electrolyte concentration (~ 0.25 mol dm^{-3}) owing to particle coagulation. In parallel, the RMM of the polymer is also barely affected by the ionic strength of the reaction medium.

Styrene polymerizations conducted at various pH values led to the conclusion that the rate of polymerization is not influenced over a wide pH range from 3 to 11, where it remains constant [45]. A drop in polymerization rate was observed in strongly alkaline media and was caused by coagulation of the latex particles. In a strongly acidic medium ($\sim$ pH 3) the following processes took places: the coagulation of particles, sensitization of the initiating rate (as a result of catalytic decomposition of peroxodisulfate) and the hydrolysis of peroxodisulfate. The particle coagulation caused the rate decrease and, conversely, the enhancement in the rate of initiation increased the polymerization rate. Due to the hydrolysis of peroxodisulfate, a considerable amount of oxygen is released, retarding the polymerization and adversely influencing the quality of the polymer.

Another important requirement, which affects the course of seed polymerization, is to keep the number of particles constant during polymerization. It is generally assumed that under stationary state conditions the particle number is constant. An analysis of the time distribution of the particles in seeded styrene polymerization showed that the formation of particles and their association occurs simultaneously during polymerization [47]. This means that measurement of a constant particle number can hide the nonstationary behaviour of the polymerization brought about by overlapping of the association and formation of new particles.

The emulsion copolymerization of styrene extends the possibilities of kinetic studies by the new facts concerning the mechanism of emulsion polymerization, especially for the copolymerization of monomers with different lipophilicity. Guillot [48, 49] studied the mechanism of emulsion copolymerization of styrene and acrylonitrile, i.e. monomers with different water solubilities. The composition of the copolymer was a function of the composition of the monomer mixture in the monomer droplets and in the monomer/polymer particles, and was not greatly influenced by the ratio of comonomers in the aqueous phase. The contribution of polymerization in the aqueous phase became more significant at the end of polymerization. Guillot found that monomer particles are consumed at about 40% conversion at which the largest monomer fraction is in the polymer particles. At 90% conversion, only acrylonitrile is in the system (in the aqueous phase) which determines the rate of polymerization.

The rate of copolymerization increased with increasing concentration of monomer, emulsifier and polymer particles. Increase in the particle size was observed up to high conversions, the biggest effect ranging from 10 to 40%. At these conversions the diameter of the polymer particles increased from 160 to 200 nm. However, the increase in the number of polymer particles has not been discussed at even higher conversions. The value of the reaction order of 0.74 was obtained by studying the dependence of the polymerization rate on emulsifier concentration. A deviation has also been recorded in an investigation of the effect of the concentration of polymer particles on the polymerization rate. The mechanism proposed for the copolymerization regards the monomer/polymer particles as the dominant centre for polymerization, determining the composition of the copolymer, the rate of copolymerization, and the particle size.

The effect of the concentration of an anionic emulsifier and of transfer agent on the emulsion copolymerization of styrene and acrylonitrile was discussed by Pavlyuchenko et al. [41]. These authors followed the course of emulsion copolymerization at emulsifier concentrations below and above the CMC (0.9×10^{-2} and 1.5×10^{-2} mol dm^{-3}) using tert-dodecylmercaptan as transfer agent (at concentrations between 1.0×10^{-3} and 5.0×10^{-3} mol dm^{-3}) and a seed poly(butadiene) latex. As the concentration of tert-dodecylmercaptan (TDM) increased, the rate of polymerization decreased. The reduction was particularly marked at higher emulsifier concentrations. The decrease in rate was ascribed to a reduction in radical concentration in a particle caused by the desorption of radicals from the particle. In the absence of micelles, on adding TDM (5×10^{-3} mol dm^{-3}), the value of average number of radicals per particle decreased from 3 to 1 and in the presence of micelles from 2.4 down to 0.19. Polymerizations conducted at submicellar emulsifier concentrations do not lead to any change in particle number. Particle nucleation was observed in the presence of micelles; on the completion of polymerization the particles were 2–3 times smaller than the original seed particles.

The emulsion copolymerization of styrene with acrylonitrile seeded on a poly(butadiene) (PB) support has given evidence for oligomeric desorption that becomes manifest

Sec. 3.4] Kinetics and mechanism

in the presence of a chain-transfer agent [50, 51]. The oligoradicals formed in the aqueous phase or resulting from desorption re-enter another particle. Chain transfer to the PB support results in the accumulation of trapped free radicals. As a consequence of the accumulation of free radicals, the overall copolymerization rate increases. Therefore, the average number of free radicals per particle exceeds the value 0.5. This implies the existence of some particles with two or more free radicals, the termination ability of which is decreased. The number of particles with a trapped free radical has been calculated from the following equation

$$N_{tr} = C_1[I]N^*, \qquad (4)$$

where [I] is the peroxodisulfate concentration, N^* is the number of 'untrapped' free radicals, and C_1 is a constant characteristic of the process of the accumulation of trapped free radicals in the PB support and expresses the probability ratios of chain transfer to the support and to the chain transfer agent, respectively. The authors reported that this copolymerization deviates from the Smith–Ewart theory.

A series of emulsion copolymerizations of styrene (St) with methyl acrylate (MA) in the presence of sodium dodecyl sulfate initiated by the redox system potassium peroxodisulfate (PS)/sodium hydrogen disulfite (SBS) was carried out at different monomer feed compositions and monomer/water (M/W) ratios [52]. The polymerization rate in the system with PS/SBS is higher than that with PS alone. The initial and final copolymer molecules are richer in the more hydrophilic monomer (MA). These results support the earlier kinetic data obtained for the emulsion copolymerization of butyl acrylate and acrylonitrile, where the initial and final copolymers formed were rich in the more water-soluble acrylonitrile. The particle sizes were smaller, and the size distributions larger, in the presence of PS/SBS than for polymerizations in the presence of PS alone. The particle number (N) tends to increase with conversion for monomer mixtures rich in MA, in contrast to St homopolymerizations in which N remains constant. Comparison of values of N and kinetic data led the authors [53] to the conclusion that the average number of radicals per particle, $\bar{n}$, remains close to 0.5. The experimental data, to some extent, enable one to quantify the competition between micellar and homogeneous nucleation mechanisms.

Nomura et al. [53] have led the investigation on the site of particle formation in the emulsion copolymerization of styrene with methyl acrylate (MA) and methyl methacrylate. For this purpose, the authors designed a new experimental method comprising the emulsion copolymerization of the sparingly water-soluble monomer, styrene (St) with each of the partially water-soluble monomers, followed by measurement of the composition of the copolymers produced at the very beginning of the reaction when the formation of polymer particles takes place. In the absence of emulsifier, polymer particles were found to be generated by the precipitation of copolymers produced by the homogeneous solution copolymerization of the monomers dissolved in the aqueous phase. It was shown that the compositions of the copolymers for the St/MA copolymers produced in the absence of emulsifier micelles are distinctly different in the very low conversion range from those of copolymers produced in the presence of micelles. This behaviour favoured the idea that the emulsifier micelle is still important as the site of particle formation even in the emulsion polymerization of considerably water-soluble monomers such as MA.

Table 15 — Kinetic parameters for emulsion copolymerization of styrene and methyl methacrylate

$n_{St}{}^a$	$R_p{}^b$ /10^{-3} mol dm^{-3} s^{-1}	$[M]_{eq}{}^c$ /mol dm^{-3}	r^d /nm	N^e /10^{15} cm^{-3}	$\bar{n}^f$	$M_n{}^g$ /10^6
1.0	3.8	5.2	39	1.9	0.16	2.1
0.69	3.5	4.8	43	1.6	0.1	2.1
0.54	4.3	4.4	43	1.7	0.1	2.3
0.39	6.9	5.4	46	1.3	0.14	2.9
0.24	9.2	5.3	47	1.2	0.18	3.6

^{a} Mole fraction of styrene in monomer feed.
^{b} Rate of polymerization in Stage II.
^{c} Equilibrium monomer concentration in polymer particle in Stage II.
^{d} Radius of final polymer particle.
^{e} Polymer particle number per cm^3 water.
^{f} The average number of radicals per particle.
^{g} Number average RMM of copolymer.

Goldwasser & Rudin [54] studied the copolymerization of styrene with methyl methacrylate. They characterized the effect of the monomer feed composition on the kinetics (Table 15). The table shows that increase of the fraction of methyl methacrylate (MMA) in the monomer feed is accompanied by an increase in the polymerization rate due to the higher reactivity of MMA and poly(MMA) radicals in the polymerization process. The RMM follows a similar trend. The inverse proportionality between the rate and the particle number has not been explained; it is probably determined by the active participation of hydrophilic MMA in the polymerizations in the aqueous phase by formation of the primary particles and their participation in particle growth via the coagulation mechanism and in desorption radical processes.

The high hydrophilicity of MMA monomer and the formation of rather small particles favours the desorption of MMA$^\bullet$ radicals from a particle. The kinetic study led to the chain transfer constants to monomer, $C_{St} = 1.9 \times 10^{-5}$, $C_{MMA} = 0.5 \times 10^{-5}$ and to the sum of the cross-transfer constant 5.8×10^{-5}, which are close to the values for homogeneous polymerization. The differences between the measured and theoretical values of the kinetic parameters were mostly explained by the inadequacy of the models used.

Nomura et al. [55, 56] studied the effect of the monomer feed composition of styrene and methyl methacrylate on the kinetic parameters of the emulsion copolymerization aimed at determining a measure of the applicability of the micellar model and the formulation of a new model which would properly describe the desorption of radicals. According to the assumptions of the micellar model, the polymerization rate increased with increasing emulsifier and monomer concentrations; monomer droplets were consumed after reaching about 40% conversion.

Investigation of the influence of particle size on adding peroxodisulfate in Stage II revealed deviations from the micellar model. The decrease in the particle size was accompanied by a reduction in the polymerization rate which was attributed to radical desorption from the particles. The polymerization rate was only slightly enhanced as the fraction of MMA in the monomer feed increased up to the limiting value of the MMA fraction (~ 0.7), then it rapidly increased. The desorption of radicals determined by the concentration of hydrophilic MMA radicals reduced the polymerization rate but it did not influence the composition of the copolymer.

Vinyl acetate is a monomer with high hydrophilicity. Its effect on the mechanism of the copolymerization with styrene has been studied elsewhere [57]. The copolymerization of styrene with vinyl acetate does not give copolymer in high yields; the copolymer has a heterogeneous composition. This phenomenon follows from the values of the copolymerization parameters which are unsuitable for these two monomers as regards the production of copolymer. The addition of tributylamine promotes the copolymerization of styrene and vinyl acetate. This behaviour results from the interaction of tributylamine with the propagating radicals, and changes their reactivity in favour of the formation of a copolymer of more homogeneous composition. The emulsion copolymerization of styrene and vinyl acetate produced a low-RMM copolymer, $M_n \sim 5 \times 10^4$, with a high polydispersity, $M_w/M_n \sim 5$. The ready hydrolysis of vinyl acetate led to the formation of a terpolymer dispersion, namely styrene/vinyl acetate/vinyl alcohol as early as during the preparation of dispersion.

α-Methylstyrene is a monomer with high lipophilicity. Its copolymerization with styrene is, under homogeneous conditions, rather difficult because of the high reversibility of the polymerization (depropagation), the negative effect of steric factors and the high degradative transfer to monomer [58]. This gives rise to a copolymer with low RMM (< 50 000) [59]. Higher copolymerization rates and higher RMMs of styrene/α-methylstyrene copolymer are obtained via emulsion copolymerization [60]. As expected, with increasing fraction of α-methylstyrene in the monomer feed, the rate of copolymerization and the RMM decrease even in emulsion polymerization. For instance, as the molar fraction of α-methylstyrene increased from 0.1 to 0.5, the rate of polymerization decreased from 1.8×10^{-3} to 0.23×10^{-3} mol dm^{-3} s^{-1} and the RMM from 2.75×10^5 to 1.32×10^5 g mol^{-1}. Mathematical analysis of the experimental data yielded transfer constants with respect to monomer $C_{St} = 1.9 \times 10^{-4}$ and $C_{MSt} = 7.2 \times 10^{-4}$. The activation energy for the transfer of styrene radical to both monomers is ca. 42 kJ mol^{-1} and for that of α-methylstyrene radical to the monomer, 71.2 kJ mol^{-1}. The degradative chain transfer to α-methylstyrene affects the rate and the RMM of the polymer significantly; however, it is not such a dominating factor as in homogeneous polymerization.

The effect of chain transfer has also been studied in the emulsion copolymerization of styrene and vinyl benzoate [61]. As before, chain transfer to vinyl benzoate caused a decrease in the polymerization rate and the RMM of the copolymer and was directly proportional to the concentration of vinyl benzoate. The copolymer was enriched with the more reactive styrene and therefore a shift in copolymer composition was observed with increasing conversion consistent with the copolymerization parameters $r_{St} = 29.6$ and $r_{VB} = 0.03$ and from the composition of the monomer mixture in the polymer particles during polymerization.

Good swelling ability of a polymer by its monomer in a polymer particle is a necessary condition for propagation in the polymer particle. This is clearly demonstrated by the emulsion copolymerization of styrene with divinylbenzene [62]. On increasing the degree of crosslinking of the copolymer by raising the mole fraction of divinylbenzene in the feed with styrene, the swelling of copolymer by monomers decreases. The rate of copolymerization decreases and the typical sigmoidal shape of the conversion curve for emulsion polymerization of styrene in the presence of divinylbenzene changes to the shape typical for the conversion curve of the solution free-radical polymerization. The lowest polymerization rate was found for the emulsion polymerization of divinylbenzene (in the absence of styrene). The decreasing ability of the polymer particles to undergo swelling by monomer leads to monomer-starved conditions in the polymer particles and to the lowering of the copolymerization rate. The formation of co-oligomer radicals and co-oligomers in the aqueous phase becomes more and more important for the overall copolymerization rate. This mechanism explains the increase of the polymer particle number up to 50% conversion and the decrease of the rate of copolymerization. The existence of trapped radicals in crosslinked polymer particles enables interparticle crosslinking in their aggregates. The agglomeration of the polymer particles is responsible for the lowering of the polymer particle number at conversions over 50%. A similar effect on the polymerization rate in free-radical emulsion polymerization is exerted by good solvents for polymers. The decrease of the polymerization rate is caused here by the competitive swelling of the polymer particles by solvent which lowers the monomer concentration in the polymer particles [63].

The emulsion polymerization of 1,4-divinylbenzene (DVB) with sodium dodecyl sulfate as emulsifier and potassium peroxodisulfate as initiator was studied kinetically by Obrecht et al. [64, 65]. It was found that the rates and polymer particles obtained were smaller ($d \sim$ 15–30 nm) than in the emulsion polymerization of styrene under the same conditions. The authors explained the difference as due to the different mechanism for the growth of crosslinked and uncrosslinked particles. The strong crosslinking of polyDVB particles reduces the swelling ability of the polymer particles or the equlibrium concentration of DVB in the particles, thus limiting polymerization in the polymer particles. The polymerization takes place essentially on the particle surface. The styrene polymerization is carried out in the monomer-swollen polymer particles and thus the resulting rate is higher. The lower polymerization rate of polyDVB particles raises the probability of the nucleation of a greater number of monomer-swollen micelles and thus increases the particle number in the system.

The effect of a crosslinked network in the poly(styrene) seed particles on the emulsion polymerization of styrene/divinylbenzene has been investigated by Sheu et al. [66]. The use of crosslinked seed particles in seeded polymerization caused phase separation in the resulting particles, i.e., the formation of nonspherical particles. As the crosslinking density of the seed particles increased, the degree of phase separation and the number of phase domains increased. A kinetic study of the phase separation revealed that it occurred before the start of polymerization and was enhanced at increasing conversions. The phase separation in the swollen network is a viscoelastic process, which is induced by the relaxation of the polymer chains resulting from increased temperature or increased swelling time. The thermodynamics of phase separation were investigated by analysis of the free-energy changes during swelling and polymerization, and the phase separation was described by a nucleation-and-growth mechanism.

The results of this study have been applied to the design and synthesis of a series of uniform nonspherical particles of different morphology, including singlets (egg-like or ellipsoidal particles), doublets (pear-shaped or dumbbell particles), and multiplets (ice cream-like and popcorn-like particles).

Emulsion polymerization is usually carried out in a macroemulsion (a conventional emulsion polymerization), an opaque system with particle sizes in the range from 200 to 1000 nm. The properties of these systems have been characterized and a variety of information is available (see above). Extension of the field towards systems with smaller particle sizes has recently been made and two new types of polymerization system can be recognized.

Ugelstad et al. [67, 68] and Chou et al. [69] have prepared emulsions with much smaller particle sizes ($\simeq$ 100–200 nm) by using ionic emulsifiers and a small amount of long-chain alkanols as coemulsifiers. These emulsions are called 'miniemulsions' to distinguish them from the much larger size 'macroemulsions' and the very small size 'microemulsions'. The latter feature a particle size of 5 to 100 nm and are transparent.

The polymerization of styrene miniemulsions prepared using a mixed emulsifier system comprising sodium dodecyl sulfate and cetyl alcohol, a water-soluble (potassium peroxodisulfate) and an oil-soluble initiator (2,2'-azobisisobutyronitrile (AIBN)) was carried out by Choi et al. [70]. From the results two major differences in the kinetic features between miniemulsion and conventional emulsion polymerization were observed. First, the particle formation stage was usually long. This was attributed to the markedly reduced rate of radical absorption by monomer droplets. Second, miniemulsion polymerizations did not exhibit the Stage II characteristic of conventional emulsion polymerization. With AIBN both R_p and N varied with only the 0.2 power of the initiator concentration, instead of the ~ 0.4 power dependence found with peroxodisulfate. Experimental results showed that the monomer droplets became the main source of polymer particle formation. This was attributed to the fact that stable emulsions with droplet diameters in the range from 50 to 15 000 nm were produced using this mixed emulsifier system.

Holdcroft & Guillet [71] have investigated the kinetics of the photopolymerization of styrene microemulsions using pulsed lasers. Conventional thermal initiation and photochemical initiation of styrene microemulsion polymerization involves steady-state initiation, propagation, and termination processes. In contrast, the polymerizations under discussion when initiated by pulsed-laser sources involve transient processes. The latter, therefore, was utilized to gain information on the dynamics of polymerization. The RMMs of poly(styrene) produced were linearly dependent on the time interval between pulses when laser pulses of lower intensity were used. In contrast, laser pulses of much higher intensity produced polymers with RMMs which were nonlinear with laser repetition rate. Furthermore, the pulsed laser initiation of styrene microemulsions yields polymers of RMMs in the range 1×10^5 to 5×10^5. The low RMMs (~ 1×10^5) observed with high intensity laser pulses (i.e., 2×10^{17} photons absorbed cm^{-3} $pulse^{-1}$) indicates that termination of the polymer chains occurs before the subsequent laser pulse. From the experimental data the following equation for the kinetic chain length was derived:

$$\gamma_p = k_p [M] \tau, \tag{5}$$

where [M] is the effective concentration of the monomer and τ is the time interval between pulses.

One of the first experimental studies of a photoredox-initiated polymerization in an oil-in-water microemulsion was reported by Grätzel et al. [72]. The course of polymerization for the two systems styrene/divinylbenzene and acrolein/divinylbenzene in the presence of hexadecyltrimethylammonium persulfate is illustrated here. The form of the polymerization rate curves is typical for a radical-initiated emulsion polymerization process. In both systems a relatively long inhibition period was observed.

In the styrene/divinylbenzene system, after an initial irradiation period of 25 min, the distribution of small particle sizes is displaced by a factor of 10 relative to the pre-irradiation sizes and there is an increase in the number of particles with diameters in the several hundred to 1000 nm range. In the DVB/acrolein system, after 20 min of irradiation there are two distributions of particle sizes, one maximizing at ~ 10 nm and the other at 600 nm.

Highly efficient polymerization was found to occur under visible light by using $Ru(bpy)_3^{2+}$ or eosin Y as a sensitizer.

The first report on the polymerization of styrene in a true ternary microemulsion system made without the addition of a coemulsifier was given by Perez-Luna et al. [73]. In the unpolymerized microemulsions two distinct sizes are detected. The smaller size is about 0.6 nm which corresponds to the dodecyltrimethylammonium bromide (DTAB) micelles. The second, larger size, varies from 2.4 to 14.8 nm and is due to the presence of styrene-containing droplets. Polymerization of these transparent microemulsions produced stable, bluish monodisperse microlatices with particle radii ranging from 20 to 30 nm, depending on styrene content. Polymerization of these microemulsions with potassium peroxodisulfate proceeds to a final conversion of approximately 60–70%, i.e. it increases with styrene content. The rate of polymerization is initially fast and decreases towards a plateau after approximately 40 or 50% conversion. Initiation takes place in the styrene swollen droplets, i.e. in the styrene-swollen micelles, and polymeric particles grow by recruiting monomer and emulsifier from uninitiated droplets and small micelles.

Microemulsion polymerizations of styrene have also been reported elsewhere [74–78].

3.4.2 Acrylic and methacrylic alkyl esters

Emulsion polymerizations and copolymerizations of acrylic and methacrylic esters have been objectives of many kinetic studies and patents. Their polymers and copolymers are interesting not only from the theoretical point of view but also from their industrial application. The wide range of properties of these polymers is determined by the different nature (hydrophilic or lipophilic) of individual monomers. Table 16 shows that the monomer becomes more lipophilic with increasing length of the alkyl chain of the acrylate or methacrylate. Variation in the type of monomer leads to the appearance of different specific features of the polymerization and properties of the final polymeric product.

Table 16 — Water solubility of vinyl monomers at 20–25°C [79–81]

Monomer	Solubility (mM)
p-$tert$-Butyl styrene	0.2[a]
n-Octyl acrylate	0.34
Dimethyl styrene	0.45
Vinyl toluene	1.0
n-Hexyl acrylate	1.2
Styrene	3.5
n-Butyl methacrylate	3.5[b]
n-Butyl acrylate	11
Chloroprene	13
Butadiene	15
Vinylidene chloride	66
Ethyl acrylate	150
Methyl methacrylate	150
Vinyl chloride	170
Vinyl acetate	290
Ethylene	200–600
Methyl acrylate	650
Acrylonitrile	1600
Acrolein	3100

[a] Ref. [80].   [b] Ref. [81].

Methyl acrylate is the acrylate monomer with the highest water solubility; the mechanism of its polymerization in emulsions was investigated by Banerjee et al. [82]. The authors studied the influence of the initiator (peroxodisulfate or Fenton's agent) and of the emulsifier system (cetyl trimethylammonium bromide — CTAB or sodium dodecyl sulfate — SDS) on the kinetic parameters of the emulsion polymerization. The dependence of the initial rate on the concentration of persulfate yielded a reaction order of 0.58 and on the concentration of Fenton's agent ($FeSO_4$ and H_2O_2) at constant $FeSO_4$ concentration it was 0.61; at a constant concentration of hydrogen peroxide the order was 0.51. The orders with respect to initiator obtained from the dependence of the RMM on the initiator concentration (Table 17) differ from the theoretical values given by the micellar model. At low conversions, the rate of polymerization is proportional to the square root of the initiator concentration

$$R_p = z\, [I]^{0.5}, \tag{6}$$

the molecular mass is proportional to reciprocal of the square root of the initiator concentration

$$M_v = p\, [I]^{-0.5} \tag{7}$$

and for the overall rate of polymerization we have

$$R_p = k_p\, (f k_d / k_t)^{0.5}\, [M]\, [S_2O_8^{2-}]^{0.5}, \tag{8}$$

where z and p are constants.

Table 17 — Variation of the reaction order x in initiator concentration with type of initiating system, obtained from $M_v \sim [I]^{-x}$ (Reprinted from ref. [82] with permission of Butterworth Heinemann Ltd., London)

% conv.	Fenton system x^a	x^b	Peroxodisulfate system x^c
5	—	—	0.41
10	0.54	0.60	0.44
20	0.43	0.61	—
30	0.33	0.52	—
40	—	0.42	—
50	—	0.35	—

[a] Recipe: MA 2% (v/v), $[H_2O_2] = 0.92 \times 10^{-5}$ mol dm^{-3}, [$FeSO_4$] varied, CTAB = 0.02% (mass/v). 20 °C.
[b] Recipe: MA 2% (v/v), [$FeSO_4$] = 2.88×10^{-4} mol dm^{-3}, CTAB = 0.02% (mass/v), [H_2O_2] varied.
[c] Recipe: MA 10% (v/v), SDS = 0.333% (mass/v), [$K_2S_2O_8$] varied. 50°C.

The results led the authors to conclude that the homogeneous polymerization significantly affects the initial step of polymerization, the initiating mechanism consisting of two steps. In the first step, oligomer radicals are formed in the aqueous phase, and, in the second step, they enter micelles swollen with monomer, where their growth and the formation of the polymer particles take place. The authors state that hydroxyl radicals formed in the redox system ($Fe^{2+} + H_2O_2$) [83, 84] and by hydrolysis of $SO_4^{\bullet-}$ radicals determine the rate of initiation. This conclusion is supported by the following facts: the existence of an energy barrier between the negatively-charged surface of the micelles and $SO_4^{\bullet-}$ radicals [85, 86] and of the high fraction of hydroxyl groups on the surface of particles prepared by polymerization initiated by peroxodisulfates [87]. They have not taken into account any possibility of hydrolysis of sulfate groups to hydroxyl during polymerization, which has been discussed elsewhere [86].

The course of the emulsion polymerization of methyl acrylate has also been studied at various mass ratios between monomer and water using the anionic emulsifier SDS and water-soluble peroxodisulfate [87]. When the mass ratio of monomer/water used was equal to or greater than 5:95, the rate in the stationary state obeyed the Eq.

$$R_p = k' [I]^{0.46} [E]^{0.3} [M]^0 \tag{9}$$

and for the ratio lower than 5:95

$$R_p = k'' [I]^{0.42} [E]^{0.25} [M], \tag{10}$$

where k' and k'' are constants and [M] is the monomer concentration in water. The values of the order with respect to initiator were in both cases lower than 0.5. The low values of the orders with respect to emulsifier were explained by the high water-solubility of the monomer.

The role of the emulsifier was in this case rather complicated. Higher concentrations of emulsifier led on the one hand to an increase in the number of particles and on the other to an increase in the efficiency of coalescence, which caused a decrease in their concentration. The emulsifier functioned as a degradative additive lowering the RMM in proportion to its concentration. In the region of lower monomer concentrations (0.1 mol dm^{-3}) the number of radicals in a particle $\bar{n}$ increased with increase in conversion; e.g. at conversions between 12 and 50%, $\bar{n}$ changed from 0.54 to 0.63. By contrast, polymerizations proceeding at higher monomer concentrations (~1.1 mol dm^{-3}) throughout the range of conversions specified, led to an increase in $\bar{n}$ from 0.66 to 0.88. The high value of the order was ascribed to the formation of large particles, in which several radicals can exit simultaneously.

In their analysis of the kinetic data of the emulsion polymerization of methyl acrylate, Guha & Palit [88] also obtained values for the orders with respect to initiator higher than 0.4 and, with respect to emulsifier, lower than 0.6. These findings were explained by the active participation of homogeneous polymerization in the mechanism for the formation of polymer particles and their growth. The reaction order in initiator, which is close to 0.5, supports the notion about deactivation of radicals by termination processes occurring in the aqueous phase.

The dependence of the rate of polymerization on the square root of the peroxodisulfate concentration is also characteristic of the polymerization of acrylonitrile in an aqueous disperse medium [89]. The contribution from homogeneous polymerization originates from the high water solubility of acrylonitrile and from its participation in addition reactions in the homogeneous medium.

Fitch & Shih [90] examined the emulsion polymerization of methyl methacrylate as a function of the concentration of emulsifier (sodium dodecyl sulfate). They found that the polymerization rate is not constantly proportional to the emulsifier concentration throughout the concentration region. The rate of polymerization was proportional to the number of particles. The drop in the number of particles at high emulsifier concentrations resulted from the properties of the emulsifier functioning as an electrolyte. The emulsifier initiated the formation of a higher concentration of primary particles and, at the same time, particle coalescence (Table 18).

Table 18 — Variations of particle size and number in emulsion polymerization of methyl methacrylate (MMA) with concentration of sodium dodecyl sulfate (SDS)

[SDS] /10^{-4} mol dm^{-3}	N /10^{16} dm^{-3}	r /nm
13.9	81.3	136
17.4	102.0	126
20.8	131.0	116
27.8	90.5	131

N is average particle number determined by electron microscopy; r is polymer particle radius.
Recipe: [MMA] = 9.51×10^{-2} mol dm^{-3}, [K$_2$S$_2$O$_8$] = 8.64×10^{-4} mol dm^{-3}, [NaHSO$_3$] = 1.3×10^{-3} mol dm^{-3}, [(NH$_4$)$_2$Fe(SO$_4$)$_2$] = 2.74×10^{-6} mol dm^{-3}. Temperature: 30 °C.

A study of the batch emulsion polymerization of methyl methacrylate (MMA) in the presence of seed poly(butyl acrylate) (PBA) particles has shown [91] that its rate of polymerization and the shape of the conversion curve depend on the mass ratio of MMA/PBA. The experimental and calculated diameters of the polymer particles agreed well for the system with a 'low' (0.632) mass ratio MMA/PBA. This result indicated that all poly(methyl methacrylate) formed was used exclusively for building up shells of poly(butyl acrylate) seed particles. For a relatively 'high' mass ratio MMA/PBA (3.160) the calculated diameter of the particles was higher than that found experimentally. At 'high' values of the MMA/PBA ratio, about 37% of poly(methyl methacrylate) formed was used for generating a crop of poly(methyl methacrylate) particles. It was shown that for the 'low' mass ratio of MMA/PBA, the 'inner' transport of monomer (from poly(butyl acrylate) particle core swollen by methyl methacrylate) to the site of propagation is decisive. For systems with a 'high' MMA/PBA ratio, the 'outer' transport of monomer from monomer droplets and micelles swollen by monomer to the propagation site is of primary importance. To explain the experimental results, a reaction mechanism for seeded emulsion polymerization was proposed [91].

Ethyl methacrylate which has been described in another kinetic study [92], is representative of hydrophilic monomers with the same water solubility as methyl methacrylate. The rate of polymerization could be expressed at low conversions by the semiempirical relation

$$R_p \sim [I]^{0.56} [E]^{0.72} \qquad (11)$$

and the molecular mass by

$$M_v \sim [I]^{-0.53}. \tag{12}$$

The reaction orders with respect to initiator again point to the role of homogeneous polymerization. As the conversion increases (Table 19), the order with respect to initiator decreases, and, conversely, that with respect to emulsifier increases. The use of high concentrations of emulsifier led to destabilization of the dispersion accompanied by a decrease in the number of particles. As for the emulsifier, the use of higher initiator concentrations leads to the lowering of the stability of the latex prepared as a consequence of the functioning of its electrolytic character.

Table 19 — Variation of reaction orders in initiator and emulsifier with conversion (Reprinted from ref. [92] with permission of Marcel Dekker Inc., New York)

% conv.	Reaction order in initiator	Reaction order in emulsifier
20	0.45	0.55
30	0.46	0.61
40	0.38	0.68
60	0.38	0.63

The effect of a bi-unsaturated monomer on the emulsion polymerization of ethyl acrylate in the presence of sodium dodecyl sulfate as emulsifier and initiated by ammonium peroxodisulfate was investigated by Capek et al. [93]. Divinylbenzene (DVB) was found to reduce both the rate of polymerization and the particle size. Addition of 1,6-hexamethylene diacrylate (HMDA) slightly increases the rate of polymerization and the particle size. The rate of polymerization in Stage II was found to be proportional to the 0.37, 0.23, and 0.5 powers of the emulsifier concentration for system A (without additive), system B (with DVB), and system C (with HMDA). The number of polymer particles is proportional to the 0.39, 0.23, and 0.57 powers of the emulsifier concentration for systems A, B, and C. All polymerization systems deviate from the predictions of micellar theory. The low reaction orders are attributed to the high hydrophilicity of the polymer formed. The low rate of polymerization in the presence of DVB was discussed in terms of retardation by the pendant vinyl groups and the decrease of the equilibrium monomer concentration in the polymer particles.

The kinetics of the emulsion polymerization of lipophilic butyl acrylate have been studied in detail by Capek et al. [94]. Increase in the emulsifier concentration was accompanied by a rate enhancement of the polymerization and an increase in the number of particles (Table 20). The values of the ratio r_{exp}/r_c point to the suitability of the micellar model for calculating the kinetic parameters of this emulsion polymerization. The trend in the average number of radicals per particle conforms with earlier kinetic studies giving evidence for the possible coexistence of several radicals in larger polymer particles. At low conversions, even with the low water-solubility of butyl acrylate, the following dependence was obtained

$$M_V \sim [I]^{-0.5}, \quad (13)$$

supporting the role of homogeneous polymerization of butyl acrylate in the system. This surmise is also supported by the dependence of the RMM on conversion, i.e. at low conversions, a polymer is formed with low RMM comparable with that formed in homogeneous polymerization. The RMM rapidly increases as reaction proceeds and reaches a maximum at about 40% conversion (Fig. 9). The RMM of the polymer obtained with an emulsifier concentration varying from 0.63×10^{-2} to 7.32×10^{-2} mol dm^{-3} decreased from 1.3×10^6 to 0.9×10^6. This decrease was attributed to chain transfer to the anionic emulsifier (Dowfax 2A1).

Table 20 — Kinetic parameters for the emulsion polymerization of butyl acrylate

$[DW]^a$ /10^{-5} mol dm^{-3}	k_t/k_p^b	$\bar{n}^c$	N^d /10^{15}	r_{exp}^e /nm	r_{exp}/r_c^f	R_p^g /10^{-2} mol dm^{-3} s^{-1}
7.32	20	0.57	6.96	33	1.18	1.74
2.87	60	0.89	3.65	41	1.21	1.05
1.18	70	1.25	1.59	55	1.22	0.73
0.63	120	1.64	0.92	75	1.39	0.55

[a] Sodium dodecylphenoxybenzene disulfonate (emulsifier).
[b] Ratio between rate constants for termination and propagation calculated for Stage II.
[c] Average number of radicals per particle.
[d] Polymer particle number per cm^3 water.
[e] Particle radius obtained experimentally (for latex with 100% conversion).
[f] Ratio between particle diameter obtained experimentally and calculation.
[g] Polymerization rate in Stage II.

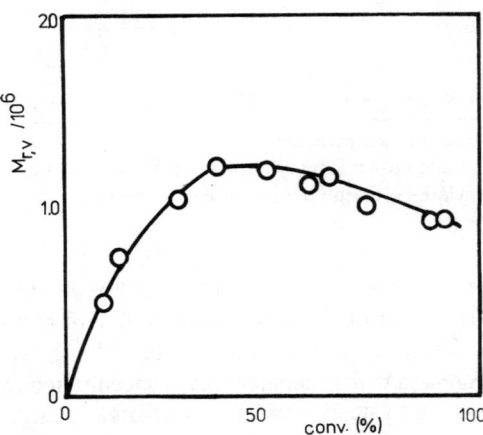

Fig. 9 — Plot of the average viscosity RMM ($\bar{M}_r$) of poly(butyl acrylate) conversion. Recipe: 140 g H$_2$O, 95.6 g BA, [DW] = 7.325×10^{-3} mol dm^{-3}, [APS] = 2.887×10^{-3} mol dm^{-3}. Temperature: 70°C. BA — butyl acrylate, DW — sodium dodecylphenoxybenzene disulfonate, APS — ammonium peroxodisulfate.

Mallya & Plamthottam [95] have determined the average number of radicals per particle, $\bar{n}$, and the termination rate constant, k_t, as a function of conversion and the size of the polymer particles for the seeded batch emulsion polymerization of n-butyl acrylate. With decreasing amounts of seed (increasing particle size), the average number of radicals per particle was reported to increase. The values of $\bar{n}$ ranged from 30 to 5 and decreased with conversion in Stage III. Contrary to expectation, the values of $\bar{n}$ were found to be unusually high and the values of the termination rate constant were found to be very low for all runs ($\sim 1 \times 10^{-4} - 1 \times 10^{-3}$ dm^3 mol^{-1} s^{-1}). Besides, k_t slightly increased with conversion. The high values of $\bar{n}$ seem to be consistent with the low rates of termination. The authors tried to explain the low values of k_t in terms of the energy barrier for rotation about the terminal C–C bond of the polymer chain end of poly(butyl acrylate), i.e. the chain end of PBA is 'dynamically rigid' due to the presence of the bulky butyl side groups.

The effect of a seed latex on the emulsion polymerization of butyl acrylate has also been kinetically investigated at different initiator, emulsifier and particle concentrations and particle domain formation [96, 97]. In the presence of poly(ethyl acrylate) seed particles (System 1), the rate of polymerization is approximately proportional to the 0.4 power of the initiator concentration. The order of the number of polymer particles is approximately 0.1 in peroxodisulfate concentration. The rate of polymerization in the presence of crosslinked poly(ethyl acrylate) seed particles by hexamethylenediacrylate (System 2) is approximately proportional to the 0.6 power of the initiator concentration and, in the presence of seed particles of poly(ethyl acrylate) crosslinked by divinylbenzene (System 3), to the power 0.56.

Table 21 — Reaction order in the initiator, emulsifier and particle concentrations

[BA] /mol dm^{-3}	$R_p \sim I^x$			$N \sim I^y$			$N \sim E^z$	
	a	b	c	a	b	c	b	c
1.16	0.37	0.55	—	0.14	0.12	—	0.20	—
2.32	0.35	0.56	0.56	0.11	0.10	0.07	0.31	0.51
3.47	0.39	0.65	—	0.11	0.22	—	0.25	—

[a] System 1 (poly(ethyl acrylate) seed particles).
[b] System 2 (poly(ethyl acrylate-co-hexamethylenediacrylate) seed particles).
[c] System 3 (poly(ethyl acrylate-co-divinylbenzene) seed particles.

The order of the number of polymer particles in System 2 is $ca.$ 0.15 in peroxodisulfate and in System 3 is 0.07. The order of the polymerization rate of butyl acrylate is 0.51 in emulsifier (sodium dodecyl sulfate) concentration (Table 21). In System 1, the rate of polymerization slightly decreases with monomer (BA) concentration in the range from 1.1 to 3.5 mol dm^{-3}. By contrast, in Systems 2 and 3 the rate slightly increases with monomer concentration. The experimental data clearly deviate from the predictions of micellar or homogeneous theory. The observed behaviour is discussed in terms of

(i) phase separation in the monomer-swollen seed particles,
(ii) the high hydrophilicity of the seed particles,

(iii) the diffusion of monomer from the monomer-swollen seed particles to the nucleated micelles,
(iv) the increase in lipophilicity of the seed particles crosslinked by the bi-unsaturated monomer,
(v) reaction with the pendant vinyl groups,
(vi) the morphology of the semi-IPNs formed, and
(vii) formation of a new generation of polymer particles.

The mechanism for the emulsion polymerization of lipophilic butyl methacrylate was investigated by Bartoň et al. [98–100]. These authors studied the effect of the oil-soluble 2,2′-azobisisobutyronitrile (AIBN) and of the water-soluble peroxodisulfate on the emulsion polymerization and discussed the special features of these systems. In the AIBN-containing systems, the orders obtained with respect to anionic emulsifier (Dowfax 2A1) were 0.56 in Stage I and 0.36 in Stage II. Similar values were obtained for systems with ammonium peroxodisulfate: for Stage I the value was 0.47 and for Stage II it was 0.43. Lower values of the orders were also obtained for the initiator concentration. At conversions of 0 to 30%, the polymerization rate was proportional to the 0.26 order in the peroxodisulfate concentration and to the 0.34 order in the AIBN concentration.

Table 22 — Rate (R_p) of the emulsion polymerization of butyl methacrylate (BMA) in Stages I and II (Reprinted from ref. [98] with permission of Hüthig and Wepf Verlag, Basel)

System	DW /g	$R_p / 10^{-4}$ mol dm^{-3} s^{-1}			
		Stage I[a]	Stage II[a]	Aqueous phase[b]	Monomer particles[c]
A	5	1.31	4.95	2.89×10^{-4}	0.32
B	1	0.53	2.79	2.89×10^{-4}	0.32
C	5	3.32	29.00	4.71×10^{-4}	[d]
D	1	1.56	14.48	4.71×10^{-4}	[d]

[a] Experimental data. Polymerization rate in Stage I was determined from slope of the conversion curve at zero time. Polymerization rate in Stage II is constant and maximal. In calculations of rates of polymerization, the value of the equilibrium concentration of BMA in the monomer/polymer particles considered was 3.83 mol dm^{-3}.
[b] Calculated from $R_p = (k_p/k_t^{0.5})R_i^{0.5}[M]_{water}$ using experimental and literature data.
[c] Calculated for monomer particles of diameter of 100 nm using equation for the polymerization rate in discrete particles: $R_p = (k_p/k_t^{0.5})R_i^{0.5}[M]_d \tanh(N_A^2 v^2 R_i / k_t)^{0.5}$, where $[M]_d$ is monomer concentration in droplets and v is volume of monomer droplets.
[d] Concentration of (NH$_4$)$_2$S$_2$O$_8$(APS) in emulsion monomer droplets is unknown. With regard to the very small probability of entry of primary SO$_4^{\bullet-}$ radicals into monomer droplets and probably zero concentration of APS in monomer droplets, the rate of polymerization in oil droplets is practically zero.

Analysis of the contributions of the polymerization rate in water, in monomer droplets, and in the polymer particles summarized in Table 22, led to the conclusion that the contribution from polymerization in the aqueous phase is very small and does not exceed 0.1% of the total value of the initial rate. Polymerization in the monomer droplets is significant under the conditions listed in Table 22. In System A it is 24% and in System B up to 60% in its contribution to the overall rate (relative to the resulting initial rate of polymerization).

Interpretation of the results showed that peroxodisulfate is much less efficient in initiation of the polymerization than AIBN. In terms of these studies, the reason for the lower initiation efficiency of peroxodisulfate is the strong electrostatic barrier between the negatively-charged particle surface and the radicals of structure $SO_4^{\bullet-}$. The high activity of the radical fragments of AIBN in initiation is determined by the ability of 2-cyanopropyl radicals to penetrate more easily into the micelles and polymer particles. The higher radical capture efficiency for radical fragments from an oil-soluble initiator, as well as the lower desorption of radicals from the particles, as explanations of the high activity of AIBN in the initiating step were supported by the authors [101].

The interaction between peroxodisulfate and emulsifier leads to acceleration of the decomposition of initiator [102]. As expected AIBN does not interact with the emulsifier and thus its decomposition is not influenced by emulsifier (Table 23). The higher RMMs of the polymer in the AIBN-initiated system were explained by the lower concentrations of primary radicals in the particles and, in parallel, by their participation in termination reactions and by the low transfer rate to AIBN.

Table 23 — Kinetic parameters for butyl methacrylate emulsion polymerization

System [a]	$[DW]^b$ $/10^{-2}$ mol dm^{-3}	$[I]^c$ $/10^{-4}$ mol dm^{-3}	M_v^d $/10^6$	R_i^e $/10^{-8}$ mol dm^{-3} s^{-1}	k_d^f $/10^{-5}$ s^{-1}
—	—	—	—	—	1.6[g]
B	0.99	2.63	7.8	2.8	1.5
A	4.94	2.63	10.5	5.0	1.4
—	—	—	—	—	0.58[h]
D	0.99	23.12	4.4	14.5	2.47
C	4.94	23.12	3.6	29.0	6.54

[a] Initial concentration of 2,2'-azobisisobutyronitrile (AIBN) initiator (Systems A and B) and peroxodisulfate (Systems C and D) is 3.8×10^{-3} mol dm^{-3}.
[b] Sodium dodecylphenoxybenzene disulfonate.
[c] Amount of initiator dissolved in water.
[d] Average viscosity RMM.
[e] Rate of initiation obtained by enumerating equation $R_i = R_p [(1/P_n) - C_m]$ using experimental values of R_p and P_n and of transfer constant to monomer $C_m = 1.4 \times 10^{-5}$ [103].
[f] Rate constant for decay of initiator obtained from equation $R_i = 2fk_d [I]$ and from experimental data for R_i and [I] and literature value $f = 0.5$ for peroxodisulfate [104] and 0.6 for AIBN [105].
[g] Ref. [105].
[h] Ref. [104].

The higher concentration of radicals in water in Systems C and D (with peroxodisulfate) led to the formation of a greater number of particles (approximately 4 times compared with the AIBN-initiated systems); in System C, the number of particles N was equal to 4.65×10^{18} cm^{-3} water, while in System A, N equalled 1.32×10^{18} cm^{-3} water. The average concentration of radicals per particle varied in Systems A and B between 0.2 and 0.3 and in Systems C and D between 0.3 and 0.5.

The applicability of the micellar model to interpreting the kinetic values of the emulsion polymerization of butyl methacrylate follows from the good agreement between the calculated and measured sizes of the particles (their ratio does not exceed 1.2).

Analysis of the radioactive poly(butyl methacrylate) formed in the system with labelled AIBN showed that each macromolecule contained on average one radioactive fragment from the initiator [100]. The most probable process of deactivation of a pair of radicals was shown to be disproportionation in this case.

Some experimental and theoretical investigations have been conducted on emulsion polymerizations initiated by oil-soluble initiators [106, 107]. The principal site of polymerization in such systems is also situated in the polymer particles.

Nomura & Fujita [107] have clarified quantitatively the factors that influence the kinetics and mechanism of a conventional emulsion polymerization initiated by oil-soluble initiators and the average number of radicals per particle. The authors modelled the emulsion polymerization process for systems in which the oil-soluble initiator used was completely insoluble in water. For this case no region was found where $\bar{n} = 0.5$. The rate of polymerization of such systems would be that of a suspension polymerization (bulk polymerization). However, this is actually not the case, because any oil-soluble initiator will be more or less soluble in water. The authors [107] therefore concluded that the emulsion polymerization initiated by oil-soluble initiators behaves rather like to a conventional emulsion polymerization initiated by water-soluble initiators, showing $\bar{n} = 0.5$, when the necessary conditions are satisfied.

The influence of the nature of the monomer (alkyl esters of acrylic acid) on the kinetics of emulsion polymerization initiated by a water-soluble initiator was investigated by Eliseeva [108], who reported that the reaction order in emulsifier (anionic types were used) decreased with increase in the hydrophilic nature of the alkyl acrylate (Table 24). The trend in the values of the reaction order was explained by the interaction between polymer and emulsifier. As the polymer becomes more lipophilic, the energy of adsorption of the emulsifier on the surface of the polymer particles, and the strength of the polymer–emulsifier interaction, both increase.

Table 24 — Kinetic and adsorption characteristics of acrylate emulsion polymerization

Monomer	Emulsifier	Reaction order in emulsifier	Adsoption energy for emulsifier /kJ m^{-2}	A_s /nm^2
Methyl acrylate	SDS	0.13	20.0	1.51
Ethyl acrylate	SDS	0.21	22.1	0.92
Ethyl acrylate	STS	0.29	23.3	0.82
Ethyl acrylate	SHS	0.33	24.6	0.74
Butyl acrylate	SDS	0.33	25.6	0.67
Hexyl acrylate	SDS	0.46	—	—

SDS — sodium dodecyl sulfate, STS — sodium tetradecyl sulfate, SHS — sodium hexadecyl sulfate, A_s — area occupied per emulsifier molecule.

The drop in the reaction order in emulsifier with increasing water-solubility of the monomer has also been observed by Breitenbach et al. [109]. The decrease in order with increasing hydrophilicity of the polymer was, in both studies, attributed to the increase in size of the polymer particles via the flocculation of particles and to the strong influence of the homogeneous polymerization.

Octadecyl methacrylate monomer, which has an extremely low solubility in water, was polymerized in an emulsion using persulfate initiation by Satpathy & Dunn [26].

The dependence of the polymerization rate on the initiator and emulsifier (SDS) concentrations, during Stage II in which it is constant, gives the following reaction orders: the initiator exponent is 0.48 and the emulsifier exponent 0.68. Emulsifier exponents below 0.6 are generally found for hydrophilic monomers and are usually attributed to the occurrence of a significant part of the polymerization in the aqueous phase (see above). The high emulsifier exponent of 0.68 is attributed to the low water solubility and high micellar solubility of octadecyl methacrylate. In addition, the monomer is proposed to be weakly surface-active and to form mixed micelles with the emulsifier.

The effect of the emulsification of monomers, such as acrylonitrile, methyl methacrylate, butyl acrylate, ethyl acrylate, and their mixtures on the stability of the monomer mixtures and polymer latices was investigated by the conductometric method [110, 111]. The dependence of the conductivity on the monomer concentration for a given emulsifier and concentration is described by a curve with a minimum and a maximum [110]. All monomer emulsions show a decrease in initial conductivity with increasing monomer concentration. In aqueous solution with acrylonitrile and methyl methacrylate, a sharp increase in conductivity was observed after the first break point. By contrast, only a slight increase in conductivity was observed in systems with ethyl acrylate/methyl methacrylate comonomer mixtures. Beyond the second break point, a small decrease in conductivity was observed. The fluctuations in conductivity are assumed to result from the solubilization of monomer in emulsifier micelles, from the release or tying up the most conducting species from the aqueous phase, from the changes in the particle surface layer and in the molecular volumes of the monomers, and from the phase inversion and transfer of emulsifier between phases. The most stable monomer emulsion was formed in systems containing the most water-soluble acrylonitrile monomer. Conversely, the stability of copolymer latices decreased with increasing acrylonitrile content in the monomer mixture. It was found that a pre-emulsification process is one of the important characteristics of the systems studied [110, 111]. In a system with a longer pre-emulsification time one observes a higher rate of polymerization, a more stable latex, and smaller polymer particles in comparison with those found in a system with a shorter pre-emulsification period.

The courses of the emulsion copolymerizations of ethyl acrylate and methyl methacrylate monomers, with the same water solubility, initiated by a water-soluble initiator in the presence of a mixed emulsifier system (anionic and nonionic emulsifier) were characterized by Capek & Tuan [112]. At low conversions (below 5%), the order in ammonium peroxodisulfate approached 0.5 and the number average RMM of the copolymer reached only 2×10^5. At medium conversions (20–50%), the order in initiator fell to 0.4. The RMM increased linearly as the polymerization proceeded and reached a maximum ($\sim 3.5 \times 10^6$) at conversions of about 40%. The average particle size was found to grow throughout all regions of conversion, most remarkably, however, at low conversions (Table 25). The increase in particle size and stability as a result of the coalescence of particles of various sizes, as well as the increase in the reaction order in initiator, resulted from the contribution of the homogeneous polymerization. The high water solubility of the monomers used and their polymerization led to the formation of primary polymer particles at even higher conversions. The fractional rate expressed by

Eq. (14) increased with the extent of polymerization owing to the increase in the number of radicals per particle:

$$R_{f,c} = \frac{R_p}{[M]_{eq}} = \frac{k_{pAA} k_{pBB} N_c \bar{n}_c (r_A + 2L + r_B L^2)}{N_0(k_{pBB} r_A + k_{pAA} r_b L)(1+L)}, \quad (14)$$

where R_p is the experimental value of the polymerization rate, the monomer feed concentration is given by the molar ratio of monomers $L = [B]/[A]$, $[M]_{eq}$ is the equilibrium monomer concentration in the polymer particles, k_{pAA} and k_{pBB} are the propagation rate constants of monomers A and B and r_A and r_B are the copolymerization parameters of the pair of monomers A and B.

Table 25 — Kinetic parameters for batch emulsion copolymerization of methyl methacrylate and ethyl acrylate

APS[a]	Conversion	$\bar{d}^b$	N^c	R_f^d	$\bar{n}_c^e$
$/10^{-3}$ mol dm^{-3}	/%	/nm	$/10^{14}$ cm^{-3}	$/10^{-4}$	particle
7.3	30	155	—	5.5	1.9
7.3	50	160	—	8.3	2.8
7.3	70	165	—	12.8	4.3
7.3	100	170	2.29	—	—
1.46	30	180	1.95	2.8	1.4
1.46	50	185	1.78	4.3	2.2
1.46	70	190	1.54	5.5	2.8
1.46	100	195	1.45	—	—

[a] APS — ammonium peroxodisulfate.
[b] Particle diameter.
[c] Polymer particle number per cm^3 water.
[d] Fractional polymerization rate.
[e] Average number of radicals per particle.
Recipe: 150 g water, 20 g methyl methacrylate, 2.5 g emulsifier (Tween 40; 97 mass %), 2.5 g emulsifier (Spolapon AOS; 40 mass %). Temperature 60°C.

The emulsion copolymerization of ethyl acrylate and methyl methacrylate has also been studied from the point of view of the influence of the composition of a mixed emulsifier system and the concentration of emulsifier on the kinetics of polymerization [113]. The dependence of the rate of polymerization on the composition of the emulsifying mixture is described by a curve with a maximum for an equimolar composition of the emulsifying mixture, $n_{Tw} \sim 0.5$ (Fig. 10). The rate enhancement of the polymerization with increase in the concentration of nonionic emulsifier in the range $0 < n_{Tw} < 0.5$ was attributed to a decrease in the repulsive forces between the surface of the polymer particles and the negatively-charged radicals entering the particles [2]. The particle size formed in systems with $n_{Tw} < 0.5$ was approximately constant, with a diameter of about 180 nm. An increase in the concentration of the nonionic emulsifier (Tween 40) above 0.5 was accompanied with a decrease in the rate of polymerization, which was ascribed to the growth of particle size.

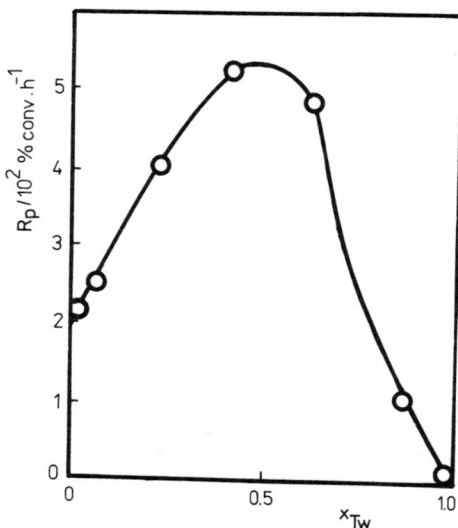

Fig. 10 — Plot of the rate R_p of batch emulsion copolymerization of methyl methacrylate (MMA) and ethyl acrylate (EA) (% conv. h^{-1}) versus composition of emulsifying mixture (Tween 40 and Spolapon AOS) n_{Tw}. Recipe: 150 g H_2O, 20 g MMA, 80 g EA, 5 g emulsifier mixture (Tween 40 and Spolapon AOS), 0.25 g APS. Temperature: 60°C.

The diameter of particles increased from 180 to 280 nm in the range of n_{Tw} between 0.5 and 1.0.

The order of reaction with respect to emulsifier (mixture of anionic and nonionic) with regard to the degree of conversion followed a trend opposite to that of the initiator. The lower value of the order in the initial steps (0.3 at 10% conversion) followed from the contribution from homogeneous polymerization. The polymer particles formed were more hydrophilic and able to adsorb only a limited number of emulsifier molecules. At medium conversions (30–40%), the course of the polymerization was determined by polymerization in the polymer particles; the polymer formed has a lipophilic character and was able to adsorb a greater fraction of emulsifier, which is why the order increased, reaching a value of 0.6. The decrease in RMM with increasing emulsifier concentration was considered as a consequence of the degradative transfer of the chain to emulsifier.

The effect of a mixed emulsifying system (Tween 40 and Spolapon AOS) on the mechanism of the emulsion copolymerization of ethyl acrylate and methyl methacrylate was studied by measuring the conductivity during polymerization [111]. The dependence of the conductance on conversion is described by a curve with a minimum at about 50% conversion. The change in conductivity was explained by the different type of association between emulsifier molecules and polymer particles and by the decrease of the total polymer particle surface. At high conversions, the type of association prevails which releases the most conducting species from the polymer particles to the aqueous phase. In addition, this sharp increase in conductance after 40 or 50% conversion goes parallel with a decrease of the polymer particle number.

Capek [114] has reported the effect of diluent agents, such as toluene and n-octane, on the kinetics of the emulsion copolymerization of ethyl acrylate and methyl methacrylate. It was found that both diluents unexpectedly strongly reduced the polymerization rate. The rate reduction increased with increasing diluent concentration, which was attributed to the desorption of monomer radicals from the polymer particles and to the suppression of the gel effect. The driving force of desorption is proposed to be the hydrophilicity of the low-RMM reactants. Toluene and n-octane, diluents which are water-insoluble, depress the polymerization in water and favour the processes of conventional emulsion polymerization. The average number of radicals per particle was found to increase with increasing conversion.

The kinetics of emulsion copolymerization of ethyl acrylate and methyl methacrylate were investigated in the presence of methanol by Capek [115]. It was observed that methanol also strongly reduced the rate of polymerization, an effect discussed in terms of the dilution of the components and of the desorption of diluent and monomeric radicals from polymer particles. Furthermore, the polymerization rate increased with conversion up to high conversions. The high polymerization rate at high conversions was attributed to the propagation of the more reactive ethyl acrylate monomer and to the gel effect. The particle size was found to increase with increasing diluent concentration. The formation of large particles in the presence of methanol was due to an increase in the nonmicellar form of the emulsifiers.

The effect of a magnetic field on the emulsion copolymerization of methyl methacrylate with vinyl acetate and acrylamide was investigated by Simionescu et al. [116]. Effecting the copolymerization in a magnetic field, or in a reaction medium pre-treated in the magnetic field, leads to an increase in the rate of polymerization. This increase was more pronounced in systems with emulsifiers as compared with those without. This behaviour was explained by the orientation or by the 'solubilization' of the reactant compounds through their treatment in a magnetic field. This study showed the latices obtained in a magnetic field (or those obtained with pre-treated compounds) are more stable, as characterized by their lower electrical conductivity and surface tension. Besides, the copolymers formed in the magnetic field-treated latices exhibit high RMMs.

Table 26 — Kinetic parameters for emulsion copolymerization of acrylonitrile (AN) and butyl acrylate (BA)^a

$n_{BA}{}^b$	$R_p{}^c$ /10^{-3} mol dm^{-3} s^{-1}	$r_0{}^d$ /nm	$r_0/r_c{}^e$	N^f /10^{15}	y^g
1.0	17.4	33	0.87	2.60	0.64
0.788	30.5	44	1.13	2.50	0.58
0.625	35.3	60	1.54	2.44	0.51
0.435	23.5	76	1.95	2.48	0.43
0.293	26.0	95	2.38	2.29	0.40

^{a}Recipe: 140 g water, 96.6 g comonomers, 0.7725 g ammonium peroxodisulfate, 7.32×10^{-2} mol dm^{-3} Dowfax 2A1. Temperature 70°C.
^{b}Mole fraction of butyl acrylate in comonomer mixture.
^{c}Polymerization rate in Stage II.
^{d}Average radius of final polymer particles obtained experimentally.
^{e}Ratio between particle radii obtained experimentally and from micellar model.
^{f}Average number of latex particles per cm^3 water.
^{g} Reaction order in emulsifier obtained from dependence $R_p \sim [DW]^y$.

Capek et al. [117, 118] examined the emulsion copolymerizations of acrylates of different water solubility, using as an example the monomer pair acrylonitrile and butyl acrylate. Increase in the acrylonitrile fraction in the monomer feed was accompanied by a decrease in the reaction order with respect to emulsifier and an increase in the ratio between the size of the particles as measured and predicted by the micellar model (Table 26). The deviation from the micellar model in systems with a high content of acrylonitrile was interpreted in terms of the active role of the homogeneous polymerization of acrylonitrile in the overall mechanism. The decrease in polymerization rate above a certain limiting concentration of acrylonitrile was caused by the increased formation of primary particles, their participation in particle growth via a coagulation mechanism, and the formation of coagulate. Investigation of the water solubility of acrylonitrile showed that the solubility of acrylonitrile is controlled by the concentration of butyl acrylate, and to a minor extent, by the emulsifier concentration. On changing the concentration of Dowfax 2A1 from 0.63×10^{-2} to 12.93×10^{-2} mol dm^{-3}, the water solubility of acrylonitrile increases from 7.3 g to 8.3 g in 100 g water; on the other hand, by adding 0.6 moles of butyl acrylate to 140 ml water and 0.36 moles of acrylonitrile, the solubility of acrylonitrile decreases from 7.3 g to 3 g in 100 g water.

In their papers, Capek & Barton [119, 120] deal with the course of homogeneous polymerization in the emulsion copolymerization of acrylonitrile and butyl acrylate. Analysis of the composition of the copolymer led to the conclusion that, in the initial steps of the polymerization (at conversions lower than 5%), the copolymer is enriched in acrylonitrile. The emulsion copolymerization of acrylonitrile and butyl acrylate using a system with a monomer feed composition n_{BA} = 0.625 (see the legend to Table 26) led to the formation of copolymer at 15% conversion containing ca. 45% of acrylonitrile; at higher conversions of ca. 40%, the acrylonitrile content decreased to 40 mol%. The theoretical value of the instantaneous composition of the copolymer obtained, using an integrated equation for calculating the copolymer composition [121], the literature data for the copolymerization parameters [122] and initial monomer feed composition, is 42 mol% of acrylonitrile and, after correction for the monomer dissolved in water, is only 37 mol%. Enrichment of the copolymer in acrylonitrile in the initial steps of polymerization was also observed in systems containing cationic cetyl pyridinium bromide and the nonionic emulsifier Slovasol 2430 [123].

Table 27 — Effect of inhibitor (p-benzoquinone — BQ and hydroquinone — QH$_2$) on rate of copolymerization (R_p) of acrylonitrile (AN) and butyl acrylate (BA) in Stage II (Reprinted from ref. [119] with permission of Hüthig and Wepf Verlag, Basel)

$[DW]^a$ /10^{-2} mol dm^{-3}	$R_p{}^b$/10^3 mol dm^{-3} s^{-1}						Inhibition period /s	
	a, c	b, c	a, d	b, d	a, e	b, e	a, d	b, d
7.32	35.3	26.4	2.4	1.4	2.2	4.1	600	600
2.87	17.0	23.4	2.6	2.5	1.9	3.6	570	540
1.18	13.3	18.1	2.9	3.5	1.5	3.2	540	500
0.63	8.7	10.1	3.3	3.7	1.4	2.5	480	480

[a] Sodium dodecylphenoxybenzene disulfonate.
[b] Polymerization rate in Stage II.
Recipe: 140 g water, 95.6 g monomers, 0.7721 g ammonium peroxodisulfate. Temperature 70°C.
a — AN/BA = 0.25; b — AN/BA = 1.0 (mass ratio); c — without inhibitor; d — [QH$_2$] = 1.72×10^{-3} mol dm^{-3}; e — [Q] = 2.27×10^{-3} mol dm^{-3}.

Addition of hydroquinone or *p*-benzoquinone suppressed the enrichment of the copolymer in acrylonitrile, reduced the polymerization rate in the stationary state, and prolonged the inhibition period [119, 120] (Table 27). The relative decrease in the rate (ratio of the rates with and without inhibitor) was proportional to the emulsifier concentration. For instance, using 0.63×10^{-2} mol dm^{-3} of Dowfax 2A1, the decrease in the polymerization rate was 3-fold, while a concentration of 7.32×10^{-2} mol dm^{-3} led to as much as a 15-fold decrease in the rate in the presence of the same amount of hydroquinone. Enhancement of the retardation efficiency of the inhibitor at higher emulsifier concentrations was explained by the increased transport and adsorption of the molecules of inhibitor at the water/particle interphase, at which the inhibitor may terminate the polymerization process.

The particle size and RMM of acrylonitrile/butyl acrylate copolymer were not substantially affected by the presence of the inhibitor. The particle size increased after consumption of the monomer droplets (at about 40% conversion). For instance, in a system with an emulsifier concentration of 0.63×10^{-2} mol dm^{-3}, the particle size (diameter) increased from 240 nm at 50% conversion to 270 nm at 80% conversion (other data are listed in Tables 26 and 27). Examination of the monomer distribution (acrylonitrile and butyl acrylate) between the individual phases gave the following results: monomer droplets are consumed at conversions of 40–50%, where, by contrast, the monomer concentration in the polymer particles reaches its maximum value. As the reaction proceeds, the concentration of acrylonitrile decreases slightly and at 80% conversion only 25% of its original amount is consumed [124]. The polymerization of acrylonitrile dissolved in water gives rise to oligomeric radicals, which, on precipitation from solution, form primary particles. These are stabilized in the presence of free emulsifier, otherwise they are subjected to coalescence with large particles. To calculate the average degree of polymerization of the oligomeric radicals, at which they precipitate from water, the authors [119, 120] used Eq. (15) [125, 126] and the literature data for $K(k_t^{0.5}/k_p)$, and R_i and φ ($k_{t,cross}/k_{t,1}^{0.5} k_{t,2}^{0.5}$) and molar concentrations of acrylonitrile (monomer 1) and butyl acrylate (monomer 2) present in the aqueous phase

$$P_{n,m} = \frac{r_1[AN]^2 + 2[AN][BA] + r_2[BA]^2}{R_i^{0.5}(r_1K_1^2[AN]^2 + 2\varphi K_1K_2 r_1 r_2[AN][BA] + r_2K_2^2[BA]^2)^{0.5}} \quad (15)$$

for initial monomer feed compositions [AN]/[BA] = 0.6 and 2.4, we obtain values for $P_{n,m}$ equal to 24 and 34, respectively, from Eq. (15). The reliability of these values was supported by that obtained experimentally for methyl methacrylate, $P_{n,m} = 65$ [127].

The emulsion polymerization of acrylonitrile and butyl acrylate in the presence of sodium dodecyl sulfate was investigated as regards the influence of the mass ratio, monomer/water (1:2, 1:4, and 1:9), of the emulsifier concentration (CMC or higher) and of the monomer feed composition on the composition of the copolymer ($f_{AN} = 0.7$ and 0.8) [128]. The disagreement observed between the copolymer compositions found experimentally and the theoretical values was caused by the high water solubility of acrylonitrile, especially at low conversions. The differences increased as the emulsifier concentration and the fraction of the aqueous phase in the reaction system increased. The dependence of the copolymer composition on the conversion was described by a curve with a minimum plateau at medium conversions. At the end of the polymerization, the copolymer was evidently enriched in acrylonitrile. For instance, for a system with a monomer feed composition of $f_{AN} = 0.7$ and with a monomer/water ratio of 1:9, the average composition of the copolymer at medium conversions was $F_{AN} = 0.55$. The dependences at low conversions have not been reported with the exception of those for the system with an emulsifier concentration equal to the CMC, where the polymerization took place under conditions $f = 0.8$, monomer/water = 1:4 and 50°C. The greatest differences between the experimental and theoretical values of the copolymer composition were observed at lowest conversions.

The emulsion copolymerizations of acrylonitrile with 2-ethylhexyl acrylate (with a solubility of 0.01 g/100 g water) led to even more remarkable shifts in the copolymer composition [128]. Surprisingly, in systems with an emulsifier concentration below the CMC, particles of the core-shell type were formed: the nucleus was enriched in acrylonitrile and the shell in 2-ethylhexyl acrylate (2-EHA). In contrast, at high emulsifier concentrations, the course of the time dependence of the copolymer composition pointed to the formation of particles with a 2-EHA-rich core and with a PAN shell. Results from the analyses of the formation of films and their properties indicate, however, the formation of core-shell particles with a large PAN fraction in the core.

A study of the effect of the concentration and type of emulsifier (anionic, cationic, nonionic) on the emulsion copolymerization of acrylonitrile and butyl acrylate led to a better understanding of the polymerization mechanism [129]. The rate of the copolymerization was enhanced with increasing concentrations of anionic Dowfax 2A1 (DW) and nonionic Slovasol 2430 (SLO) and, conversely, decreased with an increase in concentration of the cationic emulsifier, cetyl pyridinium bromide (CPB). The reaction order in anionic emulsifier was 0.41, in nonionic emulsifier 0.48, and for the cationic emulsifier a negative value of −0.42 was obtained. The values of orders less than 0.6 originate from a contribution from the homogeneous polymerization of acrylonitrile in water and, for the nonionic emulsifier, also from the solubility of emulsifier in monomer droplets [21]. The negative order follows from the interaction between cetyl pyridinium bromide and peroxodisulfate, which leads to the formation of an inactive complex. This complex does not decompose into initiating radicals, but, in contrast, consumes the active peroxodisulfate. Increase in the emulsifier concentration increases the concentration of complex and lowers the rate of initiation. This leads to a reduction in the polymerization rate even when the number of particles increases. The values of the orders of the dependence of N upon [E] were not only a function of the type of emulsifier but also of the concentration range used:

$$N \propto [DW]^{0.67-0.4}; \quad N \propto [CPB]^{0.9} \quad \text{and} \quad N \propto [SLO]^{2.7-1.9}. \tag{16}$$

The decrease in order at higher emulsifier concentrations is attributed to the active participation of primary polymer particles in the polymerization. At higher emulsifier concentrations, a larger fraction of these particles is formed and their role in propagation and coalescence increases [130].

The influence of a mixed emulsifier system (formed by anionic Dowfax 2A1 with nonionic Slovasol 2430) on the copolymerization of acrylonitrile with butyl acrylate has been determined [131]. As the fraction of anionic emulsifier increases, the polymerization rate is significantly enhanced in the regions of low and high fractions of emulsifier; on the other hand, the average particle size decreases (Fig. 11). Analysis of the kinetic data led to the following dependences when only the anionic emulsifier was taken into account:

$$R_p \propto [DW]^{0.36}, \quad N \propto [DW]^{0.62} \quad \text{and} \quad R_p \propto N^{0.55} \tag{17}$$

and for both emulsifiers, [E] (sum [DW] and [SLO])

Kinetics and mechanism

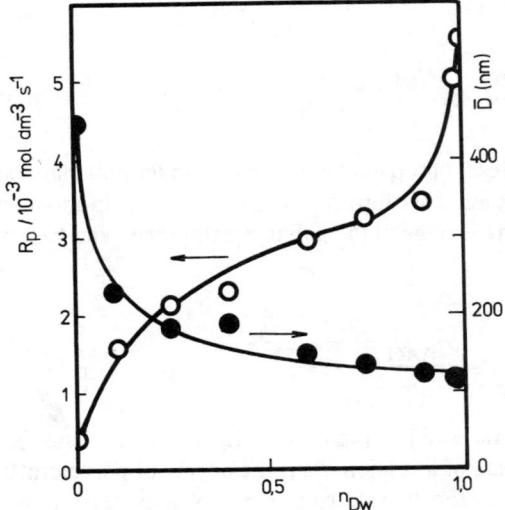

Fig. 11 — Plot of the rate of polymerization and particle size versus composition of emulsifying system, n_{DW}.
$\left(n_{DW} = \dfrac{[DW]}{[DW]+[SLO]}, \text{DW — sodium dodecylphenoxybenzene disulfonate, SLO — Slovasol 2430.} \right)$

$$R_p \propto [E]^{1.09}, \quad N \propto [E]^{1.8} \quad \text{and} \quad R_p \propto N^{0.55}. \tag{18}$$

The values for the exponents to [DW] in mixed emulsifier system are close to those obtained in the system containing only DW [119, 120]. The good agreement points to the dominant role of anionic emulsifier in directing the kinetics of the emulsion copolymerization of acrylonitrile and butyl acrylate. The higher values of the exponents to the overall concentration of emulsifier follow from the higher solubility of the nonionic emulsifier in the oil phase. If the polymerization is carried out at low emulsifier concentrations, a large fraction of the charged emulsifier is soluble in the monomer phase. In contrast, at high emulsifier concentrations, the effective emulsifier concentration is higher and a larger number of particles is generated. Therefore, this reaction order in [E] is rather empirical.

These interpretations of the kinetic data were based on simplifications relating to the neglecting of transfer of the chain to the initiator, emulsifier, and polymer. According to these interpretations, only transfer to monomer affects the RMM of the polymer [132]. These ideas have been verified in the emulsion copolymerization of acrylonitrile and butyl acrylate [133], taking account of chain transfer to monomers in terms of the scheme

$$A^{\bullet} + A \xrightarrow{k_{AA}} AH + A^{\bullet}, \tag{19}$$

$$A^{\bullet} + B \xrightarrow{k_{AB}} AH + B^{\bullet}, \tag{20}$$

$$B^\bullet + A \xrightarrow{k_{BA}} BH + A^\bullet, \tag{21}$$

$$B^\bullet + B \xrightarrow{k_{BB}} BH + B^\bullet, \tag{22}$$

where $A^\bullet$ and $B^\bullet$ represent propagating radicals with terminal monomer structural units A and B. $A^\bullet$ and $B^\bullet$ are radicals formed via chain transfer to monomer capable of further reinitiation without any change in the polymerization rate, and k_{AB}, k_{BA}, k_{AA}, and k_{BB} are transfer constants.

Using Eq. [54]

$$\frac{1}{P_n} = \frac{R_i}{n}\frac{1}{R_p} + \frac{C_{AA}r_1}{\beta} + \frac{C_{BB}r_2 L^2}{\beta} + (C_{AB} + C_{BA})\frac{L}{\beta}, \tag{23}$$

where L is the molar ratio [B]/[A], β is $r_1 + 2L + r_2L^2$, r_1 and r_2 are copolymerization parameters, R_i is the rate of initiation, P_n is the degree of polymerization, $C_{jj} = k_{jj}/k_{pjj}$, and using experimental data for the rate and degree of polymerization, we obtained transfer constants to monomer and the rate of initiation (Table 28). The transfer constants to monomer increased with increasing emulsifier concentration; this increase was ascribed to the contribution of chain transfer to the emulsifier. The transfer constants to monomer obtained in systems with low emulsifier concentrations are comparable with the literature data: for acrylonitrile $C_{AA} = 2.6 \times 10^{-5}$ and for butyl acrylate $C_{BB} = 4.0 \times 10^{-5}$ [2, 134]. In systems with low emulsifier concentrations the effect of emulsifier on the mechanism of transfer reactions is unremarkable. The value for R_i was lower than that calculated from the relationship $R_i = 2fk_d$ [I]; this result was attributed to the retarding effect of emulsifier [135].

Table 28 — Transfer constants to monomer in emulsion copolymerization of acrylonitrile (AN) and butyl acrylate (BA) at 70°C (Reprinted from ref. [133] with permission of Academia, Prague)

System	$k_{AA}{}^a$ /dm^3 mol^{-1} s^{-1}	$k_{BB}{}^a$	C_{AA} /10^{-5b}	C_{BB} /10^{-5b}	$(C_{AB}+C_{BA})^c$ /10^{-5}	$R_i{}^d$ /10^{-6} mol dm^{-3} s^{-1}
1	0.28	0.21	5.7	10.5	29	1.4
2	0.51	0.25	10.2	12.5	31	0.13
3	0.166	0.065	3.33	3.33	8	2.0

^{a} Transfer rate constant to monomer (A or B).
^{b} Relative transfer constant to monomer ($C_{AA} = k_{AA} / k_{pAA}$, $C_{BB} = k_{BB} / k_{pBB}$).
^{c} Sum of cross-transfer constants to monomer ($C_{AB} = k_{AB} / k_{pAB}$, $C_{BA} = k_{BA} / k_{pBA}$).
^{d} Rate of initiation.
Recipe: 96.6 g monomer, 140 g water.
Systems: 1 — [DW] = 7.32, [I] = 2.28,
2 — [DW] = 7.32, [I] = 1.4,
3 — [DW] = 0.63, [I] = 2.28.
Concentrations in mol dm^{-3}. DW — sodium dodecylphenoxybenzene disulfonate.

Desorption processes of radicals from the particles [136], which have also been followed in emulsion copolymerizations of acrylonitrile with butyl acrylate, are also associated with chain transfer to monomer [137]. Kinetic data for emulsion copolymerizations are discussed from the points of view of the three possible processes: chain transfer to monomer, diffusion of the monomer radical from the surface of the particle and, finally, its diffusion into the aqueous phase. Of these contributors, the dominant effect on the rate of desorption is that of chain transfer to monomer. Corrections for the desorption of radicals led to a substantial decrease in the average number of radicals per particle. In all systems at both high and low concentrations of emulsifier and at various compositions of the comonomer mixture, we obtained values for the average number of radicals below or close to 0.5. In agreement with the results of ref. [138], desorption is inversely proportional to the particle size and directly proportional to the hydrophilic nature of the monomer.

The decrease in the radical concentration in a particle was exceptionally marked in the presence of hydroquinone and p-benzoquinone [137, 138]. The decrease is caused by the formation of additive-based radicals able to desorb efficiently into water, and from the subsequent lowering in the concentration of radicals, in water and in the polymer particles [123]. The desorption of radicals was activated by the emulsifier to an extent CPB < DW < SLO. This trend results from the efficiency of chain transfer to the emulsifier and the water solubility of the emulsifier and its radical fragments.

The pair butyl acrylate/vinyl acetate is representative of monomers with different water solubilities. These monomers also show different reactivity in copolymerization (r_{BA} = 3.1 and r_{VAc} = 0.05) [139]. The kinetics of their copolymerization were investigated by Misra *et al.* [140]. They found that the latex prepared in the batch mode contained larger particles (d_n = 90 nm) with a narrower size distribution (d_w/d_n = 1.08) than the latex obtained through semicontinuous polymerization (d_n = 52 nm, d_w/d_n = 1.79). Electron micrographs of the polymer particles and dynamic mechanical spectra of the copolymers led to the conclusion that the polymer particles formed in the batch process have a core-shell morphology with a core rich in lipophilic and reactive butyl acrylate (with an average diameter of about 55 nm) and a hydrophilic shell of poly(vinyl acetate). Polymer particles prepared in the semicontinuous mode were more homogeneous, which is also evident from the smaller diameter of the poly(butyl acrylate) core (d = 24 nm). The formation of particles with a core rich in butyl acrylate is due not only to the high reactivity of butyl acrylate but also to the high content of butyl acrylate in the micelles or polymer particles. As reaction proceeds, the butyl acrylate in the particle is consumed and inevitably the fraction of vinyl acetate increases. At higher conversions, only a small amount of butyl acrylate is present, which is why it is the hydrophilic monomer that preferentially enters propagation and which then forms the shell of the polymer particle.

The polymerizations of conventional emulsions and miniemulsions of n-butyl acrylate and vinyl acetate using sodium hexadecyl sulfate as emulsifier and hexadecane as coemulsifier have shown the existence of marked differences in the polymerization kinetics of these two types of emulsion [141–145]. Analysis of the role of the coemulsifier in the miniemulsion copolymerization of 50 : 50 molar ratio butyl acrylate/vinyl acetate monomer mixture has shown that hexadecane has four different functions:

(i) it enables the formation and stabilization of submicron monomer droplets;
(ii) upon initiation of the polymerization it helps to retain the monomer originally contained in the initiated droplets;
(iii) later on, its presence in uninitiated monomer droplets reduces the equilibrium concentration of the monomers in the growing polymer particles and

(iv) its presence in the polymer particles increases dramatically its swelling capability in post-polymerization swelling experiments.

For initiator concentrations up to 10^{-2} mol dm^{-3} the rate of polymerization in the conventional emulsion copolymerization is faster than in the miniemulsion copolymerization. The particles formed in the miniemulsion copolymerization of butyl acrylate and vinyl acetate are larger in size and broader in distribution than particles formed in the conventional emulsion copolymerization under the same conditions. It has been postulated that all these kinetic differences are due to the different particle nucleation mechanisms operating. In the conventional emulsion copolymerization of butyl acrylate and vinyl acetate, where the monomer droplets are large, homogeneous nucleation in the aqueous phase is the dominant particle nucleation mechanism. Delgado *et al.* [141–145] have reported that in the miniemulsion copolymerization, polymer particles are formed by oligoradical entry/capture by the monomer droplets. Particle nucleation was found to be a slow process in the miniemulsion copolymerization where only a fraction of the monomer droplets become polymer particles. It was also shown that the coemulsifier is a key factor in the monomer transport process as a result of its profound effect on the thermodynamics of the system. The authors have developed a model, based on mass balance and equilibrium thermodynamics, which describes accurately the comonomer distribution during the copolymerization and the interdroplet mass transport process.

In the photoredox-induced copolymerization of methyl methacrylate/acrolein microemulsion, particles of diameter from 500 to 7000 nm, maximizing at *ca.* 1500 nm were prepared [72]. The methyl methacrylate/acrolein system showed no preference as to whether initiation was by monochromatic or polychromatic visible light. The form of the conversion curve is typical for a radical-initiated polymerization process. No inhibition period was observed in this system. Peroxodisulfate affords rapid photoelectron transfer when an appropriate sensitizer is excited by visible light. This photoinduced charge transfer is followed by facile initiation unencumbered by a necessary diffusion through the aqueous continuous phase. This fast initiation is followed by rapid propagation since the monomer units are readily available in the interior of the microemulsion droplets (see section 3.4.1).

The micropolymerization of acrylates and methacrylates has also been discussed elsewhere [146–150].

3.4.3 Carboxylated monomers

Carboxylated latices provide an important class of polymer materials with considerable industrial interest in many applications. Incorporation of the carboxylic monomer is mainly directed to the improvement of the ionic and mechanical stability of the latices as well as promoting interactions of the particles with various types of substrate.

Carboxylated latices are mostly prepared by the emulsion copolymerization of lipophilic monomer with acrylic and methacrylic acids and sometimes with other carboxylic monomers. The course of polymerization in strongly acid media makes many demands on the control of the process and, especially, on the choice of emulsifying system which is supposed to secure the formation of a stable emulsion. The most suitable stabilizers are alkyl sulfates and alkyl benzene sulfates.

The reproducibility of the process of preparation of carboxylated latices was studied by Blackley *et al.* [151, 152]. The authors used for this purpose the copolymerization of styrene with acrylic acid in the presence of sodium dodecyl sulfate or dodecylbenzene sulfate acid and its salt. The copolymerizations were carried out at high initiator concentrations (~ 0.5 mass % per 100 g of monomer), by using 1 mass portion of acrylic acid at 45°C. Better reproducibility of the kinetic data during polymerization was observed in the systems with sodium dodecylbenzene sulfonate. The partial neutralization of acrylic acid led in all systems (with different emulsifying systems) to an increase in the reproducibility of the polymerization process. A similar result has also been achieved on adding KCl. The authors described the fluctuations in reproducibility of the kinetics to the hydrolysis of the emulsifiers and to the active participation of the hydrolysis products in the polymerization. Sodium dodecyl sulfate, which is more sensitive to hydrolysis than other emulsifiers used, was hydrolysed to dodecyl alcohol and sodium sulfate and therefore the reproducibility of the polymerization was lower. Both hydrolysis products adversely influenced the concentration of sodium dodecyl sulfate at the interphase, the character of the electrical double layer, and the stability of the polymer particles. The neutralization of acrylic acid and addition of KCl and isopropanol lowered the extent of emulsifier hydrolysis; this raised the stability of the dispersion and also the reproducibility of the polymerization.

The rate of hydrolysis increases with increasing concentration of emulsifier micelles [153, 154] and therefore increase in the stability of the dispersion cannot be achieved by raising the emulsifier content in the system but merely by a suitable choice of emulsifying system.

The use of sodium dodecyl benzene sulfonate is suitable since it leads to the formation of stable latices even at high polymerization rates. The long nucleation period (characteristic of systems with hydrophilic monomers) somewhat complicates the investigation and modelling of the polymerization. Its shortening is achieved by adding isopropanol which raises the effectiveness of the formation of the primary particles and of the entry of the negatively-charged oligomeric radicals into the micelles and polymer particles [155]. The addition of acrylic acid reduced the rate of styrene polymerization three times on average.

The effect of the emulsifier (cationic — dodecylammonium chloride, anionic — sodium dodecyl sulfate and nonionic — based on polyoxyethylene) on the course of the emulsion copolymerization of styrene and methacrylic acid was investigated by Antonova *et al.* [20]. The highest initial rates were observed in systems with an anionic emulsifier, with rates about 3–4 times lower in systems with cationic and nonionic emulsifiers. While the shape of the curves describing emulsion copolymerization in the presence of anionic and cationic emulsifiers corresponded to that classical polymerization, in the presence of nonionic emulsifier, the conversion stopped at a certain limiting value which varied only slightly with time. Analysis of the copolymers has shown that a block copolymer is produced in the presence of an anionic emulsifier, a statistical polymer is formed with a cationic emulsifier, and a mixture of homopolymers is produced with a nonionic emulsifier. The formation of the block copolymer was explained by the entry of oligomeric radicals of the hydrophilic monomer into micelles swollen with styrene. The statistical copolymer is formed at the interphase as a result of the concentration of peroxodisulfate at the micellar surface (owing to the strong interaction between emulsifier and peroxodisulfate) [156]. The polymerization of monomers carried out in the system with nonionic emulsifier also took place separately in the particles and thus high fractions of homopolymer were formed.

The mechanism for styrene emulsion copolymerization with methacrylic acid (MAA), acrylic acid (AA), and itaconic acid (IA), monomers with different hydrophilicity and with a concentration of about 4 mass % was studied by Egusa & Makuuchi [157]. Table 29 shows that the accelerating effect of the acids on the polymerization rate decreases with increasing hydrophilicity of the carboxylic monomer. In the system with itaconic acid, a reduction in the polymerization rate was observed which follows from the great hydrophilicity of the oligomeric radicals which are not adsorbed effectively by the micelles and polymer particles. On the other hand, the addition of carboxylic monomer positively affects the formation of a larger number of polymer particles via a homonucleation mechanism. With increasing lipophilicity of the carboxylic monomer (its measure being the distribution coefficient K between styrene and the aqueous phase), the monomer concentration in the polymer particles increases in parallel with the rate of polymerization. Increase in concentration of the carboxylic monomer (AA or MAA) was accompanied by a rate enhancement of the copolymerization. On increasing the concentration of acrylic acid from 2 to 8 mass %, the rate increased from 32.5 to 50.3% conv. h^{-1}, the particle number remained, however, constant (3.3×10^{15} g^{-1} of polymer). It is therefore concluded that the increase in polymerization rate reflects rate enhancement of the decomposition of peroxodisulfate with a decrease in pH. Analysis of the copolymer has shown that acrylic acid is preferentially located in the surface particle layer (~ 60 mass %), methacrylic acid in the particle core (~ 60 mass %), and itaconic acid in the water-soluble polymer. According to ref. [157] the copolymerization reactions of styrene and carboxylic monomer take place preferentially on the particle surface. The entry of methacrylic acid into the core was explained by the adsorption of styrene molecules on the particle surface, where they polymerize and enfold the carboxyl groups.

Table 29 — nfluence of nature of carboxylic acid on kinetic parameters of emulsion copolymerization (Reprinted from ref. [157] with permission of John Wiley & Sons Inc., New York)

Carboxyl monomer	K_m / v^a	R_p^b /% conv. h^{-1}	$\bar{d}^c$ / nm	$N / 10^{15d}$
—	1	23.5	89.3	2.54
AA	0.175	38.8	82.4	3.23
MAA	1.94	46.8	82.0	3.28
IA	0.012	18.4	79.3	3.62

^{a} Distribution coefficient for carboxyl monomer between styrene and water [158, 159].
^{b} Polymerization rate in Stage II.
^{c} Polymer particle diameters.
^{d} Polymer particle number per g polymer.
Recipe: Styrene (76.8 g), carboxyl monomer (44.4 mmol), SDS (1.0 g), water (319 g). Radiation 1.7×10^4 rad h^{-1}.
AA — acrylic acid, MAA — methacrylic acid, IA — itaconic acid.

The results of ref. [158, 159] which describes the localization of acrylic and methacrylic acids in polystyrene particles, support the above conclusions relating to the structural composition of the polymer particles.

The distribution of the carboxylic monomer between the individual zones of the disperse system also depends on the size of the particles formed. In the systems containing acrylic and itaconic acids, the growth of the polymer particles was accompanied by an increase of the fraction of the water-soluble fraction of hydrophilic monomer to the detriment of the surface-bound fraction, the amount of monomer built into the interior of the particles remaining constant. The use of methacrylic acid also led to a reduction in the fraction of surface-bound carboxyl groups, but, simultaneously, their presence in the particle core increased. Methacrylic acid, a monomer with a high solubility in styrene, takes part preferentially in propagations in the core, less so on the particle surface and rarely in the aqueous phase [160–162].

According to ref. [163], increase in the lipophilicity of the carboxylic monomer positively affects the diffusion of a hydrophilic monomer into the growing polymer particles, and their presence in the copolymer as well as the rate of polymerization. Therefore, when a carboxylic monomer is used, the polymerization rate increases in the order: itaconic acid < acrylic acid < methacrylic acid.

Sakota & Okaya [164, 165] dealt with the influence of the degree of neutralization of acrylic acid and its fraction in the monomer feed on the emulsion copolymerization with styrene. They found that as the acrylic acid content in the monomer feed increases, the rate of polymerization increases on account of the growth of the polymer particle number. With increasing concentration of acrylic acid, the amount of styrene dissolved in water and its role in addition reactions, increase. The larger amount of styrene in water leads to the formation of oligomeric radicals of greater lipophilicity which increases their rate of entry into the polymer particle. The neutralization of acrylic acid leads to a decrease in rate of the copolymerization caused by the reduction of the particle number.

Coverage of the surface of the latex particles by carboxyl groups influences positively the stability of the latex (mechanical and thermal, in particular), alters the colloidal dispersions, and the nature of the polymer formed. The properties of polystyrene polymer latices are determined not only by the concentration of carboxyl groups but also by their distribution on the surface and in the particle core [166].

The classical emulsion polymerization of styrene and of carboxylic monomer in the presence of, for example, sulfate emulsifier, initiated by peroxodisulfate leads to the formation of polymer particles with both surface carboxyl and sulfate groups. The sulfate groups originate from the peroxodisulfate and from emulsifier which takes part in transfer reactions [8].

Polystyrene latices containing a large fraction of carboxyl groups can be prepared by the emulsion copolymerization of styrene and carboxylic monomer initiated by UV or γ-radiation [167]. If γ-radiation is applied, initiating hydroxyl radicals are formed; the particle surface is therefore also covered with a smaller fraction of hydroxyl groups.

The surface charge density of the polymer particles as a function of the concentration of the carboxyl groups bound on the particle surface was studied during the emulsion copolymerization of styrene and acrylic or methacrylic acids [168]. The results listed in Table 30 show that as the concentration of carboxylic monomer increases, the concentration of the surface-bound carboxyl groups also increases, especially in the system with acrylic acid. The value of the surface particle charge grows in proportion. As regards the concentrations of carboxylic monomer and peroxodisulfate used in the feed, the particle surface was covered mostly by carboxyl groups: the sulfate groups occupied only a small fraction.

Table 30 — Surface charge density of polymer particles σ

Latex	Surface charge density /μC cm^{-2}		σ
	σ_s	σ_w	$\sigma_s + \sigma_w$
	($-OSO_3^-$)	($-COO^-$)	
PS / PAA$_2$	−5.11	−33.3	−38.5
PS / PAA$_5$	−5.49	−87.0	−92.5
PS / PMAA$_5$	−6.08	−35.2	−41.3
PS	−4.52	−1.4	−5.9

Subscripts s and w denote dissociated sulfate and carboxyl groups and 2 and 5 the molar concentration of carboxyl monomer in copolymer. PS — poly(styrene), PAA — poly(acrylic acid), PMAA — poly(methacrylic acid).

The kinetics of emulsion copolymerization of butyl acrylate and methacrylic acid initiated with ammonium peroxodisulfate in the presence of sodium dodecyl sulfate functioning as an emulsifier and ammonium sulfate as a salting-out agent have been investigated by Kulikov *et al.* [169]. The polymerization rate was found to be approximately proportional to the 0.25 and 0.4 powers of the initiator and emulsifier concentrations, respectively. In the presence of ammonium peroxodisulfate (AS), the reaction order in emulsifier concentration was reduced to 0.23; this confirms the initiation of the polymerization in the aqueous phase. On adding the initiator, the concentration of methacrylic acid in water, as well as the polymerization rate and formation of polymer particles, decrease. In the absence of emulsifier, the generation of particles proceeds in water by homogeneous nucleation. If the emulsifier is present, the contribution of homogeneous nucleation decreases and, conversely, the nucleation of monomer-swollen emulsifier micelles and the contribution of the polymerization in the polymer particles increase.

Emelie *et al.* [170] studied the kinetics of the emulsion polymerization of butyl acrylate and methyl methacrylate (in equimolar ratio) in the presence of acrylic and methacrylic acids (less than 1 mass % each). Addition of the carboxylic monomer caused a decrease in the average particle size and an increase in the particle number in the system. In spite of these changes, no substantial fluctuations occurred in the rates of polymerization. Acrylic acid took part mainly in polymerization processes in water (30%) and on the particle surface (37%) and the residue entered the polymer particle core. The more lipophilic methacrylic acid mostly entered reactions in the core (72%), 23% reacted on the particle surface, and the rest (5%) formed a water-soluble polymer. This distribution of the carboxylic monomers between the individual phases of the disperse system follows from the distribution of the monomer between the oil and aqueous phases. The distribution coefficient between the oil and aqueous phases is 2.1 for methacrylic acid and only 1.3 for acrylic acid.

Emelie *et al.* [171] have prepared a series of further methyl methacrylate/butyl acrylate copolymer latices by emulsion polymerization by varying the nature of the carboxylic acid (acrylic or methacrylic acid), the type of emulsifier (sodium dodecyl sulfate (SDS) or nonyl phenyl polyoxyethylene ether (HV 25) emulsifiers), or the method of polymerization. The characterization of the latices led to the following results:

(i) The more hydrophilic acrylic acid is also here predominantly located in the aqueous phase (~ 40 mass %) or at the particle surface (~ 20–35 mass %) when SDS is used. The presence of a nonionic emulsifier (HV 25) causes an increase in the buried part.

(ii) The more lipophilic methacrylic acid is predominantly buried in the particles (~ 75 mass %). The addition of methacrylic acid drastically affects its distribution in the final latex with 65 mass % of the carboxyl groups detected at the particle surface. The fraction of methacrylic acid buried is lower with an anionic emulsifier.

Two parameters were found to perturb the distribution of the carboxylic monomer: (i) the type of initiator and (ii) the reactivity of the carboxylic monomers. SDS does not interact with the two acids but the reverse is true for the nonionic emulsifier (HV 25). Acrylic acid (co)polymerizes slowly compared to methacrylic acid. In the case of the nonionic emulsifier-stabilized latices, the stability against electrolytes, as expected, is better than for the SDS-stabilized latices, whatever the polymerization process may be. Studies of stability against different types of electrolyte and temperature showed that the surface morphology is strongly dependent upon the nature of the oligomers formed during polymerization, especially the presence of sequences rich in poly(carboxylic acid)s, and how they are anchored onto the particle. According to the type of emulsifier and polymerization process, the stability is predominantly determined either by electrostatic or steric forces.

The dependence of the copolymerization of ethyl acrylate and butyl acrylate in the presence of methacrylic acid on the nature of the emulsifying system was investigated by Eliseeva et al. [172, 173]. The emulsion copolymerizations were carried out at the monomer feed composition EA/BA/MMA = 8 : 2 : 0.3 (mass ratio) and an anionic emulsifier was used. The shapes of the conversion curves differed from that observed in classical emulsion polymerization. The maximum rate was reached in the initial stage (to 20% conversion) and, after exceeding this value, it decreased continuously as the reaction proceeded. The rates and RMMs obtained in the systems with anionic emulsifier (R_p ~ 1.8% conv. min^{-1}; M_n ~ 3×10^6) were higher than when a mixture of anionic and nonionic emulsifier was used (R_p ~ 0.8% conv. min^{-1} and M_n ~ 1.3×10^6). An increase in conversion was accompanied by growth in the size of the polymer particles and their polydispersity. At conversions lower than 5%, the average particle size given by the diameter d had a value of 100 nm with the main fraction of particle sizes lying between 50 and 150 nm. At ~ 50% conversions, the average particle size reached 250 nm, the particle size varying between 50 and 350 nm. The increase in size of the particles and their dispersity is a result of (i) the long nucleation period, (ii) the formation of primary particles over a wide range of conversions through the homonucleation mechanism, and (iii) particle growth by coagulation.

Eliseeva et al. [174] studied the effect of methacrylic acid (5 mass %) on the emulsion copolymerization of butyl acrylate and methyl methacrylate (equimolar ratio) using a nonionic emulsifier (Sulfanol). The addition of methacrylic acid led to no change in the polymerization rate although the number of particles varied; in the absence of acid, the particle number was about 3×10^{14} cm^{-3} and, on adding methacrylic acid, the number became N ~ 6×10^{14} cm^{-3}. Methacrylic acid did not greatly affect the copolymer composition but it markedly changed its RMM. In the system without carboxylic monomer, the RMM increased linearly and reached a maximum at about 80% conversions (M_n ~ 1×10^6); at 10% conversion, its value was ca. 4×10^5. In the presence of methacrylic acid, the RMM was approximately constant throughout the range of conversions, with an average ca. M_n ~ 8×10^5. The authors reported that polymer dispersions containing copolymers with lower RMMs are much more stable: this was ascribed to the fraction of –COOH groups in the polymer molecule as well as on the particle surface.

Zosel *et al.* [175] investigated the distribution of carboxylic monomer in polymer particles of the latex prepared via classical emulsion polymerization of butyl acrylate or ethyl acrylate at low and high concentrations (up to 40 mass %) of acrylic and methacrylic acids in the presence of an anionic emulsifier. The authors used electron microscopy, analytical ultracentrifugation, and mechanical spectroscopy for determining the distribution of the carboxylic monomer between the individual phases of the system. Combination of the last two methods enabled a qualitative determination of the fractions of the groups bound on the surface and located in the particle core (see Fig. 12). It was found that 50% of the acrylic acid enters the particle core, 40% the aqueous phase, and the residual 10% is located on the particle surface. Methacrylic acid in the polymer particles was distributed between the core (90%) and the particle surface (10%).

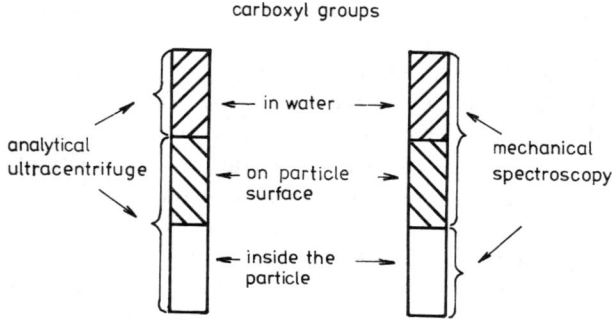

Fig. 12 — Mechanism for determining carboxyl monomer distribution in polymer dispersions by a combination of the analytical centrifuge and mechanical spectroscopy.

The copolymerization of the more hydrophilic ethyl acrylate with acrylic acid led to the formation of polymer particles with a much higher concentration of carboxyl groups in the particle core. Only 5% of the overall amount of the carboxylic monomer took part in the formation of a water-soluble polymer.

Independently of the concentration of carboxylic monomer, about 10% of the total initial concentration has always been localized on the surface of the polymer particle. This picture is supported by electron microscopy. The distribution of the carboxyl groups in a polymer particle enabled formation of a polymer particle with core/shell morphology. The particle shell made of a very thin layer rich in carboxyl monomer showed high hydrophilicity.

Some information about the structure of carboxylated polymer particles can be obtained from alkalization of the polymer latices [176, 177]. At a low content of carboxyl groups, only an increase in the effective hydrodynamic volume of polymer particles takes place, caused by the extension of their electric double layer (primary electroviscosity effect), which results in increased mutual interactions during flow (secondary electroviscosity effect). At larger contents of carboxyl groups, the volume of the dispersion particles may really increase due to swelling, or the particles may become dissolved to form a polyelectrolyte.

Quadrat *et al.* [178] have investigated the structure of ethyl acrylate/methacrylic acid latex particles. It was found that in a dispersion with an acid content up to 20 mass %, the particles only swell in aqueous ammonium hydroxide solution; at acid contents between 20 and 40 mass %, they decompose into smaller units represented by supermolecular aggregates of macromolecules, and only when appoximately starting from 40 mass % of acid does the copolymer particle become dissolved molecularly. Dispersions with a higher methacrylic acid content in the copolymer may partly dissolve during swelling; their structure is less strong during overlapping of disperse particles, or no overlapping takes place at all, because the swollen copolymer has been pre-dissolved. On the contrary, in the case of dispersions with a too low acid content (buried units) the final swelling after alkalization is probably weak. In dilute dispersions the interaction effect mentioned above does not appear, and the viscosity only increases during alkalization to a limiting extent proportional to the hydrodynamic volumes of the swollen disperse particles or mixtures of the macromolecules and their aggregates.

Microemulsion copolymerizations of acrylic acid and styrene using cetyltrimethylammonium bromide (CTAB) as emulsifier have been investigated kinetically by Puig et al. [179]. In the microemulsion polymerization of styrene, the principal site of initiation is likely to be in the monomer droplets instead of in the aqueous phase. In this system the water-soluble acrylic anion is mainly located at or close to the positively charged micelle surface. These authors [179] hypothesize that radicals derived from potassium peroxodisulfate react with acrylic acid molecules at or close to the micelle surface. The acrylic acid radicals in turn indicate reaction in the micelles where styrene begins to react. The initiated micelles can grow by recruiting styrene, acrylic acid, and emulsifier from noninitiated micelles. High conversions and reaction rates are achieved. The resulting latex is a monodisperse dispersion of spherical particles of 21 nm in radius. The copolymer backbone consists of isolated acid units randomly distributed among poly(styrene) blocks.

References

1. Chatterjee, S.P., Bandopadhyay, M. & Konar, R.S. (1976) *Indian J. Chem., A* **14** 836
2. Chatterjee, S.P., Banerjee, M. & Konar, R.S. (1978) *J. Polym. Sci., Polym. Chem. Ed.* **16** 1517
3. Alexander, A.E. & Gilbert, D.H. (1970) *Prog. Polym. Sci.* **3** 145
4. Piirma, I., Kamath, V.R. & Morton, M. (1975) *J. Polym. Sci., Polym. Chem. Ed.* **13** 2087
5. Al-Shabib, W.A.G.R. & Dunn, A.S. (1978) *J. Polym. Sci., Polym. Chem. Ed.* **16** 677
6. Van der Hoff, B.M.E. (1952) *Adv. Chem. Ser.* **34** 6
7. Hasan, S.M. (1979) *M. Phil. Thesis*, CNAA, London
8. Blackley, D.C. (1975) *Emulsion polymerization*. Applied Science Publ., London, p. 211
9. Kauffmann, H.F., Olaj, O.F. & Breitenbach, J.W. (1976) *Makromol. Chem.* **177** 939
10. Buchholz, K. & Kirchner, K. (1976) *Makromol. Chem.* **177** 935
11. Hasan, S.M. (1982) *J. Polym. Sci., Polym. Chem. Ed.* **20** 2969
12. Asahara, T., Seno, M., Shiraishi, S. & Arita, Y. (1973) *Bull. Chem. Soc. Jpn.* **46** 249
13. Lichti, G., Sangster, D.F., Whang, B.C.Y., Napper, D.H. & Gilbert, R.G. (1982) *J. Chem. Soc., Faraday Trans. 1* **78** 2129
14. Kast, H. & Funke, W. (1981) *Makromol. Chem.* **182** 1553
15. Kast, H. & Funke, W. (1981) *Makromol. Chem.* **182** 1567
16. Sakota, K. & Okaya, T. (1976) *J. Appl. Polym. Sci.* **20** 1725
17. Liu, L.H. & Krieger, I.M. (1978) In: Becker, P. & Yudenfreud, M.W. (eds) *Emulsions, latices, dispersions*. M. Dekker, New York, p. 41
18. Ohtsuka, Y., Kawaguchi, H. & Suzuki, S. (1976) *Kobunshi Ronbushu* **33** 224
19. Ohtsuka, Y., Kawaguchi, H. & Watanabe, S. (1980) *Polymer* **21** 1073
20. Antonova, L.F., Lepyanin, G.V., Zaitsev, E.E. & Rafikov, S.S. (1978) *Vysokomol. Soedin., A* **20** 687
21. Piirma, I. & Chang, M. (1982) *J. Polym. Sci., Polym. Chem. Ed.* **20** 489
22. Van der Hoff, B.M.E. (1958) *J. Polym. Sci.* **33** 487
23. Westby, M.J. (1988) *Colloid Polym. Sci.* **266** 46
24. Westby, M. (1985) Discussion contribution at the Plastics and Rubber Institute's *Polymer Latex II Conference*, London
25. Waite, F.A., private communication
26. Satpathy, U.S. & Dunn, A.S. (1988) *Br. Polym. J.* **20** 521
27. Greth, G.G. & Wilson, J.E. (1961) *J. Appl. Polym. Sci.* **5** 135
28. Kamura, S.O. & Motoyama, T. (1962) *J. Polym. Sci.* **58** 221

29. Vanderhoff, J.W., Dimonie, V.L., El-Aasser, M.S. & Settlemeyer, L.A. (1990) *J. Appl. Polym. Sci.* **41** 1549
30. Chen, K.M. & Liu, H.J. (1987) *J. Appl. Polym. Sci.* **34** 1879
31. McCartney, T.L. & Piirma, I. (1990) *Polym. Bull.* **23** 367
32. Guillaume, J.L. (1985) *Polym. Mater., Sci. Eng.* **52** 309
33. Kato, K., Kondo, H., Morita, A., Esumi, K. & Meguro, K. (1986) *Colloid Polym. Sci.* **264** 737
34. Kato, K., Kondo, H., Takeda, M., Esumi, K. & Meguro, K. (1987) *Colloid Polym. Sci.* **265** 950
35. Kato, K., Kondo, H., Yokota, K., Esumi, K. & Meguro, K. (1987) *J. Appl. Polym. Sci.* **33** 2657
36. Kato, K., Kondo, H., Esumi, K. & Meguro, K. (1987) *J. Macromol. Sci., Chem.* **24** 1391
37. Vanderhoff, J.W. (1983) In: Poehlein, G.W., Ottewill, R.H. & Goodwin, J.W. (eds) *Science and technology of polymer colloids*, vol. 1. M. Nijhoff Publ., Dordrecht, p. 1
38. Hawkett, B.S., Napper, D.H. & Gilbert, R.G. (1980) *J. Chem. Soc., Faraday Trans. 1* **76** 1323
39. Whang, B.C.Y., Lichti, G., Gilbert, R.G., Sangster, D.F. & Napper, D.H. (1980) *J. Polym. Sci., Polym. Lett. Ed.* **18** 711
40. Lichti, G., Sangster, D.F., Whang, B.C.Y., Napper, D.H. & Gilbert, R.G. (1984) *J. Chem. Soc., Faraday Trans. 1* **80** 2911
41. Pavlyuchenko, V.N., Kolosova, Z.O., Lovyagina, L.D., Kerzhovskaya, V.V., Gromov, E.V., Vylegzhanina, K.A., Egorova, E.I. & Ivanchev, S.S. (1983) *Acta Polym.* **34** 399
42. Eliseeva, V.I., Ivanchev, S.S., Kukhanov, S.I. & Lebedev, A.V. (1976) In: *Emulsion polymerization and its application in industry*. Khimiya, Moscow, p. 47 (in Russian)
43. Whang, B.C.Y., Napper, D.H., Ballard, M.J., Gilbert, R.G. & Lichti, G. (1982) *J. Chem. Soc., Faraday Trans. 1* **78** 1117
44. Penboss, I.A., Napper, D.H. & Gilbert, R.G. (1983) *J. Chem. Soc., Faraday Trans. 1* **79** 1257
45. Hasan, S.M. (1982) *J. Polym. Sci., Polym. Chem. Ed.* **20** 3031
46. Kamath, V.R. (1973) *PhD Thesis*, University of Akron
47. Bataille, P., Van, B.T. & Pham, Q.B. (1982) *J. Polym. Sci., Polym. Chem. Ed.* **20** 795
48. Guillot, J. (1981) *Acta Polym.* **32** 593
49. Guillot, J. & Guerrero, L.R. (1982) *Makromol. Chem.* **183** 1979
50. Hagiopol, C., Georgescu, M. & Deaconescu, I. (1985) *Makromol. Chem., Suppl.* **10/11** 159
51. Hagiopol, C., Deleanu,T. & Memetea, T. (1989) *J. Appl. Polym. Sci.* **37** 947
52. Ramirez-Marguez, W. & Guillot, J. (1988) *Makromol. Chem.* **189** 379
53. Nomura, M., Satpathy, U.S., Kouno, Y. & Fujita, K. (1988) *J. Polym. Sci., Polym. Lett. Ed.* **26** 385
54. Goldwasser, J.M., & Rudin, A. (1982) *J. Polym. Sci., Polym. Chem. Ed.* **20** 1993
55. Nomura, M., Yamamoto, K., Horie, I. & Fujita, K. (1982) *J. Appl. Polym. Sci.* **27** 2483
56. Nomura, M., Kubo, M. & Fujita, K. (1983) *J. Appl. Polym. Sci.* **28** 2767
57. Bataille, P. & Charifi-Sandjani, N. (1978) *J. Polym. Sci., Polym. Chem. Ed.* **16** 2527
58. Fischer, J.P. (1972) *Makromol. Chem.* **155** 211
59. Capek, I. & Bartoň, J. (1986) *Chem. Papers* **40** 443
60. Rudin, A., Samanta, M.C. & Hoff van der, B.M.E. (1979) *J. Polym. Sci., Polym. Chem. Ed.* **17** 493
61. Plavlyanich, B. & Yanovich, Z. (1981) *J. Polym. Sci., Polym. Chem. Ed.* **19** 1795
62. Bartoň, J. (1992) Paper presented at IUPAC Polymer '91 International Symposium, Melbourne. *Makromol. Chem., Macromol. Symp.* **53** 289
63. Blackley, D.C. & Haynes, A.C. (1977) *Br. Polym. J.* **9** 312
64. Obrecht, W., Seitz, U. & Funke, W. (1976) *Makromol. Chem.* **177** 1877
65. Obrecht, W., Seitz, U. & Funke, W. (1976) *Makromol. Chem.* **177** 2235
66. Sheu, H.R., El-Aasser, M.S. & Vanderhoff, J.W. (1990) *J. Polym. Sci., Polym. Chem. Ed.* **28** 629
67. Ugelstad, J., Hansen, F.K. & Lange, S. (1974) *Makromol. Chem.* **175** 507
68. Ugelstad, J. (1978) *Makromol. Chem.* **179** 815
69. Chou, Y.J., El-Aasser, M.S. & Vanderhoff, J.W. (1980) *J. Dispersion Sci., Technol.* **1** 129
70. Choi, Y.T., El-Aasser, M.S., Sudol, E.D. & Vanderhoff, J.W.(1985) *J. Polym. Sci., Polym. Chem. Ed.* **23** 2973
71. Holdcroft, S. & Guillet, J.E. (1990) *J. Polym. Sci., Polym. Chem. Ed.* **28** 1823
72. Grätzel, C.K., Jirousek, M. & Grätzel, M. (1986) *Langmuir* **2** 292
73. Perez-Luna, V.H., Puig, J.E., Castano, V.M., Rodriguez, B.E., Murthy, A.K. & Kaler, E.W. (1990) *Langmuir* **6** 1040
74. Lianos, P. (1982) *J. Phys. Chem.* **86** 1935
75. Johnson, P.L. & Gulari, E. (1984) *J. Polym. Sci., Polym. Chem. Ed.* **22** 3967
76. Gan, L.M, Chew, C.H. & Friberg, S.E. (1983) *J. Macromol. Sci., Chem.* **19** 739
77. Murtagh, J., Ferrick, M.R. & Thomas, J.K. (1987) *ACS Polym. Prepr.* **28** 441
78. Kuo, R.L., Turro, N.J., Tsang, C.M., El-Aasser, M.S. & Vanderhoff, J.W. (1987) *Macromolecules* **20** 1216
79. Vanderhoff, J.W. (1985) *J. Polym. Sci., Polym. Symp.* **72** 161
80. Dow Chemical Co. (1980) *Tech. Bull. on tert-butylstyrene monomer*
81. Bartoň, J. & Kárpátyová, A. unpublished results
82. Banerjee, M., Sathpathy, U., Paul, T.K. & Konar, R.S. (1981) *Polymer* **22** 1729
83. Baxendale, J.H., Bywater, S. & Evans, M.G. (1946) *Trans. Faraday Soc.* **42** 675

84. Banerjee, M. & Konar, R.S. (1974) *J. Indian Chem. Soc.* **51** 722
85. Chatterjee, S.P., Banerjee, M. & Konar, R.S. (1979) *J. Polym. Sci., Polym. Chem. Ed.* **17** 2193
86. Hull van den, H.J. & Vanderhoff, J.W. (1971) In: Fitch, R.M. (ed.) *Polymer colloids*. Plenum Press, New York, p. 1
87. Banerjee, M. & Konar, R.S. (1986) *Polymer* **27** 147
88. Guha, T. & Palit, S.R. (1963) *J. Polym. Sci., A-1* **1** 877
89. Dainton, F.S., & Seaman, P.H. (1959) *J. Polym. Sci.* **39** 279
90. Fitch, R.M. & Shih, L. (1975) *Prog. Colloid Polym. Sci.* **56** 1
91. Bartoň, J., Hloušková, Z. & Juraničová, V. (1992) *Makromol. Chem.* **193** 167
92. Sathpathy, U.S., Paul, T.K., Banerjee, M. & Konar, R.S. (1981) *J. Macromol. Sci., Chem.* **15** 1495
93. Capek, I., Kostrubová, J. & Bartoň, J. (1990) *Makromol. Chem., Macromol. Symp.* **31** 213
94. Capek, I., Bartoň, J. & Orolínová, E. (1984) *Chem. Zvesti* **38** 803
95. Mallya, P. & Plamthottam, S.S. (1989) *Polym. Bull.* **21** 497
96. Capek, I. (1991) *Acta Polym.* **42** 273
97. Capek, I. (1991) *Chem. Papers* **45** 481
98. Bartoň, J. & Kárpátyová, A. (1987) *Makromol. Chem.* **188** 693
99. Capek, I., Bartoň, J. & Kárpátyová, A. (1987) *Makromol. Chem.* **188** 703
100. Bartoň, J., Capek, I., Juraničová, V. & Riedel, S. (1986) *Makromol. Chem., Rapid Commun.* **7** 521
101. Lambla, M., Ali-Syed, K. & Banderet, A. (1976) *Eur. Polym. J.* **12** 263
102. Alexander, A.E. & Napper, D.H. (1970) *Prog. Polym. Sci.* **3** 145
103. Nair, A.S. & Muthana, M.S. (1961) *Makromol. Chem.* **47** 128
104. Hakoila, E. (1963) *Ann. Univ. Turk., A* **66** 7
105. Berezhnoi, C.D., Khomikovskii, P.M. & Medvedev, S.S. (1960) *Vysokomol. Soedin., A* **2** 141
106. Al-Shabib, W.A.G. & Dunn, A.S. (1980) *Polymer* **21** 429
107. Nomura, M. & Fujita, K. (1989) *Makromol. Chem., Rapid Commun.* **10** 581
108. Eliseeva, V.I. (1980) *Polymer dispersions*. Khimiya, Moscow
109. Breitenbach, J.W., Kuchner, K., Fritze, H. & Tarnowiecki, H. (1970) *Br. Polym. J.* **2** 13
110. Capek, I. (1988) *Chem. Papers* **42** 347
111. Capek, I. (1988) *Acta Polym.* **39** 221
112. Capek, I. & Tuan, L.Q. (1986) *Makromol. Chem.* **187** 2063
113. Capek, I., Bartoň, J., Tuan, L.Q, Svoboda, V. & Novotný, V. (1987) *Makromol. Chem.* **188** 1723
114. Capek, I. (1992) *Makromol. Chem.* **193** 1423
115. Capek, I. (1992) *Chem. Papers* **46** 232
116. Simionescu, C.I., Chiriac, A., Neamtu, I. & Rusan, V. (1989) *Makromol. Chem., Rapid Commun.* **10** 601
117. Capek, I., Bartoň, J. & Orolínová, E. (1985) *Acta Polym.* **36** 187
118. Capek, I. (1986) *Acta Polym.* **37** 195
119. Capek, I. & Bartoň, J. (1985) *Makromol. Chem.* **186** 1297
120. Capek, I. & Bartoň, J. (1986) *Chem. Papers* **40** 45
121. Meyer, V.E. & Lowry, G.G. (1965) *J. Polym. Sci., A-1* **3** 2843
122. Greenley, R.Z. (1980) *J. Macromol. Sci., Chem.* **14** 445
123. Capek, I., Mlynárová, M. & Bartoň, J. (1988) *Acta Polym.* **39** 142
124. Capek, I. (1987) In: *Microsymposium: Modification of Polymers*, Smolenice, p. 162
125. Melville, H.W., Noble, B. & Watson, W.F. (1947) *J. Polym. Sci.* **2** 229
126. Bijsterbosh, B.H. (1978) *Colloid Polym. Sci.* **256** 343
127. Fitch, R.M. & Tsai, C.H. (1971) In: Fitch, R.M. (ed.) *Polymer colloids*. Plenum Press, New York, p. 103
128. Rochetti, B., Guillot, J. & Guyot, A. (1984) *J. Dispersion Sci., Technol.* **5** 447
129. Capek, I., Mlynárová, M. & Bartoň, J. (1988) *Makromol. Chem.* **189** 341
130. Lichti, G., Gilbert, R.G. & Napper, D.H. (1983) *J. Polym. Sci., Polym. Chem. Ed.* **21** 269
131. Capek, I., Mlynárová, M. & Bartoň, J. (1988) *Chem. Papers* **42** 763
132. Van der Hoff, B.M.E. (1960) *J. Polym. Sci.* **48** 175
133. Capek, I. (1986) *Coll. Czech. Chem. Commun.* **51** 2546
134. Das, S.K., Chatterjee, R.S. & Palit, S.R. (1955) *Proc. R. Soc. London, A* **227** 252
135. Šňupárek, J. (1980) *Angew. Makromol. Chem.* **88** 69
136. Nomura, M. & Harada, M. (1981) *J. Appl. Polym. Sci.* **26** 17
137. Capek, I. (1987) *Chem. Papers* **41** 815
138. Litt, M., Patsiga, R. & Stannet, V. (1970) *J. Polym. Sci., A-1* **8** 3607
139. Young, L.J. (1975) In: Brandrup, J. & Immergut, E.H. (eds) *Polymer handbook*, vol. 2. J. Wiley & Sons, New York, p. 118
140. Misra, S.C., Pichot, C., El-Aasser, M.S. & Vanderhoff, J.W. (1979) *J. Polym. Sci., Polym. Lett. Ed.* **17** 567
141. Delgado, J., El-Aasser, M.S. & Vanderhoff, J.W. (1986) *J. Polym. Sci., Polym. Chem. Ed.* **24** 861
142. Delgado, J., El-Aasser, M.S., Silebi, C.A., Vanderhoff, J.W. & Guillot, J. (1988) *J. Polym. Sci., Polym. Chem. Ed.* **26** 1495

143. Delgado, J., El-Aasser, M.S., Silebi, C.A. & Vanderhoff, J.W. (1988) *Makromol. Chem., Macromol. Symp.* **20/21** 545
144. Delgado, J., El-Aasser, M.S., Silebi, C.A. & Vanderhoff, J.W. (1989) *J. Polym. Sci., Polym. Chem. Ed.* **27** 193
145. Delgado, J. & El-Aasser, M.S. (1990) *Makromol. Chem., Macromol. Symp.* **31** 63
146. Stoffer, J.O. & Bone, T. (1980) *J. Dispersion Sci., Technol.* **1** 37
147. Stoffer, J.O. & Bone, T. (1980) *J. Polym. Sci., Polym. Chem. Ed.* **18** 2641
148. Atik, S.S. & Thomas, J.K. (1981) *J. Am. Chem. Soc.* **103** 4279
149. Jayakrishnan, A. & Shah, D.O. (1984) *J. Polym. Sci., Polym. Lett. Ed.* **22** 31
150. Carnali, J.O. & Fowkes, F.M. (1985) *Langmuir* **1** 576
151. Blackley, D.C. (1983) In: Poehlein, G.W., Ottewill, R.H. & Goodwin, J.W. (eds) *Science and technology of polymer colloids*, vol. 1. NATO ASI Ser. E67. M. Nijhoff Publ., Dordrecht, p. 203
152. Blackley, D.C., Andries, S. & Sebastian, R.D. (1987) *Br. Polym. J.* **19** 25
153. Motsavage, V.A. & Kostenbauder, H.B. (1963) *J. Colloid Sci.* **18** 603
154. Muramatsu, M. & Inoue, M. (1976) *J. Colloid Interface Sci.* **55** 80
155. Shinoda, K. (1963) In: Shinoda, K., Nakagawa, T., Tamamushi, B. & Isemura, T. (eds) *Colloidal surfactants.* Academic Press, New York, p. 1
156. Antonova, L.F. & Ablyakimov, E.I. (1972) *Vysokomol. Soedin., A* **14** 881
157. Egusa, S. & Makuuchi, K. (1982) *J. Polym. Sci., Polym. Chem. Ed.* **20** 863
158. Matsumoto, T. & Shimada, M. (1965) *Kobunshi Kagaku Jpn.* **22** 172
159. Matsumoto, T. (1974) In: Lissant, K.J. (ed.) *Emulsions and emulsion technology.* M. Dekker, New York, Chapter 9
160. Sakota, K. & Okaya, T. (1976) *J. Appl. Polym. Sci.* **20** 2583
161. Vijayendran, B.R. (1979) *J. Appl. Polym. Sci.* **23** 893
162. Hen, J. (1974) *J. Colloid Interface Sci.* **49** 425
163. Ceska, G.W. (1974) *J. Appl. Polym. Sci.* **18** 427
164. Sakota, K. & Okaya, T. (1976) *J. Appl. Polym. Sci.* **20** 3255
165. Sakota, K. & Okaya, T. (1976) *J. Appl. Polym. Sci.* **20** 3265
166. Greene, B.W. & Sheetz, D.P. (1970) *J. Colloid Interface Sci.* **32** 96
167. Egusa, S. & Makuuchi, K. (1981) *J. Colloid Interface Sci.* **79** 350
168. Shirahama, H. & Suzawa, T. (1984) *Polym. J.* **16** 795
169. Kulikov, S.A., Yablokova, N.V., Nikolaeva, T.V. & Aleksandrov, Yu.A. (1989) *Vysokomol. Soedin., A* **31** 2322
170. Emelie, B., Pichot, C. & Guillot, J. (1984) *J. Dispersion Sci., Technol.* **5** 393
171. Emelie, B., Pichot, C. & Guillot, J. (1988) *Makromol. Chem.* **189** 1879
172. Eliseeva, V.I., Zharkova, N.G., Chubarova, A.B. & Zubov, P.I. (1965) *Vysokomol. Soedin.* **7** 156
173. Eliseeva, V.I., Zubov, P.I. & Malofeevskaya, V.F. (1965) *Vysokomol. Soedin.* **7** 1348
174. Eliseeva, V.I., Malofeevskaya, V.F., Gerasimova, A.S., Makarov, Yu.A., Izmailova, I.S. & Orlova, K.G. (1967) *Vysokomol. Soedin., A* **19** 730
175. Zosel, A., Heckmann, W., Ley, G. & Maechtle, W. (1987) *Colloid Polym. Sci.* **265** 113
176. Verbrugge, C.J. (1970) *J. Appl. Polym. Sci.* **14** 897
177. Verbrugge, C.J. (1970) *J. Appl. Polym. Sci.* **14** 911
178. Quadrat, O., Mrkvičková, L., Jasná, E. & Šňupárek, J. (1990) *Colloid Polym. Sci.* **268** 493
179. Puig, J.E., Corona-Galvan, S., Maldonado, A., Schulz, P.C., Rodriguez, B.E. & Kaler, E.W. (1990) *J. Colloid Interface Sci.* **137** 303.5

3.5 MODIFICATION OF POLYMER DISPERSIONS

The modification of a polymer latex is another way of preparing reactive polymer dispersions. This procedure is used in those cases where a latex of the required composition cannot be prepared via normal emulsion polymerization. The preparation of the poly(vinyl alcohol)-containing polymer latex usually obtained by hydrolysis of poly(vinyl acetate) in polymer particle microspheres provides an example [1].

The modification of polymer particles or, more correctly, their surface, is carried out through conventional organic reactions and procedures: hydrolysis, re-esterification, aminolysis, decarboxylation and oxidation or reduction. The modification results in a polymer dispersion with special physical and chemical properties.

Copolymers of lipophilic and hydrophilic monomers containing various functional or polar groups susceptible to nucleophilic attack are suitable materials for modification reactions.

3.5.1 Hydrolysis

Hydrolysis of the functional groups located on the surface of polymer particles may only be carried out under mild conditions (catalysis by hydroxyl or hydrogen ions) because of possible coagulation of the polymer latex.

Sulfate groups bound on the particle surface are particularly sensitive to hydrolysis. Even in simple operations with the latex (e. g. purification by dialysis) they are prone to hydrolysis, the rate of which increases in slightly acid and slightly basic media. The hydrolysis of sulfate groups leads to $\equiv$C–OH groups, which are readily oxidized in the presence of peroxodisulfate into carboxyl groups, as has been found in the processing of sulfated poly(styrene) latex [2].

Spontaneous hydrolysis of the alkyl esters of acrylic acid has only been observed on the surface of polymer particles containing a copolymer of vinyl acetate and butyl acrylate [3]. An increase in the concentration of carboxyl groups was observed in a system with a high concentration of alkyl ester groups at the particle surface. The hydrolysis was promoted in slightly acid media. This behaviour has also been confirmed by measurement of the surface particle charge. By applying 18 mol % of butyl acrylate in the feed, the value of the surface particle charge obtained was 0.55 μC cm^{-2}. The use of a higher concentration of butyl acrylate (68 mol %) led to a substantial increase in the surface charge up to 2.0 μC cm^{-2}.

Hydrolysis of the sulfate groups can be effectively carried out by using H$^+$ ion exchangers. The autocatalysed hydrolysis of the ester groups of sulfuric acid localized on the surface of poly(styrene) particles leads to a decrease in the surface charge

$$\sim\text{O}-\underset{\underset{\text{O}}{|}}{\overset{\overset{\text{O}}{|}}{\text{S}}}-\text{O}^-\text{M}^+ + \text{H}_2\text{O} \rightarrow \sim\text{OH} + \text{M}^+\text{HSO}_4^- \qquad (1)$$

and to an increase of the concentration of the surface-bound hydroxyl groups [4]. The hydroxyl groups transform to carboxyl groups on heating the latex in the presence of peroxodisulfate (about 15 % with respect to the polymer fraction) and of a catalytic amount of AgNO$_3$ ($\sim 10^{-5}$ M) [5].

The hydrolysis of vinyl acetate groups on the surface of the polymer particles leads to the production of hydroxyl groups. This process was observed on polymer particles containing a copolymer of butyl acrylate and vinyl acetate [6]. The hydrolysis was catalysed by the sulfate and carboxyl groups located on the particle surface as a result of the interaction between the electron-donating acetyl group and H$^+$ ions. The interaction caused the weakening of the RO–COCH$_3$ bond, its scission and the formation of acetic acid and of free hydroxyl groups remaining on the particle surface [7]. The surface sulfate groups originate from the peroxodisulfate, while the carboxyl groups result from the simultaneous hydrolysis of the butyl ester.

The copolymers of acrylamide localized on the surface of polymer particles are also modified by hydrolysis [8]. Amide groups are stable to hydrolysis at medium pH values. The surface layer of particles containing acrylamide groups is, however, subject to hydrolysis in alkaline media. The modification of copolymer latices based on acrylamide and prepared by emulsion copolymerization is normally preceded by removal of the water-soluble fraction of the polymer or copolymer [9].

One of the most important factors affecting the hydrolysis of the functional groups is the time of interaction between a nucleophilic agent and the polymer latex, which influences the degree of hydrolysis of the functional groups localized in the internal particle spheres (see section on the characterization of polymer particles).

The hydrolysis of the amide groups of poly(acrylamide) bound on the surface of the polymer particles obeys first-order kinetics. The relation between the concentration of the amide groups and the carboxyl groups being formed can be expressed as [8, 10]:

$$\log([AAm]_{s,0} - [COOH]_h) = -kt + \log[AAm]_{s,0}, \qquad (2)$$

where $[AAm]_{s,0}$ is the number of acrylamide groups in the surface layer of the polymer particles, $[COOH]_h$ is the number of carboxyl groups formed by hydrolysis, k is the reaction rate constant and t is the reaction time.

The copolymerization of the more lipophilic methacrylamide gives copolymers with a higher content of amide groups usually located in the inner particle spheres. No dependence of the type described by Eq. (2) has been observed as a consequence of hindrance to the penetration of hydroxyl ions into the particles and with regard to the lower concentration of amide groups on the particle surface. A reaction order with respect to acrylamide of 1.0 is not reached in these systems and is much lower.

3.5.2 The Hofmann reaction

In this reaction an amide group is transformed into an amine in an aqueous medium in the presence of NaOH and NaOCl [11] with yields > 98%. Poly(acryl amide) is converted to poly(vinyl amine) over a wide temperature range. The course of reaction was shown to be affected by the reactant concentrations and their relation to the acrylamide groups [12]. During the Hofmann reaction, the polymer chain undergoes scission and its RMM decreases.

The Hofmann reaction has been applied successfully to the modification of styrene/acrylamide polymer latices [13]. In this modification, it is necessary to work under milder reaction conditions to prevent the possibility of coagulation of the polymer particles. The Hofmann reaction as applied to styrene/acrylamide latex led to the isolation of a polymer latex containing both amine and carboxyl groups. The results show that the reaction yield is influenced by the concentrations of the reactants and their ratio and not by the temperature. An amphoteric polymer latex containing both amine and carboxyl groups was produced in this reaction. Its isoelectric point varied between pH 5 and 10 and was determined by the ratio of the numbers of amine and carboxyl groups on the particle surface. Similar results were obtained in ref. [14] for the isoelectric point of a latex modified by the Hofmann reaction, i.e. values of pH 4.4–9.7.

3.5.3 The Mannich reaction

In the Mannich reaction, the amide group is transformed in the presence of formaldehyde and secondary amines in an alkaline medium to a secondary or tertiary amino group which can be quaternized [15, 16]:

$$R''-CONH_2 \xrightarrow{HCHO, R_2NH} R''-CONHCH_2NR_2 \xrightarrow{R'X} R''-CONHCH_2NR_2R'^+ X^- \quad (3)$$

The yield is controlled by the character of the alkylamines used, decreasing with increasing volume of the alkyl groups. Carboxyl and hydroxymethyl amide groups are the side products of the Mannich reaction as with the Hofmann reaction.

The amide groups of the poly(acrylamide) bound at the surface of the polymer particles are hydrolysed spontaneously fairly rapidly whereas poly(acrylamide) dissolved in water undergoes only partial hydrolysis. The lower mobility of the acrylamide units on the particle surface affects the hydrolysis positively and, conversely, it diminishes the yield of the Mannich reaction.

The ratio of amine and carboxyl groups on the surface of the polymer particles formed by the Mannich reaction changes as a function both of the character of the amine and of temperature. As the amine concentration increases, the concentration of amine and carboxyl groups increases linearly up to a particular limiting concentration of the amine used, above which the number of carboxyl groups is unchanged. The yield of the Mannich reaction increases with temperature.

3.5.4 Hydroxymethylation of amide groups

Hydroxymethylation is as a preparatory step in the amination of amide groups localized on the surface of the polymer particles. The pH value of the reaction system is the controlling factor of the reaction. Table 31 shows that with increasing pH the concentration of hydroxymethyl

Table 31 — Hydroxymethylation of styrene/acrylamide latex [S] (Reprinted from ref. [8] with permission of Academic Press Inc., San Diego)

pH	Surface density of hydroxymethyl groups / nm^2
4.0	0.45
6.0	0.49
8.5	0.56
10.3	0.86
12.1	3.42

groups localized an the particle surface increases rapidly. Further study of the influence of the reaction conditions on the yield of hydroxymethyl groups led to the conclusion that a large number of the groups are formed in a strongly alkaline medium [17,18]. In this medium, however, the Cannizzaro equation also begins to participate which leads to the formation of carboxyl groups.

3.5.5 Reactions of double bonds

Double bonds on the surface of polymer particles can be saturated or altered by chemical reaction to various functional groups.

Unsaturation in the backbone of the natural rubber latex, *cis*-poly(isoprene), facilitates certain types of chemical modification to yield a range of modified polymers or composite latex particles. In natural rubber latices the all *cis*- trialkyl-substituted double bonds are electron-rich, both as a result of hyperconjugative and inductive effects. The modifications of these latex particles can be divided into two classes: low-level and high-level modifications. In the former, modification is often ca. 1 mol % while in the latter it falls in the range 20–100 mol % to endow natural rubber latex with new chemical and physical properties which effectively upgrade it into new materials [18, 19].

Use of γ-rays or high-energy electrons offers a challenging approach to initiating graft polymerization on the particle surface [20, 21].

Various procedures for altering the vinyl groups on the polymer particle surface have been summarized by Seitz [22], e.g. hydrobromination yields a hydroxyl group [23] via the following reaction mechanism:

$$\text{(P)}-CH=CH_2 \xrightarrow{BH_3} \text{(P)}-CH_2-CH_2-BH_2 \xrightarrow{NaOH, H_2O_2}$$

$$\longrightarrow \text{(P)}-CH_2-CH_2-OH, \tag{4}$$

where (P) denotes the polymer particle.

Epoxidation introduces epoxide groups onto the particle surface [22]

$$\text{(P)}-CH=CH_2 \xrightarrow[CH_2Cl_2]{m\text{-chloroperoxobenzoic acid}} \text{(P)}-\triangleleft_O \tag{5}$$

Ozonization leads to carboxylic groups [22]

$$\text{(P)}-CH=CH_2 \xrightarrow[CH_2Cl_2 / CH_3OH]{O_3} \text{'ozonide'} \xrightarrow[HCOOH]{H_2O_2} \text{(P)}-C\begin{smallmatrix}O\\OH\end{smallmatrix} \tag{6}$$

Hydrobromination is used for converting vinyl groups to halogen derivatives which can be easily modified by various nucleophilic reagents such as basic dyes or aromatic amines [22, 24].

Hydrobromination followed by addition of the basic dye can be written as

$$\text{(P)}-CH=CH_2 \xrightarrow{HBr} \text{(P)}-\underset{\underset{CH_3}{|}}{CH}-Br \xrightarrow{NH_2-dye} \text{(P)}-\underset{\underset{CH_3}{|}}{CH}-NH-dye \tag{7}$$

The functional groups obtained by reactions (4)–(7) can be further changed through oxidation and reduction processes and additions to other groups, e.g. hydroxyl, amine, aldehyde, etc.

3.5.6 Utilization of flexibility in the procedure for polymerization

Batch emulsion polymerization is used for investigating the kinetics and mechanism of the polymerization process and for preparing polymer dispersions of both theoretical and applied interests. It proceeds in three stages (see section 3.4).

The seed (two-step) emulsion polymerization is used for preparing monodisperse polymer particles, polymer dispersions with a heterogeneous structure and for studying the kinetics and mechanism of emulsion polymerizations under stationary conditions. It consists of the Stages II and III (see section 3.4).

The preparation of these polymer particles is usually accompanied by a secondary nucleation that exerts a negative effect on the process. Its effectiveness depends on several factors reported elsewhere [25, 26]. Studying the kinetics of the seed polymerization, Schmutzler [25] and Hergeth [26] came to the conclusion that as the number of seed polymer particles increases, secondary nucleation is substantially reduced, and above a concentration of 10^{13} particles cm^{-3}, is suppressed to a mininum. A secondary polymer is mostly formed on the surface of the latex particles when a high concentration of feed particles is applied during polymerization. It was confirmed that the range of the concentrations of the polymer particles from 10^{12} to 10^{13} cm^{-3} was large enough for this type of modification [26] (see section 3.4).

Both semicontinuous emulsion polymerization (i.e. monomer is distributed between the reactor and the feeding seed) and continuous (i.e. the monomer exists only in the feeding seed) are mostly applied in the industrial preparation of copolymer dispersions. Their wide application lies mainly in the possibility of controlling the exothermicity of the polymerization process [27], the rate of polymerization [28, 29], the morphology of the polymer particles [30] and, in particular, the composition of the copolymer [31, 32].

The semicontinuous emulsion polymerization and copolymerization take place in Stage III of the batch emulsion polymerization. The polymerization in these systems involves a small amount of monomer in the monomer/polymer particles. This regime is referred to as monomer-starved, which corresponds to reaction conditions under which the feed rate of monomer is lower than the corresponding rate of the batch emulsion polymerization. The rate of the semicontinuous polymerization is thus proportional to the rate of monomer dosage and is constant for a given feed rate. By accumulating the unchanged monomer in polymer particles, the rate is kept constant. This is explained by the decrease in the average number of radicals per particle as a consequence of the lowering of the internal viscosity of the polymer particles [28, 29]. Wessling & Gibbs [33] proposed a model in which the polymerization proceeds only on the particle surface. The instantaneous conversion reaches values greater than 90 % and the character of the polymer particles approaches the glassy state. Under these conditions, the model proposed describes the actual course of polymerization. The authors have found that at high feed rates of monomer, the unchanged monomer accumulates and the polymerization is uninfluenced by the feed regime. Conversely, at low feed rates the rate of polymerization is governed by the feed rate of the monomer.

Makgawinata *et al.* [34] proposed a kinetic model for semicontinuous emulsion polymerization. It provides information on the time dependence of the instantaneous conversion, on the monomer concentration in monomer/polymer particles, on the average number of radicals in a particle and the average particle size. The authors achieved good agreement between the theoretical and experimental values of the conversions and particle size in the emulsion copolymerization of vinyl acetate and butyl acrylate.

The relation between the feed rate of the reaction components, r_m, and the reaction rate, r_DM, is expressed by

$$(MW)\frac{\mathrm{d}V[\text{M}]}{\mathrm{d}t} = (r_\text{m})V - (-r_\text{DM})V, \tag{8}$$

where V is the volume of the reaction system, MW is the average RMM of the comonomer, and [M] is the monomer concentration in the system. Equation (8) can be modified to

$$\frac{d[M]}{dt} = R_a - R_{ps},\qquad(9)$$

where R_a is the feed rate of monomer and R_{ps} is the reaction rate (both in molar units).

The size of the polymer particles and its distribution are important parameters decisive for the fundamental properties of the polymer latex prepared by continuous emulsion polymerization. The modelling of the particle size distribution for these polymerizations has been the objective of several groups [35–38]. The authors have discussed the effects of the concentration of emulsifier, initiator, and monomer, of the water/monomer ratio, temperature, etc. on the particle size distribution. These are mostly complicated functions and therefore computational techniques are needed to solve them.

The distribution of emulsifier between the initial reactor charge R and the monomer emulsion M (expressed as $(R/M)_E$) and the rate of stirring the reaction mixture are important factors which affect the course of semicontinuous emulsion polymerization. Šňupárek discussed the effect of the distribution of an emulsifier between the initial reactor charge R and the monomer emulsion M on the kinetics and mechanism of semicontinuous emulsion polymerization and the copolymerization of acrylates, methacrylates, and styrene in a series of papers [39–45]. He studied the influence of the values of $(R/M)_E$ on the rate of polymerization, instantaneous conversion, copolymer composition, surface tension of the latex during polymerization and after its completion as well as the size of the polymer particles and the latex stability. At higher $(R/M)_E$, a large number of small particles were formed which had flocculated strongly during polymerization. On the other hand, at high fraction of emulsifier in the monomer feed, a latex with a low surface tension was formed which contained particles of large dimensions.

As polymerization proceeded at high emulsifier concentrations in the monomer feed ($(R/M)_E$ = 75/25, 50/50, and 25/75), the surface tension increased during feeding of the reaction mixture and, after reaching its maximum, it was kept at a plateau until the completion of polymerization. On the other hand, polymerization when $(R/M)_E$ = 0/100 was accompanied by an initial decrease of the surface tension to a plateau with much a lower value of the surface tension.

As the value of $(R/M)_E$ increases from 0/100 to 20/80 or 30/70, the average particle size decreases linearly. At higher emulsifier concentrations (> $(R/M)_E$ = 30/70), the polymer particle size remains almost constant.

The semicontinuous copolymerization gave copolymers of homogeneous composition, with small deviations from the monomer feed composition.

At high values of $(R/M)_E$, high rates of polymerization are observed; however, the latices formed show lower stability.

More vigorous stirring of the latex leads to flocculation of the polymer particles and to the formation of a coagulate [31]. The effect of stirring on the course of semicontinuous emulsion polymerization has been studied in ref. [46–48]. With increasing stirring rate, flocculation and formation of the coagulate increase and the number of polymer particles decreases simultaneously. The polymerization rate decreases in parallel. Daniel et al. [49] observed, however, a rate enhancement of the emulsion polymerization of vinyl acetate was induced by an increase in the stirring rate. By a suitable change in the stirring rate, latices can be prepared with a relatively narrow particle size distribution, with a minimum of coagulate, and with varied particle sizes [50, 51]. During the semicontinuous emulsion copolymerization of vinyl acetate and the dibutyl ester of maleic acid, Donescu *et al.* [52] observed that the instantaneous conversion is a function of the stirring rate and the time of polymerization and reaches a maximum at a particular value of $(R/M)_E$.

The continuous emulsion polymerization system consists of a series of continuous stirred-tank reactor (CSTR) systems in which all reagents are pumped continuously into the first reactor and the product latex is removed from the final reactor. One significant difference between batch and continuous reactors is the concept of a distribution of residence times. All materials have the same residence time in a batch reactor. In a single CSTR, however, material which enters at a fixed time can appear in the effluent over a broad range of times. Some will appear almost instantly while other parts will remain in the reactor for a long time before leaving the product [53]. Poehlein *et al.* [54] have suggested the following particle growth model for the latex particles in CSTR systems

$$\frac{dD'}{d\tau} = \frac{\bar{n}}{3\beta(D')^2}, \tag{10}$$

where $\beta = \pi D_0^3/6\Theta K_1 [M]_p$, $[M]_p$ is the monomer concentration in the polymer particles, K_1 is the kinetic parameter related to particle swelling with monomer and the propagation rate constant, $\bar{n}$ is the time-averaged number of radicals in a monomer-swollen particle of size D', D' is the dimensionless diameter defined as the actual particle diameter D divided by the diameter of a freshly-nucleated particle D_0, Θ is the mean residence time and τ is the dimensionless time defined as the actual time t divided by the mean residence time in the reactor.

Models to predict the particle size distribution, conversion and the kinetic parameters of continuous emulsion polymerization have been developed [55, 56]. They have been applied successfully to several continuous emulsion polymerizations to explain the behaviour of the polymerization and the latex properties [55–57].

References

1. Bauman, W.C. (1951) *U.S.* 2559529
2. Kamel, A.A., Ma, C.M., El-Aasser, M.S., Micale, F.J. & Vanderhoff, J.W. (1981) *J. Dispersion Sci., Technol.* **2** 315
3. Kong, W.C., Pichot, C. & Guillot, J. (1987) *Colloid Polym. Sci.* **256** 791
4. McCarvill, W.T. & Fitch, R.M. (1978) *J. Colloid Interface Sci.* **67** 204
5. Jayasuriya, R.M., El-Aasser, M.S., Vanderhoff, J.W. & Yue, H.J. (1985) *J. Polym. Sci., Polym. Chem. Ed.* **23** 2819
6. El-Aasser, M.S., Makgawinata, T., Vanderhoff, J.W. & Pichot, C. (1983) *J. Polym. Sci., Polym. Chem. Ed.* **21** 2363
7. Arranz, F., Fountain, J. & Ashraf, T. (1968) *Rev. Plast. Mod.* **1** 151
8. Kawaguchi, H., Hoshino, H., Anagasa, H. & Ohtsuka, Y. (1984) *J. Colloid Interface Sci.* **97** 456
9. Ohtsuka, Y., Kawaguchi, H. & Sugi, Y. (1981) In: Bassett, D.R. & Hamielec, A.E. (eds) *Emulsion polymers and emulsion polymerization.* Am. Chem. Soc. Symp. Ser., Washington DC, p. 145
10. Fitch, R.M., Gajria, C. & Tarcha, P.J. (1979) *J. Colloid Interface Sci.* **71** 107
11. Tanaka, H. & Senju, R. (1976) *Kobunshi Ronbunshu* **33** 309
12. Tanaka, H. (1979) *J. Polym. Sci., Polym. Chem. Ed.* **17** 1239
13. Kawaguchi, H., Hoshino, H. & Ohtsuka, Y. (1981) *J. Appl. Polym. Sci.* **26** 2015
14. Kawaguchi, H., Hoshino, H. & Ohtsuka, Y. (1983) *Colloids Surfaces* **6** 271
15. Nishiyama, M. (1980) *Kamipa Gikyoshi* **34** 37
16. Phillips, K.G., Ballweber, E.G. & Hurlock, J.R. (1979) *U.S.* 4179424
17. Schiller, A.M. & Suen, T.J. (1956) *Ind. Eng. Chem.* **48** 2132
18. Barnard, D. (1982) *Kautsch. Gummi, Kunstst.* **35** 747

19. Hourston, D.J. & Romaine, J. (1990) *J. Appl. Polym. Sci.* **39** 1587
20. Battaerd, H.A.J. & Tregear, G.W. (1967) *Graft copolymers.* Interscience, New York, Chapter 2
21. Egusa, S., Sasaki, T. & Hagiwara, M. (1987) *J. Appl. Polym. Sci.* **34** 2177
22. Seitz, U. (1989) *Habilitationsschrift.* Universität Stuttgart
23. Seitz, U. (1977) *Makromol. Chem.* **178** 1689
24. Wollman, D. (1978) *Dissertation.* Universität Stuttgart
25. Schmutzler, K. (1982) *Acta Polym.* **33** 454
26. Hergeth, W.D. (1983) *Thesis.* Merseburg
27. Hamielec, A.E. (1977) *Introduction to polymerization kinetics — polymer reaction engineering — intensive short course on polymer production technology.* McMaster University, Hamilton, Ontario, June
28. Gerrens, H. (1966) *ACS Polym. Prepr.* **7** 699
29. Gerrens, H. (1969) *J. Polym. Sci.*, *C* **27** 77
30. Misra, S.C., Pichot, C., El-Aasser, M.S. & Vanderhoff, J.W. (1979) *J. Polym. Sci., Polym. Lett. Ed.* **17** 567
31. Šňupárek, J. (1979) *J. Appl. Polym. Sci.* **24** 909
32. Makgawinata, T., El-Aasser, M.S., Vanderhoff, J.W. & Pichot, C. (1981) *Acta Polym.* **32** 583
33. Wessling, R.A. & Gibbs, D.A. (1973) *J. Macromol. Sci., Chem.*, *A* **7** 647
34. Makgawinata, T., El-Aasser, M.S., Klein, A. & Vanderhoff, J.W. (1984) *J. Colloid Sci., Technol.* **5** 301
35. De Graff, A.W. & Poehlein, G.W. (1971) *J. Polym. Sci., A-2* **9** 1955
36. Stevens, J.D. & Fundenburk, J.O. (1972) *Ind. Eng. Chem., Process Des. Dev.* **11** 360
37. Sundberg, D.C. (1979) *J. Appl. Polym. Sci.* **23** 2197
38. Cauley, D.A. & Thompson, R.W. (1982) *J. Appl. Polym. Sci.* **27** 363
39. Šňupárek, J. (1981) *Acta Polym.* **32** 368
40. Šňupárek, J. (1980) *Angew. Makromol. Chem.* **88** 69
41. Šňupárek, J. & Kašpar, K. (1981) *J. Appl. Polym. Sci.* **26** 4081
42. Šňupárek, J. (1972) *Angew. Makromol. Chem.* **25** 105
43. Šňupárek, J. (1972) *Angew. Makromol. Chem.* **25** 113
44. Šňupárek, J. (1975) *Angew. Makromol. Chem.* **37** 1
45. Šňupárek, J. (1977) *Acta Polym.* **28** 249
46. Dunn, A.S. & Taylor, A. (1965) *Makromol. Chem.* **83** 207
47. Kiparissides, C., MacGregor, J.F., Singh, S. & Hamielec, A.E. (1980) *Can. J. Chem. Eng.* **58** 57
48. Nomura, M., Narada, M., Eguiki, W. & Nagata, S. (1972) *J. Appl. Polym. Sci.* **16** 835
49. Daniel, N.V., Klopova, A.V. & Nikolaev, J.S. (1969) *Plast. Massy* **9** 10
50. Eliseeva, V.I. (1981) *Acta Polym.* **32** 355
51. Donescu, D., Gosa, K., Ciupitoiu, A. & Languri, J. (1985) *J. Macromol. Sci., Chem.* **22** 931
52. Donescu, D., Gosa, K., Languri, J. & Ciupitoiu, A. (1985) *J. Macromol. Sci., Chem.* **22** 941
53. Poehlein, G.W. & Dougherty, D.J. (1977) *Rubber Chem. Technol.* **50** 601
54. Poehlein, G.W., Dubner, W. & Lee, H.C. (1982) *Br. Polym. J.* **14** 143
55. Gugliotta, L.M. & Meira, G.R. (1986) *Makromol. Chem., Macromol. Symp.* **2** 209
56. Poehlein, G.W., Lee, H.C. & Stubicar, N. (1985) *J. Polym. Sci., Polym. Symp.* **72** 207
57. Lee, H.C. & Mallinson, R.G. (1990) *J. Appl. Polym. Sci.* **39** 2205

3.6 COLLOIDAL STABILITY OF POLYMER LATICES

The theory of colloidal stability developed by Derjagin & Landau [1] and Verwey & Overbeek [2] (DLVO) is based on how attractive and repulsive forces operate between particles. The surface of polymer particles covered with strongly-bound sulfate or carboxyl groups or by adsorbed molecules of polar and ionic emulsifier is electrically charged. The degree of dissociation of the functional groups controls the surface charge and is a function of the hydrogen ion concentration. Addition of sodium hydroxide lowers the concentration of protons in water and, conversely, it increases the fraction of dissociated groups on the particle surface and also, simultaneously, the surface charge, which governs the stability of the polymer dispersion.

The space distribution of the charge shown in Fig. 13 points to a high concentration of charge on the particle surface, which decreases markedly from the particle to the aqueous phase. The negatively-charged particle surface is neutralized by the cations of the Stern layer which can be covered by a negatively charged layer [3]. Both layers are in the vicinity of

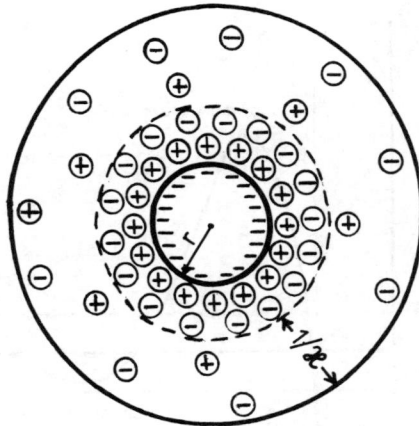

Fig. 13 — Schematic of charge distribution in a sphere surrounding a particle.

the negatively charged surface of the polymer particle and contribute to the hydrodynamic diameter of the particle. Between a particle and the water phase, there is another layer with a remarkably lower ion concentration called the Gouy–Chapman layer [4] (of a diffusional character). The thickness of this layer can be calculated from the following equation derived from the Debye–Hückel theory [5]

$$1/\kappa = \sqrt{\frac{\varepsilon_0 \varepsilon_r k T}{e^2 \sum n_i z_i^2}}, \qquad (1)$$

where e is the electron charge, k is Boltzmann's constant, T is the absolute temperature, ε_0 is the permittivity in a vacuum, ε_r is the relative permittivity, and n_i is the concentration of an ion with charge z_i. All ions occurring in the solution contribute to the expression $\sum n_i z_i^2$. Equation (1) shows that the thickness of the diffusional part of the electrical double layer (EDL) is proportional to the square root of the absolute temperature and inversely proportional to the square root of the ion concentration and charge. As the electrolyte concentration increases, the thickness of the layer decreases and, conversely, on dilution and an increase in temperature, it increases.

The overall charge of the polymer particle, Q_p is determined by measuring the particle mobility in an electric field (by electrophoretic measurements). It can be expressed mathematically

$$Q_p = 4 \zeta \varepsilon_0 \varepsilon_r a (1 + \kappa), \qquad (2)$$

where ζ is the electrokinetic potential, κ is the reciprocal of the thickness of the double layer, a is the hydrodynamic particle radius (equal to the sum of the particle radius r and the thickness of the Stern layer δ). Application of Eq. (2) to the calculation of the particle charge is limited by the fact that the hydrodynamic radius and ε_r are not exactly known very near the particle [3, 5].

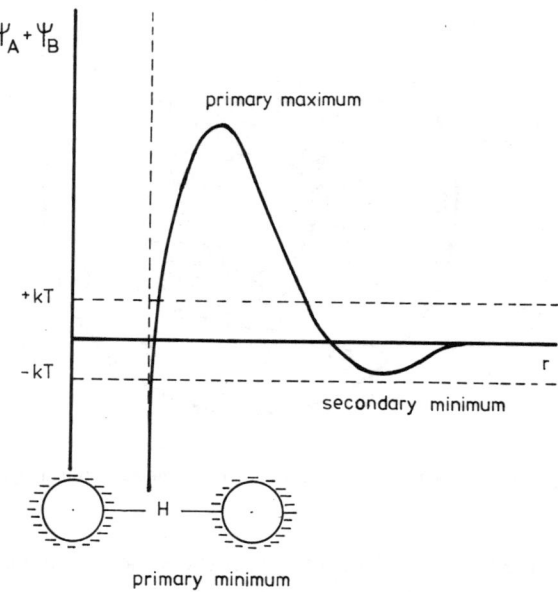

Fig. 14 — Potential of the two negatively charged polymer particles according to DLVO theory.

The mobility of polymer particles with an electrically charged surface can be expressed by using the diffusion coefficient D

$$D = \frac{1}{6}\frac{d\langle r^2\rangle}{dt} = \frac{kT}{6\pi \eta_F a} \ , \qquad (3)$$

where η_F denotes the viscosity of the liquid medium. The numerator in Eq. (3) expresses the contribution of the thermal energy and the denominator is the Stokes term for the attractive energy. During motion of the particle in a liquid medium, collisions of the particles occur which deform the diffusion particle layer and the particle adopts configurations other than spherical. Collisions and deformations result in an increase of the internal viscosity of the system, in what is called the electroviscosity effect [6].

According to DLVO theory, the potential energy of two interacting particles is given by the sum of the van der Waals attractive energy and the repulsion energy. When the mutual repulsion of the particles is strong, the polymer latex is stable. Figure 14 shows the potential of two negatively-charged particles. The thermal energy is in this case greater than the secondary minimum, and therefore stable associates are not formed. The dominant contribution from the repulsive forces keeps the distance between the particles large enough and prevents any effective influence of the attractive forces. Values of the potentials of the attractive and repulsive forces, Ψ_A and Ψ_B are obtainable from the theories of Derjagin and Landau [1] and Verwey and Overbeek [2]

$$\Psi_A = -\frac{A}{12}\left[\frac{1}{x^2-1} + \frac{1}{x^2} + 2\ln(1-x^{-2})\right] \ , \qquad (4)$$

where $x = r/2a$, r being a measure of the separation of the centres of the polymer particles and A is the combined Hamaker constant.

If $\kappa a \geq 3$, the potential Ψ_B can be calculated from Eq. (5)

$$\Psi_B = 2\pi\varepsilon_0\varepsilon_r a \Psi_\delta^2 \ln(1 + e^{-\kappa H}). \tag{5}$$

Here $H = (r - 2a)$ and is a measure of the separation of the particle surface, Ψ_δ is the potential of the diffusion layer close to the value of the ζ potential [6].

The electrokinetic potential ζ and the surface particle charge density σ are fundamental characteristics of the polymer particle in determining the stability of the polymer dispersion. Both parameters are strongly influenced by the pH of the medium. The value of ζ of a polymer particle covered by carboxyl or sulfate groups increases with increasing pH because of neutralization of the acid groups. The difference in the dependences of ζ and σ on pH follows from the character of both quantities: while σ represents the charge of *all* ionized groups on the particle surface, ζ is the potential on the shear plane of the EDL.

Polymer particles characterized by high values of σ and ζ do not aggregate on mutual collision although their mutual interactions may cause deformation of the spherical configuration of the charge or ion distribution. As the value of the surface charge decreases, the probability increases of greater overlap of the surface particle layers, and their association increases. Each latex has a particular value of the electrokinetic potential, and is stable above it but undergoes coagulation below it (Fig. 15).

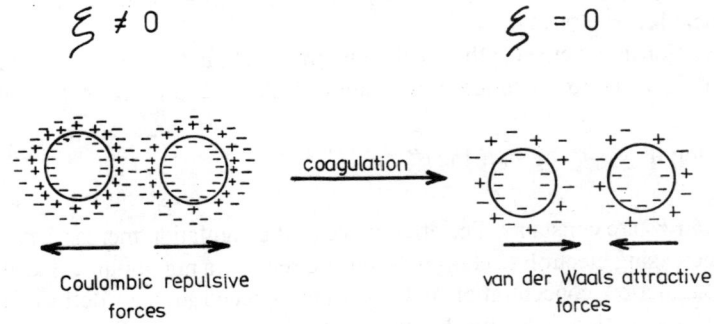

Fig. 15 — Mechanism for repulsive and attractive forces in the stabilization of polymer dispersions.

The foundations of the mechanism for coagulation, laid by Smoluchowski [7] have been supplemented later by Fuchs [8]. Fuchs introduced a new parameter W, the measure of coagulation, which is given by the ratio between rapid and slow coagulation

$$W = \frac{K_r}{K_s}. \tag{6}$$

Data about the stability of polymer dispersions are obtained from kinetic investigations of the kinetics of latex coagulation, which can be induced by adding electrolyte, by changing temperature, by ultrasound or an electric or magnetic field, by mechanical stirring or adding low-RMM or high-RMM compounds. In all cases, measurements of the breakdown of the outer structure of the polymer particles, i.e. the EDL, and the reduction in concentration of the polar groups on the particle surface, or the resistance to these effects, are made.

Association of the polymer particles can be studied, for example, from the change in the absorbance of the system taking place on adding electrolyte. The coalescence of two polymer particles is described by Eq. (1):

$$k = (1/A_0)(dA/dt)_0/N_0, \qquad (7)$$

where k is the rate constant for coagulation, A is the light absorbance, N_0 is the initial number of polymer particles and $(dA/dt)_0$ denotes the initial rate of coagulation.

The absorbance of the system increases linearly with the concentration of coagulating agent; it is associated with the increase of the efficiency of particle association which is caused by the decrease in thickness of the electrical double layer ($1/\kappa$). An investigation of the dependence of the thickness of the EDL on the concentration of the electrolyte NaCl led to the semiempirical relationship

$$1/\kappa = 3.04/[NaCl]^{0.5}. \qquad (8)$$

The increment in the solution absorbance increases up to a certain critical concentration of electrolyte above which the rate of concentration does not vary [9], which is denoted as rapid coagulation. In some systems, increase of the electrolyte concentration above its critical concentration is accompanied by a decrease of the absorbance increment (A_t/A_0, where A_0 and A_t are the absorbances at times 0 and t), given by the change in thickness of the electrical layer [10, 11].

The relationship between the Fuchs stabilization ratio W and the concentration of electrolyte C_e added to a solution of polymer particles is expressed by the equation

$$\log W = -k' \log C_e + \log k'', \qquad (9)$$

where k' and k'' are constants. The effectiveness of coagulation increases according to Eq. (9) with increasing electrolyte concentration and reaches a maximum value at the so-called critical coagulation concentration (CCC). This concentration is defined as the critical amount of coagulation agent needed to change diffusion-controlled coagulation to rapid coagulation [12]. An experimental plot of log W versus log C_e (Fig. 16) is described by the curves of slow (horizontal curve) and rapid (vertical curve) coagulation; the break point gives the value of the CCC used for characterizing the stability of the polymer dispersion [13]. A study of the stability of a poly(styrene) latex stabilized by hydroxyl, carboxyl or sulfate groups upon adding NaCl or BaCl$_2$ led to values of the CCC: 150, 280, or 400 mmol NaCl and 24, 40, or 40 mmol BaCl$_2$, respectively. Depending on the type of stabilizing functional group, the stability of the polymer latex increases in the order: hydroxyl < carboxyl < sulfate [14].

The slope of the plots of log W versus log C_e gives data about the density of the surface charge σ [9, 15]. The CCC is proportional to the surface particle charge. A change in the surface particle charge, determined by the concentration of sulfate groups of poly(styrene) latex ranging from 0.2 to 1.8 μeq m^{-2}, led to a change in the CCC from 0.15 to 0.98 mol dm^{-3} NaCl [16].

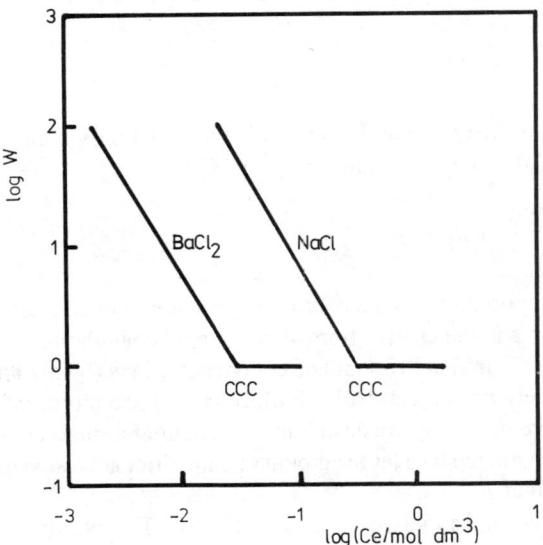

Fig. 16 — Logarithmic dependence of stability factor W on the molar concentration of NaCl or $BaCl_2$ (C_e) for a poly(styrene) latex with surface sulfate groups.

Decrease in the concentration of 'free' emulsifier and fraction of adsorbed emulsifier improves the contact of the surface of the polymer particles with the electrolyte and increases the rate of coagulation. Better interparticle contact is achieved by subjecting the polymer latex to pressure. For each latex there is a particular limiting value of the pressure above which particle coagulation occurs. Increase of pressure causes tighter packing of the particles, an increase in the number of active collisions and an increase in the effectiveness of flocculation [17–20].

Under certain conditions the hydrophilic high-RMM polymers stabilize polymer dispersions and under other conditions destabilize them. One part of the added reagent is absorbed on the polymer particle surface and the second one dissolves in the water. As the concentration of hydrophilic polymer increases, the rate of coagulation also increases and reaches a maximum at a particular concentration of the polymer, as is the case with NaCl. In contrast to the addition of low-RMM inorganic electrolytes further increase in the concentration of hydrophilic polymer causes an increase in the stability of the polymer dispersion [21, 22].

The distribution of the water-soluble polymer between the aqueous phase, where it associates with the free emulsifier molecules [23], and the polymer particles can be expressed by Eq. (10)

$$K = \frac{P_s E_i}{P_i E_s}, \qquad (10)$$

where K is the distribution coefficient, P_s and P_i are the amounts of polymer associated in solution and at the interface, respectively and E_s and E_i are the amounts of emulsifier dissolved in water and adsorbed on the particle surface, respectively.

The critical coagulation concentration of the polymer flocculant, above which flocculation of the latex is observed, is given by the sum of the components

$$\mathrm{CCC} = P_s + P_i + P_f. \tag{11}$$

Here P_f is the amount of free polymer flocculant in the system. By combining Eqs (10) and (11), the following relationship is obtained for the CCC

$$\mathrm{CCC} = \frac{KP_i}{E_i} E_s + P_i + P_f. \tag{12}$$

The CCC is also an important parameter for these systems. It characterizes the stability of the polymer dispersion and the change from slow to rapid coagulation.

Polymer particles prepared at low emulsifier concentrations (below and near the CMC) are covered ineffectively by the emulsifier molecules and are more liable to interactions with the polymer flocculant. They are destabilized on adding a small amount of flocculant. Polymer dispersions stabilized by a large amount of emulsifier are less sensitive to additions of the polymer flocculant.

Acrylamide polymers and copolymers are high-RMM efficient flocculants [24, 25]. The coagulative ability of these compounds increases with an increase in the fraction of acrylamide in the copolymer and with a decrease of the ionic strength.

The association of the polymer particles initiated by addition of a water-soluble polymer can operate via a bridging mechanism in which the polymer is adsorbed simultaneously on adjacent particles, binding them together [25, 26] or by trapping the particles into the polymer network formed by the flocculant [27–29]. The bridges reduce the particle mobility and increase their effective interactions. Only particles with functional groups and segments on their surface capable of forming strong bonds with the flocculant molecule coagulate effectively through the bridging mechanism. With the increase of the RMM of the polymer flocculant, the rate of coagulation increases; only polymers with an RMM 10^5 have been shown to be effective [30, 31]. Coagulation of the carboxyl polymer particles proceeds effectively via this mechanism; the polymer particles form bridges by forming hydrogen bonds between the carboxyl group and a polymer flocculant molecule (polyoxyethylene) [32]. The strength of the complex between the particle and flocculant is controlled by pH and is greatest at low pH.

By trapping particles into the network of a water-soluble polymer, the strength of the interparticle interactions and the rate of particle association increase. Polymer particle associates are formed with the active participation of the hydrogen bonds and electrostatic attractive forces which result from the reduced mobility of the particles localized in the network.

Flocculation by trapping the particles into the polymer network is used in the fractionation of a polymer latex and for determining the degree of polydispersity of a latex, because the participation of particles from solution is a function of their size [33]. Only polymer particles trapped in meshes of the polymer network precipitate from solution. The mesh dimensions can be determined by coagulation of polymer standards of various particle sizes [34]. Only particles with a size identical with the mesh dimensions of the polymer network are precipitated from solution.

The induction period of coagulation on addition of a flocculant is a result of the bridging or trapping of particles in the network and is a function of temperature, the concentration of free emulsifier in the system, the degree of coverage of the particles by the emulsifier and the pH [35]. It can be shortened by increase in temperature, using acids as flocculants and by reducing the emulsifier concentration [36]. The number of bridges necessary for destabilization of the dispersion is fewer at higher temperatures, which increase the number of effective particle collisions [37].

References

1. Derjagin, B. & Landau, L. (1941) *Acta Physicochim.* **14** 633
2. Verwey, E.J.W. & Overbeek, J.Th.G. (1948) *Theory of the stability of lyophobic colloids.* Elsevier, Amsterdam
3. Stauff, J. (1960) *Kolloidchemie.* Springer-Verlag, Berlin
4. Chapman, D.L. (1913) *Phil. Mag.* **25** 475
5. Sontag, H. (1977) *Lehrbuch der Kolloidwissenschaft.* VEB Deutscher Verlag der Wissenschaften, Berlin
6. Goodwin, J.W. (1981) *Colloidal dispersions.* Chem. Soc. Special Publication, London, p. 165
7. Smoluchowski von, M. (1916) *Phys. Z.* **17** 557
8. Fuchs, N. (1934) *Z. Phys.* **89** 736
9. Tsaur, S.L. & Fitch, R.M. (1987) *J. Colloid Interface Sci.* **115** 463
10. Fleer, G.J. & Lyklema, J. (1976) *J. Colloid Interface Sci.* **55** 228
11. Egusa, S. & Makuuchi, K. (1981) *J. Colloid Interface Sci.* **79** 350
12. Matijević, E., Mathal, K.G., Ottewill, R.H. & Kerker, M. (1961) *J. Phys. Chem.* **65** 826
13. Schulze, H. (1982) *J. Prakt. Chem.* **2** 25
14. Hoy, K.L. (1974) *Coatings Technol.* **51** 425
15. Hull van den, H.J. & Vanderhoff, J.W. (1971) In: Titch, R.M. (ed.) *Polymer colloids.* Plenum Press, New York, p. 1
16. Brouwer, W.M. & Zsom, R.L.J. (1987) *Colloids Surfaces* **24** 195
17. Melville, J.B., Willis, E. & Smith, A.L. (1972) *Trans. Faraday Soc.* **68** 45
18. El-Aasser, M.S. & Robertson, A.A. (1973) *Kolloid-Z. Z. Polym.* **251** 241
19. Homola, A.M. & Robertson, A.A. (1975) *Can. J. Chem. Eng.* **53** 389
20. Meijer, A.E.J., Megen van, W.J. & Lyklema, J. (1978) *J. Colloid Interface Sci.* **66** 99
21. Mabzar, L., Pefferkorn, E. & Varoqui, R. (1984) *J. Colloid Interface Sci.* **102** 380
22. Rao, I.V. & Ruckenstein, E. (1985) *J. Colloid Interface Sci.* **108** 389
23. Saito, S. & Taniguchi, T. (1973) *J. Colloid Interface Sci.* **44** 114
24. Myagchenkov, V.A., Bogdanova, L.V., Saveleva, S.D. & Frenkel, S.Y. (1974) *Khim. Tekhnol. Elementoorg. Soedin. Polim.*, Kazan 3-4 69
25. Kurenkov, V.F., Nagel, M.A. & Myagchenkov, V.A. (1984) *Eur. Polym. J.* **20** 779
26. La Mer, V.K. & Healy, T.W. (1963) *Rev. Pure Appl. Chem.* **13** 112
27. Ries, H.E. & Neyers, B.L. (1971) *J. Appl. Polym. Sci.* **15** 2023
28. Napper, D.H. (1977) *J. Colloid Interface Sci.* **58** 390
29. Packham, R.F. (1965) *J. Colloid Interface Sci.* **20** 81
30. Rusyaeva, V.A., Kukushkina, I.A., Kurenkov, V.F. & Myagchenkov, V.A. (1977) *Khim. Tekhnol. Elementoorg. Soedin. Polim.*, Kazan **6** 44
31. Evans, R. & Napper, D.H. (1973) *Nature* **246** 34
32. Ikawa, T., Abe, K., Honda, K. & Tsuchida, E. (1975) *J. Polym. Sci., Polym. Chem. Ed.* **13** 1505
33. Lindström, T. & Glad-Nordmark, G. (1984) *J. Colloid Interface Sci.* **97** 62
34. Lindström, T. & Glad-Nordmark, G. (1984) *Colloids Surfaces* **8** 337
35. Thompson, L. & McEwen, A. (1982) *J. Colloid Interface Sci.* **90** 329
36. Saito, S. & Fujiwara, M. (1977) *Colloid Polym. Sci.* **255** 1122
37. Saito, S., Taniguchi, T. & Matsuyama, H. (1976) *Colloid Polym. Sci.* **254** 882

4
Inverse emulsion (microemulsion) polymerization

4.1 CHARACTERIZATION OF INVERSE EMULSION AND MICROEMULSION SYSTEMS

The term microemulsion will, in this chapter, refer to a system composed of a nonpolar solvent (oil), water (containing a water-soluble monomer), and an amphiphilic compound. The term microemulsion was first introduced by Schulman *et al.* [1] for an optically clear mixture of oil, water, and amphiphile. These systems are of immense theoretical and practical importance (e.g. in oil production, for preparing polymers of high RMM, for enzymatic catalysis in inverse micelles) and are therefore the subjects of intensive research (see e.g. [2–7]).

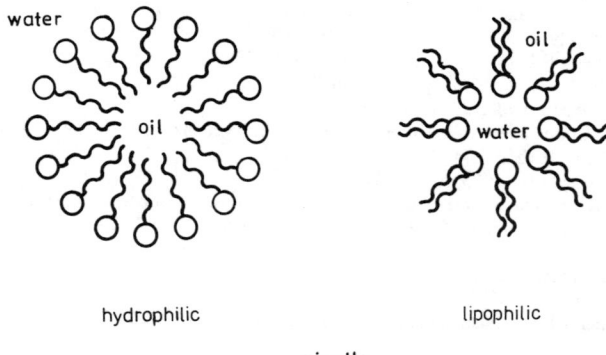

Fig. 17 — Schematic of hydrophilic and lipophilic micelles. ○ hydrophilic moiety of amphiphile (polar group), 〜 lipophilic moiety of amphiphile (alkyl chain).

Three coexisting liquid phases, namely an aqueous phase, a phase enriched in amphiphilic compound containig water and solvent, and a phase containing a high level of solvent, may be formed at low concentrations of the amphiphilic compound. In a strict sense, the phase rich in amphiphilic compound is a microemulsion (see also section 3.3.1.3). From the point of view of thermodynamics, this phase fulfils the condition of maximum mutual solubilization of oil and water. The interfacial tension between the oil and aqueous phase is then at a minimum. At low temperature the nonionic amphiphilic substance is more soluble in water than in oil, but the reverse is true at higher temperatures. The change in solubility has so far been explained only qualitatively. At a sufficiently high concentration, the amphiphilic compound forms, in water, micelles as spherical aggregates containing about 100 molecules of amphiphile per micelle. Figure 17 shows a hydrophilic (classical) micelle (oil-in-water, O/W, microemulsion) and a lipophilic micelle (inverse micelle, a W/O inverse microemulsion). On further increasing the concentration of amphiphilic compound, the spherical micelles change into cylinders (see also section 3.3.1.2., p. 87) and then, at sufficiently high concentrations of amphiphile, hexagonal and then lamellar liquid crystals are formed. This situation is depicted in Fig. 18.

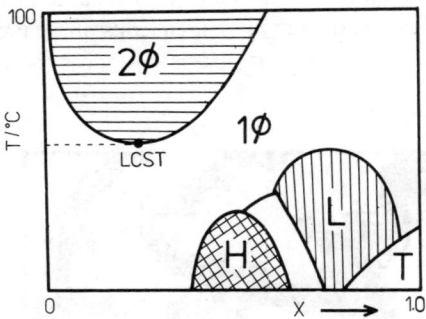

Fig. 18 — Phase diagram of water and nonionic amphiphilic compound of mass fraction x as a function of temperature. Schematic drawing. For explanations see text. (Reprinted from ref. [8] with permission of American Association for the Advancement of Science, Washington DC.)

The region 2Ø in Fig. 18 represents the coexistence of the two phases; the minimum on the curve (LCST) characterizes the value of the lower critical temperature of the mutual solubility of both components (water, oil). The symbol 1Ø covers the area of a microemulsion, i.e. a one-phase system. H and L are hexagonal and lamellar mesophases (liquid crystals), while region T denotes the existence of a solid solution. As the temperature increases, the hydrogen bonds between the hydrophilic groups of the amphiphile and water are disrupted. This lowers the water solubility of the amphiphile so that at a sufficiently high temperature, aqueous solutions of nonionic amphiphiles form two separate isotropic liquid phases (2Ø). One phase is richer in water, the other in amphiphile. With further temperature increase, the region of immiscibility broadens at first (with regard to the concentration of amphiphile in the aqueous phase, see Fig. 18), and then it gradually narrows and only when the temperatures are high enough (usually above the b.p. of water) is the system of water and amphiphile again completely miscible.

In the system water/oil/amphiphile at approximately the same mass ratios and at room temperature, the hydrophilic groups of the amphiphile form hydrogen bonds with the water molecules and draw the lipophilic groups of the amphiphile into the aqueous phase although they are strongly hydrophobic. At higher temperatures, the hydrogen bonds are weakened and the lipophilic groups of the amphiphile draw hydrophilic groups of the amphiphile into the oil-rich phase in spite of their hydrophilicity. The change in distribution of the amphiphile in the oil and water phases as a function of temperature lead to a temperature (or range of temperature) at which the amphiphile is equally soluble in water and the oil phase. At this temperature, the amphiphile prefers to be located at the interface between the oil and water phases to being dissolved in either of them. A macroscopically homogeneous mixture of water, oil and amphiphile is microscopically heterogeneous. It has a sponge-like structure (e.g. the oil phase represents the cavities, while the aqueous phase represents the skeleton of the sponge, see Fig. 19). A monolayer between the individual microscopic domains of water and oil is formed by the amphiphilic molecules. The structure undergoes remarkable fluctuational changes: the sizes and shapes of water and oil domains alter rapidly. The hydrophilic groups of the amphiphile are anchored in the aqueous phase (domains) while the lipophilic chains are directed to the domains of the oil phase. At temperature at which the properties of the amphiphile change from hydrophilic to lipophilic, the system oil/water/amphiphile splits into three phases: a lower water layer, a middle amphiphile-rich layer (microemulsion), and an upper oil-rich layer. On subsequent addition of amphiphile, the ability of the sponge to absorb both oil and aqueous phase increases, and the lower and upper layers are reduced until these two phases disappear (with only the microemulsion remaining).

Fig. 19 — Schematic of water/oil/amphiphile microemulsion (amphiphile monolayer at interfaces is not illustrated). □ aqueous phase, ■ oil phase (mass ratio of phases 1:1).

Splitting of the system into three liquid phases takes place only over a particular range of temperature which depends on the chemical structure of the oil and amphiphile. Above and below this temperature range the system forms two phases. Transfer to the one-phase system (microemulsion) can only be achieved by raising the concentration of amphiphile in the system.

An increase in hydrophilicity and lipophilicity of the amphiphile is accompanied by an increase in the strength of anchoring in the aqueous and oil phases, respectively. The monolayer formed by the molecules of amphiphile becomes stronger and the average sizes of the microdomains of water and oil increase. The structure of the microemulsion may be analysed by visible light, X-ray or neutron scattering. A measure of the size of the microdomains is the scattering peak which decreases smoothly with decreasing ability of the amphiphile to stabilize individual microdomains. With weak amphiphiles, the scattering peak is barely observed. The most recent observations show, however, that this conclusion is not fully valid [9].

Formation of the three phases for ionic amphiphiles, when the lipophilic part of the amphiphile becomes stronger, is achieved by raising the temperature or adding an electrolyte (NaCl). If the oil phase is made more lipophilic, e.g. by increasing the number of carbons in the chain or by increasing the number of lipophilic chains, the temperature of the three-phase system falls. Ionic amphiphiles show lipophilic properties at low temperatures, whereas at higher temperatures they are hydrophilic. As compared with nonionic amphiphiles, ionic amphiphiles modify the oil/water system by raising the temperature so that the oil phase is separated into the oil-rich phase and the microemulsion. On further temperature increase, the microemulsion absorbs more and more water until it merges into the aqueous phase, and an inverse microemulsion (water-in-oil) is formed. The phase diagram of the system water/oil with an ionic amphiphile is similar to that for the system water/oil with a nonionic amphiphile except that the properties and behaviour of the individual phases are reversed.

There are few ionic amphiphiles which form the three-phase system at temperatures between the m.p.'s and b.p.'s of the mixture. For instance, sodium lauryl sulfate (one lipophilic alkyl chain) forms the three-phase system below the m.p. of the mixture. When using a more lipophilic ionic amphiphile containing, for example, two sufficiently long alkyl groups, three phases appear above the b.p. of the mixture. For a mixture of suitably chosen ionic and nonionic amphiphiles (such that they form a three-phase system between the m.p. and b.p. of the mixture), the behaviour of the system is analogous to that of a nonionic amphiphile (at low fractions of an ionic amphiphile in a mixture with a nonionic) or, behaviour analogous to a system with an ionic amphiphile (at high fractions of the ionic amphiphile in a mixture with a nonionic one). At a particular ratio of both amphiphiles, which is a function of the composition of both amphiphiles and the type of oil, a microemulsion is formed over almost all the temperature range between the m.p. and b.p. of the mixture. Though weakly sensitive to the temperature change, such a microemulsion undergoes changes when the ratio of both amphiphiles and the concentration of electrolyte vary. Figure 20 shows the relationship between the phase changes of the system water/oil/amphiphile accompanying changes of the HLB (hydrophilic/lipophilic balance of amphiphile used) leading to the formation of an oil-in-water microemulsion (classical type) and an inverse water-in-oil microemulsion. The water or oil phase can be fully suppressed by raising the concentration of amphiphile (at a given HLB). A one-phase system, microemulsion O/W (a) or W/O (c) are then formed. The system O/W + W/O (bicontinuous or oil-and-water continuous regime, (b)) is schematically illustrated in Fig. 20.

The essential problem in the preparation of W/O emulsions and microemulsions as media for a radical-type polymerization reaction is their stability. The formation and stability of inverse emulsions and inverse microemulsions are affected by several parameters, mainly the type and the number of components of the system, the HLB value of the emulsifier, the

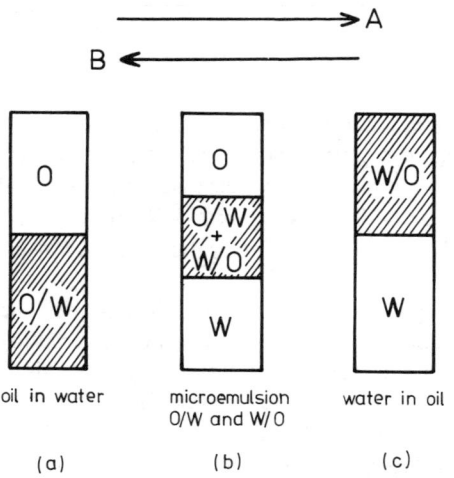

Fig. 20 — Schematic of transformation of oil-in-water microemulsion to water-in-oil microemulsion with decreasing HLB (arrow A) and vice versa (arrow B). O — oil phase, V — aqueous phase, ▨ — microemulsion. (Reprinted from ref. [10] with permission of Academic Press Inc., San Diego.)

presence of a co-emulsifier and temperature [11]. The value of the HLB necessary for the preparation of an inverse microemulsion system can easily be obtained by using a mixture of emulsifiers [12]. The concept of the cohesive energy ratio proposed by Beerbower and Hill [13] is used in the formulation of microemulsions [12]. This concept is based on the assumption that the HLB value can be determined from solubility parameters [14] of the oil component δ_{oil}, and the lipophilic part of the emulsifier(s) δ_L, of water δ_W, and of the hydrophilic part of the emulsifier(s) δ_H, according to

$$(HLB)_0 = 20\delta_L^2/(\delta_L^2 + K) ,$$

where $(HLB)_0$, is the optimum HLB if $\delta_L^2 = \delta_{oil}^2$ and K is a constant determined by

$$K = R_0 \, \delta_H^2 \, d_L/d_H .$$

Parameter $R_0 = V_L \delta_L^2 / V_H \delta_H^2$.

The symbols d_L and d_H are the densities of the lipophilic and hydrophilic parts of the emulsifier, V_L and V_H being the molar volumes of the lipophilic and hydrophilic moieties of the emulsifier. The values of the HLB and the parameters R_0 and K determine the character of the dispersion formed [13]. For instance, if the HLB of the system is in the range 4–6, $R_0 > 1$ and $K \sim 230$, then a W/O emulsion is formed. If, however, the HLB values vary between 12 and 15, $R_0 < 1$ and $K \sim 50$, then an O/W emulsion is formed. HLB values between 8 and 10, with $R \simeq 1$ and $K \sim 100$ lead to sytems which form a bicontinuous microemulsion W/O and O/W (inversion phase region). This procedure is applied in the preparation of a microemulsion of the cationic monomer of methacryloyloxyethyltrimethylammonium chloride in cyclohexane [15, 16]. The emulsifying system contained sorbitan sesquioleate (Arlacel 83, HLB = 3.7) and sorbitan monooleate with 20 ethylene oxide units (Tween 80, HLB = 15). The task was to establish the minimum amount of emulsifier necessary for the formation of a microemulsion for various concentration of cationic monomer in the system. The optimum result, i.e. the minimum amount of a mixture of emulsifier of about 6 mass % with respect to the system, was obtained for a mixture having an HLB between 12.5 and

13.0 and for monomer concentration of 2.7 mol dm^{-3} of water. At lower concentrations of cationic monomer (e. g. 2.3 mol dm^{-3} water and 1.7 mol dm^{-3} water), the requisite amount of the mixture of emulsifiers increased for the particular value of the HLB to *ca.* 15 mass % or 30 mass % with respect to the system. The result shows that the monomer functions as a co-emulsifier. A more detailed study of the structure of the dispersion showed it to be a bicontinuous microemulsion, type Winsor III (see Fig. 19, p. 188). As we have already stated, a bicontinuous microemulsion is characterized by an HLB value of between 8 and 10. The increase in value of the HLB to about 12 can be ascribed to the co-emulsifying effect of the monomer.

The presence of a water-soluble component of either polymerizable monomer or other molecule in an inverse emulsion (microemulsion) system may have a number of consequences. Detailed investigations of micellar systems by light scattering, ultracentrifugation and viscometry have shown that an admixture of acrylamide monomer extends the region of the microemulsion in the phase diagram for water–emulsifier–oil. This result also points to a coemulsifying effect of the monomer. Negative values of the second virial coefficient of the osmotic pressure in the micellar system [17] have been obtained as another consequence of the presence of acrylamide in a microemulsion system. This indicated the presence of interparticle attractive forces. In the presence of acrylamide, the conductivity of the system increases by several orders of magnitude [18]. These facts can be explained if acrylamide is partially located at the interface between the Aerosol OT (AOT) molecules (AOT: sodium bis(2-ethylhexyl)sulfosuccinate). This leads to the possibility of an increase in the attractive interactions as a result of some disorganization of the interfacial film and, consequently, rapid exchanges between water droplets [19]. Confirmatory evidence of these interactions can be obtained by small-angle neutron scattering [20]. The thickness of the shell of the inverse micelle (containing D_2O molecules but not acrylamide) is of the same order of magnitude as the length (*ca.* 12.5 A) of the lipophilic part of the AOT molecule [21]. If part of the D_2O is replaced by acrylamide, no repulsive interaction peak is observable. Introduction of acrylamide induces attractive interparticle forces resulting in percolative behaviour [17, 18]. The presence of acrylamide in the water pool of the inverse micelle reduces its size. This effect may by correlated with partial localization of the monomer at the W/O interface, resulting in a decrease of the interfacial tension [20, 22]. The presence of water-soluble additives (e. g. proteins) in the water pool of the inverse micelle gives rise to larger emulsifier aggregates than those formed in the absence of additive [23]. This finding has not been corroborated by other authors (e. g. [24]).

Changing of the [water]/[surfactant] ratio in an inverse micellar system can vary the diameter of the water pool of the inverse micelle. Assuming the droplets of water to be spherical, the radius of the water pool r_W is given by

$$r_W = 3\ V/S$$

in which V and S are the total polar volume and the total surface available, respectively. Taking the surface area per head polar group of AOT as being 60 A^2 and the water molecule volume equal to 30 A^3, the relation between the radius of the water pool and the water content w ($w = [H_2O]/[AOT]$) is

$$r_W = 1.5\ w.$$

This relation has been shown to hold up to $[H_2O]/[AOT] \leq 45$ by small-angle X-ray scattering and neutron scattering [23].

Change of the radius of the water pool of an inverse micelle by merely changing the ratio between water and emulsifier in the system provides interesting possibilities not only of introducing hydrophilic molecules of various size into the water pool but also for the preparation of polymers [18, 25]. Control of the size of the water pool, as well as of the number of water pools in disperse system (e. g. by diluting the disperse system by adding an organic disperse phase), i.e. the size and number of 'microreactors', in the particular reaction volume of the system, enables unique approaches to modelling the course of the radical polymerization. Such approaches are not provided by homogeneous polymerization systems (block or solution polymerization) and rarely even by classical emulsion polymerization systems.

We examine the questions of polymerization in inverse micellar W/O systems in other sections of this chapter. Classical dispersions of O/W systems are described in detail in Chapter 3.

References

1. Schulman, J.H., Stochenius, W. & Prince, L.J. (1959) *J. Phys. Chem.* **63** 1677
2. Borkovec, M., Eicke, H.F., Hammerich, H. & Das Gupta, B. (1988) *J. Phys. Chem.* **92** 206
3. Holmberg, K. & Österberg, E. (1986) *J. Dispersion Sci., Technol.* **7** 299
4. Friberg, S. & Podzimek, M. (1986) *J. Dispersion Sci., Technol.* **7** 57
5. MacDonald, H., Bedwell, B. & Gulari, E. (1986) *Langmuir* **2** 704
6. Robbins, M.L. & Bock, J. (1988) *J. Colloid Interface Sci.* **124** 486
7. Khmelnitskii, Yu.L., Kabanov, A.V., Klyachko, N.L., Levashov, A.V. & Martinek, K. (1989) In: Pileni, M.P. (ed.) *Structure and reactivity in reverse micelles*. Elsevier, Amsterdam, p. 230
8. Kahlweit, M. (1988) *Science* **240** 617
9. Kilpatrick, P.K., Davis, H.T., Scriven, L.E. & Miller, W.G. (1987) *J. Colloid Interface Sci.* **118** 270
10. Robbins, M.L. & Bock, J. (1988) *J. Colloid Interface Sci.* **124** 462
11. Candau, F. (1990) In: Candau, F. & Ottewill, R.H. (eds) *Scientific methods for the study of polymer colloids and their applications*. NATO Sci. Ser. D. Reidel Publ. Co., Dordrecht, p. 73
12. Candau, F. & Holtzscherer, C. (1989) *J. Chim. Phys.* **86** 2095
13. Beerbower, A. & Hill, M.W. (1971) In: *McCutcheon's detergents and emulsifier annual*. Allured Publ. Co., Ridgewood, p. 223
14. Hildebrand, J.H. & Scott, R.L. (1964) *The solubility of non-electrolytes*. 3rd ed. Dover Publ., Mineola
15. Candau, F. & Buchert, P. (1990) *Colloids Surfaces* **48** 107
16. Buchert, P. & Candau, F. (1990) *J. Colloid Interface Sci.* **136** 527
17. Candau, F., Leong, Y.S., Pouyet, G. & Candau, S.J. (1984) *J. Colloid Interface Sci.* **101** 167
18. Carver, M.T., Hirsch, E., Wittman, J.C., Fitch, R.M. & Candau, F. (1989) *J. Phys. Chem.* **93** 4867
19. Leong, Y.S., Candau, S.J. & Candau, F. (1984) In: Mittal, K. & Lindman, B. (eds) *Surfactants in solution*. Plenum Press, New York, 3: 1987
20. Holzscherer, C., Candau, F. & Ottewill, R.H. (1990) *Prog. Colloid Polym. Sci.* **81** 81
21. Sheu, E., Chen, S.H. & Huang, J.S. (1987) *J. Phys. Chem.* **91** 3306
22. Pichot, C., Graillat, C. & Revillon, A. (1987) In: *Proceedings of the XVIIth Congress AFTPV*, Nice, p. 270
23. Pileni, M.P. (1989) In: Pileni, M.P. (ed.) *Structure and reactivity in reverse micelles*. Elsevier, Amsterdam, p. 44
24. Levashov, A.V. & Khmelnitskii, Yu.L., Klyachko, N.L., Chernyak, V. Yu. & Martinek, K. (1982) *J. Colloid Interface Sci.* **88** 444
25. Candau, F., Leong, Y.S. & Fitch, R.M. (1985) *J.Polym. Sci., Polym. Chem. Ed.* **23** 193

4.2 KINETICS AND MECHANISM OF RADICAL POLYMERIZATION IN INVERSE MICELLAR SYSTEMS (WATER-IN-OIL EMULSIONS AND MICROEMULSIONS)

In the preceding section we discussed several problems associated with the preparation of microemulsions in the three-component system: water/oil/amphiphilic compound. By adding monomer to the disperse phase, i.e. to lipophilic micelles containing water [1, 2] or to the continuous (oil) phase [3, 4], a polymer can be formed via radical polymerization.

The monomer cannot, however, be considered as an inert compound which influences neither the formation nor the stability of the microemulsion. Under particular conditions, the monomer may function as a co-emulsifier (a substance, which, in the presence of a classical emulsifier, supports formation of the microemulsion; from among known cases of such pairs we can mention sodium lauryl sulfate and higher alkanols). A polymer formed on polymerization of a monomer may also lead to serious problems in view of the stability of the microemulsion. In remarkably concentrated systems, a high-volume fraction of the lipophilic micelles can limit the number of various conformations which might be occupied by a polymer in the continuous phase in the absence of a disperse phase. This limitation lowers the entropy of the polymer and leads to a remarkable increase in its Gibbs free energy [5–8] and to the phase separation of the polymer.

For instance, in their study of the effect of styrene and poly(styrene) on the properties of a microemulsion stabilized in the conventional way, i.e. by emulsifier and co-emulsifier, the authors [9] found that oligomers and polymers of styrene (of various RMM) markedly reduce the area describing the presence of the microemulsion in a phase diagram. The higher is the concentration of oligomer or polymer used and the higher their RMM, the greater is the reduction of this area.

The polymerization of methyl methacrylate (MMA) in a microemulsion composed of 41.7% MMA, 27.0% n-pentanol, 14.6% sodium lauryl sulfate, and 16.7% water at various 2,2'- azobisisobutyronitrile (AIBN) concentrations was described by the present authors [3] and again elsewhere [10]. Turbidity unavoidably occurs during polymerization even if the poly(methyl methacrylate) being formed has been dissolved in methyl methacrylate. Other authors [11, 12] also observed destabilization of a W/O microemulsion by the polymer formed. It seems to have been demonstrated in the case of n-pentanol as co-emulsifier, in which poly(methyl methacrylate) does not dissolve, that no transparent inverse latex can be prepared. On replacing n-pentanol with acrylic acid [10], the area of the microemulsion was increased. Acrylic acid is in fact a comonomer in the system (together with MMA) and is a solvent of poly(methyl methacrylate). The influence of poly(methyl methacrylate) on the destabilization of the microemulsion is not clear-cut in the presence of acrylic acid, and the preparation of a transparent microemulsion with fully polymerized methyl methacrylate using 2,2'-azobisisobutyronitrile (AIBN) or potassium peroxodisulfate as initiators is possible, though within a rather small range of concentrations of the individual components of the system. A transparent inverse poly(methyl methacrylate) dispersion (in fact a dispersion of poly(methyl methacrylate-co-acrylic acid)) was formed via polymerization of the system of water (15%), sodium lauryl sulfate (2.5%), and the mixture methyl methacrylate/acrylic acid (82.5%, mass ratio 1:1). The initiator AIBN (0.25–1.5% per monomer mixture) did not show any remarkable effect on the microemulsion. The degree of polymerization of the copolymer (inversely proportional to the initiator concentration) does not influence the stability of the microemulsion. Using water-soluble $K_2S_2O_8$, 100% conversion of the comonomers can also be achieved. However, in contrast to the system containing AIBN, the system is sensitive to change in the concentration of $K_2S_2O_8$. Selection of the concentrations of the components involved in the system for the preparation of a transparent system both during polymerization and after its completion is more restricted than for AIBN initiator.

One of the stable formulations (i.e. a system transparent before and after polymerization) was used for a kinetic study of the polymerization, viz. H_2O (10%), sodium lauryl sulfate (2%), acrylic acid (34%), and methyl methacrylate (54%). The concentration of AIBN was 0.1% per mixture of monomers (6.2×10^{-3} mol kg^{-1} monomers) or 0.5% $K_2S_2O_8$ per aqueous solution (water + emulsifier), i.e. 1.97×10^{-3} mol kg^{-1} water in the reaction system. In spite of about a 30-fold higher concentration of AIBN in the system as a whole, the rate of polymerization was only slightly greater than by using $K_2S_2O_8$. The viscosity of the system strongly increased during polymerization and at about 30% conversion the system gelated (after *ca.* 3.5 h). To affect the complete conversion of the comonomers, the system was polymerized for a total of 48 h. The RMM of the copolymers approached 10^6 g mol^{-1} while the polydispersity of the copolymers increased with conversion. The water content in the copolymers varied between 0 and 15% and the loss of water due to drying ranged between 0.5 and 5%. The complete removal of water by drying was not achieved, indicating that the water molecules are strongly bound in the copolymer matrix. The composition of the copolymers is a function of conversion. At low conversions, the copolymer is very rich in methyl methacrylate structural units ($r_{MMA} = 1.86$; $r_{acryl.acid} = 0.24$ for 50°C [12]). As conversion proceeds, the fraction of copolymerized acrylic acid increases. The authors [10] believe that the initiation of polymerization by $K_2S_2O_8$ starts in the dispersed aqueous phase with the formation of oligomers of acrylic acid or of methyl methacrylate [13]. Oligomeric radicals then migrate to the continuous oil phase, where the majority of propagations and terminations takes place.

An interesting case of polymerization in a microemulsion, where all the components except water are monomers, has been described in refs. [14] and [15]. The microemulsion was formed from a mixture of sodium acrylamideundecanoate, methyl methacrylate, acrylic acid and water. Microemulsions were polymerized using AIBN (0.5%) at 50°C. For polymerizations proceeding longer than 24 h, some polymerized microemulsions in the form of solid polymer were opaque, while others were transparent. Before polymerization the region of concentrations of the transparent microemulsion is about twice that of the polymerized microemulsion (see Fig. 32). Terpolymers containing 5–12% water were prepared in this way. At higher water concentrations, an opaque product was formed. The authors [14] named the new types of terpolymer *micellar terpolymers* to differentiate them from classical terpolymers prepared via solution terpolymerization.

The use of nonionic amphiphiles or their mixtures may be demonstrated by an example of disperse polymerization of acrylamide in a mixture of alkanes [16]. The advantage of this polymerization is that it enables formation of poly(acrylamide) of RMM > 10^7 g mol^{-1}. The polymerization system was prepared as follows: an aqueous solution of acrylamide (acrylamide 22.7 mass %, water 34.05 mass %, boron(III) oxide 0.24 mass %, and ethylenediaminetetraacetic acid 0.006 mass %)* was dispersed with vigorous stirring in a system of alkanes of b.p. between 204 and 247°C (40.53 mass %) containing sorbitan monooleate (2.47 mass %). On bubbling through nitrogen (to remove aerial oxygen) and annealing to the polymerization temperature, oil-soluble (azodimethylvaleronitrile and AIBN) or water-soluble (azobutyroamidine (A)) initiator as a solution in hydrocarbon or water) were added with a syringe:

$$A: \quad \begin{array}{c} \text{CH}_3 \quad\quad \text{CH}_3 \\ | \quad\quad\quad | \\ \text{HN}=\text{C} - \text{C}-\text{N}=\text{N}- \text{C} - \text{C}=\text{NH} \\ | \quad\quad | \quad\quad\quad | \quad\quad | \\ \text{NH}_2 \;\; \text{CH}_3 \quad\quad \text{CH}_3 \;\; \text{NH}_2 \end{array}$$

---

* Percentage relative to a mixture of all components of the disperse polymerization.

The kinetic course of disperse polymerization with water-soluble azo initiator corresponds to the course of polymerization in solution. The RMMs of poly(acrylamide) are much lower than when using oil-soluble initiator under otherwise identical conditions. The polymerization rate is proportional to the concentration of oil-soluble initiator over a wide range of concentrations. On the other hand, a square root dependence ($R_p \sim C_I^{0.5}$) of the polymerization rate on the initiator concentration was observed. The rate of polymerization decreases as a function of the emulsifier concentration with an exponent –0.2. In addition to temperature, the rate of stirring, i.e. the size of the interphase surface (as a result of the decrease in size of disperse particles at a constant ratio between water and the oil phase) affects the polymerization rate. The authors [16] propose that at low conversions, initiation of the polymerization of monomer by an oil-soluble initiator takes place via the transfer of primary radicals or oligoradicals from the oil into the aqueous phase.

It is assumed in the case of a water-soluble initiator ($K_2S_2O_8$) and at low emulsifier concentrations (stearic acid ethoxylated by 8 moles of ethylene oxide, concentration 5–6 mass % per oil phase) [17] that acrylamide polymerization begins in water droplets containing monomer. The polymerization of monomer leads to an increase in the viscosity in the water droplets, and a gel is formed, followed by phase inversion of the water-in-oil dispersion. In the later stages of polymerization, when a sufficient amount of polymer is formed in the aqueous phase, a polymer dispersion containing water is formed due to stirring-in the continuous oil phase (formation of water (+ polymer)-in-oil dispersion).

Results of a study of the disperse polymerization of acrylamide in the presence of the polyoxyethylene adduct of polyoxypropyleneethylenediamine (Tetronic 1102) and dibenzoyl peroxide are described in ref. [18]. Here it is assumed that the nucleation site of the polymer particles is a water droplet of the emulsifier lipophilic micelle. Radicals arise from decomposition of dibenzoyl peroxide dissolved or solubilized in a water droplet of the lipophilic micelle or at the emulsifier–water droplet interface. In the case of the azo initiator, radicals are formed at the emulsifier–oil interface [19].

Acrylamide is readily soluble in toluene (2% at 50°C [20]). It is therefore assumed that for acrylamide polymerization in the inverse microemulsion formed by water and toluene phases under the influence of a mixture of nonionic emulsifier (sorbitan sesquioleate Montane 83) with HLB = 4.0 and polyoxyethylene sorbitan trioleate (Montanox 85 with high HLB), the main source of radicals is AIBN initiator in the oil phase. The initial size of the dispersed water droplets in the continuous toluene phase is small (the initial droplet size is characterized by bimodal size distributions with averages of ca. 20 nm and 80–400 nm), their overall surface being therefore large. There is a high probability of trapping of primary or oligomeric radicals from the oil phase by water droplets (containing the main fraction of acrylamide). As regards nonionic emulsifiers, the barrier formed by the emulsifier at the oil–water interface is electrically neutral. In spite of this the emulsifier at the interface sterically hinders the penetration of radicals as indicated by the dependence of the rate of acrylamide polymerization on the stirring rate. It is typical that the resulting inverse latex is not monodisperse, its average particle size being little smaller than the average of the greater disperse water droplets (containing acrylamide), i. e. smaller than the range between 80 and 400 nm reported. The authors [21] came to similar conclusions.

Acrylamide polymerization in inverse micellar systems in the presence of a hydrolytic agent and of the redox initiating system $K_2S_2O_8 - Na_2S_2O_5$ is described elsewhere [22, 23].

In the presence of sodium hydroxide, partial hydrolysis of the amide groups of acrylamide takes place and sodium acrylate is formed. The polymer product is then a copolymer, i.e. poly(acrylamide-co-sodium acrylate). In contrast to poly(acrylamide) homopolymer, this copolymer is less vulnerable to imidization of its amide groups, which leads to formation of a crosslinked product during acrylamide polymerization in more concentrated solutions and at temperatures above 50°C. In the presence of NaOH, rate enhancement of the acrylamide polymerization or, better, of the copolymerization of acrylamide with sodium acrylate, was observed. Responsibility for the rate increase of the copolymerization rests with the increase in the initiation rate given by the enhanced decay rate of $K_2S_2O_8$ under the effect of NaOH or reaction intermediates of acrylamide hydrolysis. In agreement with the kinetics of radical processes, an increase in the concentration of acrylamide between 0.9 and 5.4 mol dm^{-3} (water) leads to enhancement of the polymerization rate; the reaction order with respect to acrylamide is 1.1. On the other hand, increase in the acrylamide concentration in the system leads to a decrease in the degree of hydrolysis of acrylamide. The reason is the formation of associates of acrylamide (intermolecular hydrogen bonds between the oxygen of the carboxyl group and the hydrogen of the amide group of the two adjacent acrylamide molecules). By raising the concentration of redox initiating system (at a given ratio between potassium peroxodisulfate and sodium bisulfite), the rate of polymerization is enhanced. The reaction order with respect to potassium peroxodisulfate is 1.0 which indicates the importance of monomolecular termination in the system. The monomolecular termination of macroradicals is considered to be caused by the high viscosity near the reaction site of the polymerization. The viscosity reduces the mobility of the macroradicals thus making mutual bimolecular termination impossible.

The use of the anionic emulsifier sodium bis(2-ethylhexyl)sulfosuccinate (AOT) for the polymerization of acrylamide in the inverse emulsion polymerization system toluene/water was described by Leong & Candau [24]. The system did not require any addition of coemulsifier for the formation of lipophilic micelles [25]. AIBN was used as initiator. The microemulsion contained before polymerization (mass %) AOT (17.2), acrylamide (3.5), water (10.4), toluene (68.9) and AIBN/acrylamide (0.14). The continuous phase involved toluene which has low AOT and water concentrations. Dilution of the continuous phase to the volume fraction of the dispersed phase of about 1% did not bring about any important change in the particle size of the disperse phase. The authors used for polymerization either an undiluted dispersion (a microgel of polymer is formed via polymerization) or the dispersion was diluted with 6-fold volume of toluene. They used UV radiation to decompose the initiator to radicals. During polymerization, the microemulsions were perfectly transparent and stable. Phase separation was not observed and Tyndall scattering was only observable at low ratios between acrylamide and water (< 1.3). The rate of polymerization was high, complete monomer conversion being reached in about 30 min. The viscosity of the dispersion was low (7–8 mPa s). Together with the transparency of the system, the authors consider it to be proof of the restriction of polymerization to the dispersed phase (lipophilic micelles containing water and acrylamide). The hydrodynamic radius of the particles, R_H, was determined from the dependence of the diffusion coefficient, D, of the microemulsion (before polymerization) and the dispersion (after polymerization) on the dilution of toluene according to the Stokes – Einstein equation:

$$D = kT / 6\pi \eta_0 R_H,$$

where k is Boltzmann's constant, T is the absolute temperature, and η_0 the solvent viscosity. D decreases during polymerization, indicating the change in R_H. The diffusion coefficient, D, is obtained via extrapolation of the calculated value at the given degree of dilution to zero concentration [26]. For the initial microemulsion, R_H = 3.8 nm and for the polymer dispersion, R_H = 16 nm. The RMM of poly(acrylamide) is M_w = 3.2 × 10^6 g mol^{-1} (light scattering). As regards the small hydrodynamic diameter of the polymer particles, the poly(acrylamide) chain has to form a highly compressed coil inside the micelle. The results of a study of the structure of the microemulsion before polymerization and the resulting polymer particles in the inverse micellar system acrylamide/water/toluene/AOT, by the methods of elastic and quasielastic light scattering (QELS), ultracentrifugation and viscosimetry have been described by others [26]. The hydrodynamic radius of the dispersed particles, R_H, obtained using QELS is systematically greater than the geometrical radius, R, obtained by elastic light scattering. This result, also observed by others [27, 28], may be ascribed to the penetration of toluene (the continuous phase) into hydrocarbon chains of the AOT molecules creating a micelle. At a constant H_2O/AOT ratio, the particle size increases with increasing acrylamide concentration, e.g. from a value of 3 nm at an acrylamide concentration of 2.1% to 5.7 nm at an acrylamide concentration of 5.8 %. This increase in the particle size is not a simple reflection of the monomer swelling since it is not proportional to the concentration of acrylamide. At a constant acrylamide concentration in the system, the ratio of the micelles is a linear function of the ratio H_2O/AOT (see section 4.1). The means whereby this ratio changes is unimportant (e.g. whether through an increase in the amount of the water or by decreasing the AOT concentration). The particle distribution is rather narrow and the particles are spherical or spheroidal. The micellar system gave after acrylamide polymerization (initiated by the thermal decomposition of AIBN at 45°C or its photochemical decomposition at 25°C) higher values of the hydrodynamic radius, R_H, and of the geometrical radius, R, than those measured for the microemulsion before polymerization. The dispersed particles have grown during polymerization from the original 2 nm and 6 nm to 14 nm and 26 nm, respectively. However, the values of R_H and R of the system before and after polymerization strongly depend on the composition of the original microemulsion (as mentioned above). Increase of the particle size leads to an increase in the overall interfacial surface area between the dispersed particles and the continuous phase. After polymerization, an excess of AOT molecules, with respect to the amount of AOT needed for the stabilization of the particles, is observed in the system. The excess AOT forms therefore inverse micelles in the system which are more or less hydrated by water molecules. The disperse system contains thus two types of dispersed particles (in addition to polymer particles with a diameter $d \sim 50$ nm as well as AOT micelles with a diameter $d \sim 3$ nm). With regard to this fact, the most reliable data about the size of dispersed latex particles are provided by quasielastic light scattering [26]. The number of macromolecules per particle varies as a function of the particle radius and the RMM of the polymer between 1 and 17. The transport of monomer necessary for the propagation step, i.e. for the growth of the polymer particles, is carried out via a collision mechanism (by collision of a growing particle containing a radical centre with a 'dead' particle containing monomer which serves as a source of monomer) or via a mechanism of monomer diffusion from 'dead' particles through the continuous phase into the dispersed particles comprising propagating monomer chains (by analogy with conventional emulsion polymerization) [26, 29].

The preparation of a poly(acrylamide) inverse microemulsion stabilized by sodium bis-(2-ethylhexyl)sulfosuccinate has been described [30]. Toluene was used as a continuous phase and AIBN or potassium peroxodisulfate as initiators. The rate of polymerization in the presence of AIBN is proportional to the initial monomer concentration, while in the presence of $K_2S_2O_8$, the reaction order in monomer is 1.5. The inverse proportionality observed between the RMM of the poly(acrylamide) and the emulsifier concentration indicates the participation of emulsifier in the initiation step of polymerization. This view is also supported by the fact that the concentration of initiators used has no effect upon the

RMM. The authors present as an original finding the fact that a polymer particle of the resulting polymer dispersion contains only one poly(acrylamide) molecule (*cf.*, however ref. [26]).

These results indicate the invalidity of the Smith–Ewart theory (Chapter 3) and that particle nucleation takes place throughout the polymerization. The following relationships [30] are a quantitative expression of the kinetics of polymerization. They characterize the dependence of the rate of polymerization, R_p, of acrylamide in an inverse microemulsion on the concentration of acrylamide (AM), initiator (AIBN, $K_2S_2O_8$), emulsifier (AOT) and of the viscosity RMM of the poly(acrylamide) (M_v) on the concentration of AM, AOT, AIBN, and $K_2S_2O_8$

$$R_p \sim [AM]^{1.1}$$

$$\overline{M}_v \sim [AM]^{1.1}$$

$$R_p \sim [AIBN]^{0.1}$$

$$\overline{M}_v \sim [AIBN]^{-0.03}$$

$$R_p \sim [AOT]^{-0.55}$$

$$\overline{M}_v \sim [AOT]^{-0.8} \quad \text{(systems with AIBN)}$$

$$R_p \sim [AM]^{1.5}$$

$$\overline{M}_v \sim [K_2S_2O_8]^{-0.1} \quad \text{(systems with } K_2S_2O_8\text{)}$$

Of other examples of acrylamide polymerization in inverse nonionic microemulsions, the results of ref. [31] are notable. The effects of different amounts of emulsifier, monomer, and salt (sodium acetate) on the size and stability of the polymer particles and the RMM of the polymer were studied. The authors used as the oil phase a mixture of alkanes (Isopar M) and a mixture of nonionic emulsifiers having an HLB value of 9.3. Increase of the sodium acetate concentration failed to affect either the particle size or the RMM of the poly(acrylamide). An increase in the emulsifier concentration caused a decrease in particle size and an increase in the RMM. As the AM concentration was increased, the particle size and the RMM of poly(acrylamide) increased.

The number of macromolecules per particle, N_p, is a function of the mass ratio of acrylamide and emulsifier, AM/E. The number of macromolecules, N_p, is also a function of the polymer particle diameter. For ratios of AM/E between 0.6 and 1.8, the number of macromolecules per particle varied between 5 and 45.

This dependence can be expressed as

$$N_p \sim (AM/E)^{1.6} .$$

On increasing the diameter of the polymer particles (from 65 nm to 125 nm), the values for N_p became 5 and 35. The result concerning the number of macromolecules per polymer particle disagrees with that published earlier [30], indicating the presence of only 1 macromolecule of poly(acrylamide) per polymer particle. Reference [31] reports that the value of $N_p = 1$ is a limiting case obtained when the AOT was used as emulsifier. It seems that in comparing the two microemulsion systems prepared in the presence of various emulsifiers and oil phases (Isopar M in ref. [31], toluene in [30]), and various ratios of the reactants in the feed, the assumptions required to calculate the particle number, e.g. a spherical shape of polymer particle, are not necessarily exactly fulfilled. Similarly, precise values of the particle size, monomer conversion and, above all, number average RMM of poly(acrylamide) derived from the viscosimetric value of the RMM necessary for the calculation may be unavailable. Taken as a whole, these circumstances can easily lead to the scatter in N_p values observed. This is why in consideration of the mechanism for the formation of polymer particles in an inverse emulsion system, these complications associated with the parameters of N_p have to be taken into account.

The polymerization of acrylamide in an inverse microemulsion in the presence of an ionic emulsifier (AOT) is characterized by the presence of spherical (globular) micelles of the emulsifier. The total monomer concentration in the system is about 5 mass % [24, 26]. On the other hand, when nonionic emulsifiers are used, e.g. of the sorbitan ester type [32, 33], the inverse emulsion system has a bicontinuous structure which can incorporate larger amounts of acrylamide, usually up to 25 mass %.

The high polymer content in inverse microemulsions is desirable for many industrial applications. In order to optimize the process of microemulsion formation at relatively high monomer content and at minimal emulsifier concentrations, the inverse emulsion copolymerization of acrylamide with acrylic acid was studied [34, 35]. The microemulsions were prepared by adding an aqueous solution of monomers neutralized at pH 9–10 to a mixture of nonionic emulsifiers, 2,2′-azobisisobutyronitrile and Isopar M (a mixture of alkanes). The addition of monomers considerably extends the microemulsion region in the phase diagram. A close correlation exists between the optimum HLB value of the emulsifier blend, the minimum emulsifier concentration, and the sodium acrylate content in the feed. The solubility of the polyoxyethylene group of the nonionic emulsifier in water decreases with temperature owing to weakening of the interaction (hydrogen bonding) between the water molecules and the hydrophilic moieties of the emulsifier (see also section 4.1). The addition of salt (i.e. sodium acrylate) remarkably reduces the temperature at which incipient phase separation occurs, i.e. the cloud point. Most electrolytes exert a 'salting out' or cloud-point lowering effect by dehydrating the emulsifier [36]. Thus at a particular temperature, the HLB value of the emulsifier becomes more lipophilic on addition of salt to the system. It is therefore necessary to shift the HLB value to higher values in order to counteract the decrease in solubility to optimize the emulsification. On the other hand, the progressive addition of salt, while decreasing the solubility of the emulsifier in water, tends to repel the hydrophilic moieties towards the periphery of the micelles at the W/O interface. Consequently, the interfacial layer becomes more condensed, which is a stabilizing factor. This may explain the lesser amount of emulsifier required to form an inverse microemulsion in the presence of salt. An excess of salt, which produces a large decrease in the HLB value is, of course, unfavourable towards the stabilization of microemulsions. High concentrations of emulsifier are then required, which is disadvantageous.

Some systems studied [34] were initially turbid but after polymerization became clear and stable inverse microemulsions. On the other hand, several systems, initially clear, became turbid after polymerization, and polymer particles settled out from the microemulsion. Some systems were both clear initially and remained clear after polymerization and were stable inverse microemulsions (e.g. the microemulsion composed of Isopar M 38.08, emulsifier blend 15.60, water 23.37, copolymers 22.95 mass %; the sodium acrylate content in copolymer was 42.22 mass %). The initial stage of polymerization is evidenced by the appearance of turbidity and a notable increase in the viscosity of the system. After a few minutes the system becomes clear and fluid again. This point corresponds to the end of reaction. A high degree of conversion of monomer is attained. The appearance of turbidity and the viscosity increase during polymerization is due to the bicontinuous structure of the system. The formation of macromolecules soluble in the aqueous domains of the microemulsion leads to the increase in viscosity. The formation of polymer causes a shift in the phase diagram, producing temporarily a three-phase system (see Fig. 20, scheme b) which progressively evolves towards the final globular configuration observed after the completion of polymerization. The final product is clear and stable, with a viscosity similar to that before polymerization. The diameter of the polymer particles (~ 60 nm) decreases with increasing acrylate content and emulsifier concentration.

Another example of polymerization in an inverse microemulsion system with a rather high monomer (methacryloyloxytrimethylammonium chloride) concentration was described earlier [37]. We have described details of the bicontinuous structure of this inverse microemulsion system and its composition in section 4.1 (p. 191). The size of the polymer particles increases with increasing monomer concentration in inverse microemulsion polymerization systems from ca. 55 nm for 5% of monomer to 116 nm for 25% of monomer. The concentration of the mixture of emulsifiers varied between 10 and 15 %, water between 5 and 25% and cyclohexane from 35 up to 80%. The values of the water/emulsifier ratios and the volume fractions of monomer and water in the system are not constant and interpretation of the reason for the growth in size of the polymer particles with increasing monomer concentration in the starting inverse microemulsion is difficult. The authors observed an increase in the RMM of poly(methacryloyloxytrimethylammonium chloride) with an increasing ratio between the monomer and the mixture of emulsifiers. From the diameters of the polymer particles, their number and the RMM of the polymer in the polymer particles they determined the average number of macromolecules, N_p, per polymer particle. A value of between 1 and 2 was obtained [37]. (See discussion about the problem of determining N_p, p. 199.)

The main features of inverse emulsion and inverse microemulsion processes have been analysed with emphasis given to the search for the optimal formulation of the system before polymerization [38]. Based on kinetic studies, the possible reaction mechanism of the inverse emulsion and inverse microemulsion processes have been discussed elsewhere [39]. The polymerization of water-soluble monomers in 'inverse microsuspension' (or, more commonly, 'inverse emulsion') involves the emulsification of a water-soluble monomer in a continuous organic phase. A water-in-oil steric stabilizer is used and the polymerization is carried out with either a water- or an oil-soluble initiator. Remarkable advantages of the process are that microsuspensions have low viscosities, they facilitate removal of the heat of reaction and utilize monomer concentrations higher than those possible in solution polymerization. Addition of an inverse microsuspension to water leads to its inversion and the water-swollen particles of the inverse microsuspension rapidly dissolve. In contrast, a polymer which has been dried, experiences gel blocking when added to water. When water-soluble initiators are used, all components of the inverse microsuspension (emulsion) are in dispersed monomer droplets. Each droplet functions as a small batch reactor and the kinetics are similar to those of solution polymerization [20, 40, 41]. The following locations are taken into account in inverse emulsion polymerization systems when using an oil-soluble initiator; the monomer droplets, monomer-swollen micelles, disperse (oil) phase (if the monomer is partially soluble in it), and the interphase containing the lipophilic parts of the emulsifier of the oil micelles. When using oil-soluble initiators and an aromatic continuous phase, the kinetics resemble those of emulsion polymerization with the initiation site occurring in the inverse micelles. If, instead of an aromatic compound, a paraffinic continuous oil phase is used, the site of initiation is in the monomer droplets. This polymerization resembles physically and kinetically a suspension process and is referred to as an inverse microsuspension because the average particle size is nominally 1000 nm (1 μm). The oil-soluble initiators and oil-soluble free radicals can diffuse from the oil phase through the emulsifier interface into the aqueous monomer droplets or into micelles swollen by monomer (and water) and then start the polymerization. The reverse process, i.e. the diffusion of water-soluble initiator and water-soluble free radicals into the continuous oil phase, is unimportant here because of the water-soluble monomers used for the inverse emulsion process. However, this possibility might be of some importance, as we shall see when discussing the copolymerization of oil-soluble monomer with water-soluble monomer in an inverse emulsion or microemulsion.

Depending on the initiation site, particle growth can take place by monomer diffusion from the monomer droplets and/or micelles swollen by monomer to the propagation site in a manner somewhat similar to the classical oil-in-water emulsion system. Another possibility for polymer particle growth in inverse emulsion and/or microemulsion is offered by the coagulative mechanism. The transport of monomer to a polymer particle could occur during its collisions with monomer-containing species of the system.

For the description of the kinetics of inverse emulsion systems, the kinetic laws established for classical, conventional emulsion polymerization (see Chapter 3) can be applied. However, in fact, only a few attempts have been made to apply the Smith–Ewart theory to inverse emulsion polymerization [19, 42].

Recently a mechanistic model has been developed [40] for the inverse emulsion (suspension) polymerization of acrylamide initiated by 2,2'-azobisisobutyronitrile, which includes the unimolecular termination of poly(acrylamide) radicals with emulsifier and the mass transfer of primary radicals and acrylamide oligomer radicals between the continuous oil and the dispersed water phases. The model predicts fairly well the course of conversion, the RMM of the polymer formed, and the particle characteristics. The kinetic analysis considers, in addition to the classical elementary steps of radical polymerization, i.e. initiation, propagation, transfer, and termination, another three steps: (i) reaction between a macroradical and an interfacial emulsifier, which has been found to dominate over conventional bimolecular termination reaction, (ii) a long-chain branching reaction with emulsifier terminal double bonds, and (iii) the mass transfer of primary radicals and oligomeric radicals between the organic and aqueous phases. These additional steps have a profound effect on the kinetics of the process. The polymerization rate was found [40] to depend on the initiator concentration to a power greater than one half and to be inversely proportional to the surface emulsifier concentration, i.e.

$$R_p \propto [I]^{1.0} [M]^{1.0} [E]^{-a},$$

where a is a constant equal to 0.2 for poly(acrylamide) polymerization and I, M and E denote initiator, monomer and emulsifier, respectively.

The average number of radicals per particle is very difficult to determine in the inverse emulsion process. The reason is that the different stages of the process are not well defined and the polydispersity in the size of the particles and the variation in their number is high. For water-soluble initiating systems, usually Smith–Ewart kinetics (Case 2) (average number of radicals n per particle is 0.5) are applied. In the case of oil-soluble initiators (e.g. dibenzoyl peroxide) n ranges between 0.006–0.2 in accord with the Smith–Ewart Case 1 kinetics ($n \ll 0.5$). Complications arise if the system can adopt a multicellular structure as shown for acrylamide inverse emulsion polymerization in o-xylene using the nonionic emulsifier Tetronic 1102 [18, 19]. In spite of this, the reaction kinetics could have been described by the Smith–Ewart Case 1 for an oil-soluble initiator and/or by Case 2 for a water-soluble initiator, respectively.

The success of the inverse emulsion polymerization process also depends in a complex manner on such variables as the stirring speed, ionic strength, medium pH, temperature and the type of process (batch or semicontinuous). A specific problem of many inverse emulsion polymerization systems is the formation of coagulum. The main reasons for this are reported to be the high rate of polymerization (gel effect), vigorous stirring of the reaction system, poor colloidal stability of the polymer particles and (if applicable) great difference between the densities of the dispersed polymer particles and the continuous oil phase [39].

Conductometric studies of the inverse microemulsion toluene/water/AOT at a constant AOT/toluene ratio have shown [43] that, in the presence of a particular minimum concentration of acrylamide, the conductivity of the system changes markedly in comparison with that of the system in the absence of acrylamide. For instance, after exceeding the figure of 4 vol. % of a mixture of acrylamide and water (Φ_{aw}) in an inverse emulsion system (ratio AM/water varied between 0.24 and 1.25), the conductivity rapidly increases, especially at higher AM/water ratios. On the other hand, in inverse microemulsions without acrylamide the conductivity of the microemulsion slightly decreases above $\Phi_{aw} \sim 4$ vol. %. The increase in the conductivity is attributed [43] to the formation of dynamic percolation clusters after reaching a particular ratio of components in the system, i.e. the so-called percolation threshold. Above the percolation threshold, the conductivity continues to rise with increasing Φ_{aw} until phase separation occurs. The conductivity of a multicomponent inverse microemulsion depends on the nature of the interfacial zone as well as on the total volume fraction of the particles. It may by assumed that the inverse microemulsion droplet interfaces are nonconducting and that the interfaces must temporarily open upon contact to form water channels between the particles, allowing the current to flow. It has been shown (see p. 191, 204) that acrylamide is present at the interface and can act a coemulsifier. Acrylamide augments the attraction between the particles and thus increases the time of particle contact. At the same time it increases the flexibility of the interface of the particle, allowing the particle (micelle) to open, forming a transient water channel from particle to particle.

Increase in the acrylamide concentration in inverse emulsion system, leading to an increase in the attractive interaction forces between particles, results in a lowering of the percolation threshold of the inverse microemulsion system. The percolation threshold, expressed by the volume fraction of acrylamide and water in the inverse microemulsion, is 10.6% for an acrylamide/water ratio of 0.24 and 6.3% for an acrylamide/water ratio of 1.25 [43]. The particle diameter in the percolation system at low conversion (<10%), determined by quasielastic light scattering, is ca. 160–180 nm, but falls rapidly with increasing conversion and reaches final values comparable to nonpercolating systems (30–50 nm). On the other hand, nonpercolating inverse microemulsions display a tendency for their particle sizes to decrease only slightly during polymerization. For percolating as well as for nonpercolating inverse emulsion systems at the start of polymerization (zero conversion of acrylamide), the value of the hydrodynamic diameter of the inverse emulsion particles (micelles) is 4.5–5.0 nm. In ref. [43] are discussed several reasons for the large particle diameters in the early stages of polymerization in percolating inverse emulsion systems. The most plausible explanation seems to be the idea of a rapid formation of semipermanent aggregates or clusters. The rapid drop observed in the specific conductivity with increasing polymerization parallels that of the particle diameter. This indicates a loss of the percolating network of polymer particles.

Examples of the copolymerization of acrylamide with various ionogenic monomers and their corresponding salts have been investigated in inverse emulsions [41, 44–46] and microemulsions [34, 47, 48]. Besides those factors already mentioned as influencing the polymerization kinetics in inverse emulsion and inverse microemulsion systems, the partition coefficient of the monomer between the aqueous and oil phases must be taken into account. The same reasoning is valid for the partitioning of initiator between the water and oil phases. Complications may arise from the simultaneous homopolymerization of monomer soluble in the continuous oil phase and the formation of heterogeneous polymer product (copolymer in a mixture with homopolymer).

Analysis of the microstructure of the copolymers provides some information on the copolymerization mechanism. It was shown [46] that in inverse emulsion copolymerization systems, the distribution of monomer sequences obeys a first-order Markoff model with reactivity ratios close to those obtained in solution. Such a result favours the mechanism based on the diffusion of monomer from the reservoir of unnucleated monomer droplets to the site of propagation. Another study of similar systems led to a monomer sequence distribution in the copolymer approaching Bernouillian statistics [38]. Further experiments are therefore necessary to come to definite conclusions. The average copolymer composition was found to be independent of the degree of conversion [47]. This points to an (approximately) constant comonomer concentration at the propagation site. The necessary transport of monomers to the propagation site could be realized via the above-mentioned collisional mechanism or by an interphase diffusion mechanism. The exact positioning of the propagation site (at the interface or in a water pool of the inverse microemulsion polymer particle) cannot be determined from these experiments.

The polymerization of acrylamide in an inverse emulsion system has also been studied in ref. [49]. The composition of the inverse emulsion system was chosen so that it would be able to create an inverse microemulsion at the start of polymerization. The AOT concentration used (13.1 mass %) meant that, during polymerization, by lowering the acrylamide concentration, the stability of the disperse system was gradually lowered and phase separation occurred. This was accompanied by the appearance of opalescence and finally a strong turbidity, which on standing, settled and formed a gel-like precipitate. Ammonium peroxodisulfate (APS), 2,2'-azobisisobutyronitrile (AIBN) and dibenzoyl peroxide (DBP) were used as initiators. The particle diameter was between 55–70 nm, the RMM (viscometric measurements) over 10^6 g mol^{-1}. Table 32 gives information on the conversion and average

Table 32 — Dependence of rates of homopolymerization of acrylamide (in % conv. s^{-1}) and final conversion after 330 s on polymerization time in inverse microemulsion (Reprinted from ref. [49] with permission of Hüthig and Wepf Verlag, Basel)

Initiator	Conversion / %	Rate of polym.[a] / % conv. s^{-1}
Ammonium peroxodisulfate	100	0.30
2,2'-Azobisisobutyronitrile	100	0.30
Dibenzoyl peroxide	15	0.05
Ammonium peroxodisulfate	–	0.10[b]

[a] Calculated value (final conversion : polymerization time) × 100.
[b] Value for acrylamide homopolymerization in water [60].

Reaction system: acrylamide 0.594 mol dm^{-3} of solvent (toluene + water) mixture, bis(2-ethylhexyl)sulfosuccinate (AOT) 0.337 mol dm^{-3} of solvent mixture, toluene/water mass ratio 8.4. Concentration of initiators was 4.14×10^{-3} mol dm^{-3} of solvent mixture. Temperature 60°C [49].

rate of polymerization. The solubility of acrylamide in toluene is about 2% [45]; the contribution of the acrylamide polymerization in the oil (toluene) phase to the overall rate is therefore negligible for this particular system. The different rate constants for the unimolecular decomposition of the initiators (at 60°C k_d for APS and AIBN is 0.578×10^{-5} s^{-1} [50] and 1.6×10^{-5} s^{-1} [51], respectively) predict the polymerization rate to lie in favour of the AIBN-initiated system. The solubility of AIBN in water-containing ionic emulsifier at about 20°C is *ca.* 7 mass % [52], i.e. a fraction of that of APS (100%). This means that the solubility of the initiators in the aqueous phase favours the water-soluble APS compared with the partially-soluble AIBN. Taking these facts into account, the rate of acrylamide polymerization should be considerably higher in the system with APS. The inaccuracy in the determination of the polymerization rate due to the difficulties in the precise determination of conversion at very high polymerization rates [53] cannot alter this statement. However, experiment has shown that the rate of acrylamide polymerization in the inverse emulsion system is almost the same for both initiators — APS and AIBN. This result led to the conclusion that neither the water pool of inverse micelles (in the case of AIBN) nor the continuous oil phase (for APS) can be the sites of the initiation, i.e., the solubility of APS, the sulfate anion-radical, or the oligomeric acrylamide radicals in the oil phase is negligible [54]. For the system with AIBN, the initiation site appears to be above all the interface containing lipophilic chains of the AOT emulsifier and toluene molecules. The initiation site for the APS system is, of course, the water pool of the inverse micelles. We cannot eliminate, however, the possibility that the propagation site of acrylamide oligomer radicals is the interface. In her paper supporting this idea, Candau *et al.* [30] reported that a considerable part of acrylamide is concentrated at the interface.

The polymerization of acrylamide at the interface can be therefore started very easily by both kinds of initiators, viz. AIBN and APS. In the latter case the initiating radicals (or radicals participating in propagation) are oligomeric acrylamide radicals formed in the water pools of inverse micelles owing to reactions of the primary sulfate anion radicals with acrylamide. The growing poly(acrylamide) chains at the interface are gradually drawn from the interface into the water pool. Consequently, the stability of the polymer particles being formed decreases (note that acrylamide functions as a coemulsifier in the system) and strongly compressed coils of poly(acrylamide) (in the collapsed state) are formed in the water pools of the inverse emulsion system [30]. In efforts to obtain further details about the role of the interface in radical processes in inverse microemulsions, Bartoň [55–57] applied the principle of 'partitioned' free-radical polymerization (see section 3.2.4) to study the inverse microemulsion polymerization of acrylamide initiated by water- and oil-soluble initiators. In an investigation of the effect of the water-soluble inhibitor, potassium nitrosodisulfonate (Fremy's salt), upon the inverse microemulsion polymerization of acrylamide, a microemulsion [55] was used containing (mass %) : toluene (73.17), water (7.32), AOT (17.56) and acrylamide (1.95). For initiation of acrylamide polymerization, ammonium peroxodisulfate, 2,2'-azobisisobutyronitrile and dibenzoyl peroxide were used. The course of acrylamide polymerization in this nonpercolating inverse microemulsion is characterized by the data summarized in Table 33. For the APS-initiated system, a very high polymerization rate at 50% conversion was observed. However, in the presence of Fremy's salt the rate of polymerization after the inhibition period (44 min, Table 33) is strongly retarded. In the case of the AIBN-containing system, the polymerization rate is only slightly lower compared with the polymerization rate in the APS-containing system. The retardation of the polymerization rate owing to the presence of Fremy's salt compared with the system containing APS is less pronounced. If DBP is used for the initiation of the acrylamide polymerization

Table 33 — Polymerization of acrylamide in inverse microemulsion initiated by ammonium peroxodisulfate, APS, 2,2'-azobisisobutyronitrile, AIBN, and dibenzoyl peroxide, DBP in the presence and/or absence of potassium nitrosodisulfonate, FS (Reprinted from ref. [55] with permission of Hüthig and Wepf Verlag, Basel)

Initiator	$[I]^a$ / 10^{-2} mol dm^{-3}	FS	50% conv. /min	Polym. time^b /min	Rate of polym.^{c} / % conv. s^{-1}
APS	3.03	no	2.5	9	1.333
		yes^d	17	35	0.083
AIBN	1.23	no	4.0	16	0.833
		yes	8.5	36	0.416
DBP	1.19	no	25	40^e	0.416
		yes	25	40^e	0.416

^{a} mol dm^{-3} of water (for APS), mol dm^{-3} of toluene (for AIBN and DBP).
^{b} Time needed to reach conversion ≥ 97%.
^{c} At 50% of conversion.
^{d} Inhibition period of 44 min was observed.
^{e} The rate of polymerization to 10% conversion is ca. 1/55 of the polymerization rate at 50% conversion.
Inverse microemulsion composition: toluene 73.17 mass %, water 7.32 mass %, AOT 17.56 mass %, acrylamide 1.95 mass %. Concentration of potassium nitrosodisulfonate: 3.03×10^{-3} mol dm^{-3} of water. Polymerization temperature: 60°C.

in an inverse microemulsion, the polymerization rate at 50% conversion is only half of that for the AIBN-containing system. There is no difference in the shapes of the conversion curves, regarding the presence or absence af Fremy's salt. The rate of acrylamide polymerization initiated by DBP up to a conversion of 10% is very slow; beyond 10% conversion the rate increases dramatically and is practically equal to the rate of acrylamide polymerization initiated by AIBN in the system with Fremy's salt. In comparison to the APS- and Fremy's salt-containing inverse microemulsions, the rate of acrylamide polymerization initiated by DBP is reasonably higher. There is practically no inhibition period for AIBN- and DBP-containing systems due to the presence of Fremy's salt. This means that Fremy's salt in these inverse microemulsion polymerization systems does not interfere in the initiation reaction of acrylamide polymerization. The largest difference between the times for polymerization was observed for the APS-containing system in the presence of Fremy's salt and for the APS-containing system in the absence of Fremy's salt. A slightly smaller difference between the polymerization times was observed for the AIBN-containing system. For the DBP-containing system, the difference between the polymerization times is essentially zero.

All these observations offer considerable support to the previously expressed suggestion [49] as to the initiation sites in inverse microemulsion systems for the polymerization of acrylamide initiated by DBP, AIBN and APS. Benzoyloxy radicals from the decomposition of dibenzoyl peroxide in the toluene phase initiate the polymerization of dissolved acrylamide in the toluene phase (see p. 195, 204). The polymerization proceeds with a rate $R_p = 3.15 \times 10^{-4}$ mol dm^{-3} s^{-1} (7.2% conversion h^{-1}). This is close to the rate found for acrylamide polymerization in toluene in the presence of AOT, but in the absence of water,

$R_p = 1.59 \times 10^{-4}$ mol dm^{-3} s^{-1} [58]. The polymerization rates relate to toluene and to the same concentration of acrylamide. A correction was made for the use of different DBP concentrations in both systems. The result points to a homogeneous mechanism for the polymerization of acrylamide. The oligoacrylamide radicals and acrylamide oligomers (formed by termination of oligoacrylamide radicals), after reaching their solubility limit in toluene, precipitate and are captured by AOT inverse micelles, or form aggregates which are also finally captured by inverse AOT micelles. Oligoacrylamide radicals penetrate through the interphase of the inverse micelle (i.e. formed by the toluene-solubilized lipophilic chains of AOT molecules forming the micelle) into the acrylamide-rich region in the surface layer (shell) of the water pool of the inverse micelle. The existence of an acrylamide-rich region in the shell of the water pool was documented by Candau et al. [30]. The growth of oligoacrylamide radicals continues in the acrylamide-rich region. The site of propagation for the polymerization of acrylamide in inverse microemulsion is thus initially the oil (toluene) phase but later becomes mainly the acrylamide-rich region in the surface layer of the water pool of the inverse AOT micelle. The utilization of the acrylamide-rich region as a site of propagation is responsible for the high rates in the polymerization of acrylamide beyond a conversion of 10%. Some oligoacrylamide radicals as well as acrylamide macroradicals enter the water pool and terminate with Fremy's salt (if present). The proposed mechanism of the DBP-initiated inverse microemulsion polymerization of acrylamide differs from that proposed [18] for the DBP-initiated inverse emulsion polymerization of acrylamide. In the latter case the initiation occurs by radicals formed in the aqueous monomer droplets owing to the enhanced solubility of DBP in water in the presence of acrylamide.

The mechanism proposed for the DBP-initiated inverse microemulsion polymerization of acrylamide also seems to be valid for the AIBN-initiated polymerization of acrylamide in an inverse microemulsion. The substantially higher water solubility of AIBN compared with the water solubility of DBP produces a minor modification of the proposed mechanism. AIBN present in the water pool undergoes thermal homolysis to yield free radicals which partly react with Fremy's salt and partly initiate (even in the presence of Fremy's salt) the polymerization of acrylamide. These competitive reactions for AIBN radicals proceed simultaneously. The quantitative relation between these two reactions is reflected by the relatively shorter period for the low rate of polymerization of acrylamide in the system containing Fremy's salt. The polymerization of acrylamide in the toluene phase as well as in the acrylamide-rich surface layer of the water pool is not influenced by the eventual presence of Fremy's salt in the inverse microemulsion system. This statement is supported by the shape of the conversion curves for the polymerization of acrylamide initiated by dibenzoyl peroxide in the presence and/or absence of Fremy's salt. The lower polymerization rate of acrylamide observed in the system with Fremy's salt is a consequence of the lower concentration of AIBN radicals for the initiation of the polymerization. The AIBN molecules dissolved in the water pool generate radicals which are partly consumed by Fremy's salt. These radicals are therefore unavailable for initiation of the polymerization. In the absence of Fremy's salt they can participate in the initiation of the polymerization of acrylamide in the water pool and thus increase the overall rate of polymerization.

In the case of APS as an initiator for the polymerization of acrylamide in an inverse microemulsion, the initiation site is the water pool of the inverse micelle. The idea of the acrylamide-rich region in the surface layer (shell) of the water pool as the main site of propagation agrees well with the experimental results obtained for this system. Obviously, Fremy's salt is not present in the acrylamide-rich region. If the reverse was true, i.e. in the presence of Fremy's salt in the acrylamide-rich region, an inhibition period for the AIBN-, as well as for the DBP-initiated inverse microemulsion polymerization of acrylamide, should be observed. This, however, was not confirmed (Table 33).

The kinetics and mechanism of acrylamide polymerization initiated by water-soluble ammonium peroxodisulfate and/or oil-soluble dibenzoyl peroxide in the percolating inverse microemulsion formed from toluene, water, bis(2-ethylhexyl)sulfosuccinate sodium salt and acrylamide was studied in recent papers [56, 57]. These studies confirmed the validity of the previously expressed opinion [55] as to the sites of initiation and mechanism of polymer particle formation for acrylamide polymerization initiated by APS and DBP in nonpercolating inverse microemulsion and also for percolating inverse microemulsion. The different kinetic behaviour found for percolating inverse microemulsion polymerization of acrylamide (Table 34) in comparison to the nonpercolating inverse microemulsion polymerization

Table 34 — Inverse microemulsion polymerization of acrylamide initiated by ammonium peroxodisulfate, APS and dibenzoyl peroxide, DBP at 60°C [a]

Run	Φ_{aw} [b]	t_{pol} [c] / min	t_{50} [d] / min	R_p [e] / % s^{-1}	Initiator
1[f]	8.3	9	2.5	1.33	APS
1a[f]	8.3	35	17	0.08	APS
2	12.0	14	5	0.28	APS
2a	12.0	28	12	0.10	APS
3	15.3	18	6	0.24	APS
3a	15.3	27	9	0.15	APS
4[f]	8.3	40	25	0.42	DBP
4a[f]	8.3	40	25	0.42	DBP
5	12.0	24	8	0.42	DBP
5a	12.0	23	8	0.42	DBP

[a] Toluene 15 g, AOT 3.6 g, [APS] = 3.03×10^{-2} mol dm^{-3} (based on water), [DBP] = 1.19×10^{-2} mol dm^{-3} (based on toluene). Acrylamide 0.4 g (runs 1 and 4); 0.6 g (runs 2 and 5); 0.8 g (run 3). Water 1.5 g (runs 1 and 4); 2.25 g (runs 2 and 5); 3.0 g (run 3). Runs containing Fremy's salt (3.03×10^{-3} mol dm^{-3} based on water) are indicated by small case letter a; i.e. 1a, 2a, etc.
[b] Volume fraction of acrylamide and water in initial inverse microemulsion, in %.
[c] Time needed for conversion $\geq 97\%$.
[d] Time needed for 50% conversion.
[e] Rate of polymerization at 50% conversion.
[f] Data taken from ref. [55].

of acrylamide point to changes in the partitioning of reactants in the propagation site as well as to structural changes in the inverse microemulsion. If the composition of the inverse microemulsion polymerization passes the percolation threshold [43] then the formation of 'water channels' among individual micelles of the cluster of micelles leads to creation of an 'averaged' acrylamide concentration (the acrylamide-rich region in the surface layer of the water pool of the inverse micelle is perturbed or even eliminated). Polymerization then proceeds in a system of 'apparently' lower acrylamide concentration. This can explain the lower polymerization rate of acrylamide if APS is used as initiator in a percolating inverse microemulsion system (runs 2 and 3, Table 34). The 'averaged' acrylamide concentration in the core of the water pool *de facto* is higher than that when the acrylamide-rich region is not perturbed. The higher acrylamide concentration in the core of the water pool promotes the propagation reaction of acrylamide radicals with acrylamide as compared with termination reactions of acrylamide radicals with Fremy's salt. Thus the polymerization rate of acrylamide in the system with APS and Fremy's salt is higher if the system is percolating (*cf.* run 1a versus runs 2a and 3a, Table 34). For DBP-initiated polymerization of acrylamide in an inverse microemulsion, the region of slow acrylamide polymerization in a percolating system (runs 5 and 5a, Table 34) is very short in comparison to a nonpercolating system (runs 4 and 4a, Table 34). Percolation increases not only the mass transfer between various phases of the inverse microemulsion system but very probably also the rate of capture of acrylamide oligomer radicals by inverse micelles. This seems to be the reason for the short time interval of relatively slowly DBP – initiated polymerization of acrylamide in a percolating inverse microemulsion. The rates of acrylamide polymerization initiated by DBP beyond the region of the 'slow' acrylamide polymerization, regardless of the nature of the inverse microemulsion (percolating and/or nonpercolating), have practically the same value. This points to the insignificance of the perturbation of the acrylamide-rich region for the propagation reaction in this system. The mechanism and site of the propagation reaction responsible for the rapid polymerization rate of acrylamide in this particular percolating inverse microemulsion is at present not well understood and needs further work.

The subjects of study of copolymerization in inverse microemulsions were the monomer couples acrylamide/methyl methacrylate and acrylamide/styrene [59–61]. Conversion curves of these copolymerizations display the typical course of so-called 'dead end' polymerization, i.e. polymerization with a rate which, after reaching a particular conversion (often much less than 100% conversion) is almost zero. In the acrylamide/methyl methacrylate copolymerization system, the limiting conversion for a mole fraction of acrylamide n_{AAm} = 0.477 in a mixture with methyl methacrylate was about 41%. This degree of conversion was reached in less than 6 minutes and after a further 300 min of polymerization, the conversion increased to 47%. The concentration of acrylamide structural units in the copolymer poly(acrylamide-co-methyl methacrylate) was 78 mol %. The composition of the copolymer varied only slightly even with substantial changes in the composition of the comonomer mixture. Thus for changes of n_{AAm} in the mixture from 0.25 to 0.60 the mole fraction of acrylamide structural units in the copolymer varied between 0.7 and 0.9. After completion of the period of a high rate of polymerization (after reaching the 41% conversion), the reaction product contains, in addition to the copolymer, a significant amount of poly(acrylamide) homopolymer and a small amount of poly(methyl methacrylate) homopolymer. This points to simultaneous homopolymerization and copolymerization in two different reaction sites — the water pool and the interface of the inverse microemulsion micelles [62].

The effect of the bi-unsaturated vinyl monomer, divinylbenzene (DVB), on the polymerization of acrylamide initiated by dibenzoyl peroxide in a percolating inverse microemulsion was recently studied [63]. In contrast to acrylamide polymerization in a percolating inverse microemulsion in the absence of DVB, a region of 'slow' acrylamide polymerization was observed. The rates of the 'slow' acrylamide polymerizations increased slightly, but the polymerization rates in the region of 'high' acrylamide polymerization rates (between 20–70 % conversion) decreased on increasing the DVB/AAm ratio.

The inverse microemulsion polymerization of acrylamide initiated by dibenzoyl peroxide in the presence of N,N-methylenebisacrylamide (MBAAm) is characterized [64] by the formation of polymer particles of smaller size in comparison to polymer particles prepared in the absence of MBAAm. For concentrations of MBAAm up to 5 mass % (based on AAm), the decrease of polymer particle diameter is about 25–30 %. The admixture of MBAAm is practically without influence on the polymerization rate.

References

1. Leong, Y.S, Riess, G. & Candau, F. (1981) *J. Chim. Phys.* **73** 279
2. Atik, S.S. & Thomas, J.K. (1981) *J. Am. Chem. Soc.* **103** 4279
3. Stoffer, J.O. & Bone, T. (1980) *J. Dispersion Sci., Technol.* **1** 37
4. Stoffer, J.O. & Bone, T. (1980) *J. Polym. Sci., Polym. Chem. Ed.* **18** 2641
5. Clayfield, E.J. & Lumb, E.C. (1968) *Macromolecules* **1** 133
6. Hesselink, F.T., Vrij, A. & Overbeek, J.T.G. (1971) *J. Phys. Chem.* **75** 2094
7. Napper, D.H. (1970) *Ind. Eng. Chem., Prod. Res. Dev.* **9** 467
8. Vincent, B., Luckham, P.F. & Waite, F.A. (1980) *J. Colloid Interface Sci.* **73** 508
9. Gan, L.M., Chew, C.H. & Friberg, S. (1983) *J. Polym. Sci., Polym. Chem. Ed.* **21** 513
10. Gan, L.M. & Chew, C.H. (1983) *J. Dispersion Sci., Technol.* **4** 291
11. Gan, L.M., Chew, C.H., Friberg, S. & Higashimura, T. (1981) *J. Polym. Sci., Polym. Chem. Ed.* **19** 1585
12. Gan, L.M., Chew, C.H. & Friberg, S. (1983) *J. Macromol. Sci., Chem., A* **19** 739
13. Arai, M., Arai, K. & Saito, S. (1979) *J. Polym. Sci., Polym. Chem. Ed.* **17** 3655
14. Gan, L.M. & Chew, C.H. (1984) *J. Dispersion Sci., Technol.* **5** 179
15. Candau, F. (1987) *J. Chim. Phys.* **84** 1095
16. Baade, W. & Reichert, K.H. (1984) *Eur. Polym. J.* **20** 505
17. Dimonie, M.V., Boghina, G.M., Marinescu, N.N., Cincu, C.J. & Oprescu, O.G. (1982) *Eur. Polym. J.* **18** 639
18. Vanderhoff, J.W., Di Steffano, F.V., El-Aasser, M.S., O'Leary, R., Shaffer, O.M. & Visioli, D.G. (1984) *J. Dispersion Sci., Technol.* **5** 323
19. Visioli D.G. (1984) *PhD Dissertation.* Lehigh University, Bethlehem, USA
20. Graillat, C., Pichot, C., Guyot, A. & El-Aasser, M.S. (1986) *J. Polym. Sci., Polym. Chem.* **24** 427
21. Baade, W. & Reichert, K.-H. (1986) *Macromol. Chem., Rapid Commun.* **7** 235
22. Kurenkov, V.F., Baiburdov, T.A. & Garipova, N.S. (1986) *Izv. Vyssh. Ucheb. Zaved., Khim. Khim. Tekhnol.* **29** 92
23. Kurenkov, V.F., Verizhnikova, A.S. & Myagchenkov, V.A. (1986) *Vysokomol. Soedin., A* **28** 488
24. Leong, Y.S. & Candau, F. (1982) *J. Phys. Chem.* **86** 2269
25. Eicke, H.F. (1977) In: Mittal, K.L. (ed.) *Surfactants in solution.* Plenum Press, New York, p. 429
26. Candau, F., Leong, Y.S., Pouyet, G. & Candau, S. (1984) *J. Colloid Interface Sci.* **101** 167
27. Cazabat, A.M., Langevin, D. & Pouchelon, A. (1980) *J. Colloid Interface Sci.* **73** 1
28. Cebula, D.J., Ottewill, R.H., Ralston, R.H. & Pusey, P.N. (1981) *J. Chem. Soc., Faraday Trans. 1* **77** 2585
29. Candau, F., Leong, Y.S., Pouyet, G. & Candau, S. (1985) In: Degiorgio, V. & Corti, M. (eds) *Physics of amphiphiles: Micelles, vesicles and microemulsions.* North-Holland, Amsterdam, p. 830
30. Candau, F., Leong, Y.S. & Fitch, R.M. (1985) *J. Polym. Sci., Polym. Chem. Ed.* **23** 193
31. Holtzscherer, C., Durand, J.P. & Candau, F. (1987) *Colloid Polym. Sci.* **265** 1067
32. Holtzscherer, C. & Candau, F. (1988) *Colloids Surfaces* **29** 411
33. Holtzscherer, C. & Candau, F. (1988) *J. Colloid Interface Sci.* **125** 97
34. Candau, F., Zekhnini, J. & Durand, J. (1986) *J. Colloid Interface Sci.* **114** 398
35. Candau, F., Collin, D. & Kern, F. (1990) *Makromol. Chem., Macromol. Symp.* **35/36** 105
36. Schott, H.S. (1983) *J. Colloid Interface Sci.* **43** 150
37. Candau, F. & Buchert, P. (1990) *Colloids Surfaces* **48** 107
38. Candau, F. (1989) In: El Nokaly, M. (ed.) *Polymer association structures: Microemulsions and liquid crystals.* ACS Symp. Ser. 384. Washington DC, p. 47
39. Candau, F. (1990) In: Candau, F. & Ottewill, R.H. (eds) *Scientific methods for the study of polymer colloids and their applications.* NATO ASI Ser. D. Reidel Publ. Co., Dordrecht, p. 73
40. Hunkeler, D., Hamielec, A.E. & Baade, W. (1989) *Polymer* **30** 127
41. Kurenkov, V.P., Verizhnikova, A.S., Kuznetsov, E.V. & Myagchenko, V.A. (1982) *Izv. Vyssh. Ucheb. Zaved., Khim. Khim. Tekhnol.* **25** 221
42. Vanderhoff, J.W., Tarkowski, H.L., Schaffer, J.B., Bradford, E.B. & Wiley, R.M. (1962) *Adv. Chem. Ser.* **34** 32
43. Carver, M.T., Hirsch, E., Wittmann, J.C., Fitch, R.M. & Candau, F. (1989) *J. Phys. Chem.* **93** 4867
44. Baade, W., Hunkeler, D. & Hamielec, A.E. (1987) *PMSE Prepr., Am. Chem. Soc. Div.* **57** 850
45. Glukhikh, V., Graillat, C. & Pichot, C. (1987) *J. Polym. Sci., Polym. Chem. Ed.* **25** 1127
46. Pichot, C., Graillat, C. & Llauro, M.F. (1985) *Polymer latex II.* Plastics and Rubber Institute, London, p. II/1

47. Candau, F., Zekhnini, Z. & Heatley, F. (1986) *Macromolecules* **19** 1895
48. Candau, F., Zekhnini, Z., Heatley, F. & Franta, F. (1986) *Colloid Polym. Sci.* **264** 676
49. Vašková, V., Juraničová, V. & Bartoň, J. (1990) *Makromol. Chem.* **191** 717
50. Berezhnoi, G.D., Khomikovski, P.M. & Medvedev, S.S. (1960) *Vysokomol. Soedin.* **2** 141
51. Blackley, D.C. (1975) *Emulsion polymerization: Theory and practice.* Applied Science Publ., London, p. 163
52. Bartoň, J. & Kárpátyová, A. (1987) *Makromol. Chem.* **188** 163
53. Bartoň, J., unpublished results
54. Lambla, M., Ali Syed, K. & Banderet, A. (1976) *Eur. Polym. J.* **12** 263
55. Bartoň, J. (1991) *Makromol. Chem., Rapid Commun.* **12** 675
56. Bartoň, J. (1991) Lecture presented at the *International Symposium on Polymeric Microspheres*, Fukui, Japan. Preprints, p. 19
57. Bartoň, J. (1993) *Polym. Int.* **30** 151
58. Bartoň, J., unpublished results
59. Vašková, V., Juraničová, V. & Bartoň, J. (1990) *Makromol. Chem., Macromol. Symp.* **31** 201
60. Vašková, V., Juraničová, V. & Bartoň, J. (1991) *Makromol. Chem.* **192** 989
61. Vašková, V., Juraničová, V. & Bartoň, J. (1991) *Makromol. Chem.* **192** 1339
62. Bartoň, J. (1992) *Makromol. Chem., Macromol. Symp.* **53** 289
63. Vašková, V., Stillhammerová, M. & Bartoň, J. (1992) Paper presented at the *34th IUPAC Symposium on Macromolecules*, Prague, Book of Abstracts, p. 1
64. Vašková, V., Stillhammerová, M. & Bartoň, J., *Makromol. Chem.*, in press

4.3 SPECIAL CASES OF RADICAL POLYMERIZATION IN WATER-IN-OIL MICELLAR SYSTEMS

In the period following the pioneering work of Vanderhoff *et al.* [1], a number of studies dealing with radical polymerization in inverse dispersion systems have appeared (see section 4.2). Discussion about the subject will be concluded by a brief mention of some particular procedures applied in preparing polymer dispersions in inverse micellar systems.

The photoinitiated radical polymerization of vinyl monomers in homogeneous reaction systems in the presence of sensitizers of either organic or inorganic nature, charge-transfer complexes or donor-acceptor type complexes has been the subject of numerous papers [2]. In classical dispersion systems, e.g. conventional emulsion polymerization systems, photoinitiation is impossible because of optical nontransparency caused by light scattering by particles of the dispersed phase throughout the reaction system. Reduction of the size of the dispersed particles to the order of nanometres, common in the so-called microemulsion systems, enables photochemical initiation. The radical-photoinitiated polymerization of acrylamide in inverse micelles described elsewhere [3] provides an example. The microemulsion was formed by the system *n*-decane/sodium bis(2-ethylhexyl)sulfosuccinate/water/acrylamide. A benzophenone (BP) derivative was the water-soluble sensitizer located in the dispersed aqueous phase

$$\text{Ph-C(=O)-C}_6\text{H}_4\text{-R},$$

where R = $CH_2SO_3^- Na^+$ (BP^-) or
$CH_2N^+(CH_3)_3Cl^-$ (BP^+).

On excitation (λ = 366 nm) the BP transforms to its excited singlet state

$$BP \xrightarrow{h\nu} {}^1BP^*,$$

which, on intersystem crossing (ISC), gives BP in the triplet state (biradical)

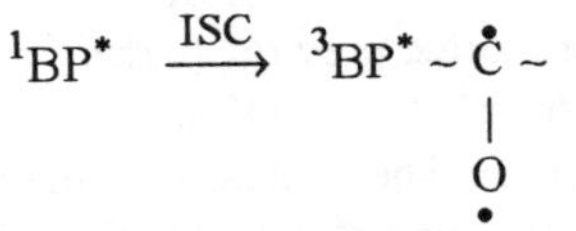

Special cases of radical polymerization

$$^1BP^* \xrightarrow{ISC} {}^3BP^* \sim \overset{\bullet}{\underset{\underset{\bullet}{O}}{C}} \sim$$

The triplet benzophenone derivative reacts with a hydrogen donor molecule (AOT, n-decane)

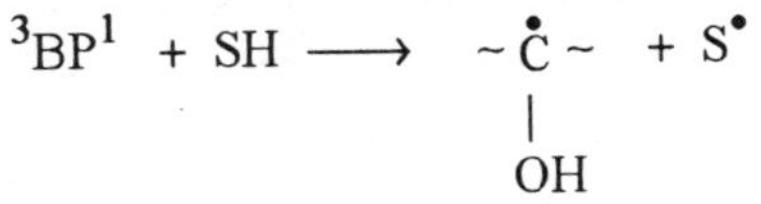

$$^3BP^1 + SH \longrightarrow \sim \underset{OH}{\overset{\bullet}{C}} \sim + S^{\bullet}$$

with the formation of a weakly-reactive ketyl radical

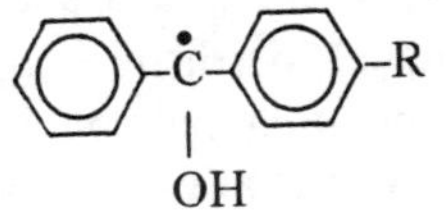

⟨phenyl⟩–Ċ(OH)–⟨phenyl⟩–R

and AOT or n-decane radical ($-\overset{\bullet}{C}-$ radical).

BP$^+$ was shown to be more effective in initiating acrylamide polymerization. The rate of polymerization is about hundred times higher than that of the polymerization in a homogeneous system. The complete conversion of acrylamide was achieved in tens of seconds. The rate of polymerization is proportional to the square root of the light intensity. The dependence of the rate of polymerization on acrylamide concentration is defined by the equations

$$R_p \sim [AM]^{1.3} \quad \text{(for BP}^{\overset{\bullet}{+}}\text{)},$$

$$R_p \sim [AM]^{1.4} \quad \text{(for BP}^{\overset{\bullet}{-}}\text{)}.$$

On replacing n-decane by toluene, the polymerization rate decreased, which is attributed to the low initiation activity of benzyl radical

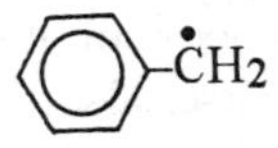

⟨phenyl⟩–ĊH$_2$

as compared with the alkyl radical formed via the transfer reaction of the benzophenone derivative in its triplet state with an n-decane molecule.

The kinetics of the photoinitiated polymerization of acrylamide in toluene/AOT/water inverse microemulsion have also been examined for systems initiated by 2,2'-azobisisobutyronitrile or a dye/triethanolamine redox system [4]. Here AIBN was considered to be an oil-soluble initiator while the dye/triethanolamine redox system was seen as a water-soluble initiator. Before polymerization the microemulsion contained (mass %) toluene 75, AOT 17, water 4 and acrylamide 4. For the redox system dye/triethanolamine, a methylene blue/eosin mixture was used. Light scattering, viscometry and ultracentrifugation experiments have shown that the system contains two kinds of particles in equilibrium after polymerization: polymer particles of diameter smaller than 50 nm and small AOT micelles of diameter $ca.$ 3 nm. The polymerized microemulsion exhibited a narrow size distribution, and despite the small size of the polymer particles, a high RMM (10^6– 10^7 g mol^{-1}). Each of the polymer particles contained a single polymer molecule. Unlike previous data [3] the rate of polymerization is first-order with respect to monomer concentration at least up to conversions of 60 %. This is in accord with the results reported previously [5]. For a constant incident light intensity, variation in the AIBN concentration led to a first-order dependence of the polymerization rate on the initiator concentration. The radical termination is thus first-order in radical concentration. The first-order dependence of the rate of polymerization on the initiator concentration was suggested to be due to participation of the emulsifier [6, 7] or monomer [8–10] in transfer reactions with radicals. The postulated [4] reaction scheme considers as terminations in the polymer particle of the inverse microemulsion system, the following steps:

$$\sim\sim M_{n-1} M^\bullet + TH \rightarrow \sim\sim M_{n-1} MH + T^\bullet,$$

$$\sim\sim M_{n-1} M^\bullet + MH \rightarrow \sim\sim M_{n-1} MH + M^\bullet,$$

where $\sim\sim M_{n-1} M^\bullet$ is a poly(acrylamide) radical and TH and MH are toluene and acrylamide monomer, respectively. In the next step the monomer radical $M^\bullet$ diffuses out from the polymer particle and reacts with the TH molecule thus regenerating the monomer MH and the radical derived from the transfer agent $T^\bullet$ or yields a relatively stable (unreactive) benzyl radical $C_6H_5CH_2^\bullet$ in a transfer reaction with toluene in the continuous phase. The monoradical mechanism of poly(acrylamide) radical termination, with respect to biradical termination, is possible only if the concentration of transfer agents is high. However, this is not the case because of the high RMMs of the poly(acrylamide) formed. The supposed dominance of transfer reactions in the system should substantially decrease the RMMs of the poly(acrylamide) formed. On the other hand, biradical termination can be physically inhibited by the fact that the polymer particle contains only one growing polymer chain and that the primary radicals are very unlikely to enter a previously-nucleated particle to terminate propagation in the particle, as is the case for classical emulsion polymerization (cf. Chapter 3). The presence of unnucleated AOT micelles after complete conversion of the monomer in the microemulsion supports this argument.

The explanation of closely similar efficiencies of water-soluble redox initiating systems and AIBN for acrylamide polymerization in inverse microemulsion is based on the assumption [4] of entry of an initiating radical, formed via decomposition of an AIBN molecule in the continuous toluene phase, into the acrylamide-containing AOT micelle. On the other hand, in the redox initiating system, radicals are generated in the water pool of the inverse AOT micelle containing acrylamide. A similar situation is, of course, also the case for ammonium peroxodisulfate as initiator. As regards the concentration of the methylene blue/eosin complex used, about 100–1000 AOT micelles are present compared with one molecule of the complex; it is therefore improbable that the AOT micelles would contain more than one molecule of the complex. Accepting the assumption that there is only one entry of a radical into an inverse AOT micelle, both initiating systems, in spite of the considerable difference in their solubilities in water and oil, provide the same mechanistic picture in the inverse emulsion polymerization of acrylamide. However, the authors do not take into account any possibility of the occurrence of AIBN in the water pools of inverse AOT micelles. Although the water solubility of AIBN is rather low (ref. [52]; p. 204), the possibility of formation of initiating radicals via decomposition of AIBN molecules in the water pools of inverse AOT micelles cannot be ruled out. The mechanism considered cannot obviously extend to the water-soluble initiator ammonium peroxodisulfate, where the presumption of only one radical per AOT micelle is unreal. The problem of bimolecular [3] versus monomolecular [4] mechanisms of the termination of poly(acrylamide) radicals in inverse emulsion systems was studied with respect to the roles of emulsifier, monomer and solvent [11]. At 10°C, nondegradative chain transfer to monomer appears to be insignificant. There is almost no dependence of the polymerization rate on the AOT concentration. The experiments were, however, inconclusive. It is difficult to change the AOT concentration without destabilizing the microemulsion while trying to keep the monomer concentration constant. If toluene was replaced by benzene, the exponent of the incident light intensity was between 0.5 and 0.6. The rate of polymerization is faster in a benzene-containing inverse microemulsion compared with a toluene-containing inverse microemulsion. In benzene, which has no readily-available hydrogen atoms for a transfer reaction, the termination of poly(acrylamide) radicals is a biradical process. In heptane, the exponent of the incident light intensity ranged between 0.7 and 0.8, indicating the possibility of monoradical and biradical mechanisms of termination operating simultaneously.

Polymerization of ordered monomer molecules can be achieved in several ways [12–14], e.g. by the polymerization of mesomorphic monomers (section 6.5). Here we shall discuss the polymerization of a monomer which is inserted before polymerization into a lyotropic liquid-crystalline structure. In the preceding section (4.1) we discussed the formation of bicontinuous microemulsions. These isotropic acrylamide-containing bicontinuous microemulsions lead, after polymerization, to stable and transparent inverse microemulsions of poly(acrylamide) particles dispersed in the oil phase [15]. Similarly, polymerization of acrylamide in the swollen lamellar mesophase formed by a mixture of nonionic emulsifiers (Arlacel 83 and G 1086) showing long range smectic order and high emulsifier concentration (~ 22 mass %), lead to a stable inverse poly(acrylamide) microemulsion. This gives clear evidence of a polymerization-induced structural change from an initial swollen lamellar phase to a final dispersion of spherical polymer particles. The polymerization of acrylamide promoted phase separation and the transfer of monomer from the organized mesophase to a single disordered isotropic phase [15].

Reference [16] reports the preparation of solid porous polymeric materials via the polymerization of water-in-oil microemulsions as well as through polymerization of the so-called middle phase (the bicontinuous phase formed by applying an emulsifying system with an HLB value between the HLBs of the emulsifying systems for the formation of oil-in-water and water-in-oil microemulsions). The microemulsions were prepared from a mixture of styrene, water, sodium dodecyl sulfate (SDS) and 2-pentanol as a cosolvent. The compositions of the microemulsions before polymerization were as follows:

Microemulsion with oil continuous phase (system A):
    styrene   55 %
    water   10 %
    2-pentanol   25 %
    SDS   10 %

Microemulsions of middle phase (system AB):
    styrene   40 %
    water   25 %
    2-pentanol   25 %
    SDS   10 %

Microemulsions with aqueous continuous phase (system B):

styrene	5 %
water	60 %
2-pentanol	25 %
SDS	10 %

The properties of polystyrene prepared in the microemulsion systems A, AB and B are presented in Table 35.

Table 35 — Second-order glass transition temperature T_g and weight average RMM of poly(styrene) prepared in microemulsion (Reprinted from ref. [16] with permission of John Wiley & Sons Inc., New York)

Microemulsion	T_g / °C	M_w / g mol^{-1}
A	133 ± 2	709 000 ± 7000
AB	128 ± 1.2	274 000 ± 1000
B	118 ± 2	284 500 ± 1500

The investigation of inverse microemulsion system with conventional oil phases and conventional reaction conditions (temperature, pressure) has been extended to study of inverse microemulsion systems formed by supercritical fluids [17–20]. These studies provide the basis for subsequent polymerization experiments and are the first reports published on the properties of inverse micelles of nonionic emulsifiers in supercritical fluids [21]. The inverse microemulsion polymerization of acrylamide in a near-critical or supercritical alkane (e.g. ethane/propane mixture) continuous phase seems a promising route to polymers with novel physical properties [21]. It is assumed that the high pressure and low viscosity of the supercritical system can increase the polymerization rate significantly [22]. These studies open up new avenues of research into inverse microemulsion polymerization.

References

1. Vanderhoff, J.W., Tarkowski, H.L., Shaffer, J.B., Bradford, E.B. & Wiley, R.M. (1962) *Adv. Chem. Ser.* **34** 32
2. Bartoň, J. & Borsig, E. (1988) *Complexes in free-radical polymerization.* Elsevier, Amsterdam
3. Fouassier, J.P., Lougnot, D.J. & Zuchowicz, I. (1986) *Eur. Polym. J.* **22** 933
4. Carver, M.T., Dreyer, U., Knoesel, R., Candau, F. & Fitch, R.M. (1989) *J. Polym. Sci., Polym. Chem. Ed.* **27** 2161
5. Candau, F., Leong, Y.S. & Fitch, R.M. (1985) *J. Polym. Sci., Polym. Chem. Ed.* **23** 193
6. Kurenkov, V.F. & Myagchenkov, V.A. (1980) *Eur. Polym. J.* **16** 1229
7. Kurenkov, V.F., Verizhnikova, A.S. & Myagchenkov, V.A. (1984) *Dokl. Akad. Nauk SSSR* **278** 1173
8. Gilbert, R.G. & Napper, D.H. (1983) *J.Macromol. Sci., Rev. Macromol. Chem.* **23** 127
9. Hawkett, B.H., Napper, D.H. & Gilbert, R.G. (1980) *J. Chem. Soc., Faraday Trans. 1* **76** 1323
10. Lansdowne, S.W., Gilbert, R.G., Napper, D.H. & Sangster, D.F. (1980) *J. Chem. Soc., Faraday Trans. 1* **76** 1344
11. Carver, M.T., Candau, F. & Fitch, R.M. (1989) *J. Polym. Sci., Polym. Chem. Ed.* **27** 2179

12. Paecht-Horowitz, M. (1977) In: Elias, H.G. (ed.) *Polymerization of organized systems.* Gordon and Breach Sci. Publ., New York, p. 89
13. Krentsel, B.A. (1977) In: Elias, H.G. (ed.) *Polymerization of organized systems.* Gordon and Breach Sci. Publ., New York, p. 117
14. Barrall, E.N. & Johnson, J.F. (1979) *J. Macromol. Sci., Rev. Macromol. Chem.* **17** 137
15. Holtzscherer, C., Wittmann, J.C., Guillon, D. & Candau, F. (1990) *Polymer* **31** 1978
16. Haque, E. & Qutubuddin, S. (1988) *J. Polym. Sci., Polym. Lett. Ed.* **26** 429
17. Gale, R.W., Fulton, J.L. & Smith, R.D. (1987) *J. Am. Chem. Soc.* **109** 920
18. Fulton, J.L. & Smith, R.D. (1988) *J. Phys. Chem.* **92** 2903
19. Blitz, J.P., Fulton, J.L. & Smith, R.D. (1988) *J. Phys. Chem.* **92** 2707
20. Fulton, J.L., Blitz, J.P., Tingey, J.M. & Smith, R.D. (1989) *J. Phys. Chem.* **93** 4198
21. Beckmann, E.J. & Smith, R.D. (1990) *J. Phys. Chem.* **94** 345
22. Ogo, Y. (1984) *J. Macromol. Sci., Rev. Macromol. Chem.* **24** 1

5

Dispersion polymerization in nonaqueous media

5.1 BASIC PRINCIPLES OF FORMATION AND STABILIZATION OF NONAQUEOUS POLYMER DISPERSIONS

In the majority of polymerizations in disperse systems, water is used as the dispersing medium because of its low price, noncombustibility and lack of threat to the environment. Aqueous polymer dispersions are also suitable for a variety of technological processes. Aqueous polymer dispersions have also some shortcomings which limit their applications. For instance, aqueous dispersions cannot be processed at temperatures higher than 100°C or below the f.p. of water. The high latent heat of water evaporation means that much energy is required to obtain polymers in the solid (powder) state from the dispersions. The surfactants used to stabilize the polymer particles of an aqueous dispersion can adversely affect the properties of the particle or polymer after removal of the disperse medium.

The technique of disperse polymerization in nonaqueous media is quite new. This method can be used to prepare a variety of stable polymer dispersions with controllable particle size in that medium most suitable as regards the practical application. The technique of disperse polymerization was originally developed for the dye and paint industries. Currently its applications are much more extensive.

Radical disperse polymerization in nonaqueous media* is a process in which the monomer polymerizes in an organic liquid which acts as a solvent for the monomer but as a precipitant for the polymer being formed. The mixture of monomer and solvent (or diluent) forms the disperse medium (continuous phase) which contains the required amount of initiator and stabilizing additive to prevent the polymer formed from coagulation in the disperse medium. The composition of the disperse medium changes gradually during polymerization because of the transformation of the monomer to polymer. The resulting reaction product is a polymer dispersion with a certain (polymer) particle size in a disperse medium, which is, of course, the solvent (diluent) used. In the absence of a stabilizer of the primary polymer particles, produced by the aggregation of macromolecules with a

---

* Dispersion polymerization in aqueous media is described in Chapter 3 (emulsion polymerization) and in Chapter 7 (suspension polymerization).

rather low degree of polymerization (of oligomers),* their coagulation starts and results in formation of the polymer as a coagulate (precipitate). A similar phenomenon has also been observed without any precipitant for the polymer, if the monomer does not function as a solvent for its polymer. The process of polymer preparation, in which a polymer insoluble in the reaction medium is deliberately produced, is the essence of precipitation polymerization (e.g. the block polymerization of acrylonitrile or vinyl chloride, i.e. in the absence of a solvent for the poly(acrylonitrile) or poly(vinyl chloride) being formed). In the polymerization of a monomer which is a solvent of its polymer, the viscosity of the reaction medium gradually increases because of the increase in concentration of the polymer at the expense of the monomer. No precipitation of polymer thus occurs. The resulting product, the polymer, contains a small amount of unchanged monomer (e.g. the block polymerization of methyl methacrylate in the preparation of organic glass-like materials).

Disperse polymerization is a special case of precipitation polymerization, which, by adding a stabilizer, prevents the formation of the coarse coagulate (particles of precipitated polymer of uncontrolled size). In aqueous polymer dispersions (Chapter 3), the polymer particles were stabilized by adding low-RMM compounds, e.g. emulsifiers or tensides forming a barrier (mostly electrostatic) on the surface of the particles which hinders their coagulation. In polymer dispersions in nonaqueous media, which are usually alkanes, stabilization of the particles by an electric charge does not protect them against coagulation [1].

Barrett [2] described the basic features of preparation of a nonaqueous polymer dispersion using as an example the disperse polymerization of methyl methacrylate. The monomer, initiator and steric stabilizers are dissolved in an alkane or mixture of lower alkanes and are heated with stirring to the reflux temperature. At the start of polymerization, the transparent mixture goes through a fine opalescence to the stage of formation of a white opaque latex. No visual changes are observed during the later course of polymerization. The conversion curve of a disperse polymerization in a nonaqueous medium has the typical sigmoidal character. The initially slow rate enhancement of the polymerization is followed by a rapid increase in the rate (the gel effect) to reach the maximum rate over a wide range of conversions. Then the polymerization rate gradually decreases because of consumption of monomer. The time needed to achieve *ca.* 99% conversion is about 1 h. The RMM of the poly(methyl methacrylate) is *ca.* 10^6 g mol^{-1}. A polymer with lower RMM is obtained in the presence of a chain transfer agent, e.g. a mercaptan. The concentration of the steric stabilizer used determines the particle size of the resulting polymer dispersion. As a rule, several per cents of stabilizer (with respect to monomer) are used. Dispersion polymerization is for most monomers a much quicker process than their polymerization in solution.

A model of the dispersion polymerization of vinyl monomers in *n*-alkanes is based on the assumption that the site of initiation is a disperse phase containing dissolved initiator and monomer molecules. The first phase of dispersion polymerization thus proceeds as a solution radical polymerization.

---

* The degree of polymerization of the oligomer, at which the aggregation of oligomers takes place and the primary (polymer) particles are formed, is a function of the type of monomer, the reaction medium temperature and the presence of other compounds. It depends on a group of effects which increase or reduce the ability of the reaction medium to solvate (dissolve) the polymer being formed. This condition follows from the definition of disperse polymerization, namely the continuous phase has to act as the precipitant for the polymer with the required RMM. Otherwise this would be an example of polymerization in bulk or in a solvent.

The RMM of the macromolecules at a given temperature and in the given reaction system is decisive for the production of primary polymer particles by coagulation (agglomeration) of the macromolecules (oligomers of rather low degree of polymerization) in the disperse phase. At 298 K in n-heptane, precipitation takes place [3] if the RMM of the oligomers is $ca.$ 2000 g mol^{-1}. At 333 K, oligomers precipitated if their RMM $\geq$ 9000 g mol^{-1}. The RMMs at which styrene oligomers precipitate indicate the considerable power of n-heptane to solvate styrene macromolecules. After precipitation, the agglomeration of the primary particles (flocculation, coagulation) is hindered by the stabilizer present.

Unstabilized polymer (colloidal) particles in disperse liquid media coagulate rapidly owing to the effect of van der Waals attractive forces between individual particles. Two ways of weakening the attractive forces between the particles, and thus hindering their coagulation, are known, i.e. (i) an electric double layer (EDL) or (ii) a steric barrier is formed on the particle surface. Both effects impede the necessary mutual approach of individual polymer particles. The electrostatic repulsion of particles with the same electric charge is very effective even in aqueous (polar) media; in media with a low relative permittivity (i.e. in nonpolar media), the steric stabilization of particles is more efficient.

The idea of a steric barrier brings us only a little closer to the actual situation. A more precise explanation of the mechanism for steric stabilization of the polymer particles follows from an analysis of the changes in Gibbs free energy as a function of distance between the two sterically stabilized polymer particles. As the polymer particles approach, at a certain distance, compression (mutual penetration) of solvated (solubilized) parts (segments) on the surface of the particles of anchored molecules of steric stabilizer takes place. The effective concentration of the solubilized particles of steric stabilizer in the region between the polymer particles increases; at the same time, the chemical potential of the disperse medium decreases (Fig. 21). The solvent therefore diffuses into this condensed space in order to reduce the concentration of the

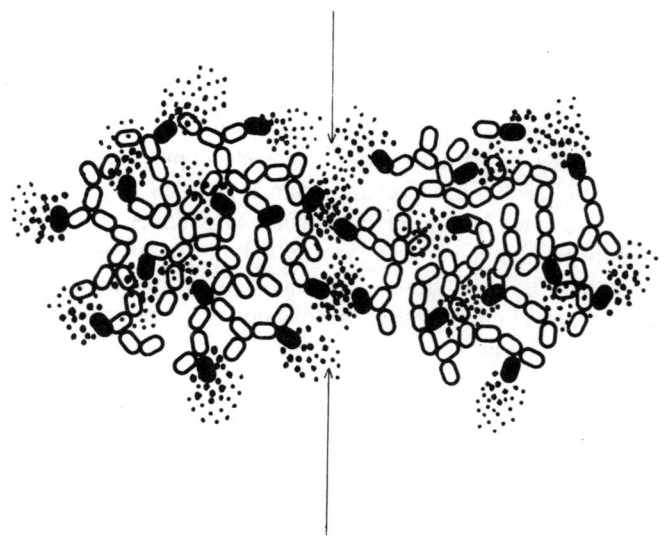

Fig. 21 — Schematic of compression (mutual penetration) of solvated parts of steric stabilizer anchored on surface of polymer particles during mutual approach (collision). •:•:• solvated segment of steric stabilizer. Arrows show direction of solvent diffusion (into zone of higher concentrations of solvated segments).

Basic principles

solubilized molecular segments of the steric stabilizer. This process causes mutual repulsion of the polymer particles until the penetration of the solubilized molecular segments of steric stabilizer anchored on the surface of both polymer particles is broken.

As the two polymer particles with the steric stabilizer molecules anchored on their surface in the disperse medium, which functions as the solvent for the soluble segment of the steric stabilizer, approach, the Gibbs energy G increases (ΔG is positive) during penetration or compression of the solvated segments of the dispersing agent molecules. The equation

$$\Delta G = \Delta H - T\Delta S$$

shows that a positive value of ΔG, as a condition of formation of a thermodynamically-stable polymer dispersion, can be obtained in three ways [4]:

(i) If $\Delta H > T\Delta S$ and both ΔH and ΔS are positive (enthalpic stabilization). The dispersion coagulates on heating.
(ii) If $\Delta H < T\Delta S$ and both ΔH and ΔS are negative (entropic stabilization). Dispersion coagulates on cooling.
(iii) If ΔH is positive and ΔS negative (combined enthalpic and entropic stabilization). The dispersion is stable over a wide temperature range.

The heat of solution of polymers is generally positive and ΔH is negative. Entropic stabilization is therefore more important in nonaqueous media. The main origin of particle repulsion in sterically-stabilized polymer dispersions in nonaqueous (nonpolar) media are entropy changes. The negative change in entropy causes a decrease in the number of possible conformations of the soluble dispersant segment after anchoring to the polymer particle. The change in enthalpy caused by the reorganization of solvent molecules during solvation of the soluble dispersant segment contributes to particle repulsion in nonpolar media less significantly than in polar media [5]. In typical athermal systems, the value of the resulting particle repulsion is determined by the difference between the number of conformations of the soluble dispersant segment in the final volume unperturbed by interactions with other solvated segments present in the system, and the number of conformations of the soluble dispersant segment in a system in which the soluble segments of the dispersant influence each other.

A more detailed discussion on the problems of steric stabilization of polymer dispersions is available [6].

To achieve permanent stabilization of the polymer particles, it is necessary to anchor strongly the dispersing agent onto the surface of the polymer particle, i.e. not to allow it to be released from the particle surface, or, eventually, to be shifted on the particle surface. Dispersants which only weakly or reversibly adsorb on the low-energy particle surface (i.e. on a surface with a low Hamaker constant [7] which is, for example, for poly(methyl methacrylate) 6.3×10^{-20} J) are less efficient stabilizers. Homopolymers and random copolymers of vinyl monomers are an example. Dispersing agents successfully applied to disperse pigments, the preparation of dispersions of alkyd resins and copolymers of vinyl monomers with polar substituents (i.e. compounds with a Hamaker constant greater than that of poly(methyl methacrylate), have been shown to be unsuitable for preparing stable dispersions of most commercially significant vinyl and acrylate polymers. An increase in the interaction between the particle surface and the dispersant can be achieved by introducing suitable functional groups into the stabilizer molecule or into the macromolecules producing polymer particles. This method was used to prepare stable polymer dispersions but has not found wider application.

In addition to the need for the strong anchoring of the dispersant on the particle surface, the surface of the particle has to be sufficiently covered with adsorbed molecules of the steric stabilizer. Thus the condition that the fraction of steric stabilizer soluble in the disperse medium should form a barrier around the polymer particle dense enough to protect it from aggregation is fulfilled.

Several reviews and monographs dealing with the adsorption of polymeric steric stabilizers (dispersants) onto the polymer particle surface have been published [8–11]. Homopolymers adsorb on such a surface using several chain segments; a homopolymer is thus able to form loops, entanglements and free ends [10] (Fig.22).

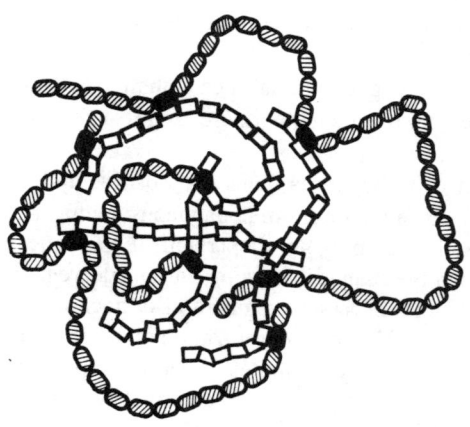

Fig. 22 — Schematic of adsorption of homopolymer (statistical copolymer) on polymer particle. ⟅⟆ homopolymer chain with 1 anchoring, ⟅⟆ dispersed polymer.

Provided they are amphiphilic, block copolymers adsorb on the surface in a different way. Diblock copolymers adsorb through one (so-called anchoring) block; the second block, being solvated by the solvent, is directed away from the surface into the solvent [11, 12] (see Fig. 21). Adsorption of a copolymer onto the particle surface is complicated by the possibility of it forming micelles in solution. Problems associated with the formation of micelles and the adsorption of diblock copolymers on the surface of polymer particles have been described [13–20].

By analogy with the CMC, the critical adsorption concentration is the dispersant concentration above which almost all the copolymer is adsorbed. The critical adsorption concentration decreases with increasing attraction between the particle surface and the stabilizer block, being determined by the energy of adsorption, with an increasing degree of incompatibility of the anchoring block with solvent, and with decreasing RMM of the solvated (soluble) block. The thickness of the adsorbed layer increases with increasing attractive interaction between the surface and the anchoring block and with an increase of the RMM of the solvated block copolymer.

Adsorption of the block copolymer, as compared with its micellization, is preferred if one of its segments (blocks) shows a strong affinity towards the particular surface or if the solvent of the soluble block is poor. This competition between micellization and adsorption is very important as regards the choice of stabilizer of the polymer dispersion. By varying the solvent, systems are obtainable in which the adsorption of a given copolymer onto a particular surface will be preferred to its micellization.

A block copolymer AB with incompatible blocks A and B forms microdomains in the absence of solvent in which a homopolymer composed of the structural units, e.g. block A, can be solubilized. Analogously, in the presence of a solvent to particular blocks, e.g. B, spherical micelles of the copolymer AB are formed, with a core composed of A blocks and shell B blocks which can solubilize a homopolymer in its core with structural units of block A [21–23]. There is experimental evidence that solubilization of homopolymer A in microdomains of the copolymer AB in the absence of solvent [24, 25], as well as in concentrated solutions [26], takes place if the RMM of the homopolymer is lower than that of block A in the copolymer AB and if the concentration of homopolymer does not exceed a certain critical concentration. If the RMM of the homopolymer is higher than that of the blocks forming the micelle core of the block copolymer, then the homopolymer is insufficiently solubilized in the micelles and the disperse system is not perfectly stabilized. These conclusions are also valid for triblock copolymers ABA. The micelles of a copolymer such as styrene/butadiene/styrene have a core of butadiene structural units (B). The solubilization of the poly(butadiene) homopolymer in the micelle core of the triblock copolymer only takes place if the solubilized poly(butadiene) has a lower RMM than block B in the micellar core. Macromolecules of copolymer of the same type can also be solubilized in the micelles of ABA on condition that the composition (length of blocks) of the ABA copolymer is different [27].

The size of the core of the spherical micelle of the AB copolymer containing, for example, blocks B, is influenced by the ratio of the RMMs of the anchoring (B) and solvated (A) blocks and also by the RMM of the anchoring block B.

Electron microscopy, light scattering, viscometry, small-angle-X-ray scattering [28–31] and small-angle neutron scattering [32] have been used to determine the size of the micelles and the number of molecules in a micelle of the block copolymers styrene/dimethylsiloxane, styrene/butadiene and styrene/isoprene in n-alkanes. By way of illustration, the following data are available for the AB block copolymer of poly(styrene-b-dimethylsiloxane), the RMM of block B being 2×10^4 g mol^{-1} and of block A 3300 g mol^{-1}, i.e. the ratio between the anchoring and the solvated blocks is 6.1 in n-heptane at 371 K. The number of molecules per micelle is 1350, and the micellar core diameter 44.0 nm. The surface area covered by one chain of the solvated block A is 4.4 nm^2. If the ratio between the anchoring and solvated blocks differs much from 1, being for example 18 or more, then the solvated blocks with regard to the size of the micellar core do not cover the whole area of the micelle. The result is an aggregation of micelles and the formation of nonspherical aggregates [31].

Attention has also been paid to copolymers with the same lengths of both blocks as well as to copolymers with an anchoring block much shorter than the solvated block. These studies led to the conclusions that the CMC increases if the RMMs of both the anchoring block and the solvent* (low- RMM or high-RMM, i.e. polymer) decreases. Increase in the solubility of the anchoring block leads to an increase in the CMC of the diblock copolymer. The size of the micelle increases if the solubility of the anchoring block decreases, even if the number of molecules per micelle decreases. As the CMC of a diblock copolymer increases, the number of molecules forming a micelle normally increases.

The thickness of the micellar shell, and hence the size of the micelle, increase if the incompatibility between the anchoring block of the copolymer (e.g. B) and the solvent S increases, its measure being the product $\chi_{BS}N_B$.** At a given degree of incompatibility expressed by this product, spherical micelles are formed preferentially if the RMM of the insoluble anchoring block B of the copolymer is lower than that of block A and if the RMM of the solvent is low compared with that of the copolymer. When both copolymer blocks have approximately the same RMM and when the solvent is a large molecule (polymer), then lamellar micelles are produced.

The micellar model of diblock AB copolymers in solution assumes a spherical geometry of the micelle composed of the three regions (layers) [33]. The micellar core (first region) is composed of B blocks for which the solvent used is poor. The core shell (the second region) consists of the A segment. The inequality $\chi_{AS} \ll \chi_{BS}$ of the interaction parameters of blocks A and B with respect to solvent S holds for this case. The surroundings of the micelle contains molecules of solvent S and of molecularly dispersed stabilizer (the third region). The core and shell of the micelle are separated by a narrow layer forming an interphase, its size being a function of the type and composition of the block copolymer and its concentration in solution. The calculated thickness of the interphase for the block copolymer of poly(styrene-b-butadiene) is about 0.77 nm at 25°C, the radius of gyration of the poly(styrene) core 8 nm and that of the micelle 20 nm. The micelle is composed of 179 copolymer molecules [13, 23].

The effectiveness of adsorption of the anchoring segment onto the polymer particle will markedly increase if there is specific affinity between the dispersing agent and the polymer forming the particle. This possibility is, under certain conditions, the only means of preparing a stable polymer dispersion. In aqueous polymer dispersions, the association (anchoring) of the anchoring dispersant segment (in the case of the lipophilic part of a surface-active compound, the emulsifier) on the surface of a polymer particle is usually a reversible process (an associated dispersant molecule can be replaced by a dispersant molecule in a disperse medium). Moreover, the chemical composition of the anchoring segment of a dispersant and of the polymer forming the particle can differ considerably. Despite this, good stabilization of the dispersion is achieved. This is due to the fact that the permanent desorption of dispersant from the particle surface requires considerable energy to break the structure of the water associates in the disperse medium. The penetration of the desorbed dispersant molecules into the disperse medium in organic solvents requires much less energy than in the aqueous disperse phase because not much energy is needed to break the structure because water associates are absent. To achieve good stabilization of the polymer particle in a disperse nonpolar medium, the desorption must be negligible. The

---

*   References [18, 19] define the RMM of the solvent by the ratio $\beta = N/N_S$, N_S being the number of interacting segments of a molecule of solvent S and N the number of segments of the AB copolymer, i.e. $N = N_A + N_B$; β varied between 0 and 50. A polymer solvating one of the copolymer blocks should also be considered to be solvent.

** χ_{BS} is Flory's dimensionless interaction parameter defining the interaction between block B of the copolymer and solvent S. It is a measure of the effective interaction energy for the B block. The product of χ_{BS} and N_B therefore represents the total interaction energy for the B blocks in the given system. It is also a measure of the incompatibility of the system [14].

anchoring (adsorption) of the dispersant segment on the surface of the polymer particle has to be an irreversible process characterized by a considerable energy release (needed to carry out desorption of the dispersant molecule from the polymer particle surface). The degree of irreversible anchoring of the dispersant on the polymer particle depends on the dispersant and the particle surface, and on the number of individual anchorings (or entanglements of the anchoring dispersant segment with the polymer chains forming the particle) on the surface or inside the polymer particle [34, 35]. The strength of anchoring can be increased by the formation of a covalent chemical bond, hydrogen bond or donor–acceptor interaction between dispersed polymer and the anchoring dispersant segment. Another possibility consists of the crosslinking of the anchoring segments of the dispersant molecules after adsorption onto the particle and thus the formation of a crosslinked shell of anchoring dispersant segments on the surface of the polymer particle. These features are taken into account especially when the disperse medium is able, to a degree, to swell the dispersed polymer. The reduction of the degree of swelling of the dispersed polymer can be achieved by varying the disperse phase as well as by crosslinking the dispersed polymer (e.g. by adding a small amount of multifunctional monomer) [34]. The problem of the effect of solvation of the dispersed polymer particles by the disperse medium (by its own monomer) can partly be reduced by carrying out the disperse polymerization in two stages. In the first stage, part of the monomer is polymerized and in the second stage the residual part of the monomer is gradually added. This may lead, in addition to an increase in stability of the dispersion during its preparation, also to a decrease in the size distribution of the resulting polymer particles [3].

If there is a possibility of motion of the anchored segment of the steric stabilizer on the surface of the polymer particle and if, at the same time, the number of adsorbed molecules of steric stabilizer does not fully saturate the particle surface, the movements of the anchored segments of steric stabilizer on the particle surface are not associated with any significant mutual repulsion due to a polymer–polymer interaction. The polymer particles can approach each other sufficiently closely since the energy needed to overcome the repulsive forces between the two particles is negligible (the increase in concentration of the solvated dispersant blocks in the region between the particles as a result of the movement of anchored segments is small). A low stability of the polymer dispersion results.

In systems where the desorption of stabilizer from the particle surface is important, there is always an equilibrium between the adsorbed dispersant and the dispersant in the disperse medium. If micellization of the desorbed steric stabilizer takes place in the continuous phase, the energy obtained by micellization must be subtracted from the energy needed for desorption. That means that if the micellization energy is the same or greater than the energy obtained via adsorption of a dispersant onto the particle surface, then the London attractive forces between the polymer particles will considerably exceed the energy needed to maintain equilibrium between adsorption and desorption of the stabilizer molecules, which causes particle coagulation. Since, however, micellization of the stabilizer molecules must be preceded by their desorption from the surface of the polymer particles, the energy needed to carry out desorption forms a barrier which must be overcome before interaction (collision) of the two polymer particles takes place. The system is thermodynamically unstable, but for practical purposes it is sufficiently resistant to collisions caused by the Brownian motion of the polymer particles and collisions of the particles brought about by different force fields. The stabilizer is constantly anchored on the surface of the polymer particles and the repulsion between the particles is conditioned by the compression of the solvated soluble segments of the steric stabilizer [34].

The kinetic aspect of the process is significant [6]. If the relaxation time of an adsorbed molecule of the steric stabilizer is longer than the time of Brownian contact of two polymer particles, then desorption of the stabilizer molecule from the particle surface is unlikely even though the process of desorption is energetically favoured.

A number of papers on disperse radical polymerization in nonaqueous media have described the preparation of polymer particles stabilized by block copolymers. The block AB and ABA copolymers prepared via anionic 'living' polymerization have exactly-defined block lengths and a narrow RMM distribution (section 5.2). The polymer particles thus stabilized have therefore on their surface a layer of constant thickness formed by the anchoring blocks of the polymer dispersant. For instance, diblock AB copolymers of poly(styrene-b-dimethylsiloxane) were used to stabilize nonaqueous dispersions of poly(styrene) [3]. The ratio of the RMMs of the anchoring block (poly(styrene) segment) and the soluble block (poly(dimethylsiloxane) segment) varied between 0.5 and 21.5. The RMM of the poly(styrene) of the anchoring block ranged from 1 to 5×10^4 g mol^{-1}. The radical polymerization of styrene initiated by azoisobutyronitrile, dibenzoyl peroxide, and bis(4-$tert$-butylcyclohexyl)peroxodicarbonate in the presence of the block copolymer is very slow. The shape of the conversion curve recalls that of the conventional solution polymerization of styrene [36]. In contrast, the shape of the conversion curve of methyl methacrylate polymerization under the same reaction conditions is sigmoidal [3, 36]. The higher rate of polymerization results from the higher value of the ratio $k_p/k_t^{0.5}$ of the propagation and termination rate constants of this radical polymerization compared with styrene. The acceleration of the polymerization rate at a certain degree of conversion is attributed to the gel effect (Trommsdorf's effect).

A recent study [37] has shown that the rate of dispersion polymerization is slow and the RMM of the final polymers is of the order of 10^4 when more polar dispersion media are used. The mechanism was entirely one of solution polymerization, since no, or only a weak, gel effect was observed. Monodisperse latex particles in the size range of 3–9 μm were produced. The dispersion polymerization of styrene in ethanol initiated by AIBN in the presence of a polymer particle stabilizer, poly(N-vinylpyrrolidone), is characterized by two polymerization sites, namely the monomer-swollen polymer particles and the continuous (ethanol + styrene) phase [38]. At low conversions, the polymerization in solution dominates. At high conversions, the heterogeneous pathway becomes dominant. The slow rate of polymerization and low RMM of the final polymer produced by this process are due to the low monomer concentration in the polymer particle, the low particle number density, and high average number of oligomeric radicals per particle. The number average polymer particle diameter is in the range 2.45–4.68 μm and the polymer particle density between $(4.7 - 16.3) \times 10^9$ cm^{-3}. The RMMs of dispersion polymers increase with increasing conversion and decrease with increasing initiator concentration.

References

1. Napper, D.H. (1970) *Ind. Eng. Chem., Prod. Res. Dev.* **9** 467
2. Barrett, K.E.J. (1973) *Br. Polym. J.* **5** 259
3. Dawkins, J.W. & Taylor, G. (1979) *Eur. Polym. J.* **15** 453
4. Napper, D.H. & Hunter, R.J. (1972) In: Kerker, M. (ed.) *Surface chemistry and colloids*, vol. 7. Butterworths, London, p. 241
5. Ottewill, R.H. (1973) In: *Colloid sciences*, vol. 1. Specialist periodical reports. Chemical Society, London, p. 173
6. Osmond, D.J.W. & Waite, F.A. (1975) In: Barrett, K.E.J. (ed.) *Dispersion polymerization in organic media*. Wiley–Interscience, London, p. 9

7. Hamaker, H.C. (1936) *Rec. Trav. Chem.* **55** 1015, **56** 727
8. Goddart, E.D. & Vincent, B. (eds) (1984) *Polymer adsorption and dispersion stability.* American Chemical Society, Washington DC
9. Tadros, T.F. (ed.) (1982) *The effect of polymers on dispersion properties.* Academic Press, London
10. Ottewill, R.H., Rochester, C.H. & Smith, A.L. (eds) (1983) *Adsorption from solution.* Academic Press, London
11. Buscall, R., Corner, T. & Stageman, J.F. (eds) (1985) *Polymer colloids.* Elsevier, London
12. Napper, D.H. (1983) *Polymeric stabilization of colloidal dispersions.* Academic Press, London
13. Noolandi, J. & Hong, K.M. (1983) *Macromolecules* **16** 1443
14. Leibler, L., Orland, H. & Wheeler, J.C. (1983) *J. Chem. Phys.* **79** 3550
15. Whitmore, M.D. & Noolandi, J. (1985) *Macromolecules* **18** 657
16. Roe, R.J. (1986) *Macromolecules* **19** 728
17. Roe, R.J. & Rigby, D. (1984) *Macromolecules* **17** 1778
18. Munch, M.R. & Gast, A.P. (1988) *Macromolecules* **21** 1360
19. Munch, M.R. & Gast, A.P. (1988) *Macromolecules* **21** 1366
20. Gast, A.P. & Munch, M.R. (1988) *Colloids Surfaces* **31** 47
21. Krause, S. (1970) *Polym. Prepr. (Am. Chem. Soc., Div. Polym. Chem.)* **11** 568
22. Meier, D.J. (1977) *Polym. Prepr. (Am. Chem. Soc., Div. Polym. Chem.)* **18** 340
23. Tuzar, Z. & Kratochvíl, P. (1976) *Adv. Colloid Interface Sci.* **6** 201
24. Inoue, T., Soen, T., Hashimoto, T. & Kawai, H. (1970) In: Aggarwal, S.L. (ed.) *Block polymers.* Plenum Press, London
25. Molau, G.E. & Wittbrodt, W.M. (1968) *Macromolecules* **1** 260
26. Skoulios, A., Helffer, P., Gallot, Y. & Selb, J. (1971) *Makromol. Chem.* **148** 305
27. Tuzar, Z., Bahadur, P. & Kratochvíl, P. (1981) *Makromol. Chem.* **182** 1751
28. Price, C. & Woods, D. (1973) *Eur. Polym. J.* **9** 827
29. Price, C., McAdam, J.D.G., Lally, T.P. & Woods, D. (1974) *Polymer* **15** 228
30. Pleštil, J. & Baldrián, J. (1975) *Makromol. Chem.* **176** 1009
31. Dawkins, J.W. & Taylor, G. (1979) *Makromol. Chem.* **180** 1737
32. Higgins, J.S., Dawkins, J.W. & Taylor, G. (1980) *Polymer* **21** 627
33. Noolandi, J. & Hong, K.M. (1982) *Macromolecules* **15** 482
34. Walbridge, D.J. (1975) In: Barrett, K.E.J. (ed.) *Dispersion polymerization in organic media.* Wiley–Interscience, London, p. 45
35. Gennes de, P.G. (1987) *Adv. Colloid Interface Sci.* **27** 189
36. Flory, P.J. (1953) *Principles of polymer chemistry.* Cornell University Press, Ithaca, Chapter IV
37. Lok, K.P. & Ober, C.K. (1985) *Can. J. Chem.* **63** 207
38. Lu, Y.Y., El-Aasser, M.S. & Vanderhoff, J.W. (1988) *J. Polym. Sci., Polym. Phys.* **26** 1187

5.2 DISPERSANTS (STERIC STABILIZERS) AND DISPERSE MEDIA OF NONAQUEOUS DISPERSIONS OF POLYMERS

A suitable choice of steric stabilizer is a requirement for the successful preparation of a nonaqueous dispersion of a polymer. This choice requires knowledge of the mechanism of influence of the steric stabilization of polymer dispersions (see section 5.1). The choice of steric stabilizer is closely connected with the selection of disperse medium. Both selections have to be optimal in their relation to the chosen monomer, i.e. to the polymeric component of the resulting dispersion.

Suitable dispersants for the efficient stabilization of polymer dispersions are those with the chemical structure and configuration of the polymer chain that satisfy the demand for functions of both anchoring and a solvated block (segment). Block and graft copolymers best meet this requirement. Using the starting monomers A and B, the following structures of block copolymers can be obtained: A_nB_m, $A_nB_mA_n$, $B_mA_nB_m$, $(A_nB_m)_x$, where m, $n \gg 1$; $x = 1, 2, \ldots$. Figure 23 shows structures that may be taken into account for graft copolymers prepared using monomer A (homopolymer A) and monomer B (homopolymer B) as starting monomers.

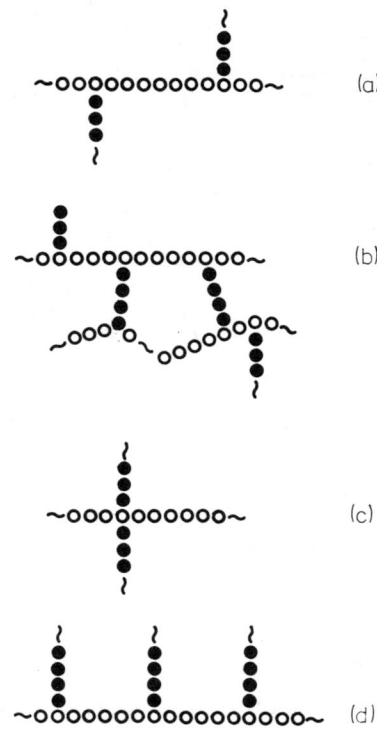

Fig. 23 — Schematic of possible structures of graft copolymers prepared by radical polymerization. ~OOOOOO~ chains of polymer to be grafted, ~●●●●●●~ chains of graft branches. (a) — graft copolymer with two graft branches, (b) — crosslinked graft copolymer (crosslinking bonds between chains of starting polymer form graft branches), (c) — graft copolymer prepared by introducing double bond in chain of starting polymer, (d) — comb graft copolymer.

The literature dealing with the preparation of block and graft copolymers is extensive [1–13]. The methods used for preparing block and graft copolymers are based on ionic and radical polymerization mechanisms. An anionic polymerization technique for the synthesis of block polymers has been reported in [6–13]. Seymour et al. [14–16] described the preparation of block copolymers via the radical mechanism. His method is based on the experimental finding that propagating macroradicals whose contact with other radicals is hindered (termination), e.g. by their precipitation from solution, can be preserved as living entities and can then be used to initiate polymerization of another monomer. However, the more suitable from the applications point of view was found to be living anionic polymerization [6–13,17–22], which is currently the only way of preparing structurally exactly-defined block copolymers.*

A particularly interesting possibility for the preparation of block copolymers is provided by a combination of macroions first described by Kučera et al. [23]. He used anionic and cationic poly(dimethylsiloxane) for this purpose. In the preparation of block copolymers we can make use of our knowledge of the transformation of the propagating centres. Detailed information has appeared in ref. [22]. Walbridge [24] gives practical instructions for the preparation of polymer dispersants based on block and graft copolymers.

Transfer reactions of radicals with a polymer are mainly used for the preparation of graft copolymers. In the transfer reaction (mainly of hydrogen), a radical centre ($P^{\bullet}$) is formed on a polymer chain (PH) which, in the presence of the radical-polymerizing monomer, initiates its polymerization

$$R^{\bullet} + PH \rightarrow P^{\bullet} + RH,$$

$$P^{\bullet} + nM \rightarrow PM_{n-1}M^{\bullet}.$$

According to the site and number of transfer reactions of the polymer macromolecule, various structures can be prepared (Fig. 23). The reaction product is usually a combination of several structures which can rarely be exactly defined. In spite of this, and taking account of their simple preparation, graft copolymers prepared by radical polymerization display a greater variety of combinations of monomers as compared with those made by ionic polymerization. They are applied successfully as polymer dispersion stabilizers and represent a significant group of polymeric dispersants of polymer particles. The introduction of a reactive side group, e.g. –SH, –CCl$_3$, into the polymer chain can considerably increase the effectiveness of grafting, with the preparation of graft copolymers with a greater number of more-or-less regularly located grafted branches. A graft copolymer is also formed by adding a radical to a double bond of the polymer chain followed by the addition of monomer to the macroradical formed

$$\sim CH_2-CH=CH-CH_2\sim \xrightarrow{R^{\bullet}} \sim CH_2-CH(R)-\overset{\bullet}{C}H-CH_2 \sim \xrightarrow{nM}$$

$$\longrightarrow \sim CH_2-CH(R)-CH-CH_2 \sim$$
$$|$$
$$M_{n-1}$$
$$|$$
$$M^{\bullet}$$

---

* By this term we understand ionic or radical polymerization as long as it is characterized by the absence of a termination step and transfer reactions.

During termination of the growing branches, three-dimensional structures can be formed (in the case of termination by the recombination of macroradicals),

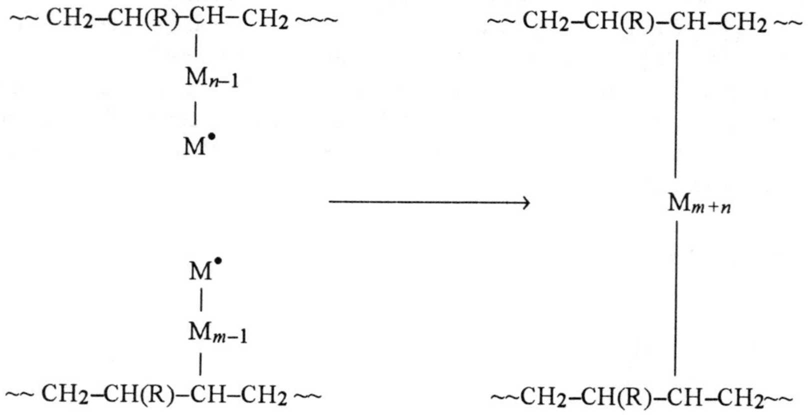

$$\sim\sim CH_2-CH(R)-CH-CH_2 \sim\sim$$
$$|$$
$$M_{n-1}$$
$$|$$
$$M^\bullet$$

$$M^\bullet$$
$$|$$
$$M_{m-1}$$
$$|$$
$$\sim\sim CH_2-CH(R)-CH-CH_2 \sim\sim$$

$$\longrightarrow$$

$$\sim\sim CH_2-CH(R)-CH-CH_2 \sim\sim$$
$$|$$
$$M_{m+n}$$
$$|$$
$$\sim\sim CH_2-CH(R)-CH-CH_2 \sim\sim$$

which can also arise by adding a propagating macroradical of the grafted branch to the double bond of another macromolecule

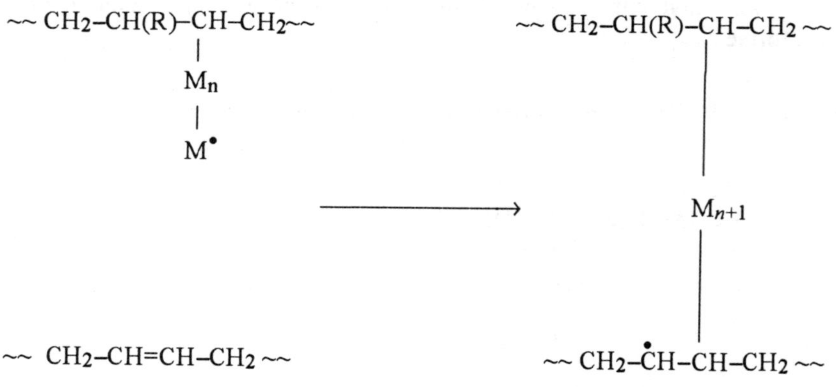

$$\sim\sim CH_2-CH(R)-CH-CH_2 \sim\sim$$
$$|$$
$$M_n$$
$$|$$
$$M^\bullet$$

$$\sim\sim CH_2-CH=CH-CH_2 \sim\sim$$

$$\longrightarrow$$

$$\sim\sim CH_2-CH(R)-CH-CH_2 \sim\sim$$
$$|$$
$$M_{n+1}$$
$$|$$
$$\sim\sim CH_2-\overset{\bullet}{C}H-CH-CH_2 \sim\sim$$

A comb graft copolymer can be prepared via polymerization of a monomer in the presence of a polymer with a reactive double bond at one end (a so-called macromonomer).

Dispersants

```
    R'•                          R'
                                   >CH₂
    R–CH=CH₂                    CH–R
                                   >CH₂
    ~~CH=CH₂                    ~~ CH
                                   >CH₂
    R–CH=CH₂        ⟶          CH–R
                                   >CH₂
    ~~CH=CH₂                    ~~ CH
                                   >CH₂
    R–CH=CH₂                    CH–R
                                   >CH₂
    ~~CH=CH₂                    ~~ CH
                                     •
    etc.                         etc.
```

Such a macromonomer can be prepared by esterification of the terminal carboxyl group of poly(12-hydroxystearic acid) by glycidyl methacrylate. The copolymerization of the macromonomer with acrylic or vinyl monomers gives a comb graft copolymer. Polymer dispersants based on comb graft copolymers are prepared by radical polymerization of a mixture of macromonomer (the potential solvated segment of the polymer dispersant), monomer (the potential anchoring segment of the polymer dispersant), and a radical initiator in the solvent (usually esters) at the reflux temperature of the reaction mixture. The RMM of the comb graft copolymer is normally in the region of M_w = 20 000 to 35 000 g mol^{-1}, M_n = 6000 to 15 000 g mol^{-1} [24]. Modification of the composition of poly(hydroxy acid) by copolymerization with various hydroxy acids or amino acids makes possible the preparation of a variety of soluble segments of the polymer dispersant, depending on the polarity of the disperse medium (alkanes, aromatics, esters, etc.).

The chemical composition and structure of the polymer dispersant (localization of the solvated and anchoring dispersant segments) are of fundamental importance in achieving the efficient steric stabilization of nonaqueous polymer dispersion. The data in Table 36 show that in the dispersion polymerization of 4-vinylpyridine only a dispersant based on diblock poly(styrene-b-tetrahydrofuran) copolymer was efficient. Multiblock copolymers based on styrene and tetrahydrofuran were inefficient. A change in the order of the solvated and anchored segments in triblock copolymers causes a change in efficiency of their stabilization of polymer dispersion. The triblock copolymers poly(isoprene-b-styrene-b-isoprene) show good stabilization efficiency of poly(methyl methacrylate) dispersions. Di- and triblock copolymers of styrene with dimethylsiloxane also show good stabilization efficiency. In contrast, poly(styrene-b-isoprene-b-styrene) copolymer with anchoring styrene segments in boundary positions of the triblock copolymer is largely inadequate for the stabilization of a poly(methyl methacrylate) dispersion in n-heptane (N.B.: the isoprene structural units were completely hydrogenated).

The results with poly(styrene-b-ethylene oxide) copolymers as stabilizers in styrene polymerization indicated that, for efficient anchoring, the block length need not be more than 10 monomeric units and that a poly(ethylene oxide) with weight-average-RMM M_w = 3000 is as effective a stabilizer as a block with poly(ethylene oxide) of M_n = 9000 [29].

The suitability of application of dispersants based on comb graft copolymers follows

Table 36 — Efficiency of block copolymers during stabilization of nonaqueous polymer dispersions

Monomer	Dispersant	Block type	Calculated RMM of segment /g mol^{-1}		Number of blocks	Efficiency[a]	Ref.
			solvated A	anchoring B			
Vinylpyridine[b]	Poly(styrene-b-tetrahydrofuran)	AB	5 600	7 200	2	poor	[25]
		(AB)$_2$	5 500	7 500	4	very low	[25]
		(AB)$_2$	5 300	1300	4	very low	[25]
		A(BA)$_2$BA	5 600	1 300	7	very low	[25]
Methyl methacrylate[c]	Poly(siloxane-b-tetrahydrofuran)	A(BA)$_4$ BA	4 100	3 600	11	very low	[26]
	Poly(isoprene-b-styrene-b-isoprene)[d]	ABA	2 × 7 400	6 100	3	good	[26]
			2 × 16 600	13 700	3	good	[26]
			2 × 41 400	34 200	3	good	[26]
			2 × 62 300	51 400	3	good	[26]
	Poly(styrene-b-isoprene-b-styrene)	BAB[e]	38 700	2 × 7 155	3	very low	[26]
		BAB[f]	60 300	2 × 11 150	3	very low	[26]
	Poly(dimethylsiloxane-b-styrene-b-dimethyl-siloxane)	ABA	—	—	3	good	[27]
Methyl methacrylate[g]	Poly(dimethylsiloxane-b-styrene)	AB	—	—	2	good	[28]
Styrene[g]	Poly(dimethylsiloxane-b-styrene)	AB	—	—	2	good	[28]

[a] Evaluated from point of view of formation of stable dispersion.
[b] Disperse medium: toluene.
[c] Disperse medium: n-heptane.
[d] Ratio of RMMs of anchoring and solvated blocks: 0.41.
[e] Craton® G 1650 (hydrogenated poly(isoprene) segments).
[f] Craton® G 1652 (hydrogenated poly(isoprene) segments).
[g] Disperse medium: Freon 113.

from Table 37. The higher the second-order transition temperature, T_g, of the anchored dispersant block, the higher is the reaction temperature needed to reach efficient stabilization of the dispersion polymer particles. Similarly, when associates are formed (e.g. because of hydrogen bonding) between the anchoring segments, the polymer dispersion has to be prepared at a higher temperature to achieve the required stability of the polymer particles. The reaction temperature or the power of the solvent (disperse medium) must enable the establishment of equilibrium between the dispersant micelles and the polymer dispersant molecularly scattered in the disperse phase (molecules and not dispersant micelles are adsorbed on the particle surface).

Table 37 — Efficiency of comb graft dispersants[a] of nonaqueous polymer dispersions in aliphatic hydrocarbon (Reprinted from ref. [24] with permission of Imperial Chemical Industries Ltd., Wilton)

Monomer	Anchoring dispersant segment	Reaction temperature / °C	Particle size /nm
Vinyl acetate	Methyl methacrylate/ methacrylic acid (98:2)	80	200
		50	coagulation
	Vinyl acetate	50	100–200
	Ethyl acrylate /methyl methacrylate (50:50)	50	800–1000
Methyl methacrylate	Methyl methacrylate / methacrylic acid (98:2)	80	100
		50	coagulation
	Methyl methacrylate	50	coagulation
	Ethyl acrylate/methacrylic acid (50:50)	50	200–300

[a] Solvated dispersant segment formed by grafted chains of poly(hydroxy acid). Homopolymers and statistical copolymers of varying chain polarity and flexibility formed anchoring component of polymer dispersant. Ratio between solvated and anchoring blocks of polymer dispersant equal to 1.

Table 38 gives data about the use of dispersants based on block and graft copolymers for the dispersion polymerization of vinyl monomers in addition to the preparation of dispersants [24].

Table 38 — Examples of use of dispersants based on block and graft copolymers in disperse polymerization of vinyl monomers

Monomer	Dispersant	Disperse phase	Ref.
Methyl methacrylate	Poly(isoprene-*b*-styrene-*b*-isoprene)	*n*-Heptane	[26]
	Poly(2-ethylhexyl acrylate-*g*-vinyl acetate)	Alkane	[30]
Lauryl methacrylate	Poly(lauryl methacrylate-*b*-methacrylic acid)	Ethanol	[30]
Styrene	Poly(2-ethylhexyl acrylate-*g*-methyl methacrylate)	Alkane	[31]
	Poly(butadiene-*g*-methacrylic acid)	Ethanol	[32]

A special feature of polymer particle stabilization is the grafting mechanism of stabilization. In the polymerization of styrene in a polar solvent, the steric stabilizer (e.g. hydroxypropylcellulose) becomes grafted and ends up on the polystyrene particle surface [32]. The effect of solvent seems to be related to the solubility properties of the grafted stabilizer — the greater its solubility, the larger the particle size.

The choice of disperse medium (continuous phase) for disperse polymerization should meet the need for the polymer to be insoluble in the disperse medium. On the other hand, the disperse medium has to be a good solvent for the soluble dispersant segment. Data about the theta solvents of polymers are one criterion for the choice of disperse medium [34]. To choose the disperse medium and the necessary composition of the soluble and anchoring dispersant segments, one can use data on the solubility parameters of polymers and solvents [35, 36].

A good solvent for a given polymer is that with a solubility parameter close to that of the polymer. Information about theta temperatures and theta solvents together with data on solubility parameters in relation to the choice of disperse medium for a particular polymer and a suitable dispersant, create a quite reliable basis for defining a system for disperse polymerization. The optimum selection, however, can normally be done only on the basis of experimental verification.

Table 39 — Critical flocculation volume of precipitant V_{cr} for solvated segment of steric stabilizer and theta composition of disperse phase V for poly(methyl methacrylate) dispersion in n-heptane (Reprinted from ref. [37] with permission of Royal Society of Chemistry, Cambridge)

Steric stabilizer	Precipitant	T_Θ /K	V_{cr} /%	V /%
Poly(12-hydroxy-stearic acid)	Ethanol	274	39.5	39
	Ethanol	297	50.5	51
	Ethanol	313	58	59
	n-Propanol	274	61	65
	n-Propanol	297	78	84
	n-Butanol	274	74	81
Poly(lauryl methacrylate)	Ethanol	249	31	33
	Ethanol	274	38	39
	Ethanol	297	44	45
	Ethanol	313	51	51
	n-Propanol	274	56	52
	n-Propanol	297	70	67

Table 39 indicates the relation between the theta composition of the disperse medium and the critical flocculation volume, V_{cr}, of the precipitant for the solvated segment of the steric stabilizer, at which the polymer dispersion starts to precipitate. The beginning of precipitation of the polymer dispersion lies near the theta composition of the disperse phase.

A deterioration in the solvation properties of a solvent (disperse phase) as regards the solvated segment of the polymer dispersant can be achieved by gradual addition of a precipitant for the solvated segment. Good agreement was found between the critical flocculation volume and the theta composition for ethanol used as solvent.

Table 40 — Theta temperature T_Θ and critical flocculation temperature T_{cr} of poly(methyl methacrylate) dispersion in n-heptane (Reprinted from ref. [37] with permission of Royal Society of Chemistry, Cambridge)

Steric stabilizer	T_Θ /K	T_{cr} /K
Poly(methyl methacrylate-g-12-hydroxystearic acid)	240	260
Poly(vinyl alcohol-g-12-hydroxystearic acid)	240	250
Poly(methyl methacrylate-g-lauryl methacrylate)	175	215

With aliphatic alcohols, the dispersion precipitated before reaching the theta composition of the disperse medium (stabilizer poly(12-hydroxystearic acid)) or after exceeding the theta composition of the disperse medium (poly(lauryl methacrylate)). Good agreement was observed between the theta temperature and the critical flocculation temperature (Table 40). The critical flocculation temperature is seen to be about 10–20 K above the theta temperature of the solvated blocks of the steric stabilizer in the particular disperse medium. The dispersion thus precipitates even before reaching the theta temperature. These examples show that the dependence of the beginning of precipitation of the polymer dispersion on the theta conditions of the system can be used in choosing the composition of the solvated segment of the steric stabilizer for the chosen disperse medium. We know from the theory of polymer solutions that the theta temperature, critical flocculation volume and critical flocculation temperature should not be dependent on the RMM of the solvated segment of the steric stabilizer. This assumption was fully confirmed for aqueous sterically-stabilized dispersions [38, 39]. This conclusion is also expected to be valid for dispersions in nonaqueous media [24].

References

1. Burlant, W.J. & Hoffman, A.S. (1960) *Block and graft copolymers*. Reinhold, New York
2. Ceresa, J.R. (1962) *Block and graft copolymers*. Butterworths, London
3. Allport, D.C. & Janes, W.H. (1973) *Block copolymers*. Applied Science Publ., London
4. Noshay, A. & McGrath, J.E. (1977) *Block copolymers*. Academic Press, New York
5. Riess, G., Periard, J. & Bandert, A. (1971) In: Molau, G.E. (ed.) *Colloidal and morphological behaviour of block and graft copolymers*. Plenum Press, New York, p. 173
6. Szwarc, M. (1968) *Carbanions, living polymers and electron transfer processes*. Wiley–Interscience, New York
7. Morton, M. & Fetters, L. (1967) In: Peterlin, A., Goodman, M., Okamura, S., Zimm, B.H. & Mark, H.F. (eds) *Macromolecular reviews*, vol. 2. Wiley–Interscience, New York, p. 71
8. Morton, M. (1970) In: Aggarwal, S.L. (ed.) *Block polymers*. Plenum Press, New York
9. Bywater, S. (1976) In: Bamford, C.H. & Tipper, C.F.M. (eds) *Comprehensive chemical kinetics*, vol. 15. Elsevier, Amsterdam
10. Saam, J.C., Gordon, D.J. & Lindsey, S. (1970) *Macromolecules* **3** 1
11. Zilliox, J.G., Roovers, J.E.L. & Bywater, S. (1975) *Macromolecules* **8** 573
12. Marsiat, A. & Gallot, Y. (1975) *Makromol. Chem.* **176** 1641

13. Davies, W.G. & Jones, D.P. (1971) *Ind. Eng. Chem., Prod. Res. Dev.* **10** 168
14. Seymour, R.B. & Kincaid, P.D. (1973) *J. Paint Technol.* **45** 33
15. Seymour, R.B., Stahl, G.A. & Wood, H. (1975) *Appl. Polym. Symp.* **26** 249
16. Seymour, R.B., Kincaid, P.D. & Owen, D.R. (1973) *Adv. Chem. Ser.* **129** 230
17. Fetters, L.S. (1969) *J. Polym. Sci., C* **26** 1
18. Zelinski, R. & Childers, C.W. (1969) *Rubber Chem. Technol.* **41** 161
19. Rempp, P. & Franta, E. (1972) *Pure Appl. Chem.* **30** 229
20. Greber, G. (1967) *Makromol. Chem.* **101** 104
21. Braun, D., Neumann, W. & Arcache, H. (1968) *Makromol. Chem.* **112** 97
22. Kučera, M. (1992) *Mechanism and kinetics of addition polymerizations*. Academia, Prague; Elsevier, Amsterdam
23. Kučera, M., Jelínek, M., Lániková, J. & Veselý, K. (1961) *J. Polym. Sci.* **53** 311
24. Walbridge, D.J. (1975) In: Barrett, K.E.J. (ed.) *Dispersion polymerization in organic media*. J. Wiley & Sons, London
25. Šušoliak, O. (1983) *PhD Thesis*. Slovak Academy of Sciences, Bratislava
26. Šušoliak, O. & Bartoň, J. (1985) *Chem. Papers* **39** 379
27. Everett, D.H. & Stageman, J.F. (1977) *Colloid Polym. Sci.* **255** 293
28. Higgins, J.S., Dawkins, J.W. & Taylor, G. (1980) *Polymer* **21** 627
29. Piirma, I. (1990) *Makromol. Chem., Macromol. Symp.* **35/36** 467
30. Brit. 1122397 (1968)
31. Brit. 1101983 (1964)
32. Brit. 941305 (1963)
33. Paine, A.J. (1990) *Macromolecules* **23** 3109
34. Elias, H.G., Adank, G., Dietschy, H., Etter, O., Gruber, U. & Ibrahim, F.W. (1966) In: Brandrup, J. & Immergut, E.H. (eds) *Polymer handbook*. Wiley–Interscience, New York, p. IV-163
35. Burrel, H. & Immergut, B. (1966) In: Brandrup, J. & Immergut, E.H. (eds) *Polymer handbook*. Wiley–Interscience, New York, p. IV-341
36. Hoy, K.L. (1970) *J. Paint Technol.* **42** 76
37. Napper, D.H. (1968) *Trans. Faraday Soc.* **64** 1701
38. Napper, D.H. (1970) *J. Colloid Interface Sci.* **32** 106
39. Napper, D.H. (1972) *Polym. Lett.* **10** 449

5.3 KINETICS AND MECHANISM OF RADICAL DISPERSION POLYMERIZATION IN NONAQUEOUS MEDIA

The basic condition for dispersion polymerization is the presence of an inert diluent, which dissolves the monomer but precipitates the polymer, and a stabilizer able to stabilize the polymer particles efficiently and thus prevents formation of a mass of precipitated polymer (see sections 5.1 and 5.2). On satisfying these conditions, any method of polymerization can be used to prepare a polymer dispersion: radical polymerization (used most often) or ionic polymerization, polymerization involving ring opening or various polycondensation reactions. Details about the problems of nonradical dispersion polymerizations are available in the specialist literature [1–3].

The first step of a radical dispersion polymerization can be characterized as that of a typical solution polymerization. Thermal or catalysed decomposition of an initiator I produces radicals $R^\bullet$, part of which reacts with the monomer M to form monomer radical $M^\bullet$. The monomer radicals add to other monomer molecules, which causes a gradual increase in the RMM of the reaction product, i.e.

$$I \rightarrow 2R^\bullet$$
$$R^\bullet + M \rightarrow RM^\bullet$$
$$RM^\bullet + nM \rightarrow RM_nM^\bullet$$

The product at a certain degree of polymerization n (usually $10 < n < 50$), depending on the type of monomer and reaction medium, is insoluble in the medium and precipitates as a new phase. This marks the end of the first stage of dispersion polymerization. Since it is insoluble in the reaction medium, an oligomeric radical of a given degree of polymerization (or an oligomer formed by interaction of a radical centre of the oligomer with radical R$^{\bullet}$ or with another oligomer radical or by transfer reaction) acts as the locus for the formation of polymer particles. With the formation of polymer particles, the second stage of the dispersion polymerization starts, marked by a gradual enhancement of the polymerization rate. The extent of this acceleration depends at this stage on the type of monomer. The final stage is characterized by a gradual decrease in the rate of polymerization [4].

To evaluate the threshold RMM of the oligomers where they precipitate (phase separation) from the reaction medium of a dispersion polymerization, the Flory–Huggins theory of polymer solutions can be applied [5–7]. The polymer solubility (measured by the Flory–Huggins interaction parameter χ)* rapidly decreases in a given solvent with increasing RMM and with an increasing value of interaction parameter. To guarantee the miscibility (solubility) of a polymer with a solvent and for any RMM of the polymer, the value of χ must be less than 0.5. That means that for a solvent with a molar volume of 100 cm^3 mol^{-1}, the difference in values between the solubility parameters for solvent and polymer must be less than 1.7; the figure depends on the value of μ_s. The interaction parameter χ depends on temperature; the temperature at which the interaction parameter has a value of 0.5 is the theta temperature [7]. The individual components of the reaction system (polymer and solvent molecules) and their configurations have, under theta conditions, a random arrangement. Data on the values of the Flory–Huggins interaction parameters χ for selected polymers and solvents are given in the literature [9–13].

The particle size and distribution of the polymer dispersion are determined by the number of new particles formed during the disperse polymerization as well as by their further growth, and the eventual mutual aggregation (flocculation) of the polymer particles produced. The solvent power of the disperse medium towards the polymer particles and the type, structure and concentration of the dispersant used, are the key factors for these processes (see sections 5.1 and 5.2). In the dispersion polymerization of methyl methacrylate in an alkane (i.e. in a diluent which is a typical nonsolvent for poly(methyl methacrylate) particles) the stage of nucleation of polymer particles is complete in a very short time, i.e. the order of several tens of seconds from the start of the polymerization. As polymerization proceeds, only a negligible quantity of new particles is formed as long as the solvent power of the disperse medium does not change (increase) or a further amount of dispersant is not added. The rate of polymer particle formation decreases (i.e. the duration of their formation is prolonged) if the solvent power of the disperse medium increases. The formation of larger polymer particles is also characteristic of disperse polymerization under these conditions. The change in solvent power with respect to the dispersed polymer is, of course, reflected by the ability

---

* The interaction parameter χ, which expresses the interaction between polymer and solvent, is defined [8] by the relationship

$$\chi = \mu_s + \frac{V_1}{RT}(\delta_1 - \delta_2)^2,$$

where μ_s is the entropic component of the heat of mixing (which, for a good solvent of the given polymer, is 0.2–0.4), δ_1 and δ_2 are the solubility parameters of the solvent and polymer, respectively and V_1 is the molar volume of the solvent (cm^3 mol^{-1}), R is the universal gas constant and T the absolute temperature.

of the dispersant to stabilize the polymer particles. This problem has been covered in some depth in earlier sections.

The concentration of the dispersant plays an important role in particle formation and influences the size of the polymer particles. There have been few systematic examinations of the power law dependence of the particle size on the reaction parameters. Most comprehensive data on the effects of the concentration and RMM of the dispersant on dispersion polymerization were obtained for systems generating particles of size < 1000 nm in hydrocarbon solvents [14, 15]. An increase in the dispersant concentration was accompanied with an increase in the particle number and a decrease in the average particle size. The data of Table 41 illustrate this general dependence. The dependence on the dispersant concentration of the polymer particle size is expressed by the relation

$$d = K [D]^{-1},$$

where d is the polymer particle diameter (nm), D is the dispersant, and K is a constant, which characterizes the particular system. The relation holds for a certain concentration region of the dispersant and after satisfying further conditions. For instance, at high monomer concentrations the particle size depends only slightly on the dispersant concentration. If, at the same time, the concentration of the dispersant in the reaction system is rather high, a quite narrow particle size distribution is obtained. When the dispersant concentration is low, the polymer particles display a rather wide size distribution.

Table 41 — Dependence of particle size of poly(methyl methacrylate) dispersion in n-heptane on dispersant concentration. Polymerization temperature: 50°C. Initiator: dibenzoyl peroxide [15]

Dispersant	Dispersant concentration / g dm^{-3} monomer	d^a / nm	d^b / nm
Poly(isoprene-b-styrene-b-isoprene)^c	100	480	465
	80	515	–
	60	525	530
	40	575	–
	20	730	–
	10	–	850
Poly(isoprene-b-styrene-b-isoprene)^d	100	830	820
	80	870	–
	60	970	950
	20	–	1200

[a] Determined by light-scattering technique.
[b] Electron microscopy.
[c] RMM of poly(isoprene) blocks: 2 × 41 400; of poly(styrene) block: 34 200 (g mol^{-1}).
[d] RMM of poly(isoprene) blocks: 2 × 16 000; of poly(styrene) block 13 700 (g mol^{-1}).

For the dispersion polymerization of styrene in ethanol, the exponent for the dispersant poly(N-vinylpyrrolidone) in the equation relating the dispersant concentration to polymer particle size is about –0.3 [14]. This value of the exponent is substantially smaller than one would expect for the simplest adsorption mechanism of stabilization by poly(N-vinylpyrrolidone), where the amount of surface area stabilized should be proportional to the dispersant concentration, predicting a slope of –1. The theoretically predicted [16] value of the exponent is –0.5. Values from 0.2 to 1.0 have been reported in [14, 15] although some of these exponents refer to block copolymers or comb-stabilized dispersion polymerizations, where a grafting mechanism of stabilization may not be operative and much smaller particles were obtained.

In contrast to the observed marked influence of the RMM of the dispersant on the size of poly(methyl methacrylate) particles in the dispersion polymerization of methyl methacrylate in n-heptane [15] (cf. Table 41), during styrene polymerization in ethanol, using poly(N-vinylpyrrolidone) as dispersant, there is only a small effect on the poly(styrene) particle size. This result may represent a balance between the improved stabilization by longer poly(N-vinylpyrrolidone) chains and the poorer adsorption of grafts containing longer poly(N-vinylpyrrolidone) chains onto the particle surface. One reason why a graft copolymer containing longer poly(N-vinylpyrrolidone) chains might be less strongly adsorbed is that the PVP/PS ratio is greater, hence the graft copolymer could be more soluble in the continuous phase, particularly at the beginning of the reaction when a significant amount of styrene is present [14]. Similarly, there is only a marginal effect on the polystyrene particle size as the RMM of (hydroxypropyl)cellulose is varied from 64 000 to 300 000 g mol^{-1}, although slightly larger particles were formed at an intermediate RMM of 100 000 [14].

However, no noticeable effect on the poly(styrene) particle size as a function of the RMM of poly(acrylic acid) was observed from 3500 to 10^6 g mol^{-1} [17].

In most cases of dispersion polymerization, an inverse correlation between the particle size and the RMM of the polymer in a polymer particle is noted: the larger particles have the lowest RMM and *vice versa*. When the particle size is infinite (coagulate formed in the absence of dispersant), the RMM is very close to that expected for solution polymerization. The reason for the higher RMM from smaller particles lies in the site of polymerization, which is in both the continuous phase and the monomer-swollen particle phase [18]. The RMMs from particle phase polymerization are usually higher than those from solution polymerization because the termination rate is reduced by the viscosity of the monomer/polymer particle medium. There are two contributors to particle phase polymerization: that caused by initiation and polymerization inside the particles and the other caused by the capture of solution-initiated oligomeric radicals. For the same volume fraction of polymer, the former is not particle-size-dependent, whereas the latter depends on particle size. In the case of small particles, a higher proportion of particle-phase polymerization occurs because solution-initiated oligomeric radicals are captured more efficiently by smaller particles which have a greater surface area.

On raising the initial styrene concentration in ethanol from 5 to 40 vol. %, the particle size obtained increases. A comparison of this effect to that of added toluene suggests that the primary effect of monomer concentration is a solvent effect: the more soluble the poly(styrene) in the reaction medium, the larger are the poly(styrene) particles [18].

The temperature at which the polymer dispersion is prepared exerts little influence on the polymer particle size. Since the dissolving ability of a disperse medium increases with temperature even with respect to the anchoring segment of the polymer dispersant, a growth in the particle size might be expected. However, the experimental results show the reverse (Table 42).

Table 42 — Dependence of particle size of poly(methyl methacrylate) dispersion on polymerization temperature. Dispersant: poly(isoprene-*b*-styrene-*b*-isoprene). [a] Reaction time: 4 h [15]

T / °C	50	55	65	70	75	80	85	90
d / nm	465	450	455	450	455	450	425	430

[a] RMM of poly(isoprene) blocks: $2 \times 41\,400$; poly(styrene) blocks $34\,200$ (g mol^{-1}). Dispersant concentration: 100 g dm^{-3} monomer.

Table 43 — Dependence of RMM (M_v) of poly(methyl methacrylate) on RMM of anchoring segment and concentration of dispersant poly(isoprene-*b*-styrene-*b*-isoprene) [15]. Disperse medium: *n*-heptane. Initiator: dibenzoyl peroxide. Temperature: 50 °C

RMM of anchoring poly(styrene) block[a] /g mol^{-1}	Dispersant concentration /g dm^{-3} monomer	$M_v / 10^5$
34 200	100	6.08
	80	4.95
	60	5.42
	40	5.02
	20	4.82
	10	4.56
6 100	80	3.19
13 700	80	3.50
34 200	80	4.95
51 400	80	10.80

[a] Ratio between the RMMs of anchoring poly(styrene) and soluble poly(isoprene) segments is 0.41.

The type of polymer dispersant and its concentration also influence the RMM of the dispersed polymer. The dependence of the RMM of poly(methyl methacrylate) on dispersant concentration is documented in Table 43. The RMM of poly(methyl methacrylate) decreases with decreasing dispersant concentration and with the RMM of the dispersant, i.e. with increase of the polymer particle size (*cf.* Table 41). Since the relative RMM of poly(methyl methacrylate) is a function of the rate of dispersion polymerization, we can regard this result to be indirect evidence of the effect of the particle size of the polymer dispersion on the rate of dispersion polymerization. In contrast to data published earlier [1], direct evidence is provided by the results presented in Table 44 and in Figs 24 and 25. We can infer from the results that the rate of dispersion polymerization in a nonaqueous medium is to an observable extent affected by the concentration and the RMM of the dispersant [15]. The shape of the conversion curves of a dispersion polymerization depends on the polarity of the disperse phase. The conversion curve of the dispersion polymerization of styrene initiated by dibenzoyl peroxide shows a typical gel effect in the strongly polar ethanol/water disperse phase. In the less polar disperse phase ethanol/2-methoxyethanol, the shape is typical of solution polymerization. The different forms of the conversion curves are explained [19] by the increased diffusion of styrene and dibenzoyl peroxide into nonpolar polymer particles in the strongly polar dispersion, i.e. by the increase in concentration of the reactants at the reaction site. This partitioning effect is less pronounced in the less polar disperse phase.

Table 44 — Dependence of conversion of methyl methacrylate in dispersion polymerization in presence of poly(isoprene-b-styrene-b-isoprene) dispersant on dispersant concentration. Disperse medium: n-heptane. Reaction time: 4 h. Temperature: 50°C [15]

RMM of dispersant segments /g mol^{-1}		Dispersant concentration[c] /g dm^{-3}					
PS[a]	PI[b]	100	80	60	40	20	10
6 100	2 × 7 400	—	58.8	59.4	—	—	—
13 700	2 × 16 600	83.0	81.8	80.4	77.5	73.6	—
34 200	2 × 41 400	86.2	86.8	80.3	83.4	78.5	74.7
51 400	2 × 62 300	89.8	89.4	84.9	82.2	78.0	—

[a] Poly(styrene) segment.
[b] Poly(isoprene) segment.
[c] Dispersant concentration with respect to methyl methacrylate.

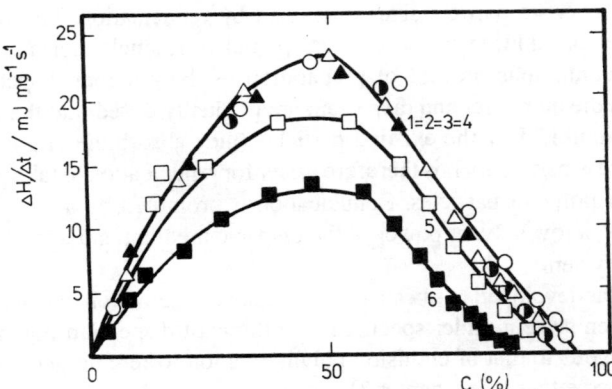

Fig. 24 — Rate of release of reaction heat $\Delta H/\Delta t$ in the dispersion polymerization of methyl methacrylate (MMA) in heptane as a function of MMA conversion. Dispersant: poly(isoprene-b-styrene-b-isoprene); RMM of poly(styrene) segment: 34 200 g mol^{-1}: poly(isoprene) segments 2 × 41 400 g mol^{-1}. Concentration of dispersant with respect to methyl methacrylate (g dm^{-3}). Temperature: 70°C [15].

1 — 100 (○)   2 — 80 (◐)   3 — 60 (△)
4 — 40 (▲)   5 — 20 (□)   6 — without dispersant (■)

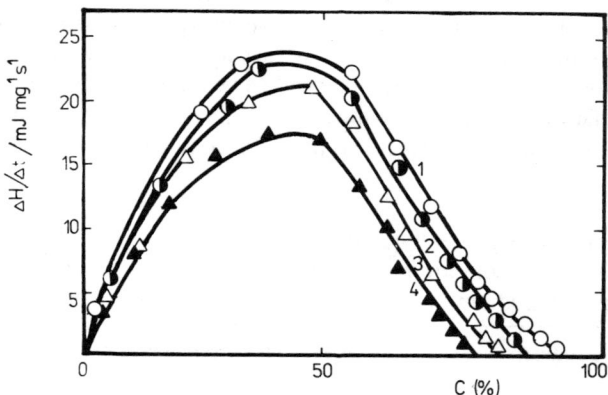

Fig. 25 — Rate of release of reaction heat $\Delta H/\Delta t$ in disperse polymerization of methyl methacrylate (MMA) in *n*-heptane as function of MMA conversion for poly(isoprene-*b*-styrene-*b*-isoprene) dispersant with different RMMs of poly(styrene) and poly(isoprene) segments [15].
RMMs of poly(styrene) and poly(isoprene) segments:
1: 51 400; 2 × 62 300 (g mol^{-1}) (○)
2: 34 200; 2 × 41 400 (g mol^{-1}) (◑)
3: 13 700; 2 × 16 600 (g mol^{-1}) (△)
4: 6 100; 2 × 7 400 (g mol^{-1}) (▲)

The particle size and particle number as well as the particle size distribution are strongly influenced by aggregation (flocculation) of the polymer particles which takes place in the absence of any effective stabilizer of the polymer particles, or at low concentrations (see section 5.2). A similar influence but in the opposite sense (particle number increase, reduction of the average particle size) is exerted by renucleation, i.e. formation of new polymer particles in addition to the existing particles. Renucleation mainly takes place during the semicontinuous process of preparation of the polymer dispersion, i.e. in the process in which the monomer and dispersant are gradually dosed into the reaction system. Renucleation is inhibited by the existing particles (they absorb the oligomers before they are able to form new particles). It is therefore easier for renucleation to take place in a system containing larger polymer particles. Renucleation is promoted by a low concentration of polymer particles, a low solvent power of the disperse medium and a low concentration of monomer in the system.

No theory so far developed has been able to predict quantitatively the number of polymer particles in a given system under specified conditions of dispersion polymerization. This situation is analogous to that of emulsion polymerization, where no generally valid theory has been developed either (see Chapter 3).

Several mechanisms have been proposed for the production of polymer particles. The mechanism of self-nucleation [20] is based on the idea that each propagating oligomeric radical moves freely in the disperse medium until it reaches a certain critical RMM, when it is separated in the form of condensed phase.

The model predicts that the propagating oligomeric radical does not interact with other oligomeric radicals present in the reaction system. The idea of aggregative nucleation is based on the association of growing oligomeric radicals in the system. The degree of association increases both with the concentration of oligomeric radicals and their RMM.

Kinetics and mechanism of radical polymerization

The aggregates formed are at first unstable and the oligomeric radicals associate only reversibly. On reaching a certain critical size, the aggregates become stabilized and gradually change to polymer particles. The model of aggregative nucleation accords with the theory of homogeneous nucleation [21, 22]. The final possibility, the model of nucleation from dispersant micelles, is important for emulsion polymerization (Chapter 3); the formation of polymer particles in the dispersant micelles is improbable in nonaqueous media in which the monomer is usually highly soluble.

The first two mechanisms for particle formation are therefore the most important for dispersion polymerization in nonaqueous media. The two mechanisms are complementary rather than in competition. Assuming that no competitive processes take place, nucleation ought to proceed throughout the polymerization until all the monomer is consumed. In practice, the rate of formation of new particles usually decreases to a negligible level very early (p. 235). As soon as a larger number of particles is formed, the generation of new particles is strongly inhibited. The oligomeric radicals produced in the reaction medium are assumed to be absorbed by the existing particles before new ones can appear [20].

The model of diffusion absorption due to Fitch and Tsai [20] starts from an assumption that each oligomeric radical is, on collision with a polymer particle, irreversibly absorbed by that particle before it reaches the critical size needed for nucleation. The rate of absorption of oligomeric radicals is proportional to the particle surface area and their total number. The difference between the rates of initiation and absorption of oligomeric radicals is governed by the nucleation rate of the polymer particles. The finding that in real dispersion polymerization systems many more polymer particles are formed than required by the postulate of irreversible absorption of oligomeric radicals by polymer particles represents conflict between theory and practice. The model of equilibrium absorption of oligomeric radicals attempts to resolve this contradiction [1]. This model assumes that oligomers at low degrees of polymerization, which are strongly solvated by the disperse medium, need not be absorbed by polymer particles on collision with them. The possibility thus arises for oligomeric radicals to nucleate new particles. As the solvent power changes during polymerization (with the loss of monomer, which gradually transforms to polymer) the reduction in the degree of solvation of the oligomeric radicals probably reduces their ability to 'survive' collision with the polymer particle and not be absorbed. The result is a steady decrease in the number of new particles being formed and the state is reached in which new particles no longer appear.

Polymer particles adsorb monomer from the continuous disperse phase. The monomer-swollen polymer particle becomes the main site of the propagation step. The high viscosity in the polymer particle reduced the proportion of macroradicals undergoing termination which causes a rapid increase in rate of the polymerization. The corresponding kinetic relations were derived [4] from the following consideration: If the polymer particles occupy, at a given time, a volume fraction V_p of the total volume of the dispersion, and the rate of initiation is R_i, then the rate of initiation in the polymer particles is

$$R_{ip} = R_i/V_p.$$

If the monomer concentration in the polymer particles is $[M]_p$, then the rate of polymerization in the polymer particles (R_{pp}) (which is equivalent to the expression for the kinetics of radical polymerization in a block or in a solvent)

$$R_{pp} = k_p[M]_p (R_{ip}/k_t)^{0.5} = k_p [M]_p (R_i/k_t V_p)^{0.5},$$

where k_p and k_t are the propagation and termination rate constants in dm^3 mol^{-1} s^{-1}. Since the polymerization occurs in polymer particles with an overall volume V_p, the total polymerization rate is

$$R_p = V_p R_{pp} = k_p [M]_p (V_p R_i/k_t)^{0.5}.$$

In dispersion polymerization in nonaqueous media, the monomer is miscible with the disperse phase and its concentration in the polymer particles depends on its partition coefficient α between the polymer and disperse medium. The overall rate of dispersion polymerization is therefore given by the expression

$$R_p = \alpha [M]_d k_p (V_p R_i/k_t)^{0.5},$$

where $[M]_d$ is the monomer concentration in the continuous phase.

The kinetic model of dispersion polymerization described has two experimental variants. The first characterizes the situation where α and V_p are low and $[M]_d$ can be considered to be almost equal to $[M]_0 K V'_p$, $[M]_0$ being the initial monomer concentration, K is the degree of conversion (%/100) and V'_p is the volume of polymer formed from one mole of monomer. It then follows that

$$R_p = \alpha [M]_0 K^{0.5} (1 - K)([M]_0 R_i V'_p)^{0.5} k_p/k_t^{0.5}.$$

This equation defines satisfactorily the dispersion polymerization of vinyl acetate and methyl methacrylate.

The second situation occurs when α is large and almost all the monomer is rapidly transported into polymer particles. V_p is then approximately equal to $[M]_0 V_M$, where V_M is the molar volume of monomer and $[M]_p$ is roughly equal to $(1 - K)/V_M$. Then it follows:

$$R_p = (1 - K)([M]_0 R_i/V_M)^{0.5} k_p/k_t^{0.5}.$$

This equation is in principle equivalent to that for the rate of a suspension polymerization. It defines quite well the dispersion polymerization of acrylonitrile and also seems to be applicable to other polar monomers, e.g. acrylic acid and acrylamide.

The average degree of polymerization of the polymer P_n for termination by disproportionation is given by *

$$1/\nu = 1/P_n = k_t R_p/k_p^2 [M]^2 + C_M + C_T [T] / [M],$$

---

* The kinetic length of the chain ν and its degree of polymerization P_n is defined by the relation $\nu = kP_n$. The coefficient of proportionality k has, for termination by disproportionation, a value of 1, while for termination by recombination it is equal to 2. For the mixed termination of macroradicals (disproportional and recombination) its value lies between 1 and 2.

where ν is the kinetic length of the polymer chain, k_p and k_t are the propagation and termination rate constants, C_M is the transfer constant to monomer, C_T the transfer constant to a transfer agent and T and M are the transfer agent and monomer, respectively.

In the dispersion polymerization of some monomers, transfer to monomer (if C_M is high) is a dominant factor which limits the RMM of the polymer. Although the RMM can in principle be controlled by a change in the polymerization rate (e.g. by a change in initiator concentration), a more marked change in P_n is achieved by addition of chain transfer agents. An important factor affecting the functioning of a chain transfer agent is its ability to distribute itself between the polymer particles and the disperse medium in a suitable proportion. From this point of view, alkyl mercaptans with an alkyl chain length of C_8–C_{12} are excellent regulators of RMMs for disperse polymerizations in alkanes.

The particle size for a poly(styrene) dispersion in ethanol stabilized by poly(N-vinylpyrrolidone) is proportional to the 0.39 power [14] or 0.40 power [23] of the AIBN concentration. It is believed that larger particles are obtained at higher initiator concentrations because lower-RMM poly(styrene) is formed, making the grafted PVP-PS more soluble and thus less effective as a stabilizer. In the presence of a transfer agent (butanethiol), polydisperse particles were obtained and the exponent for the AIBN concentration decreased to 0.19. The GPC chromatograms of these particles contained several broad peaks and the ratio M_w/M_n was ca.100. Like the monomer and initiator, the chain transfer agent may have distributed itself between the continuous and polymer phases, influencing the kinetics in both. Additionally, the lowest-RMM poly(styrene) might have been formed at the beginning of the reaction, with the result that the initial PVP-PS graft could be a very poor dispersant.

The effect of initiator concentration on the rate of dispersion polymerization in a polar medium was studied for the dispersion polymerization of styrene in ethanol. The polymerization rate increases with 4,4'-azobis(4-cyanopentanoic acid) concentration [18] at low conversions. However, the polymerization rate is independent of initiator concentration for conversions beyond 50%. At low conversions the continuous phase is rich in monomer and initiator, and hence solution polymerization dominates the reaction. At high conversions, the heterogeneous polymerization mechanism dominates and the rate of polymerization is then given by

$$R_p = k_p [M]_p (\rho_A V_p/2k_t)^{0.5},$$

where ρ_A is the rate of capture of oligomeric radicals by polymer particles and V_p is the volume fraction of polymer particles.

A simple mechanistic model for prediction of the polymer particle size in dispersion polymerization has been described recently. The key components of the model [16] are a multibin kinetic model for unstabilized particle coalescence, a grafting mechanism of stabilization, and the radius of gyration of the grafted stabilizer chains. This model was developed for dispersion polymerization in polar media (e.g. ethanol) and for a dispersant which is a homopolymer (which is transformed during dispersion polymerization to a graft copolymer). The model does not apply to other types of dispersion polymerization where pregrafted comb or block copolymer dispersants are involved [15, 24, 25], for which the particles are often much smaller because much more true dispersant is available. Recent achievements in dispersion polymerization in nonaqueous media as regards the effect of the dispersant, the reaction medium and initiation rate on the size of the polymer particles are discussed elsewhere [26].

In spite of some common features with conventional emulsion polymerization, dispersion polymerization in nonaqueous media shows several remarkable differences. In contrast to emulsion polymerization involving monomer droplets as a reservoir of monomer, in dispersion polymerization in nonaqueous media, the monomer is dissolved in the disperse phase. This situation results in the fact that the polymer particles prepared via nonaqueous dispersion polymerization contain much smaller amounts of monomer compared with the polymer particles of emulsion polymerization. The resulting polymer dispersion prepared in a nonpolar medium normally has a polymer particle diameter between 100 and 1000 nm. On the other hand, aqueous dispersions of polymers prepared via polymerization in an emulsion have an upper limit of the polymer particle diameters which is lower, i.e. about 500 nm, and a wide particle distribution. By using a polar solvent, the particle diameter lies between 1000 and 20 000 nm. Under favourable conditions monodisperse large size particles can be obtained in a single step using dispersion polymerization.

References

1. Barrett, K.E.J. & Thomas, H.R. (1975) In: Barrett, K.E.J. (ed.) *Dispersion polymerization in organic media.* J.Wiley & Sons, London, Chapter IV
2. Stampa, G.B. (1970) *J. Appl. Polym. Sci.* **14** 1227
3. Barrett, K.E.J. & Thompson, M.W. (1975) In: Barrett, K.E.J. (ed.) *Dispersion polymerization in organic media.* J. Wiley & Sons, London, Chapter IV
4. Barrett, K.E.J. & Thomas, H.R. (1969) *J. Polym. Sci., A-1* **7** 2621
5. Flory, P.J. (1942) *J. Chem. Phys.* **10** 51
6. Huggins, M.L. (1942) *J. Am. Chem. Soc.* **64** 1712
7. Flory, P.J. (1953) *Principles of polymer chemistry.* Cornell University Press, Ithaca
8. Huggins, M.L. (1941) *J. Chem. Phys.* **9** 440
9. Boyer, R.F. & Spencer, R.S. (1948) *J. Polym. Sci.* **3** 97
10. Maron, S.H. & Nakajima, N. (1960) *J. Polym. Sci.* **42** 327
11. Bristow, G.M. & Watson, W.F. (1958) *Trans. Faraday Soc.* **54** 1742
12. Orofino, T.A. & Flory, P.J. (1957) *J. Chem. Phys.* **26** 1067
13. Fox, T.G. (1962) *Polymer* **3** 111
14. Paine, A.J., Luymes, W. & McNulty, J. (1990) *Macromolecules* **23** 3104
15. Šušoliak, O. & Bartoň, J. (1985) *Chem. Papers* **39** 379
16. Paine, A.J. (1990) *Macromolecules* **23** 3109
17. Corner, T. (1981) *Colloids Surfaces* **3** 119
18. Lu, Y.Y., El-Aasser, M.S. & Vanderhoff, J.W. (1988) *J. Polym. Sci., Polym. Phys.* **26** 1187
19. Lok, K.P. & Ober, C.K. (1985) *Can. J. Chem.* **63** 209
20. Fitch, R.M. & Tsai, C.H. (1971) *Polymer colloids.* Plenum Press, New York
21. Volmer, M. (1939) *Kinetik der Phasenbildung.* Steinkopf-Verlag, Dresden
22. Becker, R. & Doring, W. (1935) *Ann. Phys.* **24** 719
23. Tseng, C.M., Lu, Y.Y., El-Aasser, M.S. & Vanderhoff, J.W. (1986) *J. Polym. Sci., Polym. Chem.* **24** 2995
24. Barrett, K.E.J. (ed.) (1975) *Dispersion polymerization in organic media.* J. Wiley & Sons, London
25. Dawkins, J.W., Shakir, S.A. & Croucher, T.G. (1987) *Eur. Polym. J.* **23** 173
26. Ober, C. (1990) *Makromol. Chem., Macromol. Symp.* **35/36** 87

5.4 PROPERTIES AND APPLICATIONS OF NONAQUEOUS POLYMER DISPERSIONS

The basic characteristics and methods for investigating polymer dispersions, such as particle size and particle size distribution, the concentration of polymer particles of the dispersion, the stability of the dispersion, etc. are discussed in Chapters 3 and 8. Conclusions from these studies can also be applied in most cases to nonaqueous polymer dispersions.

Sec. 5.4] Properties and applications of nonaqueous polymer dispersions

Of the special methods for investigating the properties and structure of polymer dispersions, and, particularly, their polymeric components, the fluorescence technique provides interesting possibilities. It offers data on molecular orientation and motion in amorphous polymers [1–5], about interpenetrating polymer chains [6], the thermodynamics of polymer blends [7], and solution properties [1–5]. The use of fluorescence for examining nonaqueous polymer dispersions can be illustrated as follows [8]. A chromophore-containing polymer dispersion was prepared by the radical copolymerization of methyl methacrylate with 1-naphthylmethylmethacrylate in the presence of a graft copolymer poly(isobutylene-g-methyl methacrylate) in cyclohexane. The naphthalene chromophore is located on the core surface of the stabilized polymer particles. The core diameter of the polymer particles was 2000 nm and the thickness of the core surface layer containing naphthalene was estimated to be 5 nm. The polymer particle of the statistical copolymer poly(methyl methacrylate-co-1-naphthyl methyl methacrylate) is stabilized by the graft copolymer; its short grafted branches, composed of methyl methacrylate structural units, are anchored on the surface of the polymer particles of the copolymer, and the soluble part of the chain, containing isobutylene structural units, is oriented towards the disperse phase. The emission spectrum of the polymer particles of the dispersion (mole ratios PIB:PMMA:P-NMMA were 13:100:11 and 13:100:2) is typical of derivatives of 1-alkylnaphthalenes. On adding anthracene to the dispersion and exciting the system at λ_{exc} = 280 nm, the fluorescence spectrum is observed as a superposition of those of naphthalene and anthracene. The excitation of anthracene molecules occurs predominantly by energy transfer from the excited naphthalene, and only in part via direct excitation of anthracene at 280 nm. The required distance between an anthracene molecule and an excited naphthalene molecule can be determined by measuring the decay curves of the naphthalene and anthracene fluorescence using the Förster dipole-dipole mechanism for energy transfer between the donor excited naphthalene and the acceptor ground-state anthracene [9,10] and from the spectroscopic properties of the participating molecules for the critical distance of a pair of nondiffusing chromophores. The distance in this particular case is 3–4 nm, which indicates that the transfer of energy from the naphthalene chromophore on the core surface of a polymer particle to anthracene in solution occurs without penetration of the naphthalene into the particle core of the statistical copolymer. However, the fact that the lifetime of the excited naphthalene shortens with increasing anthracene concentration in solution leads to the conclusion that nonpolar anthracene penetrates into the polar polymer particles of the statistical copolymer. This is consistent with the observation of a remarkable decrease in the second-order glass transition temperature, T_g, of polystyrene particles found on exposure to n-decane acting as a precipitant for poly(styrene) [11].

Fluorescence-quenching experiments carried out on methanolic and aqueous suspensions of dispersion-polymerized poly(styrene) particles stabilized with pyrene-labelled (hydroxypropyl)cellulose [12] have shown that most of the label is in an accessible (hydroxypropyl)cellulose layer of the surface of the polymer particles. This result corroborated electron microscopic observations. Some fraction of between 25 and 50% of the pyrene labels is apparently inaccessible to solution-based quenchers and can be trapped inside the particle. The precipitated particles have a morphology practically identical with those from which they originate. This similarity may suggest that the inaccessible fraction of dispersant is somewhat nearer the particle surface rather than buried deep inside. The result indicates that dispersion polymerization may involve a precipitation/accretion mechanism of growth of the polymer particle [12].

The size of polymer particles prepared by the nonaqueous dispersion polymerization is influenced by the physical character of the disperse medium (organic solvent or mixture of solvents).* The characteristic feature of nonaqueous dispersion polymerization is that, at the beginning of polymerization, all components of the system produce one phase. The polymer formed, which is insoluble in the disperse phase (otherwise it would be a solution polymerization), forms a new microphase, namely polymer particles dispersed in the continuous disperse phase. The stability of the polymer particles in the disperse phase is ensured by a suitable dispersant. The fact that the polymer being produced is insoluble in the disperse phase does not mean that the dispersed polymer phase and the disperse medium do not interact. The polymer particles are always more-or-less swollen by the solvent (disperse phase), part or sometimes most of which is composed of monomer which can act as a solvent for its polymer. This is of particular importance to the so-called seeded or batch dispersion polymerization.

Lok & Ober [13] have given an example of the effect of solvent power (disperse phase) on the course of dispersion polymerization and on the polymer particle formation. Their results obtained from a study of the radical-initiated dispersion polymerization of styrene in the binary solvents ethanol/2-methoxyethanol, ethanol/tetrahydrofuran, ethanol/water in the presence of (hydroxypropyl)cellulose functioning as a steric stabilizer of the polymer particles, can be summed up as follows. The dissolving ability of the disperse phase (related to the poly(styrene) being formed) controls the particle size of the polymer dispersion since the formation of the polymer particles is restricted to the initial phase of the dispersion (batch) polymerization.

The resulting size and size distribution of the polymer particles also depend on the degree of coalescence (flocculation) of the polymer particles. The size of the polymer particles decreases with increasing polarity of the disperse phase. It was confirmed that the degree of solvation of the polymer particles by the disperse phase is not the only factor affecting the size of the polymer particles. Transfer reactions of the macroradicals with the molecules making up the disperse phase, as well as the solubility of the stabilizer of the polymer particles in the disperse phase, also exert an influence on the resulting polymer particle size.

The size of the polymer particles also depends on the amount of initiator used being proportional to the reciprocal of the initiator concentration. The temperature mainly influences the degree of polydispersity of the particles, i.e. it affects the stage of polymer particle formation; its influence on the growth stage is only marginal. Application of a lower temperature in the stage during which polymer particles appear, i.e. from zero to *ca.* 10% conversion, causes a narrower particle size distribution.

Thermodynamic investigations [14–17] of sterically stabilized nonaqueous dispersions have shown that the colloidal stability is governed by an interaction between the solvated steric stabilizer anchored on the particle surface and the disperse medium. As the solvent power towards the solvated part of the steric stabilizer molecule increases, the repulsion between the sterically stabilized polymer particles also increases. The thermodynamic quality of the disperse medium also influences the conformation of the steric stabilizer at the interface between polymer and solvent; this leads to a significant effect on the rheological characteristics of the polymer dispersion. For instance, temperature increase effects an increase in solvent power and a resultant increase in viscosity which results from an increase of the effective volume of the polymer particles (more precisely, the particle-stabilizing

---

* Water may also be a component of the disperse phase [13].

layer of the adsorbed steric stabilizer). In dilute sterically-stabilized nonaqueous polymer dispersions, the steric stabilizer (usually a high-RMM compound) is in an equilibrium configuration. By increasing the concentration of polymer particles in the dispersion, the mutual interaction of the steric barriers of individual particles of dispersion begins. Further increase in particle concentration leads to the mutual entanglement of the solvated chains of the steric stabilizer of individual polymer particles or eventually to the compression of individual steric barriers. The number of possible configurations of the stabilizer molecule will thus decrease. This entropic effect will be manifested as an elastic component in the rheological response of the system to external forces affecting the dispersion [18]. The rheological properties of polymer dispersions have often been described [19, 20]. Three concentration regions of the polymer dispersion are of particular interest. Dispersions with a volume of the polymer phase of only several per cent are so dilute that the transfer of linear momentum in particle collisions during flow is negligible. The viscosity of the dispersion, η, is proportional both to that of the continuous phase, η_0, and the volume fraction, Φ, of the polymer particles present, i.e.

$$\eta/\eta_0 = 1 + \alpha_0 \Phi,$$

where α_0 is the Einstein coefficient, whose value for this particular case is 2.5.

In dispersions in which the concentration of polymer particles varies between 4 and 40 vol. %, the number of collisions between the particles is no longer negligible. At the usual shear rates developing in the flow, the dispersion also maintains its Newtonian character. In dispersions in which the concentration of polymer particles approaches the critical concentration at which the dispersion polymer particles are arranged, marked deviations are observed from Newtonian behaviour at shear rates from 1 to 5000 s^{-1}. Either an increase in the viscosity of the dispersion with increasing shear rates, or, conversely, a decrease, is observable. These conclusions hold strictly for solid spherical particles with ideal colloidal stability and a negligible volume formed by the stabilizer on the particle surface. These requirements are fulfilled by large particles; with particles at the submicron level the influence of the shell of dispersant on the particle surface on the flow properties of the dispersion can no longer be neglected.

The chief merit of polymer dispersions is their long lasting stability especially as regards practical applications. This problem was discussed in sections 5.1 and 5.2.

Film formation in polymer dispersions, seemingly a trivial phenomenon, is in fact a serious problem both theoretical and practical. All components of the polymer dispersion, i.e. the liquid diluent (disperse phase), polymer particles and dispersant, participate in formation of the film and its resultant properties. The diluent can be very volatile during film preparation but it can also contain less volatile components which then remain in the film produced. The composition, structure and size of the polymer particles strongly influence the properties of the film.

The nature and concentration of the dispersant are reflected in the final properties of the film. The interaction of the components of the polymer dispersion during film formation cannot be neglected (a change in compatibility of the polymer component with respect to the disperse phase, interaction between the polymer and stabilizer). The film resulting from a polymer dispersion has the structure of a composite material whose properties are not simply the sum of those of its components [21–24].

The major application of nonaqueous polymer dispersions is in the preparation of film-forming products for paints. Their applications are, however, much wider. Thus they are used for the preparation of adhesives and impregnated additives in textile and fibre processing. The extent and the methodology of their application are determined by the properties of the dispersion (more precisely of the polymer particles). Crosslinked polymer particles with a low T_g can function as an elastic filling upon insertion into a polymer particle, imparting an increased impact strength to the composite. Polymer dispersions in alkanes can be dispersed in various liquid lipophilic materials, e.g. in paraffin wax, fats and vegetable oils. The products are applied in the preparation of various cosmetics and polishes.

Nonaqueous polymer dispersions are applicable to the encapsulation of materials insoluble in alkanes with the possibility of using diverse substances, i.e. inorganic pigments, insecticides, metal salts, etc. The compounds encapsulated by the polymer dispersion can in this way be protected from external effects or, conversely, they can be control-released into the environment (e.g. the release of pesticides). Nonaqueous dispersion polymerization can be used with advantage to prepare fine polymer powders with a second-order glass transition temperature above room temperature. Such a polymer powder can be used for injection moulding. The flow properties of polymers prepared via dispersion polymerization are better than those of polymers prepared by polymerization in bulk.

The development of applications of nonaqueous polymer dispersions predicted in the 60s was hindered by the price increase of oil products (alkanes and other oil-based organic solvents) and also later by the increased efforts to reduce environmental pollution. Special applications of dispersion polymerization in nonaqueous media which do not have large materials requirements offer the best developmental possibilities at present.

References

1. Morawetz, H. (1979) *Science* **203** 405
2. Beavan, S.W., Hargreaves, J.S. & Phillips, D. (1979) *Adv. Photochem.* **11** 207
3. Schryver de, F.C. (1979) *Makromol. Chem., Suppl.* **3** 85
4. Valeur, B. & Monnerie, L. (1976) *J. Polym. Sci., Polym. Phys. Ed.* **14** 11
5. Rigler, R. & Ehrenberg, M. (1976) *Q. Rev. Biophys.* **9** 1
6. Morawetz, H. (1981) *Ann. N.Y. Acad. Sci.* **366** 404
7. Frank, C.W. & Gashgari, M.A. (1981) *Ann. N.Y. Acad. Sci.* **366** 387
8. Pekcan, O., Winnik, M.A., Egan, L. & Croucher, M.D. (1983) *Macromolecules* **16** 699
9. Birks, J. (1971) *Photophysics of aromatic molecules.* J. Wiley & Sons, New York
10. Berlman, I.B. (1973) *Energy transfer parameters of aromatic compounds.* Academic Press, New York
11. Veksli, Z., Miller, W.G. & Thomas, E.L. (1976) *J. Polym. Sci., Polym. Symp.* **54** 299
12. Winnik, F.M. & Paine, A.J. (1989) *Langmuir* **5** 903
13. Lok, K.P. & Ober, C.K. (1985) *Can. J. Chem.* **63** 209
14. Silberberg, A. (1978) *Faraday Discuss. Chem. Soc.* **65** 194
15. Croucher, M.D. & Hair, M.L. (1978) *Macromolecules* **11** 874
16. Croucher, M.D. & Hair, M.L. (1979) *J. Phys. Chem.* **83** 1712
17. Croucher, M.D. & Hair, M.L. (1980) *Colloids Surfaces* **1** 349
18. Milkie, T. H., Lok, K. & Croucher, M.D. (1982) *Colloid Polym. Sci.* **260** 531
19. Osmond, D.W.J. & Wagstaff, I. (1975) In: Barrett, K.E.J. (ed.) *Dispersion polymerization in organic media.* J. Wiley & Sons, London, Chapter VI
20. Sherman, P. (1970) *Industrial rheology.* Academic Press, London, Chapter 3
21. Vanderhoff, J.W. (1970) *Br. Polym. J.* **2** 146
22. Boutman, L.J. & Krock, R.H. (eds) (1967) *Modern composite materials.* Addison-Wesley Publ. Co., Reading
23. Ashton, J.E., Halpin, J.C. & Petit, P.M. (1969) *Primer on composite materials: Analysis.* Technomic Publ. Co., Stamford
24. Voyutskii, S.S. (1958) *J. Polym. Sci.* **32** 528

6

Special cases of polymerization in disperse systems

6.1 SYNTHESIS OF POLYMER DISPERSIONS WITH A HETEROGENEOUS PARTICLE STRUCTURE

In the emulsion copolymerization of comonomers which remarkably differ strongly in their water solubility, the structure of the resulting polymer particles is compositionally heterogeneous. The chemical composition of a polymer particle in an imaginary section made through the centre of the particle changes more or less evenly from the core of the particle to its surface. The chemical composition of the polymer inside the particle is enriched in structural units of the more lipophilic (less polar) monomer; conversely, the particle surface contains polymer enriched with structural units of the more polar monomer. This situation does not depend on the mode of polymerization — whether feed polymerization (all components of the polymerization system are present from the very beginning) or semicontinuous (monomers are added successively at a feed rate higher than, or almost the same as, the polymerization rate, i.e. so as to prevent accumulation of the monomer in the feed). The composition of the copolymer in these cases is controlled by the monomer concentrations at the site of the reaction which is approximately determined by the solubilities of the mononomers in the aqueous phase and by the degree of solvation of the polymer being formed by the pair of comonomers. In this context, the copolymerization parameters are of minor significance.

The introduction of polar or ionizing groups into the polymer chains of the polymer dispersion improves some properties, such as its stability against mechanical effects and low temperatures. This improvement of the properties of the dispersion is closely associated with the distribution of those groups bound chemically, particularly on the surface of the polymer particles [1]. The distribution of the carboxyl groups of carboxylated copolymer dispersions has been studied by many investigators [2–6]. Knowledge of the structure of the polymer particles significantly contributes to our understanding of the polymerization mechanism and, by the same token, our understanding of the mechanism enables the preparation of polymer dispersions with predetermined properties. In conventional emulsion polymerization, the polar groups on the surface of the latex stem from the decomposition products of the initiators ($SO_4^{\bullet}$ from $K_2S_2O_8$). This fact complicates investigation of the properties of dispersions prepared via copolymerization in the presence of carboxyl-containing comonomers. One way of avoiding this complication is to employ a means of initiation of the emulsion polymerization which does not entail the incorporation of polar groups into the polymer particles. This possibility is provided by using ^{60}Co γ-rays as the initiator of emulsion polymerization [7]. The ȮH radical and H atom formed during this means of polymerization function as the initiating species [8]. This approach has been used in the emulsion copolymerization of styrene with itaconic, acrylic, and methacrylic acids. The partition coefficients between the styrene and aqueous phases for these three acids are 0.012, 0.0175 and 1.94, respectively [9]. If the assumption is made that the chief factor governing the rate of copolymerization is the transfer rate of styrene molecules from monomer droplets (styrene) to growing polymer particles through the aqueous phase [10], then the transfer rate of styrene through the aqueous phase is enhanced in the presence of a substance able to distribute itself between the oil (styrene) and aqueous phase. If the transfer rate of styrene through the aqueous phase increases with increasing partition coefficient, then to a first approximation, the copolymerization rate is expected to increase in the order itaconic acid < acrylic acid < methacrylic acid. This expectation was confirmed experimentally [7]. The different lipophilicities of these monomer acids is reflected in the finding that, by using methacrylic acid, the polymer particles contain structural units of this acid not only on its surface but also in its core. On the other hand, with the highly hydrophilic itaconic acid, the carboxyl groups were located only on the surface of the particles. Although the copolymerization of the monomer pairs specified, r_{St} = 0.26 and r_{IA} = 0.12; r_{St} = 0.68 and r_{AA} = 0.14; r_{St} = 0.21 and r_{MAA} = 0.55 [11] indicate that copolymers with a regular structure (alternating copolymers) should be formed, the differences in monomer reactivity do not obviously operate in this emulsion copolymerization. The process of copolymerization is controlled by the distribution of monomer between the oil and aqueous phases.

The incorporation of carboxyl groups into the polymer chain opens up the possibility of similar modifications of the polymer [12–14]. The properties of the resulting copolymer and polymer dispersion are determined by the monomer feed composition, the nature of the monomers, and the reaction conditions [15]. If polymer particles with a high content of carboxyl groups bound on the surface are required, then acrylic acid (AA) is used as one component of the polymerization system; if a polymer with regular distribution of carboxyl groups is required, then a combination of methacrylic acid (MA) with a lipophilic monomer is used (see Table 45). The use of the highly hydrophilic itaconic acid (IA) leads to a water-soluble polymer with a high content of carboxyl groups.

Table 45 — Influence of concentration and type of carboxyl monomer on its localization in polymer dispersion[a] (Reprinted from ref. [5] with permission of John Wiley & Sons Inc., New York)

Localization of carboxyl groups in dispersion	% Carboxyl group in three sites					
	1% carboxyl monomer			2% carboxyl monomer		
	IA	AA	MAA	IA	AA	MAA
Free in serum	59	45	34	52	46	5
Surface-bound	40	51	31	41	22	22
Incorporated in particle	1	4	35	7	32	73

[a] Dispersion prepared via copolymerization of styrene with itaconic acid (IA), acrylic acid (AA), and methacrylic acid (MAA).

A negative feature in the preparation of polymer particles with a heterogeneous structure is the formation of a large fraction of the water-soluble polymer or copolymer, mainly in the copolymerization of hydrophilic monomer, as found in the styrene/acrylic acid system [16].

Increase in the lipophilicity of a monomer leads to the formation of polymer particles with a higher content of carboxyl groups bound on the surface and inside the particles. This finding is illustrated by the emulsion copolymerizations of methyl methacrylate [17] and ethyl acrylate [18] with carboxyl monomers.

Egusa & Makuuchi [19] prepared carboxylated latices via the emulsion terpolymerization of n-butyl acrylate, acrylic acid, and hydroxyethyl methacrylate in the presence of an anionic emulsifier, and used various means of initiation. The experimental results obtained reflect the dependence on the mode of initiation of the locations of the carboxyl groups. When the redox peroxodisulfate system was used, a larger fraction of carboxyl groups was located on the particle surface. In contrast, initiation by γ-rays led to a preferred location of carboxyl groups inside the particles at the expense of carboxyl group containing polymer dissolved in water. The peroxodisulfate-initiated polymerization yielded a polymer dispersion with a distribution of the carboxyl-containing polymer between water, the surface, and the particle interior in the ratio 125:156:41; whereas with γ-radiation, a dispersion was obtained with the corresponding ratio 44:151:87 (the values denote microequivalents of carboxyl groups per 5 g latex). Latices prepared using γ-radiation failed to hydrolyse the ester groups of n-BMA and HEMA both during polymerization and on completion. Acid groups on the surface of particles (HSO_3^-) prepared using peroxodisulfate/hydrogen sulfite as initiator catalysed this hydrolysis effectively.

The introduction of an amide group in polymer particles is achieved by the emulsion copolymerization of lipophilic monomers with acrylamide or its derivatives, e.g. in the copolymerization of styrene and acrylamide and its derivatives [20, 21]. The polymerization can be divided into three stages. In the first (low conversions) the polymerization of acrylamide in water is preferred. In the second stage (medium conversions) in the presence of primary polymer particles, the copolymerization proceeds mainly in these particles. Finally, at high conversions, the polymerization of residual acrylamide is completed in water. Of the total amount of acrylamide used, only a small part was copolymerized on the particle surface. The reason for the formation of a large fraction of the water-soluble polymer, which represents about 75 % of acrylamide used, is the great hydrophilicity of this monomer. With an increase in the lipophilic character of monomers in the order acrylamide $\simeq N$-hydroxymethyl acrylamide < methacrylamide < N,N-dimethylacrylamide, given by the increase in the partition coefficient of these compounds between styrene and water (0.093 $\simeq$ 0.089 < 0.159 < 0.293), the content of acrylamide structural units both on the surface of the particles and inside them, increases. The incorporation of acrylamide groups into the particle surface layers increases the stability of the dispersion. Formation of a stable dispersion was observed in the absence of an emulsifier. The incorporation of acrylamide units into the particles was also adversely influenced by copolymerization parameters for the individual comonomer pairs greater than 1.

A further study of the emulsion copolymerization of acrylamide and styrene [20, 22] also led to the conclusion that acrylamide groups are localized only in the outermost regions of the polymer particle; they form a hydrophilic polymer layer. Conductometric and viscometric measurements confirmed the presence of the water-soluble surface layer, the thickness of which increased linearly with increasing acrylamide content.

Sulfate groups can be introduced into the polymer particles through the emulsion copolymerization of monomers containing these groups after the decomposition of peroxodisulfate by propagating radicals; surface-active oligomeric radicals are formed which enter micelles or latex particles to initiate growth of the polymer chain [23, 24]. A certain limiting fraction of sulfate groups is thus introduced into the particless. A larger presence of sulfate groups on the particle surface is achieved by the copolymerization of ionic and surface-active monomers [25–28]. This approach was used to prepare model colloidal systems with a known particle size and surface charge density. The model systems are used for examining colloidal stability, interactions between particles and the surfaces of other materials, measurement of diffusion coefficients and the rheological properties of the dispersion.

By applying surface-active monomers, one can prepare monodisperse poly(styrene) latices, namely by the modification of feed latices by a monomer or a comonomer pair containing an ionic monomer [26]. The polydispersity of the latices prepared in this way varied between 1.005 and 1.015 and the particle size reached values of 90 to 120 nm. The surface charge density depended on the comonomer feed composition and the type of initiator used; it also increased with increasing concentration of ionic monomer. The use of water-soluble initiators led to the formation of particles with a higher surface charge density than for the oil-soluble initiators (benzoyl peroxide and lauroyl peroxide). For a series of initiators (lauroyl peroxide, benzoyl peroxide, diacetyl peroxide, and potassium peroxodisulfate) the levels of incorporation of a polar comonomer on the particle surface were reported to be 10%, 15.5%, 52.3% and 54.6%, respectively. The surface sulfate groups are oxidation-resistant and behave as electrolytes. The degree of hydrolysis of these groups is independent of the pH of the medium and these particles show a unique stability towards acids.

The classical peroxodisulfate-initiated emulsion copolymerization of styrene and sodium styryl sulfonate in the presence of anionic emulsifier is very rapid [29–31], i.e. much faster than styrene polymerization. The rate enhancement is ascribed to the operation of the gel effect inside the polymer particles due to strong intermolecular interactions. The addition either of an internal plasticizer or acrylamide reduces the degree of association and hence the very fast course of polymerization. An ionic monomer promotes the formation of a water-soluble polymer as well as of the polymer latex particle as expected. A linear relationship is observed between the monomer feed composition and the copolymer up to the limiting concentration of the ionic monomer (10 mol %).

Parallel production of a water-soluble polymer is observed above this concentration. A maximum of *ca.* 15 mol % of styryl sulfonate monomer was buried inside the particles, mostly on their surface, as a result of emulsion polymerization. An increasing number of bound sulfate groups positively affects the particle stability.

The emulsion copolymerization of water-soluble monomers, such as *N*-hydroxymethyl acrylamide or methacrylamide, with lipophilic monomers (styrene) [21] can be used for preparing copolymers containing hydroxymethyl groups. The amounts of these groups in the polymer particles, as well as their presence in the water-soluble polymer, are influenced by the nature of the ionic monomer and its distribution between the oil and water phases.

2-Hydroxyethyl methacrylate (HEMA) is an important monomer often used for the preparation of hydroxylated particles. Emulsion copolymerization of this monomer with lipophilic monomers, especially with styrene, in both the presence and the absence of emulsifier, gives polymer particles with a high concentration of surface-bound hydroxyalkyl groups [32–34]. The water-soluble monomer HEMA enters propagations mainly at lower conversions. After the development of turbidity due to the formation of primary particles, and after the diffusion of styrene into these particles, styrene mainly takes part in propagation. The resulting particles contain copolymers rich in hydrophilic and lipophilic monomers. The monomer feed composition influence the shape of the polymer particles. At HEMA concentrations of between 5 and 10 mol % in the feed, particles having a raspberry-like surface are formed. Higher concentrations of HEMA (15–20 mol %) cause the formation of confetti-like particles [33, 35]. The polymerization of HEMA with lipophilic monomers can in principle be classified as a two-stage polymerization and not as a typical copolymerization.

Recent information on the preparation and characterization of structured and functionalized polymer particles prepared by free-radical emulsion polymerization has been given in a review by Pichot [36]. Two main methods have been considered and discussed: first, the functionalization of latices by copolymerization of a basic monomer with addition of small amounts of monomer bearing a chemically reactive group, with special emphasis on the kinetics, the distribution of the functional monomer, the surface morphology of the latex particles and the optimization of processes; secondly, the use of surface-active monomers in those reactions which may confer potential advantages (low levels of water-soluble polymers, control of the surface charge density) provided further information can be obtained on their kinetic behaviour in emulsion polymerization [37].

Shirahama & Suzawa [38] prepared monodisperse copolymer latices via the emulsion copolymerization of styrene and HEMA. The polymerization of styrene led, under these conditions, to the formation of polymer particles with diameters $ca.$ 530 nm. The copolymerization of styrene with HEMA (5 mol % in feed) gave a latex with a particle diameter of 510 nm and particles of only 490 nm diameter were obtained using 10 mol % of HEMA.

The emulsion copolymerization of styrene with maleic acid in the presence of lecithin gave an optically active copolymer [39]. The copolymers obtained were richer in styrene than those prepared in the homogeneous phase. The optical activity of the products results from the chirality of lecithin. Lecithin forms an interphase in which the propagation of styrene and maleic acid proceeds. This interphase affected substantially both the rate of copolymerization and the reactivity of both monomers in the formation of the optically active copolymer.

Amphoteric latices are examples of special polymer dispersions which can be prepared via emulsion copolymerizations. The latices are obtained by the copolymerization of monomers containing acid and basic groups, e.g. that of acrylic acid with 2-alkylaminoethyl acrylate [40]. These copolymerizations also give, in addition to polymer latices, a water-soluble copolymer in high yield. A large fraction of water-soluble copolymer was also observed after addition of a lipophilic monomer (styrene) with ionic monomers (acrylic acid, methacrylic acid and 2-diethylaminoethyl acrylate) [41]. The concentration of basic groups, which exerts a considerable effect upon the surface of the polymer particles, is influenced adversely by hydrolysis of the cationic monomer.

Homola & James [42] prepared amphoteric latices based on dimethylaminoethyl or diethylaminoethyl methacrylate and carboxylic acids. The latices prepared had an isoelectric point dependent on the concentration, type and composition of the reaction mixture. The surface particle charge, particle size and the degree of polydispersity were controlled by a suitable choice of reaction conditions or by adding lipophilic monomer (styrene).

Harding & Healy [43] prepared amphoteric latices by the terpolymerization of styrene, methacrylic acid and dimethylaminoethyl methacrylate, varying only the amount of carboxylic monomer in the feed. The change in value of the isoelectric point was proportional to the reciprocal of the carboxyl monomer concentration. The polydispersity of the polymer latices was mostly less than 1.01 and, by keeping the content of basic monomer in the polymer particles constant, it did not vary.

A high content of water-soluble polymer adversely affects the means of characterization and the possibility of exploiting amphoteric polymer latices. Kawaguchi *et al.* [44] therefore modified the method of preparation of the amphoteric latex by replacing the water-soluble peroxodisulfate by the lipophilic azo initiator N,N-azoisopropylamidine hydrochloride. This procedure substantially increased the fraction of ionic monomers in the polymer particles. The decrease in average particle size observed simultaneously was explained as due to the increased rate of initiation and hence the rate of formation of primary particles. The higher rate of the polymer particle formation was thus accompanied by the generation of a larger number of particles able to adsorb more ionic monomer.

In addition to the statistical monomer, block or graft copolymers may also be products of the copolymerization of a lipophilic monomer with a strongly hydrophilic monomer in view of the configuration of the structural units of the chain.

The formation of a block copolymer in the emulsion copolymerization of vinyl chloride with acrylamide was expected in ref. [45]. The initiator radicals initiate acrylamide polymerization in the aqueous phase. The acrylamide macroradical formed attacks a vinyl chloride molecule in emulsifier micelles. The propagation of the macroradical in the micelle gives a copolymer of poly(acrylamide-*b*-vinyl chloride). The block copolymer cannot be, however, the sole product of this system. Termination reactions of poly(acrylamide) radicals in the aqueous phase, as well as the possible transfer to monomer in the micelles or to the polymer, induce the formation of the homopolymers of poly(acrylamide) and poly(vinyl chloride) or of poly(acrylamide-*g*-vinyl chloride) graft copolymer. The structure of the polymer particles formed is therefore rather compositionally heterogeneous.

Inhomogeneous polymer particles are prepared via classical batch polymerizations as has been reported by Vanderhoff *et al.* [46]. The emulsion copolymerization of butyl acrylate and vinyl acetate gave polymer particles with a butyl acrylate-rich core and a vinyl acetate-rich shell. The different reactivities of the monomers, which follow from the copolymerization parameters (r_{VAc} = 0.02 and r_{BA} = 3.49), also contribute to changes in the composition of the copolymer during polymerization. In the initial stage, the more reactive butyl acrylate is incorporated into the copolymer which forms the particle core. As reaction proceeds, the comonomer mixture is enriched with vinyl acetate which, after polymerization, enriches the outer layers of the particle.

The formation of particles with a heterogeneous structure is observed in the emulsion copolymerization of monomer pairs with different reactivities, e.g. in the methyl methacrylate/butyl acrylate system, where as much as 90% of the butyl acrylate is localized in the particle shell layer [47]. Particles with a heterogeneous structure are also formed in the styrene/acrylonitrile [48] and butyl acrylate/acrylonitrile [49, 50] emulsion copolymerizations.

A useful method for achieving a compositionally heterogeneous structure of the polymer particles is that of the seed latex. A monomer polymerizes in an emulsion system containing the usual components (water, initiator, emulsifier, transfer agent) in the presence of a polymer dispersion prepared in advance (i.e. two-stage emulsion polymerization). The polymer particles are swollen by the monomer, the degree of swelling of a given polymer particle by a given monomer being a function of the particle diameter (the surface layers are swollen more than the core). The seed polymer particles thus trap the particles being formed or oligomers. The polymerization of a monomer in a polymer particle takes place largely in the surface particle layer. A particle shell is thus formed with a composition different from that of the original polymer particle. Because of the diffusion of the monomer into the particles, the structure of the particle does not have a very precise boundary between the core and shell. The boundaries are usually diffusion expanded (unless referring to an exclusively immiscible polymer system) over a considerable part of the particle diameter.* The core-shell morphology of the polymer particles in two-stage emulsion polymerization is based on the prediction of the enrichment of the polymer particle surface layers with monomer [51–58]. Criticizing this model, Gardon [59, 60] postulated the possibility of polymerization of the monomer on the particle surface. Although such a possibility is admissible, it is questionable whether this model would provide sufficient explanation of all the special morphological features of two-stage emulsion copolymerization under various experimental conditions and the miscibility of the components of the polymer particle [61]. Objections have also been raised on the basis of diffusion theory [62].

In the two-stage (seed) emulsion polymerization, the structure of the polymer particles depends on the resulting ratio of the polymer components. In styrene-butadiene (SB) copolymerization in the presence of a seed poly(styrene) latex at a 30:70 ratio of poly(styrene)/styrene-butadiene copolymer, the copolymer SB separates as microdomains within the poly(styrene) matrix. On raising the fraction of copolymer SB, i.e. to a 20:80 ratio, the copolymer SB becomes the continuous phase in the polymer particle. This means, however, that the phases have become inverted [61] at a poly(styrene)/copolymer SB ratio between 30:70 and 20:80 (Fig. 26).

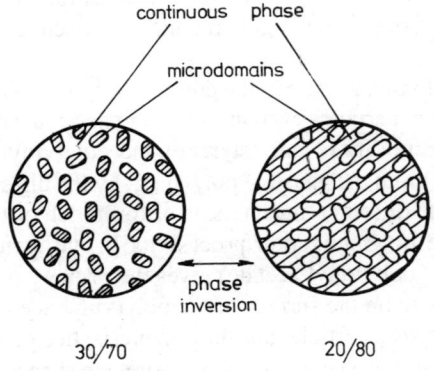

Fig. 26. — Schematic of phase inversion of polymer particles containing microdomains of immiscible polymers.

The particle morphology of the two-stage polymer dispersions of poly(styrene)/styrene-butadiene copolymer and butadiene-styrene copolymer/poly(styrene), depending on the ratio of the polymer phases, the miscibility of the polymers, their RMMs and the type of

---

* In the case of particle flocculation of the polymer dispersion owing to the effect of another polymer dispersion, the aggregates formed have much more defined boundaries between the individual particles which form the agglomerates. However, an agglomerate has neither a precise spherical structure nor precisely defined composition.

polymer used in the first stage (preparation of seed latex), is very different. The morphology of the resulting polymer particles in the limiting case can be characterized by the core-shell structure or by a structure of complete separation (Fig. 26).

Conditions for the preparation of concentric core-shell polymer particles of various vinyl monomers obtained by the two-stage emulsion polymerization technique are discussed in [63, 64]. As a rule, the following conditions need to be met in order to obtain a core-shell structure of the particles:

(i) the use of water-soluble initiator,
(ii) addition of monomer under monomer-starved conditions,
(iii) sufficiently high seed particle number,
(iv) lipophilicity of the core in comparison with the water solubility of the second monomer, and
(v) incompatibility of the core and shell polymers.

The nature of the seed polymer is another influential factor, as shown in the study of Hergeth & Schmutzler [65] on the seeded emulsion polymerizations of styrene in the presence of seed poly(vinyl acetate) latex, and vinyl acetate in the presence of seed poly(styrene) latex. In the polymerization of the lipophilic styrene in the presence of the hydrophilic poly(vinyl acetate) latex, the covering of the seed latex polymer particles was not even regular but adopted a lens-like structure which appeared on the surface, with the lens diameter increasing with increasing monomer concentration. The whole surface of the seed particles was covered only when higher monomer concentrations (> 2.3 mass %) were used.

The polymerization of vinyl acetate in the presence of poly(styrene) seed particles led to core-shell regular spherical particles even at low monomer concentrations. This result was explained by the incompatibility of poly(styrene) and poly(vinyl acetate), by the limited penetrating of monomer 1 into the phase of polymer 2, by the different water-solubilities of both polymers, and by the different mechanisms of formation of the primary particles and their participation in the polymerization process [66]. The emulsion polymerization of styrene in the presence of poly(vinyl acetate) gives the primary particles in water. Styrene polymerizes simultaneously on the surface of the poly(vinyl acetate) particles. Adsorption of the primary particles by seed particles and the polymerization of monomer on their surface facilitate the production of those aggregates and aggregate shapes that occupy the smallest contact area with the primary polymer. An increase in the monomer concentration leads to the growth of these 'islands' and, on reaching a certain critical size, to their mutual connection and the formation of a continuous polymer surface phase. Conversely, the use of a lipophilic seed latex and hydrophilic monomer leads to the formation of a continuous surface polymer layer on the seed polymer particles from the very beginning of polymerization.

The morphology of polymer particles prepared by seeded emulsion polymerization has been discussed recently by Okubo [67].

The separation of microphases in two-stage seeded emulsion polymerization has also been examined for the application of poly(ethyl acrylate) particles (seed) in styrene polymerization [68]. Dynamic mechanical studies can be used to ascertain the change in structure of the polymer particles [69, 70].

A literature survey offers the following conclusions [71] for the particle morphology in two-stage seeded emulsion polymerization:

(i) if polymer A of the seeded polymer dispersion is insoluble in monomer B, then polymer B forms surface layers on the polymer particles A of the seeded dispersion; (poly(vinylidene chloride)/poly(methyl acrylate) is a typical example);
(ii) if polymer B is miscible with polymer A and if both have the same hydrophilicity, then core-shell polymer particles are formed. The amount of polymer B in the particle shell will be greater than that of polymer A (examples: polymer particles of the two-stage emulsion polymerization of poly(methyl acrylate)/poly(methyl methacrylate), poly(ethyl acrylate)/poly(methyl methacrylate);
(iii) if monomer B swells polymer A, but polymer B is immiscible with polymer A, then phase separation of the polymers in the particle takes place, resulting in structures typical, for example for poly(styrene)/styrene-butadiene copolymer [61];
(iv) if polymer B is more hydrophilic than polymer A, a core-shell structure can be formed, the shell being mostly formed by polymer B;
(v) if polymer A is crosslinked, then polymer B forms two interpenetrating, continuous phases surrounded by polymer B-rich shells;
(vi) if polymer A is more hydrophilic than polymer B, then polymer B forms separated phases in polymer A. As a result many different structures of the polymer particles are formed.

Several techniques are available to characterize polymer particles e.g. (i) density measurements, (ii) flocculation tests and (iii) microelectrophoresis after purification of the latices by microfiltration in combination with solubility experiments in organic solvents, and (iv) chemical analysis of the particles using pyrolysis gas chromatography and X-ray photoelectron spectroscopy after drying the latex particles below the glass transition temperature of the polymer [72].

The microstructure of poly(acrylate)/poly(styrene) two-stage latices has been investigated in the frozen state by a transmission electron microscope through differences in the response of their component polymers when irradiated by an electron beam [73].

An example of the effects of the principal process parameters — monomer/polymer ratio, emulsifier concentration, size of seed particles, temperature and mode of monomer addition — on the polymerization, kinetics, and particle morphology is the styrene/acrylonitrile emulsion copolymerization in the presence of poly(styrene) seed particles was studied [74]. The seeded copolymer latex system comprises two types of particles: the smaller size of the new crop and the larger size of the grown seed. These particles differ in composition and morphology. The smaller size particles were S–AN copolymer, whereas the larger were core-PS and S–AN copolymer shells with PS-grafted-S–AN copolymer in between. These results indicate the aqueous phase as the dominant site of initiation. Polymerization within the PS seed leads to the formation of the surface zone of the interpenetrating S–AN free and graft copolymers with PS. The magnitude of this zone is determined by the polymerization temperature. Above the glass-transition temperature (65°C) of the swollen S–AN copolymer, the amount of graft core-PS is much higher than at lower temperatures [74].

The presence of unpolymerized monomer (methyl methacrylate) in seed poly(methyl methacrylate) particles influences the formation of the graft copolymer and/or statistical copolymer at the interface between the core and shell of the polymer particle during the seeded emulsion copolymerization of butyl acrylate and methacrylic acid. The yield of graft copolymer increases with reduction in the monomer content of the particles of the seed latex. An increase in the content of unpolymerized monomer in the seed polymer particles leads to the formation of a statistical copolymer at the interface. The different mechanical properties of latex films are thus directly associated with the structure of the interphase of the core-shell polymer particle [75].

The effect of various concentrations of an anionic emulsifier of the type alkyl phenol ether sulfate on the copolymerization of acrylonitrile vinylidene chloride in the presence of crosslinked (4% by mass of ethylene glycol dimethacrylate to n-butyl acrylate) poly(n-butyl acrylate) particles was investigated [76]. At high emulsifier concentrations (>1.0 g dm^{-3}) the acrylonitrile/vinylidene chloride copolymer formed a 'second crop' of latex particles. However, at low emulsifier levels (≤ 1.0 g dm^{-3}) only core-shell particles were present in the product.

The problem of the nonuniform distribution of free radicals in polymer particles has been discussed for emulsion radical homopolymerization [77]. The main results were also extended to emulsion copolymerization, i.e. for compositionally-heterogeneous polymer particles obtained by two-stage (or seeded) emulsion polymerization. The latter is characterized by a nonuniform polymer particle morphology when two or more different monomers are polymerized *at different times* such that separate polymer phases are formed due to the incompatibility of the polymer pairs or the sequence of introducing the monomers. In addition to the concepts of a nonuniform distribution of free radicals in the polymerizing polymer particles, this new approach also takes into account the diffusion-controlled termination and propagation steps in describing the gel effect and limiting conversions [78]. This concept has been incorporated into the development of a kinetic model for grafting reactions, and enables the prediction of grafting efficiency as a function of reaction conditions. It can also be used for evaluating the rate constants for grafting reactions [79].

Two-stage emulsion polymerization offers an interesting possibility for the preparation of interpenetrating polymer networks (IPNs) [80]. Monomer B polymerizes in the presence of a crosslinking agent (multifunctional monomer) and the seed latex A under radical initiation. Since no other emulsifier is added to the system, the formation of new particles is expected to be rather limited. The polymerization proceeds either on the surface of, or inside, the seed latex polymer particles already produced (see above). This leads to the formation of polymer particles containing networks of both polymers. Polymers that form IPNs have to be mutually miscible. With immiscible polymers, phase separation can take place before it can be prevented by development of the crosslinked structure [69]. As regards partially miscible components of IPNs, a shift of the maximum of the plot of loss angle tan δ versus temperature is observed under dynamic mechanical stress of the dispersion polymer component. No shift of the maximum of the loss angle tan δ on the temperature axis with respect to the tan δ values for poly(ethyl acrylate) and poly(*tert*-butyl acrylate) occurs for a clearly incompatible system (e.g. poly(ethyl acrylate)/poly(*tert*-butyl acrylate) [81].

In later studies [82, 83], the effect is discussed of altering the composition of the IPN formed from glassy and rubbery polymer components, e.g. poly(ethyl acrylate)/poly(ethyl methacrylate) and poly(ethyl acrylate)/poly(*tert*-butyl acrylate), the reverse order of IPN synthesis (polymer 1, monomer 2 and polymer 2, monomer 1) and swelling of polymer 1 (2) by monomer 2 (1). Hardness and stress-strain measurements showed that although the polymer formed second dominated the final mechanical properties, the compatibilities of the polymers played an important role. Dynamic mechanical analysis for the inverse systems showed evidence of enhanced mixing. Enhanced mixing of the IPN networks was also observed if the contact time between seed polymer and monomer before polymerization was sufficiently long (of the order of tens of hours).

A special type of poly(styrene)/poly(styrene) latex IPN was prepared by seeded emulsion polymerization of styrene/divinylbenzene mixtures in crosslinked monodisperse poly(styrene) seed latices [84, 85]. The resulting latices were uniformly nonspherical, e.g. ellipsoidal and egg-like singlets, symmetric and asymmetric doublets, and ice cream cone-like and popcorn-like multiplets. The nonspherical particles were formed by separation of the second-stage monomer from the crosslinked seed network during swelling and polymerization. The degree of phase separation increased with an increasing degree of crosslinking of the seed particles, the monomer/polymer swelling ratio, polymerization temperature, and seed particle size, and with decreasing divinylbenzene concentration in the swelling monomer.

Okubo et al. [86] drew attention to the existence of another structure of the polymer particles obtained from styrene emulsion polymerization in the presence of poly(methyl methacrylate) latex. Poly(styrene) domains were formed during polymerization within seed particles. They reported this phenomenon to be more marked in poly(methyl acrylate) particles than poly(methyl methacrylate).

Lee & Ishikawa [71] have also described the formation of particles with a domain structure, preparing these particles via styrene/butadiene (60:40) copolymerization in the presence of soft (seed copolymer latex (ethyl acrylate/methyl methacrylate; 90:10)) and by styrene polymerization in the presence of hard terpolymer latex (ethyl acrylate/styrene/acrylic acid, 50:40:10). In the soft polymer systems the formation of the inverted structure of the polymer particle was total (Fig. 27b).

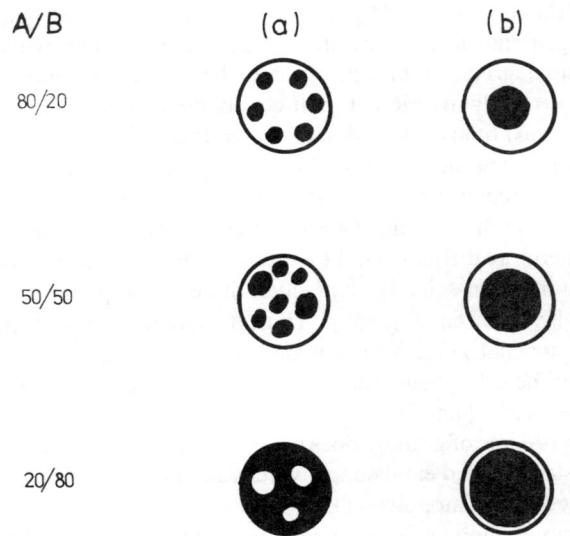

Fig. 27 — Schematic of structure of latex particles formed in seed emulsion polymerization. A — hydrophilic polymer, B — lipophilic polymer. Light regions represent A and dark regions B. (a) Seed polymer has higher RMM and is less hydrophilic; (b) seed polymer is more hydrophilic.

The polymer particle comprises a lipophilic core formed by polymer B while polymer A of the seed particle creates the shell of the modified particle. Such a structure was produced at various RMMs of A and B polymers. In the case of the hard polymer, the degree of inversion of the core-shell particle morphology was dependent on the RMMs of both seeded and formed polymers. Decreases in the RMM of polymer A and/or polymer B, or both, affected the degree of particle inversion positively.

Particles of the structure shown in Fig. 27a develop as a function of the ratio A/B if a seed polymer of higher RMM is used and polymer B is highly lipophilic.

The morphology of the particles formed in the presence of seed polymer particles also depends on the mode of polymerization as reported by Min *et al.* [87]. The anchoring of monomer (styrene) into the seed latex (poly(butyl acrylate)) may be carried out in several ways:

(i) by equilibrium swelling — the monomer is added to the latex which is swollen by monomer through gentle stirring for several days; this is followed by addition of initiator and polymerization;
(ii) semibatch — the monomer is added continuously to the seed-latex and initiator-containing reaction mixture;
(iii) batch method — the monomer is added to the reaction mixture all at once.

Method (i) gave poly(butyl acrylate) particles with the highest content of styrene polymer: a smaller content was formed in the batch and the smallest in semibatch polymerization. The effectiveness of seed polymer grafting by styrene varied in parallel. The minimum grafting efficiency observed in semibatch polymerizations was explained by the low concentration of styrene occurring in the surface layers of the polymer seed particles. The higher styrene concentrations in seed particles enable confer mobility to the monomer in the particles which increases its ability to participate in grafting.

The kinetics of the process [88–90] enable us to formulate some general rules. Observance of the rules guarantees the preparation of a graft copolymer by two-stage emulsion polymerization. The degree of grafting, which is the measure of preparation of the graft copolymer, is defined by the fraction of graft copolymer and the total amount of polymerized monomer. The quantitative detection of the graft copolymer fraction in a mixture of starting (seed) polymer and newly-formed homopolymer is difficult mainly because the graft copolymer shows remarkable emulsifying properties, especially for a combination of polymers of different hydrophilicity (see Chapter 5). Its separation from a mixture of homopolymers is therefore difficult. Of the experimental methods used to separate the graft copolymer, we can mention selective extraction or selective precipitation (turbidimetric titration) and thin-layer chromatography. The effectiveness of grafting decreases with increasing initiator concentration. The degree of seed latex grafting decreases with monomer conversion. On the other hand, the degree of grafting increases with increase in the polymer particle surface/volume ratio.

In most cases, the degree of grafting does not usually exceed 0.2 (i.e. 20% of the polymer formed in the two-stage seeded emulsion polymerization is part of the graft copolymer, the other 80% comprises new homopolymer).

Seeded emulsion polymerization and copolymerization can be used for preparing graft copolymers which can increase the compatibility of two immiscible polymers. Increase in the fraction of graft polymer on the surface of the particle core improves the miscibility of the polymer that forms the core with the polymer which forms the shell of the particle. The polymerization of styrene with poly(butyl acrylate) (PBA) latex provides an example: the graft PBA/PS copolymer improves the miscibility and adhesion of the core to the shell. In the case of the PS/PBA blend, the mass fraction of the homogeneously mixed polymers assumes only a very low value for both homopolymers (below 10^{-5}) [90]; after mixing the homopolymers, separation occurs and two homogeneous phases are formed [91]. The PS molecules generated in the mass of poly(butyl acrylate) aggregate into micro- and then to macrodomains to form a separated homogeneous phase (Fig. 26).

Dispersions with a heterogeneous particle structure

The graft copolymer improves the compatibility of the two immiscible polymers. This follows from the emulsifying function of the graft polymer which reduces the interfacial tension when two continuous polymer phases are in contact [92]. The emulsifying and compatibilization properties of graft copolymers change substantially the properties of the mixed polymer system prepared by seeded emulsion polymerization (e.g. the styrene–methyl methacrylate pair) [93]. The copolymers prepared in this way yield homogeneous polymer films not produced by the poly(styrene)–poly(methyl methacrylate) blend prepared via homopolymerization. Yamazaki [94, 95] characterized the physical properties of seed copolymers of ethyl methacrylate and methyl methacrylate and showed that films prepared from the core-shell polymers show as much as fourfold tensile strength with respect to the films obtained from a mixture of the corresponding homopolymers.

Sundberg *et al.* [96] described the grafting mechanism for seeded polymerization. They have shown that the grafting mechanism is substantially affected by several factors: temperature, concentration of initiator (peroxodisulfate), concentration of transfer agent (CCl_4), the mode of seeded polymerization and the ratio between the monomer content and the content of seed latex. The rate of polymerization rapidly increased with temperature in batch polymerization; by contrast, in the case of semicontinuous polymerization, only a slight increase was observed under the same conditions. The dependence of the grafting efficiency on temperature and the mode of polymerization was the reverse. The authors explained the increase in the fraction of graft copolymer with temperature by the more frequent entry of initiating radicals into the particles, where they attacked more vigorously the unsaturated double bonds to increase the initiation rate of grafting. The temperature enhancement was accompanied by a decrease in the equilibrium monomer concentration in the particles and an increase in the polymer/monomer ratio, which also led to an increase in grafting efficiency. The grafting efficiency increased with peroxodisulfate concentration and, conversely, it decreased with increase in concentration of a transfer agent.

The grafting of polymer in seeded emulsion polymerization occurs on the entry of radicals from the aqueous phase into the polymer particle, where they add to a double bond or abstract hydrogen from an alkyl group [97]. By using a seeded poly(butadiene) latex (PBH), the dominant reaction which starts grafting is α-hydrogen abstraction from the hydrocarbon chain [98]. The grafting mechanism characterized by reactions (1)–(6) is:

$$\text{initiator dissociation} \quad I \xrightarrow{k_d} 2\,R^\bullet, \tag{1}$$

$$\text{homopolymerization initiation} \quad R^\bullet + M \xrightarrow{k_{i,M}} P_1^\bullet, \tag{2}$$

$$\text{grafting initiation} \quad R^\bullet + PBH \xrightarrow{k_{i,B}} PB^\bullet + RH, \tag{3}$$

$$PB^\bullet + M \xrightarrow{k_{i,M}} P^\bullet, \tag{4}$$

$$\text{propagation} \quad P_n^\bullet + M \xrightarrow{k_p} P_{n+1}^\bullet, \tag{5}$$

chain transfer to monomer $P_n^{\bullet} + M \xrightarrow{k_{tr,M}} P_n + P_1^{\bullet}$. (6)

As is evident from this mechanism, the grafting consists of several successive steps. Homopolymerization is a side reaction reducing the extent of grafting and whose participation depends on the nature of the seed polymer. Cross-terminations of macroradicals observed in homogeneous systems are not found in heterogeneous emulsion polymerization and are therefore excluded from the reaction mechanism. The propagation of the graft polymer chain is governed by the reactivity and rate of monomer diffusion or the monomer concentration near the reaction centre.

The use of a lipophilic initiator (azo compound), and increase in its concentration caused a decrease in the grafting efficiency of poly(butadiene) latex by methyl methacrylate [89]. The grafting efficiency decreased with increasing conversion and increased with the particle surface area. Increase in the initiator concentration led to a decrease in the RMM of poly(methyl methacrylate), but it did not influence the RMM of the dissolved fraction of grafted poly(butadiene). These results led the authors to the conclusion that grafting takes place on the particle surface only via the chain transfer mechanism. This view was also supported by the findings that the RMM of the grafted polymer decreased with increasing particle surface area and the fraction of the graft polymer fell with an increase in the ratio between monomer and seed polymer. Investigation of seeded emulsion polymerization showed that, under various polymerization conditions, the fraction of homopolymer formed is greater than that of the graft copolymer. On the basis of these results, the authors concluded that the transfer reactions are first-order whereas the terminations of poly(methyl methacrylate) radicals are second-order reactions.

The grafting of the seed polymer was also affected by the emulsifier concentration as has been observed in kinetic studies of styrene/acrylonitrile emulsion copolymerization in the presence of seed poly(butadiene) latex [99].

If the polymerization proceeded at emulsifier concentrations lower than the CMC, the copolymer obtained was localized and grafted in polymer particles. Higher emulsifier concentrations (above the CMC) caused the formation of new particles obtained from acrylonitrile and styrene, which brought about a substantial decrease in grafting efficiency. Of the various factors influencing grafting (e.g. the mode of polymerization, concentration and type of initiator, ratio monomer/seed polymer, emulsifier concentration, and temperature), it was the emulsifier concentration that played the decisive role. The same conclusions were also drawn by Gasperowicz *et al.* [100].

Specific physical and chemical properties result from seed polymer grafting [101]: e.g. through grafting of poly(vinyl acetate) latex by bifunctional unsaturated acids (maleic, fumaric, citric, etc.), the viscosity of the latex changes, and the adhesive properties of the film and its transparency, drying and preparation are improved. Depending on the reaction conditions, e.g. initiator and emulsifier concentrations, the grafting efficiencies varied from 2 to 10% indicating low grafting [102].

Emulsion polymerization can be used for preparing heterogeneous compositions based on organic polymers and inorganic compounds. Several procedures can be used to achieve this aim. The polymerization of monomer in the double layer of an emulsifier adsorbed on the surface of an inorganic pigment (i.e. by analogy with the polymerization of a monomer in the double layer of an emulsifier adsorbed on the surface of aluminium(III) oxide, section 6.5, p. 286) leads to encapsulation of the pigment particles by polymer. The mutual flocculation of some encapsulated particles gives rise to a particle with several buried pigment particles surrounded by a polymer shell [103]. Another way to prepare encapsulated inorganic substances is emulsion polymerization without any conventional emulsifier (section 6.2), e.g. the emulsion polymerization of methyl methacrylate in the presence of powder $BaSO_4$ or $CaCO_3$. The kind of initiator strongly affected the extent of coverage of the inorganic particles by poly(methyl methacrylate). Optimal properties of the polymer film coating were achieved by using the redox initiating system $NaHSO_3$/air oxygen and by the surface modification of the inorganic particles by an emulsifier before the start of polymerization. This modification is thought to support the formation of initiating centres of the polymerization on the surface of the inorganic particles, followed by film formation on the encapsulated particles [104].

An important factor in the preparation of latex particles is the chemical nature of the emulsifier used, which determines the location of the reaction centre [104]. The polymerization of methyl methacrylate on the surface of TiO_2 particles was proportional to the concentration of emulsifier, the emulsifier being adsorbed on the particle surface. The higher adsorption ability of ionic emulsifiers (sodium dodecyl sulfate — SDS and dodecyltrimethylammonium bromide — DTAB) compared with a nonionic emulsifier (Tween 80) increased the rate of polymerization and stimulated formation of a larger amount of polymer on the TiO_2 particle surface. The inorganic particles formed the core and the polymer the shell of the latex particle.

The surface charge and the corresponding amount of adsorbed emulsifier can be controlled by changing the pH; as a result of the presence of surface hydroxyl groups [105]. Elimination of the effect of the emulsifier micelles and the formation of homogeneous particles are achieved via polymerization at emulsifier concentrations lower than the CMC. The low CMC values of nonionic emulsifiers prevent the effective covering of the surface of inorganic particles by emulsifier molecules and hence the polymerization is ineffective on the surface. If the initiator and emulsifier have the same charge, an increase in the RMM of the polymer is observed owing to the infrequent entry of radicals into the particles, caused by the mutual electrostatic repulsion of the initiator radicals by emulsifier bound to the particle surface. A combination of $K_2S_2O_8$ and SDS leading to the formation of a polymer with $M_n = 3.1 \times 10^5$ provides an example; on the other hand, the system $K_2S_2O_8$ plus DTAB gives a polymer with $M_n = 1.4 \times 10^5$.

The effect of small silicon or titanium particles on the kinetics of styrene emulsion polymerization initiated by sodium peroxodisulfate was determined by Hasan [106, 107]. The dispersed particles were stabilized by SDS. Increase in the concentration of inorganic additive by as much as 1–2 orders of magnitude did not change the polymerization rate. At higher concentrations of additive, spherical well-separable particles were formed. Agglomeration of these particles and the formation of irregular associates were observed for styrene polymerization at higher concentrations of Si or Ti particles. Increase in the SDS concentration positively affected the growth of polymer on the surface of the particles and the particle stability, indicating the probable nucleation of the new homogeneous polymer particles. Increase in the number of seed Si particles also caused an increase in the number of final heterogeneous particles. Si particles of diameter 14 nm grow, on completion of polymerization, to ca. 110–113 nm diameter as a function of the initial number of silicon particles and the reaction time. Comparison of the sizes of the silicon and final particles shows that the particle shell consists of a thick layer of polymer.

Additional information on the characterization of polymer composites containing quartz powder particles as filler and prepared by emulsion polymerization is available [108].

Yamaguchi et al. [109–112] studied the effect of microparticular additives on the emulsion polymerization of vinyl monomers. The microparticles were gradually encapsulated by the developing polymer. The mechanism of encapsulation of hard microparticles of graphite and other inorganic materials includes initiation of the polymerization on the particle surface resulting from redox reactions. The radicals thus formed initiate polymer chain growth in water; the chains produced are adsorbed on the particle surface and formed the nuclei of latex particles. In the following stage, monomer diffusing from monomer droplets onto the particle surface swell the polymer layer, and participate in propagation which causes particle growth.

Mikulášová et al. [113] investigated the mechanism of action of inorganic additives (the synthetic zeolites Potasit and Calsit, γ-alumina, and metal oxides — MgO, NiO, CuO, Al_2O_3, TiO_2 and PbO_2) as heterogeneous initiators in the emulsion polymerization of styrene and methyl methacrylate. The emulsion polymerization of these monomers did not take place without the additives. The addition of microparticles (Potasit, Calsit, γ-alumina) in the presence of triethylenetetramine and emulsifier induced formation of a polymer dispersion. Activation of the additives by oxygen led to an increase in polymerization rate (from 4% conv. h^{-1} to 8% conv. h^{-1}). Of the oxides combined with activators ($FeSO_4$–ethylenetetraacetic acid or triethylenetetramine), only ZnO and PbO_2 and, in part CuO and MgO, were shown to be effective, but in these systems activation by oxygen did not cause any increase of styrene polymerization. However, metal oxides in combination with activators were very effective in the emulsion polymerization of methyl methacrylate. The polymerization was sensitized effectively by light. The rates of polymerization varied between 10 and 100% conv. h^{-1}. One can assume that in these polymerizations, composite materials can also be formed from the polymer and inorganic additive.

Moustafa et al. [114, 115] reported that metal oxides (CuO and MnO_2) combined with $NaHSO_3$ initiate efficiently the polymerization of methyl methacrylate. Increase in the concentration of oxides was accompanied by an increase in rate and a decrease in the RMM of poly(methyl methacrylate). Higher polymerization rates were obtained in systems with CuO, although in both cases the polymers produced had comparable RMMs. With increasing concentration of the co-initiator ($NaHSO_3$), both the polymerization rate and the RMM increased. Within $NaHSO_3$ concentrations between 0.02 and 0.05 mol dm^{-3} in the presence of CuO, the rate of polymerization grew from 23×10^{-5} to 30×10^{-5} mol dm^{-3} s^{-1} or with MnO_2, from 3.9×10^{-5} to 13.9×10^{-5} mol dm^{-3} s^{-1}. As reported elsewhere [116, 117], polymerization proceeds not only on the particle surface (formed by the metal oxides) but polymer is also formed in the micelles and monomer droplets. The polymerization is accelerated by the presence of aerial oxygen and by increase in temperature.

A composite material can be prepared by heterocoagulation of an organic polymer dispersion with dispersed spherical silica gel particles. The dispersion of silica gel particles, with the polymer particles functioning as a shell (the silica gel particle diameter is ca. 1 to 10 times greater than that of the polymer particle) is very stable in the presence of electrolytes both in acid and basic media. The surface charge of the composite particles of the dispersion (charge sign and size) can be controlled by changing the pH of the disperse phase [118].

The prepagation of heterogeneous organic and inorganic materials via heterocoagulation has been described by Matijevič et al. [119–121], who examined the formation of disperse colloidal systems during mixing polymer latices with inorganic particles. A new colloidal system is formed by the settling of small polymer particles on the surface of the inorganic particles. Its properties depend on the conditions of preparation, the ratio between polymer and inorganic phase, the latex particle number, the surface charge, temperature, and pH. Stable dispersions with regular particles were produced only at higher ratios of the polymer and silicon phases, by applying smaller polymer and larger inorganic particles, at greater differences in the surface charge of the polymer and inorganic particles and at pH values enabling the presence of particles of opposite charge.

Furusawa & Anzai [122] described the preparation of stable disperse particles on the basis of polymer and silicon phases. A stable system containing spherical particles formed by a core (silicon particle) and a surface shell (smaller amorphous latex particles) was produced at pH between 4 and 6. In this medium, the core and the shell bore opposite charges. Particle formation was influenced by the ratio between the polymer and the silicon phase. Stable dispersions were formed only at high ratios of these two phases, i.e. above 150:1. In this case, regular particles with a narrow size distribution were formed. Increase in size of the silicon particles positively influenced the formation of regular particles and the stability of the dispersion. Regular composite particles were formed by using particles of diameter d ca. 960 nm, whereas smaller particles (with d ca. 460 nm and 240 nm) initiated formation of irregular associates. The amount of polymer particles adsorbed on the surface is expressed by the fractional covering, F_A, given by the ratio A/A_S (A being the amount of adsorbed polymer phase and A_S the maximum number of latex particles which can be adsorbed on the surface of a silicon particle) [120]. F_A decreases with decreasing size of the inorganic seed particles.

References

1. Greene, B.W. & Sheetz, D.P. (1970) *J. Colloid Interface Sci.* **32** 96
2. Ceska, G.W. (1974) *J. Appl. Polym. Sci.* **18** 427
3. Hen, J. (1974) *J. Colloid Interface Sci.* **49** 425
4. Sakota, K. & Okaya, T. (1976) *J. Appl. Polym. Sci.* **20** 1735; (1977) ibid. **21** 1035
5. Vijayendran, B.R. (1979) *J. Appl. Polym. Sci.* **23** 893
6. Hoy, K.L. (1979) *Coatings Technol.* **51** 27
7. Egusa, S. & Makuuchi, K. (1982) *J. Polym. Sci., Polym. Chem. Ed.* **20** 863
8. Hochanadel, C.J. (1960) In: Burton, M., Kirby-Smith, J.S. & Magee, J.L. (eds) *Comparative effects of radiation*. J. Wiley & Sons, New York, Chapter 8
9. Matsumoto, T. (1974) In: Lissant, K.J. (ed.) *Emulsions and emulsions technology*. M. Dekker, New York, Chapter 9
10. Harkins, W.D. (1947) *J. Am. Chem. Soc.* **69** 1428
11. Greenley, R.Z. (1980) *J. Macromol. Sci., Chem.* **14** 445
12. Muroi, S., Hosoi, K. & Ishikawa, T. (1967) *J. Appl. Polym. Sci.* **11** 1963
13. Green, B.W. (1973) *J. Colloid Interface Sci.* **43** 449
14. Green, B.W. (1973) *J. Colloid Interface Sci.* **43** 462
15. Sakota, K. & Okaya, T. (1976) *J. Appl. Polym. Sci.* **20** 2583
16. Fordice, R.G. & Ham, G.G. (1947) *J. Am. Chem. Soc.* **69** 695
17. Matsumoto, T. & Shimada, M. (1965) *Kobunshi Kagaku* **22** 172
18. Muroi, S. (1966) *J. Appl. Polym. Sci.* **10** 713
19. Egusa, S. & Makuuchi, K. (1981) *J. Colloid Interface Sci.* **79** 350
20. Ohtsuka, Y., Kawaguchi, H. & Sugi, Y. (1981) *J. Appl. Polym. Sci.* **26** 1637
21. Kawaguchi, H. Sugi, Y. & Ohtsuka, Y. (1981) *J. Appl. Polym. Sci.* **26** 1649
22. Tamai, H., Iida, A. & Suzawa, T. (1984) *Colloid Polym. Sci.* **262** 77
23. Vanderhoff, J.W. & Hul van den, H.J. (1968) *J. Colloid Interface Sci.* **28** 36
24. Goodall, A.R., Wilkinson, M.C. & Hearn, J. (1977) *J. Polym. Sci., Polym. Chem. Ed.* **15** 2193
25. Juang, M.S. & Krieger, I.M. (1976) *J. Polym. Sci., Polym. Chem. Ed.* **14** 2089
26. Tsaur, S.L. & Fitch, R.M. (1987) *J. Colloid Interface Sci.* **115** 450
27. Green, B. P., Sheetz, B.P. & Filer, T.D. (1970) *J. Colloid Interface Sci.* **32** 90
28. Liu, L.J. & Krieger, I.M. (1981) *J. Polym. Sci., Polym. Chem. Ed.* **19** 3013
29. Weiss, R.A., Turner, S.R. & Lundberg, R. D. (1985) *J. Polym. Sci., Polym. Chem. Ed.* **23** 525
30. Weiss, R.A., Lundberg, R.D. & Turner, S.R. (1985) *J. Polym. Sci., Polym. Chem. Ed.* **23** 549
31. Turner, S.R., Weiss, R.A. & Lundberg, R.D. (1985) *J. Polym. Sci., Polym. Chem. Ed.* **23** 535
32. Kamei, S., Okubo, M. & Matsumoto, T. (1986) *J. Polym. Sci., Polym. Chem. Ed.* **24** 3109
33. Okubo, M., Katsuta, Y. & Matsumoto, T. (1980) *J. Polym. Sci., Polym. Lett. Ed.* **18** 481
34. Okubo, M., Yamada, A. & Matsumoto, T. (1980) *J. Polym. Sci., Polym. Chem. Ed.* **16** 3219

35. Okubo, M., Katsuta, Y., Inoue, K., Nakamae, K. & Matsumoto, T. (1980) *J. Adhesion Soc. Jpn.* **16** 278
36. Pichot, C. (1987) *Bull. Soc. Chim. Fr.* p. 725
37. Pichot, C. (1990) *Makromol. Chem., Macromol. Symp.* **35/36** 327
38. Shirahama, H. & Suzawa, T. (1984) *J. Appl. Polym. Sci.* **29** 3651
39. Doiuchi, T. & Minoura, Y. (1977) *Macromolecules* **10** 260
40. Homola, A. & James, R.O. (1977) *J. Colloid Interface Sci.* **59** 123
41. Tamai, H., Hamada, A. & Suzawa, T. (1982) *J. Colloid Interface Sci.* **88** 378
42. Homola, A. & James, R.O. (1976) *Austr. Appl. PC* 6431
43. Harding, I.H. & Healy, T.W. (1985) *J. Colloid Interface Sci.* **107** 382
44. Kawaguchi, H., Hoshino, H., Amagasa, H. & Ohtsuka, Y. (1984) *J. Colloid Interface Sci.* **97** 465
45. Lambla, M., Valentin, B., Guerrero, S. & Banderet, A. (1977) *J. Macromol. Sci., Chem.* **11** 1439
46. Misra, S.C., Pichot, C., El-Aasser, M.S. & Vanderhoff, J.W. (1979) *J. Polym. Sci., Polym. Lett. Ed.* **17** 567
47. Emelie, B., Pichot, C. & Guillot, J. (1984) *J. Dispersion Sci., Technol.* **5** 393
48. Rios, L., Pichot, C. & Guillot, J. (1980) *Makromol. Chem.* **181** 677
49. Capek, I., Bartoň, J. & Orolínová, E. (1985) *Acta Polym.* **36** 187
50. Capek, I., Mlynárová, M. & Bartoň, J. (1988) *Acta Polym.* **39** 142
51. Grancio, M.R. & Wiliams, D.J. (1970) *J. Polym. Sci., A-1* **8** 2617
52. Grancio, M.R. & Wiliams, D.J. (1970) *J. Polym. Sci., A-1* **8** 2733
53. Williams, D.J. (1973) *J. Elastoplast.* **3** 187; **5** 6
54. Keusch, P. & Williams, D.J. (1973) *J. Polym. Sci., Polym. Chem. Ed.* **11** 143
55. Williams, D.J. (1973) *J. Polym. Sci., Polym. Chem. Ed.* **11** 301
56. Keusch, P., Prince, J. & Williams, D.J. (1973) *J. Macromol. Sci., Chem.* **7** 623
57. Williams, D.J. (1974) *J. Polym. Sci., Polym. Chem. Ed.* **12** 2123
58. Keusch, P., Graff, R.A. & Williams, D.J. (1974) *Macromolecules* **7** 304
59. Gardon, J.L. (1973) *J. Polym. Sci., Polym. Chem. Ed.* **11** 241
60. Gardon, J.L. (1974) *J. Polym. Sci., Polym. Chem. Ed.* **12** 2133
61. Lee, D.I. (1981) *Am. Chem. Soc., Symp. Ser.* 1965, Washington DC, p. 405
62. Napper, D.H. (1971) *J. Polym. Sci., A-1* **9** 2089
63. Hergeth, W.D., Bittrich, J., Eichhorn, F., Schleuker, S., Schmutzler, K. & Steinau, U.J. (1989) *Polymer* **30** 1913
64. Hergeth, W.D., Schmutzler, K. & Wartewig, S. (1990) *Makromol. Chem., Macromol. Symp.* **31** 123
65. Hergeth, W.D. & Schmutzler, K. (1985) *Acta Polym.* **36** 472
66. Kakuschke, R. (1982) *Thesis*. Technische Hochschule C. Schorlemmer, Merseburg
67. Okubo, M. (1990) *Makromol. Chem., Macromol. Symp.* **35/36** 307; Okubo, M., Katayama, Y. & Yamamoto, Y. (1991) *Colloid Polym. Sci.* **269** 217; Okubo, M., Shiozaki, M., Tsujihiro, M. & Tsukuda, Y. (1991) *Colloid Polym. Sci.* **269** 222
68. Matsumoto, T., Okubo, M. & Shibao, S. (1976) *Kobunshi Ronbunshu, Eng. Ed.* **5** 784
69. Sperling, L.H., Chin, Tai-Woo, Hartmann, C.P. & Thomas, D.A. (1972) *Int. J. Polym. Mater.* **1** 331
70. Sperling, L.M., Chin, Tai-Woo & Thomas, D.A. (1973) *J. Appl. Polym. Sci.* **17** 2443
71. Lee, D.I. & Ishikawa, T. (1983) *J. Polym. Sci., Polym. Chem. Ed.* **21** 147
72. Brouwer, W.M. (1989) *Colloids Surfaces* **40** 235
73. Silverstein, M.S., Talmon, Y. & Narkis, M. (1989) *Polymer* **30** 416
74. Dimonie, V., El-Aasser, M.S., Klein, A. & Vanderhoff, J.W. (1984) *J. Polym. Sci., Polym. Chem. Ed.* **22** 2194
75. Eliseeva, V.I., Shapiro, Yu.E., Titova, N.V. & Budanov, N.A. (1989) *Vysokomol. Soedin., A* **31** 263
76. Brown, R.A., Price, V., Randall, P.D. & Satgurunathan, R. (1989) *Polym. Commun.* **30** 349
77. Chern, Chorng-Shyan & Poehlein, G.W. (1987) *J. Polym. Sci., Polym. Chem. Ed.* **25** 617
78. Chern, Chorng-Shyan & Poehlein, G.W. (1990) *J. Polym. Sci., Polym. Chem. Ed.* **28** 3055
79. Chern, Chorng-Shyan & Poehlein, G.W. (1990) *J. Polym. Sci., Polym. Chem. Ed.* **28** 3073
80. Sperling, L.H. (1981) *Interpenetrating polymer networks and related materials*. Plenum Press, New York
81. Hourston, D.J., Satgurunathan, R. & Varma, H. (1986) *J. Appl. Polym. Sci.* **31** 1955
82. Hourston, D.J., Satgurunathan, R. & Varma, H. (1987) *J. Appl. Polym. Sci.* **33** 215
83. Hourston, D.J., Satgurunathan, R. & Varma, H. (1987) *J. Appl. Polym. Sci.* **34** 901
84. Sheu, H.R., El-Aasser, M.S. & Vanderhoff, J.W. (1990) *J. Polym. Sci., Polym. Chem. Ed.* **28** 629
85. Sheu, H.R., El-Aasser, M.S. & Vanderhoff, J.W. (1990) *J. Polym. Sci., Polym. Chem. Ed.* **28** 653
86. Okubo, M., Yamada, A. & Matsumoto, T. (1980) *J. Polym. Sci., Polym. Chem. Ed.* **16** 3219
87. Min, T.I., Klein, A., El-Aasser, M.S. & Vanderhoff, J.W. (1983) *J. Polym. Sci., Polym. Chem. Ed.* **21** 2845
88. Bartoň, J., Vašková, V. & Juraničová, V. (1986) *Makromol. Chem.* **187** 257
89. Merkel, M.P., Dimonie, V.L., El-Aasser, M.S. & Vanderhoff, J.W. (1987) *J. Polym. Sci., Polym. Chem. Ed.* **25** 1755
90. Somani, R.H. & Shaw, M.T. (1981) *Macromolecules* **14** 1549
91. Krause, S. (1971) In: Molau, G.E. (ed.) *Colloidal and morphological behaviour of block and graft copolymers*. Plenum Press, New York, p. 223
92. Noolandi, J. & Hong, K.M. (1982) *Macromolecules* **15** 482
93. Hughes, L.J. & Brown, G.L. (1961) *J. Appl. Polym. Sci.* **5** 580

94. Yamazaki, S. (1976) *Kobunshi Ronbunshu* **33** 663
95. Yamazaki, S. (1977) *Shikizai Kyokaishi* **50** 266
96. Sundberg, D.C., Arndt, J. & Tang, M.Y. (1984) *J. Dispersion Sci., Technol.* **5** 433
97. Dinges, K. & Schuster, H. (1967) *Macromol. Chem.* **101** 200
98. Brydon, A., Burnett, G.M. & Cameron, G.G. (1974) *J. Polym. Sci., Polym. Chem. Ed.* **12** 1011
99. Bucknall, C.B. (1978) *Toughtened plastics.* Applied Science Publ., London, p. 95
100. Gasperowicz, A., Kolendowicz, M. & Skowronski, T. (1982) *Polymer* **23** 839
101. Schildknecht, C.E. & Skeist, I. (1977) *Polymerization processes, vol. 29, High Polymers.* J. Wiley & Sons, New York, p. 228
102. Verma, S.K. & Bisarya, S.C. (1986) *J. Appl. Polym. Sci.* **31** 2675
103. Meguro, K., Yabe, T., Ishioka, S., Kato, K. & Esumi, K. (1986) *Bull. Chem. Soc. Jpn.* **59** 3019
104. Hasegawa, M., Arai, K. & Saito, S. (1987) *J. Polym. Sci., Polym. Chem. Ed.* **25** 3231
105. Lewis, K.E. & Praffit, G.D. (1966) *Trans. Faraday Soc.* **62** 204
106. Hasan, S.M. (1982) *J. Polym. Sci., Polym. Chem. Ed.* **20** 2969
107. Hasan, S.M. (1979) *M. Phil. Thesis.* CNAA, London
108. Hergeth, W.D., Steinau, U.J., Bittrich, H.J., Simon, G. & Schmutzler, K. (1989) *Polymer* **30** 254
109. Yamaguchi, T., Ono, T. & Ito, H. (1973) *Angew. Makromol. Chem.* **32** 177
110. Yamaguchi, T., Ono, T. & Saito, Y. (1974) *Chem. Ind.* (London) p. 783
111. Yamaguchi, T., Ono, T., Saito, Y. & Ohara, S. (1975) *Chem. Ind.* (London) p. 748
112. Yamaguchi, T, Ono, T., Saito, Y. & Ohara, S. (1976) *Angew. Makromol. Chem.* **53** 65
113. Mikulášová, D., Chrástová, V., Citovický, P. & Hudec, I. (1983) *Chem. Zvesti* **37** 475
114. Moustafa, A.B. & Abd-el-Hakim, A.A. (1977) *J. Appl. Polym. Sci.* **21** 905
115. Moustafa, A.B., Ghanem, N.A. & Abd-el-Hakim, A.A. (1976) *J. Appl. Polym. Sci.* **20** 2643
116. Mukherjee, A.R., Ghosh, P., Chandha, S.C. & Palit, S.R. (1967) *Makromol. Chem.* **80** 208
117. Moustafa, A.B. & Abd-el-Hakim, A.A. (1976) *J. Polym. Sci., Chem.* **1** 433
118. Furusawa, K. & Anzai, T. (1987) *Colloid Polym. Sci.* **265** 1
119. Gherardi, P. & Matijevič, E. (1986) *J. Colloid Interface Sci.* **109** 57
120. Hansen, F.K. & Matijevič, E. (1980) *J. Colloid Sci., Faraday Trans. 1* **76** 1240
121. Bleier, A. & Matijevič, E. (1978) *J. Colloid Sci., Faraday Trans. 1* **74** 1346
122. Furusawa, K. & Anzai, C. (1987) *Colloid Polym. Sci.* **265** 882

6.2   EMULSIFIER-FREE EMULSION POLYMERIZATION

Polymer dispersions prepared via emulsion polymerization in the absence of a conventional emulsifier have several advantages over those prepared by using an ionic or a mixture of ionic and nonionic emulsifiers. The properties of the dispersion polymer are influenced by the presence of rather high (usually 2–5 mass % per monomer) emulsifier concentrations; the emulsifier remains, of course, as a constituent of the polymer film during its application for making paints or as part of the emulsion polymer after removal of the aqueous phase by drying (preparation of emulsion PVC). Polymer dispersions prepared without emulsifiers are suitable for various applications. 'Clean' monodisperse polymer dispersions are used for example for calibration of various measuring instruments and techniques, e.g., electron microscopy, light-scattering instruments, ultracentrifuges and aerosol-particle counting instruments. 'Clean' polymer dispersions are an important component applied in diagnostic tests and determination of the pore size of filters, sorbents, and biological membranes [1]. They are a suitable model for various studies of colloidal properties of disperse systems, and have found wide application as carriers in the immobilization of enzymes and in research on cell *fagocytosis* [2].

Despite practical advantages of emulsion polymerization without a conventional emulsifier, it has certain shortcomings, e.g., the resulting dispersions contain a rather small fraction of the polymer component. It should be noted, however, that if we speak of the emulsifier-free technique, it does not mean that in the preparation of stable polymer dispersion one can dispense with the function of the emulsifier. The necessary stabilization (prevention from coagulation) of the polymer particles in the emulsifier-free emulsion polymerization is achieved by some component of the system, e.g., an ionizable initiator (e.g. ammonium peroxodisulfate [3, 4], azobis(alkylamidine)hydrochloride [5–7], potassium peroxodiphosphate [8]), amphiphilic polymer or oligomer [9] (for the preparation of nonaqueous polymer dispersions, see Chapter 5), hydrophilic comonomers in copolymerization with lipophilic monomer, e.g. carboxyl group-containing monomers [10–12], acrylamide and its derivatives [13], ionic monomers like sodium salt of 2-sulfoethyl methacrylate [14] or 4-vinyl-N-methylpyridinium methyl sulfate [15].

By using sodium peroxodisulfate, the necessary electrostatic stability of the polymer particles is secured by $-OSO_3^-$ groups on the particle surface (for details see Chapter 3). The presence of solvated hydrophilic groups on a polymer chain composed of comonomer molecules on the particle surface stabilize the polymer particles in disperse aqueous media [10–13]. Polymer particles are electrostatically stabilized by using an ionic comonomer as a component of the polymerization system. The ionizable groups of a copolymer on the particle surface (such as $-OSO_3^-$) create an electrostatic barrier to particle aggregation. Using amphiphilic, hydrophilic or ionizable monomers (ionomers) as stabilizing copolymerization components of the emulsion polymerization system, the polymer particles formed are compositionally heterogeneous. The particles structure is more-or-less of core-shell type depending on the degree of hydrophilicity of the comonomer (see section 6.1). This can be in some cases to the detriment of further application of polymer dispersion, since the polymer product prepared from such a dispersion is in principle a composite material with a different spatial distribution of its chemical composition at the molecular and supermolecular levels.

The kinetics of emulsion polymerization without a conventional emulsifier have been relatively little studied. Kinetic studies of the initial phase (Stage 1) of styrene emulsion polymerization [16], where the particle number rapidly decreases due to coagulation, have shown that the primary particles are generated by agglomeration of styrene oligomers with RMM *ca.* 1000 (degree of polymerization *ca.* 10). The particles become destabilized as they grow (the greater the surface, the lower the value the surface charge density) and partly coagulate. The particles formed contain both high-RMM and low-RMM polymer [17]. As the emulsion polymerization proceeds (Stage 2), the particle number is stabilized. This phase was analysed [18] using the Smith–Ewart models [19], using Smith–Ewart case 3 and applied to the core-shell model [20], Wessling's polymerization model including the surface phase [21] and Gardon's model [22]. The models were derived for emulsion polymerization in the presence of a conventional emulsifier. The experimental results [18] could have been interpreted optimally assuming polymerization on the particle surface. The observed bimodal distribution of the RMMs of the polymer, which cannot be theoretically justified by applying any of these models, is probably a result of polymerization in two different reaction zones.

Several models for emulsifier-free emulsion polymerization and polymer particle nucleation have been published recently [23, 24]. Song & Poehlein [25] proposed a two-stage model of particle formation for emulsifier-free emulsion polymerization of styrene initiated by potassium peroxodisulfate. In Stage 1 a large number of micelle-like oligomer particles are formed. At the end of Stage 1 the oligomer particles lose their stability and undergo particle coagulation when the surface charge density decreases due to the production of high-RMM polymer inside the particles. Stage 2 begins with a decrease in polymer particle number as a result of significant particle coagulation. A dynamic equilibrium between particle nucleation and particle coagulation is established in Stage 2. The two-stage model takes into account a critical chain length which varies with time during Stage 1.

In order to predict variations of conversion, particle size and RMM with time, the effects of monomer concentration, initiator (potassium peroxodisulfate) concentration, reaction temperature and agitator speed in the emulsifier-free emulsion polymerization of styrene were studied [26]. It was found that the polymerization rate and particle number increased with increasing initiator concentration and the reaction temperature. The particle size distribution became broad as the initiator concentration decreased. Increase in the initial monomer concentration led to a decrease in polymerization rate. For a given conversion, the polymer particle size increased with increasing initial monomer concentration. At high stirring speeds significant coagulation was observed and a smaller number of large polymer particles with a broad size distribution were formed. The experimental findings can be plausibly explained in terms of the two-stage model for emulsifier-free emulsion polymerization [25] after introducing some modifications [26].

A considerable number of papers have been devoted to copolymerization of lipophilic (rather slightly water-soluble) monomers with remarkably water-soluble monomer, e.g. styrene/acrylamide [13, 27].

As regards the reaction site, the emulsion copolymerization in the system styrene/acrylamide can be divided into three stages. In the initial stage of copolymerization, acrylamide is polymerized preferentially in the aqueous phase. After polymer particle formation, styrene polymerization takes place in the polymer particles although there is still abundant acrylamide in the aqueous phase. After disappearance of the styrene droplets, the reaction site returns back to the aqueous phase. Temperature increase and reduction of the pH of the system prolongs the time of polymerization in the aqueous phase. The particle size and compositional homogeneity decreases with increasing acrylamide concentration in the reaction system. When acrylamide is replaced by N-hydroxymethyl acrylamide or N,N-dimethyl acrylamide for the copolymerization with styrene, the observed deviations of the copolymerization process can be explained by changes in the hydrophilic-lipophilic properties of acrylamide derivates on the one hand as well as by the different copolymerization parameters of these monomers with styrene [13].

Reference [27] details the emulsifier-free emulsion polymerization of styrene in the presence of minor amounts of acrylamide as water-soluble comonomer. It was confirmed that the primary particles are formed by a mechanism of homogeneous coagulative nucleation and, for particle growth in the post-nucleation stage, the shell growth mechanism was established. The average number of radicals in the polymer particle was found to be between 2 and 6 and the diameter of the polymer particles between 100 and 200 nm. The high value of the radical number per particle is responsible for the lower RMM of the polymer product, i.e. 10^4–10^5. The distribution of monomer in the particles was considered to be uniform throughout the growth period. Reference [27] reported that the shell region, where the polymerization occurs, had a thickness equal to the root-mean-square end-to-end distance of the chains of the growing polymers (about 10–40 nm).

The emulsifier-free emulsion polymerization of styrene in the presence of about 0.33–2.7% (relative to styrene) of the water-soluble comonomer 2-hydroxyethyl methacrylate has been reported in [28]. The potassium peroxodisulfate initiator played a dominant role in the particle nucleation process, since the number density of the polymer particles was dependent on the 0.97 power of the initiator concentration. On the other hand, the nucleation ability of 2-hydroxyethyl methacrylate was weak, since the particle number density depended only on the 0.17 power of the 2-hydroxyethyl methacrylate concentration. The particle nucleation stage ended quite early (before 1% conversion) leading to nearly monodisperse polymer particles. It was [28] suggested that polymer particle nucleation is due to the homogeneous nucleation mechanism. The core-shell structure mechanism (shell region polymerization) is responsible for the growth of the polymer particles. The authors [28] see support for their opinion in the polymer particle size, which ranges from 150 to 600 nm.

Monodisperse reactive polymer dispersions prepared in the absence of emulsifier on the basis of the styrene/glycidyl methacrylate copolymers are described in ref. [29]. Oxirane groups on the particle surface can be transformed to aldehydes according to the scheme:

$$\text{)–OCO–CH}_2\text{–CH–CH}_2 \xrightarrow{\text{H}_2\text{O}/\text{H}^+} \text{)–OCO–CH}_2\text{–CH–CH}_2 \longrightarrow$$
$$\text{O (epoxide)} \qquad\qquad \text{OH OH}$$

$$\longrightarrow \text{)–OCO–CH}_2\text{–CH=O}$$

This scheme is applicable to the binding of various substrates. The number of oxirane groups on the particle surface does not correspond to the concentration of polymerized glycidyl methacrylate, and some oxirane groups are part of the copolymer in the particle core. The particle morphology, particle size and particle size distribution are functions of the initiator ($K_2S_2O_8$) concentration and of the ionic strength of the aqueous phase. Changes in the concentrations of the monomers and their mutual ratio exert a negligible effect on the polymer particles. The initial stage of the emulsifier-free copolymerization of styrene with glycidyl methacrylate has features typical of solution polymerization until oligomers become separated from the continuous aqueous phase. At this point, the degree of polymerization, P_n, of the oligomers is given by [30, 31]

$$P_n = \frac{(r_1[\text{GMA}]^2 + 2[\text{GMA}][\text{S}] + r_2[\text{S}]^2)}{R_i^{1/2}\,(r_1^2\delta_1^2[\text{GMA}]^2 + 2\Phi\delta_1\delta_2 r_1 r_2[\text{GMA}][\text{S}] + r_2^2\delta_2^2[\text{S}]^2)^{1/2}},$$

where R_i is the rate of initiation.

On substituting the values $\delta_1^2 = k_t/k_p^2 = 39.6$ mol dm^{-3} s^{-1} [32], $r_1 = 0.53$, $[\text{GMA}]_0 = 1.5 \times 10^{-4}$ mol dm^{-3}, $\delta_2^2 = k_t/k_p^2 = 156$ mol dm^{-3} s^{-1} [33], $r_2 = 0.44$ and $[\text{S}]_0 = 5 \times 10^{-3}$ mol dm^{-3}, $\Phi = k_{t,1,2}/(k_{t,1}\,k_{t,2})^{0.5} = 1$, which means that for an initiator concentration of ca. 4×10^{-3} mol dm^{-3} that the degree of polymerization of the oligomers on their precipitation from the aqueous phase should be ca. 4. The experimentally-determined P_n is about 7. In the emulsifier-free emulsion polymerization, the precipitation of oligomers takes place (and thus particle nucleation) after reaching a substantially lower degree of polymerization than in the conventional emulsion polymerization. Explanation for this has been made elsewhere [29].

Kinetic studies of the emulsifier-free copolymerization of styrene with the surface-active comonomer 2(11-undecanoyloxy)-1-ethane sodium sulfonate

$$CH_2=CH(CH_2)_8-\overset{\overset{O}{\|}}{C}-O-CH_2CH_2SO_3^-\ Na^+$$

using concentrations higher than the CMC of this monomer, have shown that the particle number and the rate of copolymerization in the stationary region depend on the comonomer concentration with exponent 1 and on the concentration of potassium peroxodisulfate with an exponent equal to 0.5. This result indicates particle formation via a mechanism of homogeneous nucleation. The particle distribution is rather narrow, the polydispersity being nearly 1 at 30% conversion. Addition of an inorganic salt to increase the ionic strength of the aqueous phase causes the formation of micelles which can also be a centre of later polymer particles. Change in the ionic strength leads therefore to a change in particle number as a consequence of the appearance of two nucleation mechanisms [34].

Another example of the emulsifier-free emulsion polymerization of styrene in the presence of an ionomer was reported earlier [14]. The emulsion polymerization of styrene (S) without conventional emulsifier initiated by $K_2S_2O_8$ leads to the formation of polymer particles of diameter 500 to 1000 nm. On adding a small amount (0.5%) of ionomer of sodium p-styrene sulfonate (NaSS)

$$CH_2=CH-\!\!\!\left\langle\bigcirc\right\rangle\!\!\!-SO_3Na$$

the particle size decreases (d = 150–400 nm). The particle diameter is a function of the ratio between the ionic strength and the ionomer concentration (exponent –0.64), initiator concentration (exponent –0.20), and monomer concentration (exponent of 0.46).

The particle size can be controlled to some extent by adding an inhibitor (*tert*-butyl catechol) as confirmed in the polymerization of styrene with 1% sodium 2-sulfoethyl methacrylate [14]. From the kinetic data, the authors [14] propose that the copolymerization proceeds in the aqueous phase. Nuclei of the polymer are generated via homogeneous nucleation. The particles of the polymer dispersion are stabilized by the sulfonate group of the comonomer and by sulfate end groups of the initiator

$$(^-O_3SO\text{--}\sim)\ .$$

Compared with styrene, vinyl acetate is highly water-soluble. Its emulsifier-free polymerization initiated by potassium peroxodisulfate gives a stable polymer dispersion [35–38], and its kinetics, over a wide range of concentrations of monomer and initiator at 50°C have been described elsewhere [39]. At constant initiator concentration (1.25 g dm^{-3} water) and with varying vinyl acetate concentration (20–500 g dm^{-3} water), the conversion curves have a sigmoidal shape with a distinct gel effect. The particle number quickly reaches a constant value as early as the initial stage of polymerization. The particle number is proportional to the reciprocal of the monomer concentration; it decreases with increasing monomer concentration and the polymerization rate also decreases. The log-log relationship between the particle number and the monomer concentration is linear with a slope of –0.75. This result has not been theoretically justified [36, 39]. At constant monomer concentration ([VAc] = 200 g dm^{-3} water), the rate of polymerization increases with increasing initiator concentration (0.14–7.5 g dm^{-3} water). The rate of polymerization increases smoothly with conversion up to about 80 % conversion and, in contrast to the conventional emulsion polymerization of VAc, no region of constant polymerization rate is observed. After rapid stabilization at the beginning of polymerization, the particle number remains virtually unchanged with conversion. The initiator concentration does not affect the particle number, which contradicts an earlier paper [40]. The conclusions in ref. [39] point out that the monomer dissolved in the aqueous phase influences not only the monomer concentration in the polymer particles (see Chapter 3) but also the rate constant for termination in the polymer particles.

In ref. [41] there are reported the surface-active properties of particles of polymer dispersions prepared without a conventional emulsifier using as examples the emulsion polymerization of styrene, methyl methacrylate, vinyl acetate, acrylamide, methacrylic acid and 2-hydroxyethyl methacrylate. The authors established the conditions for homo- and heterocoagulation. Solvated polymer chains on the surface of polymer particles protect the dispersion from homo- and heterocoagulation (steric stabilization, Chapter 5). On the other hand, the bridging effect encourages coagulation as documented by the effect of a dispersion of styrene/acrylamide copolymer on the precipitation of the other dispersions mentioned.

A special preparation of a polymer dispersion based on water-soluble or partially water-soluble monomers is the synthesis of monodisperse aqueous dispersions based on mixtures of acrylamide, N-isopropyl acrylamide and N,N'-methylenebisacrylamide [42]. Stable latices can be prepared with solid contents of 2.5%, the optimum concentration of acrylamide being $ca.$ 1%. The diameter of the particles of the polymer dispersion crosslinked by N,N'-methylenebisacrylamide is $ca.$ 1000 nm and the particles are highly swollen (if the particles were not crosslinked, a solution of copolymer would be formed). On raising the temperature above the lower critical solution temperature (LCST), contraction of the particles occurs by a volume factor of about 10. In the presence of 0.1 M $CaCl_2$, the critical temperature of coagulation increased from 33°C to 44°C as the amount of acrylamide in the reaction mixture was increased from 0 to 1.24 g dm^{-3}. At temperatures lower than the LCST, the latices are stabilized by the combined steric and electrostatic effects. At temperatures higher than the LCST, the particles are stabilized by the electrostatic effect.

A survey of emulsifier-free emulsion polymerization has recently been reported by Eliseeva & Aslamazova [43]. The survey treats the problem as regards the nature of the monomer(s) used and the stability of the latices formed. Several possible applications of emulsifier-free polymer dispersions are discussed.

References

1. Goodall, A.R., Wilkinson, M.C. & Hearn, J. (1977) *J. Polym. Sci., Polym. Chem. Ed.* **15** 2193
2. Bangs, L.B. (1987) *Uniform latex particles.* Seradyn Inc., Indianapolis
3. Matsumoto, T. & Ochi, A. (1965) *Kobunshi Kagaku* **22** 481
4. Kotera, A., Furusawa, K. & Takeda, Y. (1970) *Kolloid-Z. Z. Polym.* **239** 677
5. Liu, L.J. & Krieger, I.M. (1978) In: Becker, P. & Yudenfreund, M.N. (eds) *Emulsions, lattices and dispersions.* M. Dekker, New York, p. 11
6. Goodwin, J.W., Ottewill, R.H. & Pelton, R. (1979) *Colloid Polym. Sci.* **257** 61
7. Sakota, K. & Okaya, T. (1976) *J. Appl. Polym. Sci.* **20** 1725, 3133
8. Goodall, A.R., Hearn, J. & Wilkinson, M.C. (1978) *Br. Polym. J.* **10** 141
9. White, W.W. & Jung, H. (1976) *J. Polym. Sci., Polym. Symp.* **45** 197
10. Ceska, G.W. (1974) *J. Appl. Polym. Sci.* **18** 427
11. Ceska, G.W. (1974) *J. Appl. Polym. Sci.* **18** 2493
12. Sakota, K. & Okaya, T. (1976) *J. Appl. Polym. Sci.* **20** 1725, 1745, 2583, 3133, 3265
13. Kawaguchi, H., Ohtsuka, Y. & Sugi, Y. (1981) *J. Appl. Polym. Sci.* **26** 1637, 1649
14. Juang, M.S. & Krieger, I.M. (1976) *J. Polym. Sci., Polym. Chem. Ed.* **14** 2089
15. Liu, L.J. & Krieger, I.M. (1981) *J. Polym. Sci., Polym. Chem. Ed.* **19** 3013

16. Goodall, A.R., Wilkinson, M.C. & Hearn, J. (1980) In: Fitch, R.M. (ed.) *Polymer colloids*, vol. 2. Plenum Press, New York, p. 629
17. Hearn, J., Wilkinson, M.C., Goodall, A.R. & Chainey, M. (1985) *J. Polym. Sci., Polym. Chem. Ed.* **23** 1869
18. Cox, R.A., Wilkinson, M.C., Creasey, J.M., Goodall, A.R. & Hearn, J. (1977) *J. Polym. Sci., Polym. Chem. Ed.* **15** 2311
19. Smith, W.V. & Ewart, R.H. (1948) *J. Chem. Phys.* **16** 592
20. Ugelstad, J. & Hansen, F.K. (1976) *Rubber Chem. Technol.* **49** 536
21. Wessling, R.A. & Harrison, I.R. (1971) *J. Polym. Sci., A-1* **9** 3471
22. Gardon, J.L. (1970) *Br. Polym. J.* **2** 1
23. Feeney, J.P., Napper, D.H. & Gilbert, R.G. (1987) *Macromolecules* **20** 2922
24. Rawlings, J.B. & Ray, W.H. (1988) *Polym. Eng. Sci.* **28** 237
25. Song, Z. & Poehlein, G.W. (1989) *J. Colloid Interface Sci.* **128** 501
26. Song, Z. & Poehlein, G.W. (1990) *J. Polym. Sci., Polym. Chem. Ed.* **28** 2359
27. Chen, Show-An & Lee, Song-Tai (1991) *Macromolecules* **24** 3340
28. Chen, Show-An & Chang, Herng-Show (1990) *J. Polym. Sci., Polym. Chem. Ed.* **28** 2547
29. Žurková, E., Bouchal, K., Zdeňková, D., Pelzbauer, Z., Švec, F. & Kálal, L. (1983) *J. Polym. Sci., Polym. Chem. Ed.* **21** 2949
30. Melville, H.W., Noble, B. & Watson, W.F. (1947) *J. Polym. Sci.* **2** 229
31. Bijsterbosch, B.H. (1978) *Colloid Polym. Sci.* **256** 343
32. Kokhnov, I.M. & Sorokin, M.F. (1964) *Vysokomol. Soedin.* **6** 791
33. Bamford, C.H., Barb, W.G., Jenkins, A.D. & Onyon, P.F. (1958) *The kinetics of vinyl polymerization by radical mechanism*. Butterworths, London
34. Chen, Show-An & Chang, Herng-Show (1985) *J.Polym. Sci., Polym. Chem. Ed.* **23** 2615
35. Napper, D.H. & Parts, A.G. (1962) *J. Polym. Sci.* **61** 113
36. Dunn, A.S. & Taylor, P.A. (1965) *Makromol. Chem.* **53** 207
37. Netschey, A. & Alexander, A.E. (1970) *J. Polym. Sci., A-1* **8** 399
38. Netschey, A. & Alexander, A.E. (1970) *J. Polym. Sci., A-1* **8** 407
39. Nomura, M. & Sasaki, S. (1978) *J. Appl. Polym. Sci.* **22** 1043
40. Dunn, A.S. & Chong, L.C.H. (1970) *Br. Polym. J.* **2** 49
41. Tamai, H., Fujii, A. & Suzawa, T. (1987) *J. Colloid Interface Sci.* **116** 37
42. Pelton, R.H. & Chibante, P. (1986) *Colloids Surfaces* **20** 247
43. Eliseeva, V.I. & Aslamazova, T.R. (1991) *Russ. Chem. Rev.* **60** 206

6.3 POLYMERIZATION OF MONOMERS WITH EMULSIFYING PROPERTIES (EMULSIFYING MONOMERS, SURFACE-ACTIVE MONOMERS, INTERNAL EMULSIFIERS, SURFOMERS)

The use of emulsifiers is necessary to produce stable dispersions with high solid contents. By altering the chemical nature and concentration of the emulsifiers, control of the polymerization kinetics, particle size and other properties of the emulsion polymerization system become possible. On the other hand, disadvantages are caused by the presence of emulsifiers during the application of emulsion polymers. Adsorptively-bonded emulsifiers can desorb from the polymer particle surface, which leads to destabilization of the polymer dispersion. The emulsifier can migrate during polymer film formation on the polymer surface. As a consequence, special effects such as 'blooming' or 'blushing' of the polymer film appear [1]. An active area of research in emulsion polymerization is how to overcome these problems while maintaining the advantages of the technique.

One possible way of solving the problem of emulsifier desorption from the polymer particles is to use polymerizable (or copolymerizable) emulsifiers, i.e. monomers with surface-active properties [2]. A report concerning the synthesis and application of this new class of emulsifiers for emulsion polymerization has been published recently [3].

We are currently witnessing increased efforts in research on the applications of monomeric emulsifiers in both the homopolymerization and their copolymerization with vinyl monomers. The goal of this research is the preparation of new types of polymer dispersions for various applications.

A monomeric emulsifier can generally be characterized as an amphiphilic compound containing a radically-polymerizable double bond, which can be either in its lipophilic or hydrophilic moiety. The double bonds can thus be either in the surface layer or inside the micelles. An important example of monomeric emulsifiers is given by derivatives of quaternary ammonium salts, e.g.:

$$CH_2=C(CH_3)-C(=O)-O-CH_2CH_2N^+(R)_2R' \quad Br^-$$

(R = methyl, R' = dodecyl)

which can be readily prepared from N,N-dimethylaminoethyl methacrylate by reaction with a higher alkyl bromide (R'Br) [4, 5].

Another remarkable group of monomeric emulsifiers are vinylpyridinium

[structure: vinylpyridinium with CH$_2$=CH– on pyridine ring with –CH$_3$, N$^+$–CH$_2$–CO–OR, Br$^-$] (R = C$_{10}$H$_{23}$)

and vinylimidazolium salts [6, 7]. As regards other examples [8] we can mention sodium 4-[11-(methacryloylamino)undecanoylamino]benzenesulfonate

$$CH_2=C(CH_3)-CO-NH-(CH_2)_{10}-CO-NH-C_6H_4-SO_3^-Na^+$$

2-acryloylamino-2-methyldecane sulfonic acid

$$CH_2=CH-CO-NH-C(CH_3)(CH_2-(CH_2)_6CH_3)-CH_2-SO_3H$$

and sodium 2-[11-(methacryloylamino)undecanoylamino]-1-ethanesulfonate

$$CH_2=C(CH_3)-CO-NH-(CH_2)_{10}-CO-NH-CH_2-CH_2-SO_3^-Na^+.$$

Monomeric emulsifiers form micelles in aqueous solution [9–13]. The CMC for aqueous solutions of quaternary ammonium salts prepared by reaction of N,N-dimethylaminoethyl

Sec. 6.3] Polymerization of monomers with emulsifying properties

methacrylate with alkyl bromide (alkyl: C_8, C_{12}, C_{14} or C_{18}) is given by

$$\log \text{CMC} = 1.46 - 0.31 N,$$

where N is the number of alkyl carbon atoms [5]. This relation recalls an analogous equation derived for ionic emulsifiers [14].

A polymer generated by radical polymerization in water (initiator — ammonium peroxodisulfate) precipitates from solution as a granulate, but in the AIBN-initiated polymerization in benzene, gelation of the system occurs. The rate of polymerization in aqueous solution increases with alkyl chain length. Since the concentrations of monomer emulsifiers were above their CMC, lengthening of the alkyl group should cause an increase in the fraction of the oriented molecules of monomeric emulsifier in the micellar solution. It is therefore assumed that monomeric emulsifier molecules bound into the micelles take part in the polymerization. The polymerizations thus lead to the formation of polymerized micelles of the monomeric emulsifier.

In styrene polymerization with N-decylacetato-2-methyl-5-vinylpyridinium bromide (about 1% per aqueous phase, the ratio between phases styrene:water = 1:2) the polymerization rate is comparable with that of styrene in the presence of the cation-active emulsifier cetylpyridinium bromide. The stability of the polymer dispersion formed using the monomeric emulsifier compared with cetylpyridinium bromide is, however, much higher.

The influence of the initiator $K_2S_2O_8$ on the rate of styrene polymerization (at concentrations $(1.2-24.0) \times 10^{-3}$ mol dm^{-3}) and on the particle size of the polymer dispersion is negligible. The monomeric emulsifier affects both the rate of polymerization (the rate increases with the increasing concentration of emulsifier) and the particle size of the polymer dispersion (the particle size decreases with increasing emulsifier concentration).

The authors [6] assume the polymer-monomer particles to be formed from microemulsion droplets. The radical formed in the aqueous phase reacts with the double bond of the monomeric emulsifier located on the surface of the styrene microdroplets giving a monomer emulsifier radical. The radical formed produces a homopolymer of the monomeric emulsifier via repeated addition to its double bonds. The microdroplet thus changes gradually into a polymer particle. This mechanism does not require diffusion of the monomer between individual microdroplets. The particle size is, in principle, given by the dimensions of the original (unpolymerized) microdroplets.

As has been shown recently [15] radical polymerization in micelles of surface-active monomers might provide a route to polymer dispersions with small particles as conductors of drugs, proteins and other biologically and/or catalytically active substances. Again, this may be a method for fixation of micelles. Polymerization studies of the surface-active monomer

$$\left[\begin{array}{c} \text{CH}_3 \\ | \\ \text{CH}_2 = \text{C} \qquad \text{CH}_3 \\ | \qquad\qquad | \\ \text{CO-OCH}_2\text{CH}_2-\text{N}^+-\text{CH}_2-\text{CO-O-C}_n\text{H}_{2n+1} \\ | \\ \text{CH}_3 \end{array} \right] \quad X^-$$

$X = \text{Cl, Br}$
$n = 4-16$

photoinitiated (AIBN, 300 K) or thermally initiated (AIBN, 343 K or $K_2S_2O_8$, 333 K) in aqueous solution have shown [15] that the polymerization rate decreases with increasing surfomer concentration at concentrations up to the CMC of the surfomer (region I). This is a result of the increasing degree of association of the ionic surfomer molecules and of their propagating radicals, which leads to a decrease in the ratio $k_p/k_t^{0.5}$ of the rate constant for propagation and termination, respectively [16]. In region II, which is characterized by the presence of isotropic spherical micelles (i.e. the concentration of surfomer lies between CMC_1 and CMC_2), the rate of polymerization is practically constant, i.e. it does not depend on the surfomer concentration. Increasing the surfomer concentration over CMC_2 (region III) leads to a decrease in the polymerization rate. This is a result of the decreasing fraction of surfomer in the system which undergoes polymerization. It is supposed that polymerization proceeds in isotropic (spherical) micelles but not in anisotropic ones (i.e. in rod-like micelles). At very high surfomer concentrations (over CMC_3, region IV) the formation of surfomer lamellar structures leads to an increase in the polymerization rate. Reference [15] reports an investigation of region II in detail and proposes a reaction mechanism of polymer particle (polymerized micelles) formation. The initiation, propagation and termination steps of the free-radical polymerization of the surfomer occur in the outer shells of the micelles, where the methacrylic groups of the surfomer are located. For detailed discussion see the original paper [15].

Another example of the polymerization of micelle-forming amphiphiles, sodium 10-undecenoate and sodium 8-nonenoate, was introduced by Arai [17], who also included short but illustrative recent literature references.

Emulsifiers with polymerizable double bonds can be used for the preparation of polymerized monolayers (see also section 6.5). A paper by Fendler [18] exemplifies this possibility. Under UV irradiation in a nitrogen atmosphere, the chloroform solution of the surfomer di-n-octadecylmethyl(p-vinylbenzyl) ammonium chloride

$$[(C_{18}H_{37})_2N(CH_3)(CH_2C_6H_4CH=CH_2)]^+ \quad Cl^-$$

spread on aqueous NaCl solution undergoes polymerization. This is an example of production of polymerizable Langmuir–Blodgett films at the gas–water interface.

The principal aspects of the polymerization of micelle-forming emulsifiers in the aggregated state based on papers published up to 1988 are discussed in the review of Paleos & Malliaris [19]. The general aspects of micellar polymerization, e.g. synthesis and physicochemical behaviour, are highlighted by the examples of micellar polymerization of vinylpyridinium salts, vinylimidazolium salts, sodium 10-undecenoate, allyldimethyldodecylammonium bromide and 2-methacryloylethyldimethylalkylammonium halides.

As a special application, monomeric emulsifiers can be used for the surface modification of polymers [20]. Addition of a small amount (1%) of polymerizable surfomer

$$F(CF_2)_8 - \underset{\underset{O}{\overset{\|}{S}}}{\overset{O}{\|}} - \underset{C_2H_5}{N} - CH_2 - CH_2 - O - \overset{O}{\overset{\|}{C}} - \underset{CH_3}{CH} - CH = CH_2$$

to a curable acrylic lacquer gives a surface concentration of 50% of the monomeric emulsifier at the interface towards air. Emulsifier-modified surface layers are usually thicker than monolayers made by the Langmuir–Blodgett technique.

An example of surfomer copolymerization with vinyl monomer has been described earlier [21]. The surfomer nonylphenoxypoly(etheroxy)ethyl acrylate

$$C_9H_{19}C_6H_4-(O-CH_2-CH_2)_n-O-CO-CH=CH_2 \;; \quad n = 2-18$$

can be easily copolymerized with acrylamide to form soluble copolymers, i.e. poly(acrylamide)-co-nonylphenoxypolyetheroxyethyl acrylate which contain low levels (≤ 5 mol %) of emulsifier moieties. The RMM of the copolymer tends to decrease with increasing levels of surfomer. This result suggests some chain transfer to emulsifier macromonomer or concomitant alcohol (residue from the synthesis of the surfomer). The acrylamide surfomer copolymers containing even small concentration of surfomer moieties, i.e. ≤ 0.5 mol %, show interesting solution properties. The limiting viscosity number and Flory–Huggins constant depend on the extent of the intramolecular and/or intermolecular chain association. At low surfomer concentrations in the copolymer (below 1000 ppm), intramolecular chain association leads to lower values of the limiting viscosity numbers and an elevated Flory–Huggins constant (collapsed coils), while for concentrations of the surfomer above 1000 ppm (polymer overlap concentration), the copolymer solution shows a substantially higher viscosity than a solution of acrylamide homopolymer. The higher viscosity of the copolymer solution is due to the larger hydrodynamic radius of the intermolecularly-interacting chains. The surfomer copolymerization described is an example of the preparation of hydrophobically-associating acrylamide copolymers (e.g. copolymers of acrylamide and long-chain N-alkyl acrylamides). Since acrylamide and long-chain alkyl acrylamides are mutually incompatible, specialized copolymerization techniques are required. The application of these techniques often leads to formation of copolymers of low RMM and/or large usage of external emulsifier. This latter drawback can be avoided by the incorporation of amphiphilic structures into the polymer chain, i.e. by the use of an appropriate surfomer [21].

References

1. Dickstein, J. (1986) *Polym. Prepr. (Am. Chem. Soc., Div. Polym. Chem.)* **27** 427
2. Tauer, K., Goebel, K.H., Kosmella, S., Stähler, K. & Neelsen, J. (1990) *Makromol. Chem., Macromol. Symp.* **31** 107
3. Tauer, K., Goebel, K.H., Kosmella, S. & Neelsen, J. (1988) *Plaste Kautsch.* **35** 373
4. Luskin, L.S. (1974) In: Yocum, R.H. & Nyquist, E.B. (eds) *Functional monomers*. M. Dekker, New York, Chapter 3
5. Nagai, K., Ohishi, Y., Inaba, H. & Kudo, S. (1985) *J. Polym. Sci., Polym. Chem. Ed.* **23** 1221
6. Malyukova, E.B., Egorov, V.V., Zubov, V.P., Gritskova, I.A., Pravednikov, A.N. & Kabanov, V.A. (1982) *Dokl. Akad. Nauk SSSR* **265** 375
7. Egorov, V.V. *et al.* (1980) *USSR* 713868
8. Rohm and Haas, GmbH (1984) *Ger.* 3239527
9. Winter de, W. & Marien, A. (1984) *Makromol. Chem., Rapid Commun.* **5** 593
10. Jungermann, E. (1970) *Cationic surfactants*. M. Dekker, New York
11. Kabanov, V.A. (1967) *Pure Appl. Chem.* **15** 391
12. Mielke, I. & Ringsdorf, H. (1972) *Makromol. Chem.* **153** 307
13. Salamone, J.C., Israel, S.C., Taylor, P. & Snider, B. (1973) *Polymer* **14** 639
14. Rosen, M.J. (1978) *Surfactants and interfacial phenomena*. J. Wiley & Sons, New York
15. Egorov, V. (1990) *Makromol. Chem., Macromol. Symp.* **31** 157
16. Ringsdorf, H. & Thunig, D. (1977) *Makromol. Chem.* **178** 2205

17. Arai, K. (1990) *Makromol. Chem., Macromol. Symp.* **31** 227
18. Rolandi, R., Paradiso, R. & Fendler, J.H. (1989) *Colloids Surfaces* **35** 343
19. Paleos, C.M. & Malliaris, A. (1988) *J. Macromol. Sci., Rev. Macromol. Chem. Phys.* **28** 403
20. Torstensson, M., Ranby, B. & Hult, A. (1990) *Macromolecules* **23** 126
21. Schulz, D.N., Kaladas, J.J., Maurer, J.J., Bock, J., Pace, S.J. & Schulz, W.W. (1987) *Polymer* **28** 2110

6.4 EMULSION POLYMERIZATION IN THE PRESENCE OF COMPOUNDS OF THE INITIATOR-EMULSIFIER TYPE (SURFACE-ACTIVE INITIATORS)

Many well-known characteristics of emulsion free-radical polymerization are directly connected with the nature of the free-radical initiator used. The main function of the initiator is the generation of free radicals which start the polymerization of the radically-polymerizable monomer. Completely water-soluble initiators of the peroxodisulfate type are mostly used for emulsion polymerization. Owing to their water solubility they are located in the continuous aqueous phase of the classical emulsion polymerization system. Initiators of the peroxodisulfate type are also oxidizing agents and can start various redox reactions in the system. As a result, hydroxyl and/or carboxyl groups are introduced into the final polymeric product. By contrast, azo initiators are not able directly to oxidize the polymer (hydrocarbon) chain. Highly water-soluble azo initiators are commercially available only to a limited extent [1]. The most typical azo initiator AIBN (2,2'-azobisisobutyronitrile) is considered to be oil soluble. However, as has been shown [2], its water solubility cannot be neglected. The presence of an initiator in the oil phase (e.g. in monomer droplets) can lead to complications during the synthesis of emulsion polymers (i.e. a simultaneously proceeding polymerization in the polymer/monomer particles and in monomer droplets). These drawbacks of classical initiators in emulsion polymerization systems, and recognition of the role of oligomeric radicals having a hydrophilic/lipophilic nature, have generated the idea of making initiators with surface-active properties. A surface-active compound will tend to localize itself at the interface between the oil and water phases in the emulsion polymerization system.

To achieve deliberate localization of the initiating step of radical polymerization at the interface, certain compounds were synthesized; their molecules had two functions, (i) that of an initiator (i.e. the presence of an unstable chemical bond, able to dissociate homolytically), and (ii) that of a surfactant (aliphatic lipophilic chain combined with a polar hydrophilic group).

Fixing the reagents at the interfacial surface usually affects the reaction kinetics. In the case of an oil-in-water emulsion polymerization in the presence of initiator-emulsifier compounds, the thermally unstable chemical bonds are located on the surface of, or inside, the micelles (by analogy with the localization of the double bonds of a monomeric emulsifier) according to whether the thermally unstable chemical bond is in the lipophilic or hydrophilic moiety. The lipophilic part of the initiator-emulsifier compounds is normally represented by a longer alkyl chain or a chain of the lipophilic polymer (macromolecular initiator-emulsifier). Of the known cases given in the literature [3], perester I and hydroperoxide II are illustrative)

$$C_{16}H_{33}(OCH_2CH_2)_{20}-O-\overset{\overset{O}{\|}}{C}-CH_2CH_2-\overset{\overset{O}{\|}}{C}-O-O-C(CH_3)_3 \qquad (I)$$

and hydroperoxide II

$$C_{16}H_{33}(OCH_2CH_2)_{20}-O-\underset{\underset{O}{\|}}{C}-\underset{\underset{\underset{O}{\|}}{\underset{R_1-C}{|}}}{\underset{CH_2}{|}}{CH} - \underset{\underset{\underset{OOH}{|}}{\underset{R_2C-OC_2H_5}{|}}}{\underset{CH_2}{|}}{CH}-\underset{\underset{O}{\|}}{C}-OH \quad \text{(II)}$$

(R_1 and R_2 are H or CH₃), their preparation being protected by patent [4].

A polymeric initiator-emulsifier can be conveniently prepared via the emulsion copolymerization of styrene with the methacrylic ester of α-oxyethyl-*tert*-butyl peroxide III

$$CH_2=\underset{\underset{O}{\|}}{\underset{\underset{CH_3}{|}}{C}}-\underset{\underset{CH_3}{|}}{C}-O-CH-O-O-C(CH_3)_3 \quad \text{(III)}$$

by subjecting it to a redox initiating system [5].

The micellar properties of compounds I and II are characterized by very low CMCs. For (I), the CMC is 2×10^{-5} mol dm^{-3} [6] and for II it is only 4×10^{-6} mol dm^{-3} [4]. These data show that the majority of initiator-emulsifier molecules (I and II) are concentrated in the micelles. This fact was demonstrated by the increase in rate constant for the decomposition of the compound II, which is *ca*. 2.5 times that of its analogue, i.e. IV

$$C_2H_5-O-\underset{\underset{O}{\|}}{C}-\underset{\underset{\underset{O}{\|}}{\underset{R_1-C}{|}}}{\underset{CH_2}{|}}{CH} - \underset{\underset{\underset{OOH}{|}}{\underset{R_2C-OC_2H_5}{|}}}{\underset{CH_2}{|}}{CH}-\underset{\underset{O}{\|}}{C}-OH \quad \text{(IV)}$$

By contrast, the decomposition rate of (I) differs only slightly from that of common peresters in aliphatic hydrocarbons. This is assumed to be due to localization of the perester groups in the micellar core [7], which eliminates the role of micellar catalysis during the decomposition of the perester group of compound I.

As had been expected, a most remarkable effect was achieved during the decomposition of polymeric dialkyl peroxides (of copolymers (III)) with styrene and *n*-butyl acrylate (the concentration of compound (III) in these copolymers was 6.5 mass %).

The calculated values [3] of the rate constants for unimolecular decomposition of the peroxide groups of polymeric dialkyl peroxides are of the order of 10^{-5} s^{-1} at 333 K. As the particle size of the resulting copolymer latex (III + *n*-butyl acrylate) increases, the rate constant for the decomposition of the peroxide groups slightly decreases ($k_d = 6.5 \times 10^{-5}$ s^{-1} for $d = 60$ nm; $k_d = 3.7 \times 10^{-5}$ s^{-1} for $d = 193$ nm). The accelerated decomposition cannot be ascribed to the effect of the polymer since in the organic solvent (chlorobenzene) the rate constant for decomposition is comparable to that for low-RMM dialkyl peroxides. The decomposition rate of the dialkyl peroxide can be attributed to the fact that the peroxide groups in the surface layer of the micelle decompose much more rapidly than those in its core. It was found that after consuming the peroxide groups in the surface layer (shell) of the micelle, the decomposition of peroxide groups slows down remarkably [3]. Rather low initiation efficiencies f (for compound II: $f = 0.019$; for the copolymer of n-butyl acrylate with compound III: $f = 0.08$; for the copolymer of styrene with compound III: $f = 0.015$) are probably a result of the high decomposition rate of peroxide groups in the rather small volume of the reaction zone (the surface layer of micelles). A considerable fraction of the primary radicals being formed terminates before their reaction with a monomer starts. In the case of a polymeric initiator-emulsifier, one of the pair of radicals formed is fixed on the surface zone of the micelle, which makes its eventual diffusion from the cage more difficult.

The possible effect of the emulsifier adsorption layer on the propagation steps of emulsion polymerization reported earlier [8, 9] has not been proved directly. The results published more recently [3] show that the adsorption layer on the surface of a polymer-monomer particle of the emulsion polymerization affects the propagation neither in the core of the particle nor in its shell. The particle shell cannot be exchanged for, or identified with, the adsorbed layer of emulsifier on the polymer particle surface. The kinetics of termination depends on the properties of the emulsion polymerization system as a whole, i.e. also on the particle size and on the effect of the colloidal and chemical parameters of the reaction system on the conformational entropy of the propagating macroradicals. The low values of the termination rate constants (10^2–10^4 dm^3 mol^{-1} s^{-1}) result from the higher value of the activation energy for the termination steps of emulsion polymerization in comparison with that for terminations in homogeneous polymerization systems.

Some other compounds were proposed [10] to function as initiator-emulsifiers for emulsion polymerization based on the azobisisobutyronitrile (AIBN) derivatives

$$\begin{array}{c} CH_3 \quad CH_3 \\ | \quad\quad | \\ R-C-N=N-C-R \\ | \quad\quad | \\ CH_3 \quad CH_3 \end{array}$$

R: $-C\begin{array}{l} \diagup N-CO-NH-\!\!\bigcirc\!\!-CH_3 \\ \diagdown O-C_2H_5 \quad\quad NH-CO-NH-(CH_2)_{10}-COO^- \end{array}$

R: $-CH-O-(\!\!-R'-\!\!)_n\,$; $\quad n = 13 \pm 2$

R': $-CH_2-CH_2-O-$; $\quad -CH_2-CH_2-CH_2-O-$

The polymerization behaviour of the surface-active initiator prepared from poly(ethylene glycol) and AIBN according to a published procedure [11] was tested [10] on the system styrene (10 g), water (90 g) and initiator (0.5–4 g). A change in concentration of the surface-active initiator exerts only a slight influence on the latex particle size. It was found, however, that with increasing surface-active initiator concentration, the polymer particle size distribution becomes progressively narrower. The number average RMM of the polymer decreases with increasing concentration of surface-active initiator. The linear dependence of the reciprocal degree of polymerization on the square root of the surface-active initiator concentration indicates that the normal free-radical polymerization mechanism holds. The particle sizes of the final latices are much more strongly influenced by alteration of the monomer concentration than the initiator concentration. The polymerizations were examined under the same conditions as mentioned above, but with a constant surface-active initiator concentration of 2 g per run and variable amounts of monomer (10–50 g) and water (100–60 g). (The total amount of monomer and water was always 110 g).

AIBN-based surface-active initiators, compared with peroxodisulfates, offer the possibility of preparing polymers of substantially higher RMM.

A special merit of surface-active initiators is the possibility of preparing latices with a very low electrolyte content and reduced foam formation [10].

Finally, if these substances should, in addition to their function as initiators, act as stabilizers of the polymer dispersion, they must be added (relative to the initiator concentrations used in classical emulsion polymerization) in rather large amounts. Strengthening of the function of the stabilizer of a dispersion when using 'normal' initiator concentrations can also be obtained by adding normal emulsifiers in amounts needed to achieve good stability of the disperse system.

References

1. Bedford, M. (1982) *Paint Resin* **52** 22
2. Bartoň, J. & Kárpátyová, A. (1987) *Makromol. Chem.* **188** 693
3. Ivanchev, S.S., Pavlyuchenko, V.N. & Byrdina, N.A. (1987) *J. Polym. Sci., Polym. Chem.* **25** 47
4. Ivanchev, S.S., Syrov, A.A., Pavlyuchenko, V.N., Lesnikova, N.N. & Rozhkova, D.A. (1981) *U.S.* 4283553
5. Ivanchev, S.S., Pavlyuchenko, V.N., Pessina, Z.M. & Vasieva, E.D. (1982) *U.S.* 4342676
6. Pavlyuchenko, V.N., Ivanchev, S.S., Rozhkova, D.A., Dikaya, N.N, Domnicheva, N.A. & Budtov, V.P. (1978) *Kolloidn. Zh.* **45** 64
7. Medvedev, S.S. (1968) In: Kargin, V.A. (ed.) *Kinetics and mechanism of the formation and reaction of macromolecules*. Nauka, Moscow, p. 5 (in Russian)
8. Urzhenko, A.I. & Vilshanskii, V.A. (1963) *Dokl. Akad. Nauk SSSR* **148** 1145
9. Urzhenko, A.I. & Vilshanskii, V.A. (1970) *Dokl. Akad. Nauk SSSR* **190** 616
10. Tauer, K., Goebel, K.H., Kosmella, S., Stahler, K. & Neelsen, J. (1990) *Makromol. Chem., Macromol. Symp.* **31** 107
11. Walz, R., Böhmer, B. & Heitz, W. (1977) *Makromol. Chem.* **178** 2527

6.5 POLYMERIZATION IN MONO- AND MULTILAYERS. OTHER SPECIAL CASES OF EMULSION POLYMERIZATION

Radical polymerization is mostly carried out under isotropic conditions, i.e. in a chaotic, random arrangement of the components of the polymerization system. The arrangement of the monomer molecules, i.e. of the basic component needed to form a polymer, is achieved in various ways, the degree of their organization being of course different. Among the known arrangements of monomer molecules we can mention the orientation of a monomer at the liquid-liquid interface [1], the use of thermotropic liquid-crystalline media [2, 3] of micellar systems (see Chapters 3 and 4) and of systems characterized by the formation of mono- and multilayers [4, 5].

Radical polymerization in an organized system tries to mimic natural biological reactions, which usually proceed in organized systems [4]. Moreover, the arrangement of the monomer molecules can influence the kinetics of polymerization, the structure of the polymer, and its microstructure in particular [6]. For these reasons, considerable scientific and technological interests have been directed to the preparation of liquid crystalline polymers and polymer emulsifiers (see section 6.4) and to polymerization in mono- and multilayers. These problems will also be discussed in the following notes.

The polymerized mono- and multilayer aggregates preserve the structure of their starting unpolymerized aggregates and since they are sufficiently stable, are considered to be a good model of biological membranes. The ultimate aim to develop a stable model of a cell has thus been almost successfully realized [7]. The models are used for studying membrane properties [4], transport processes in membranes [5, 8] in human and veterinary medicine as drug carriers [9–11], and in research on artificial photosynthesis [12, 13].

The interfacial surface is the area of contact of two immiscible liquids. The properties of molecules at the interfacial surface differ from those in the continuous phase, the ability to undergo orientation [14, 15] being one of the most important. An amphiphilic molecule forms a monolayer on the interfacial surface as seen in the scheme in Fig. 28.

Fig. 28 — Schematic of monolayer of amphiphilic compound (○ polar moiety, ∧∧∧ nonpolar moiety) at contact of two immiscible phases (e.g. liquid – liquid or liquid – gas).

The polymerization of 4-vinyl-N-methylpyridinium methyl sulfate (I) provides an example of polymerization in a monolayer

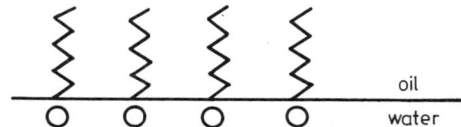   (I)

on the interfacial water/toluene surface [16, 17]. The quaternary nitrogen atom of the monomer is solvated in this case by the aqueous phase. Vigorous stirring causes the formation of toluene and water droplets in individual continuous phases. The polymerization system is depicted in Fig. 29, illustrating monomer molecules aggregated in a cylindrical form and dispersed into a water or toluene phase.

The polymer produced by polymerization at the interfacial surface has a greater number of syndiotactic triads than polymer prepared by isotropic polymerization. This fraction is to some extent distorted by the fact that polymerization is primarily conducted in the continuous aqueous phase (isotropic polymerization) because the monomer used is highly hydrophilic. The formation of micelles by self-organizing monomer molecules into aggregates in the aqueous phase was hindered by use of the monomer concentrations below the critical concentration for the formation of monomer aggregates [18, 19].

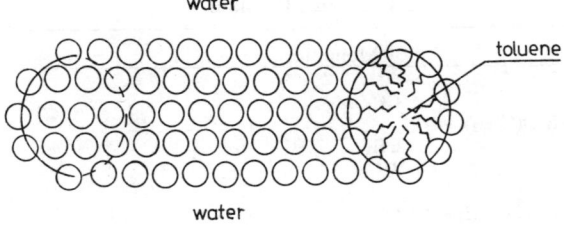
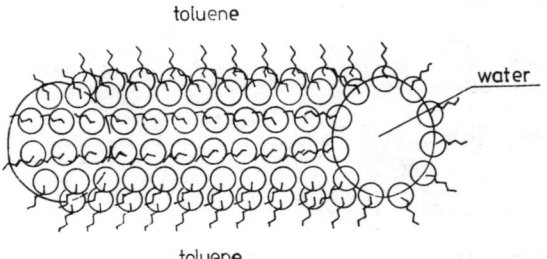

Fig. 29 — Schematic of a cylinder-like micelle in aqueous or toluene continuous phases. ∧∧∧ nonpolar moiety, ◯ polar moiety.

Liposomes [4, 5] and multilayers [4, 5, 20, 21] are smectic mesophases of the closed structure of phospholipids and synthetic surface-active compounds. They form spherical or elliptical multilayer arrangements, where the number of layers can be altered by subjecting them to ultrasound. Single bilayer shells 30 to 50 nm in diameter are called vesicles.

The molecular structural characteristics of vesicle-forming amphiphiles are usually governed by the presence of *two* long alkyl chains (in contrast to the one alkyl needed for micelle formation) and a polar group, namely ammonium, carboxylate, sulfate, sulfonate or phosphate [20, 22]. For the production of multilayer structures, an amphiphile with one [23] or three [24] chains, as well as other compounds with the structure of thermotropic liquid crystals combined with a flexible chain and a polar group [25, 26], can also be used.

The polymerizable double bond of amphiphiles able to form multilayer vesicles does not as a rule perturb either the structure of the amphiphile or its ability to form organized structures. Table 46 summarizes examples of monomers forming multilayer structures (vesicles).

Table 46 — Examples of monomers suitable for formation of polymerized multilayer structures (vesicles)

Monomer	Ref.
$CH_2=CH-C_6H_4-NHCO(CH_2)_{10}-N^+(CH_3)_2(C_{16}H_{33}) \; Br^-$	[5]
$HOOC-(CH_2)_8-C\equiv C-C\equiv C-(CH_2)_8-COOH$	[26]
$CH_2=C(CH_3)-CO-O-(CH_2)_{10}-N^+(CH_3)_2(CH_2-(CH_2)_{14}-CH_3) \; Br^-$	[27]
$CH_3-(CH_2)_9-C\equiv C-C\equiv C-(CH_2)_9-O-P(=O)(OH)-O-(CH_2)_9-C\equiv C-C\equiv C-(CH_2)_9-CH_3$	[28]
$CH_2=CH-(CH_2)_8-CH_2-O-P(=O)(OH)-O-CH_2-(CH_2)_8-CH=CH_2$	[29]
$CH_2=C(CH_3)-CO-(CH_2)_{11}-O-CO-CH_2-CH(O-CO-(CH_2)_{14}CH_3)-CH_2-O-P(=O)(O^-)-O-(CH_2)_2-N^+(CH_3)_3$	[30]
$CH_2=C(CH_3)-CO-O-CH_2-CH(O-CO-(CH_2)_{16}-CH_3)-CH_2-O-CO-(CH_2)_{16}-CH_3$	[26]
$CH_3-(CH_2)_{12}-C\equiv C-C\equiv C-(CH_2)_8-CO-O-CH_2-CH(O-CO-(CH_2)_8-C\equiv C-C\equiv C-(CH_2)_{12}-CH_3)-CH_2-O-P(=O)(O^-)-O-(CH_2)_2-N^+(CH_3)_3$	[31]

The investigation of polymerizable monomer amphiphiles able to form vesicles [5, 26–31] has mainly been necessitated by efforts to obtain aggregates of enhanced stability and controllable size and permeability. The structure and properties of the polymerized multilayer structures depend on the site, number and nature of the groups capable of polymerizing (vinyl, acrylic, diacetylene groups). In some polymerized multilayer structures (vesicles), the mobility of the so-called head-group is preserved, the chain mobility being preserved in the others. A schematic representation of the situation is given in Fig.30.

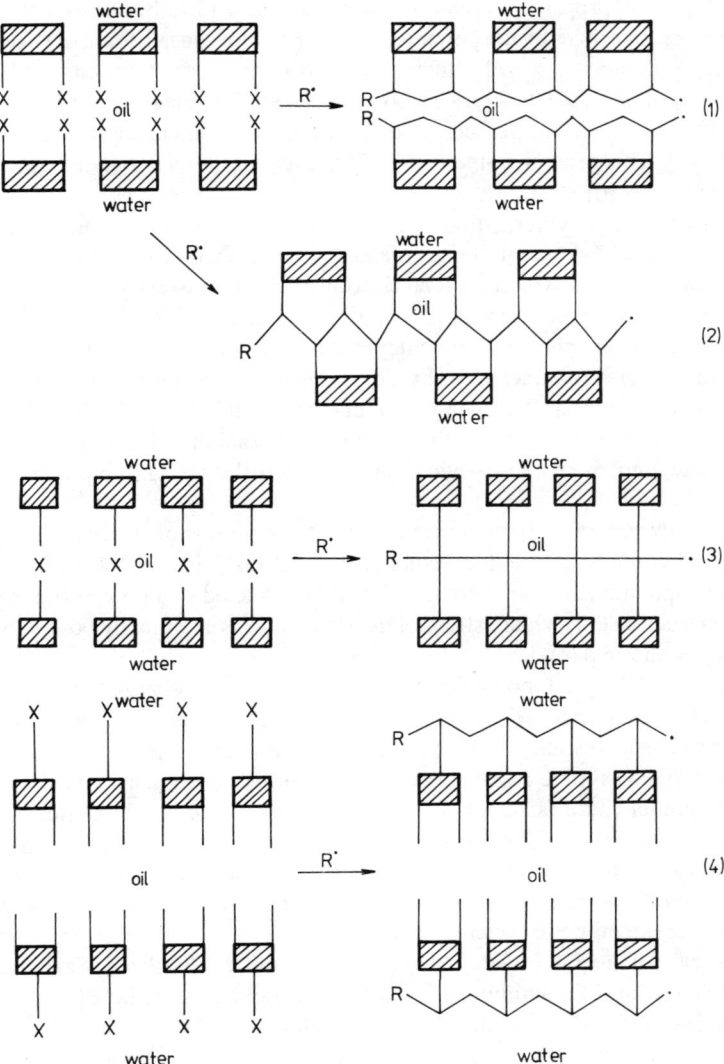

Fig. 30 — Schematic of formation of polymerized multilayers preserving mobility of head group — (1–3) and the chain mobility — (4). x — group containing double or triple bond; ▨ head group of amphiphile, — amphiphile chains

A good, critical literature survey on the polymerization of monomeric vesicles to their polymerized counterparts is due to Paleos [32]. These first investigations were followed by more elaborate studies on the kinetics and mechanisms of polymerization coupled with cooperative characterization of monomeric and polymeric vesicles. One example is the elucidation of the photochemical polymerization of styrene-bearing quaternary ammonium salts and the finding that the polymerized vesicles consist of several polymer emulsifiers assembled to form aggregates [33, 34]. These aggregates are not exclusively formed under topochemical conditions, i.e. from monomers organized into monomeric vesicles, but they may also be formed by sonication of polymeric emulsifiers obtained by the homogeneous polymerization of appropriate vesicle-forming monomers [35]. Studies were also done on the stabilization of vesicles by the polymerization of counterions, leading to vesicles covered by protective polymers [36, 37]. Studies devoted to the formation, characterization and utilization of polymerized vesicles for biological applications have been reviewed by Ringsdorf et al. [38, 39]. Fendler has reviewed the photochemical applications of stabilized vesicles [5, 40]. The general framework of polymerization in organized media has been discussed by Paleos [6].

Research into the polymerization of monomers of amphiphilic character able to form multilayer structures has mainly been oriented to the preparation and characterization of these products. The kinetics of the polymerization process is currently still in the background of scientific activity. For instance, the preparation of an encapsulated hydrophilic poly(acrylamide) polymer in an phospholipid bilayer has been described [41]. The size of the phospholipid bilayers is characterized by a diameter of ca. 650 nm; the bilayer size does not change on polymerization. Addition of the classical emulsifier (TRITON X-100) before polymerization causes a decay of the bilayers (UV-radiation-initiated polymerization). Addition of the emulsifier after polymerization affects the size of the polymer-containing bilayer only slightly.

A recent study reports [42] on the emulsion polymerization of styrene in the presence of a phospholipid containing two double bonds, i.e. 1,2-bis[5-(4-vinylphenyl)pentanoyl]-*sn*-glycero-3-phosphatidylcholine. Here the phospholipid acted as a polymerizable emulsifier (surfomer, cf. section 6.3) which, after polymerization, was chemically bound to the surface of the latex polymeric particles.

The kinetics of the photopolymerization of the negatively charged emulsifier bis[2-(*n*-hexadecanoyloxy)ethyl]methyl (*p*-vinylbenzyl)ammonium chloride in monolayers, bilayer lipid membranes and vesicles were reported by the Fendler group [43]. In sharp contrast to the monolayers, the average degree of polymerization of the emulsifier vesicles was found to be much smaller (by a factor of ca. 50). The formation of much shorter polymers is a consequence of the absence of puckering and of much looser emulsifier packing in vesicles as compared to monolayers. This confirms that molecular packing and organization is an important parameter which influences polymerization-dependent organizational changes in different membrane mimetic systems.

A special case of the polymerization of a vinyl monomer solubilized in a bilayer adsorbed on the particle surface of aluminium(III) oxide was described elsewhere [44]. The adsorbed bilayer consisted of sodium dodecyl sulfate molecules. The polymerization led to the formation of a thin film on the surface of the aluminium(III) oxide particles. The authors introduced the term of 'polymerization in an adsorbed micelle', i.e. in the adsorbed layer on the Al_2O_3 surface and/or the polymerization in a two-dimensional solvent, since the bilayer thickness in which the polymerization proceeds is much smaller than the area given in principle by the surface of Al_2O_3 (see also p. 20).

A special emulsion polymerization is that in a concentrated emulsion. The system is characterized by a high monomer concentration. If the volume ratio of the oil (monomer) and water phase is greater than 0.74 [45–47], the particles of the dispersed phase are separated by a thin film of the continuous phase. This is why the shape and size of the polymer particles are almost unchanged in the concentrated emulsion during polymerization, which is substantially different from the conventional low-concentration emulsion polymerization. This procedure can provide an entirely new possibility for the control of the particle size of polymer dispersion. In the emulsion polymerization of styrene at a ratio of the phases equal to 0.96, the polymerization rate and RMM of the polystyrene were higher than for polymerization in bulk [48].

The copolymerization of styrene (37 cm^3) and methacrylic acid (1 cm^3) was carried out by the concentrated emulsion polymerization method using sodium dodecyl sulfate (0.4 g), AIBN (0.3 g) in the presence of water (6 cm^3). The concentrated emulsion, which has the appearance of a gel, was prepared at room temperature and polymerized at 40°C. The size of the copolymer particles was affected by the internal phase ratio (here 0.93) and was in the range 2000–3000 nm in diameter. The copolymerization rate in the concentrated emulsion is higher relative to the rate of the copolymerization in bulk. The observed significant gel effect can be ascribed to the restriction of monomer mobility in the small droplets which are protected by the stiff emulsifier layers. A plot of polymer conversion versus. internal phase ratio (volume fraction of the dispersed phase) for a given polymerization time gives a curve with a maximum for the internal phase ratio of about 0.82. No significant role for concentrated emulsion polymerization systems compared with bulk polymerization on the composition of the copolymers has been established [49].

The concept of concentrated emulsion polymerization can be used for preparing polymer composites [50]. If the volume fraction of the dispersed phase is very large, say 0.99, the volume of the continuous phase is correspondingly very small. A comonomer containing an initiator constitutes the dispersed phase and a dilute solution of emulsifier in water constitutes the continuous phase. Two concentrated emulsions containing different monomers (e.g. styrene, butyl acrylate, butyl methacrylate) were separately polymerized (up to 5% conversion) and then mechanically mixed. The mixture was subjected to additional polymerization. The microstructure of the polymer composites was investigated by scanning electron microscopy. Mixing of the nonprepolymerized concentrated emulsion led to the formation of almost a copolymer and not to polymer composites.

The basic principle of concentrated emulsion polymerization can also be made use of in inverse emulsion polymerization (see Chapter 4). In inverse polymerization, the continuous phase is usually formed by an alkane, the hydrophilic monomer being normally dispersed by a suitable emulsifier in the aqueous phase. The preparation of poly(acrylamide) latex by concentrated (inverse) emulsion polymerization is described in ref. [48]. In contrast to conventional inverse emulsion polymerization [51], in the preparation of the polymer latex, much smaller amounts of organic solvent are required as is seen from the recipe [48]: acrylamide: 12 g; water: 28 g; Span 80 (sorbitan monooleate): 1 cm^3; decane: 2.5 cm^3; $K_2S_2O_8$: 2.5×10^{-4} g. The monomer dissolved in water is added gradually with vigorous stirring to a reactor containing decane and emulsifier. Before adding the monomer solution into the reactor, an aqueous solution of initiator is injected into the monomer solution. The polymerization is run in an inert atmosphere for about 3 hours. The shape of the conversion curve is similar to that of solution polymerization without any hint of the gel effect. The RMM of polyacrylamide prepared under these conditions is a little lower than that prepared via solution polymerization. The reduction in RMM of poly(acrylamide) prepared in the emulsion system is caused by transfer reactions of the poly(acrylamide) radicals with emulsifier.

The theoretical and practical aspects of concentrated colloidal dispersions and a discussion of modern techniques for examining particle–particle interactions in dispersion systems have been reviewed in a recent article of Goodwin and Ottewill [52].

Plasma initiation can be used to prepare oil-in-water and water-in-oil polymer dispersions [53]. Exposure to 60 s (100 W) and several days' postpolymerization at 25°C is sufficient to initiate the emulsion polymerization of isoprene and to reach almost complete conversion. Emulsion polymerization did not take place, however, when butadiene, styrene, α-methylstyrene, butyl vinyl ether, isobutyl vinyl ether, and vinyl acetate were used. Water-soluble monomers, e.g. sodium methacrylate, 2-hydroxyethyl methacrylate and acrylamide rapidly polymerize under the effect of a plasma in a water-in-oil emulsion.

References

1. Millich, F. & Carracher, C.E., Jr. (eds) (1977) *Interfacial synthesis*, vols I, II. M. Dekker, New York, Basel
2. Barral, E.M. & Johnson, J.F. (1979) *J. Macromol. Sci., Rev. Macromol. Chem.* **17** 137
3. Kelker, H. & Hatz, R. (1980) *Handbook of liquid crystals*. Verlag Chemie, Weinheim
4. Fendler, J.H. (1982) *Membrane mimetic chemistry*. J. Wiley & Sons, New York
5. Fendler, J.H. (1984) *Acc. Chem. Res.* **17** 3
6. Paleos, C.M. (1985) *Chem. Soc. Rev.* **14** 45
7. Mark, H.F. (1981) *Angew. Chem., Int. Ed.* **20** 303
8. McNeil, R. & Thomas, J.K. (1980) *J. Colloid Interface Sci.* **73** 522
9. Gregoriadis, G. & Allison, A.C. (eds) (1980) *Liposomes in biological systems*. Wiley–Interscience, Chichester, New York
10. Ryman, B.E. & Tyrrell, D.A. (1981) In: Campbell, P.N. & Marshall, R.D. (eds) *Essays in biochemistry*. Academic Press, New York, p. 49
11. Gross, L., Ringsdorf, H. & Schupp, H. (1981) *Angew. Chem., Int. Ed.* **20** 305
12. Grätzel, M. (1980) *Ber. Bunsenges. Phys. Chem.* **84** 981
13. Kaneko, M. & Yamada, A. (1984) In: *Advances in polymer science*, vol. 55. Springer-Verlag, Berlin, New York, p. 2
14. Menger, F.M. (1972) *Chem. Soc. Rev.* **1** 229
15. Menger, F.M. (1979) *Pure Appl. Chem.* **51** 99
16. Paleos, C.M. (1977) *J. Polym. Sci., Polym. Lett. Ed.* **15** 535
17. Paleos, C.M., Evangelatos, G.P., Dais, Ph. & Kipouros, G. (1979) *J. Polym. Sci., Polym. Chem. Ed.* **17** 1611
18. Kabanov, V.A. (1967) *Pure Appl. Chem.* **15** 391
19. Morgan, P.W. (1962) *Adv. Chem. Ser.* **34** 191
20. Kunitake, T. (1979) *Macromol. Sci., Chem.* **13** 587
21. Valer, E.W., Murthy, A.K., Rodriguez, B.E. & Zasadzinski, J.A.N. (1989) *Science* **245** 1371
22. Kunitake, T. & Okahata, Y. (1978) *Bull. Chem. Soc. Jpn.* **51** 1872
23. Hargreaves, W.S. & Deamer, D.W. (1978) *Biochemistry* **17** 3759
24. Kunitake, T., Kimizuka, N., Higashi, N. & Nakashima, N. (1984) *J. Am. Chem. Soc.* **106** 1978
25. Kunitake, T. & Okahata, Y. (1980) *J. Am. Chem. Soc.* **102** 549
26. Akimoto, A., Dorn, K., Gross, L., Ringsdorf, H. & Schupp, H. (1981) *Angew. Chem., Int. Ed.* **20** 90
27. Regen, S.L., Czech, B. & Singh, A. (1980) *J. Am. Chem. Soc.* **102** 6639
28. Lopez, E., O'Brien, D.F. & Whitesides, T.H. (1982) *J. Am. Chem. Soc.* **104** 305
29. Paleos, C.M., Christias, C., Evangelatos, G.P. & Dais, Ph. (1982) *J. Polym. Sci., Polym. Chem. Ed.* **20** 2565
30. Regen, S.L., Singh, A., Oehme, G. & Singh, M. (1982) *J. Am. Chem. Soc.* **104** 791
31. Hupfer, E., Ringsdorf, H. & Schupp, H. (1981) *Makromol. Chem.* **182** 247
32. Paleos, C.N. (1990) *J. Macromol. Sci., Rev. Macromol. Chem. Phys.* **30** 379
33. Reed, W., Guterman, L., Tundo, P. & Fendler, J.H. (1984) *J. Am. Chem. Soc.* **106** 1897
34. Serrano, J., Murino, S., Millan, S., Reynoso, R., Facugauchi, L.A., Reed, W., Nome, F., Tundo, P. & Fendler, J.H. (1985) *Macromolecules* **18** 1999
35. Elbert, R., Laschewsky, A. & Ringsdorf, H. (1985) *J. Am. Chem. Soc.* **107** 4134
36. Ringsdorf, H. & Schlarb, B. (1988) *Makromol. Chem.* **189** 299
37. Fukuda, H., Diem, T., Stafely, J., Kezdy, F.J. & Regen, S.L. (1986) *J. Am. Chem. Soc.* **108** 2321
38. Ringsdorf, H., Schlarb, B. & Venzmer, J. (1988) *Angew. Chem., Int. Ed.* **27** 114

39. Bader, H., Dorn, K., Hupfer, B. & Ringsdorf, H. (1985) *Adv. Polym. Sci.* **64** 1
40. Fendler, J.H. (1985) *Isr. J. Chem.* **25** 3
41. Torchilin, V.P., Klibanov, A.L., Ivanov, N.N., Ringsdorf, H. & Schlarb, B. (1987) *Makromol. Chem., Rapid Commun.* **8** 457
42. Yamaguchi, K., Watanabe, S. & Nakahama, S. (1989) *Makromol. Chem.* **190** 1195
43. Rolandi, R., Paradiso, R., Xu, S.Q., Palmer, C. & Fendler, J.H. (1989) *J. Am. Chem. Soc.* **111** 5233
44. Wu, J., Harwell, J.H. & O'Rear, E.A. (1987) *J. Phys. Chem.* **91** 623
45. Lissant, K.J. (1966) *J. Colloid Interface Sci.* **22** 492
46. Princen, H.M. (1979) *J. Colloid Interface Sci.* **71** 55
47. Princen, H.M. (1983) *J. Colloid Interface Sci.* **91** 160
48. Kim, K.J. & Ruckenstein, E. (1988) *Makromol. Chem., Rapid Commun.* **9** 285
49. Ruckenstein, E. & Kim, K.J. (1989) *J. Polym. Sci., Polym. Chem. Ed.* **27** 4375
50. Ruckenstein, E. & Park, J.S. (1990) *Polymer* **31** 2397
51. Dimonie, M.V., Boghina, C.M., Marinescu, N.N., Marinescu, M.M., Cincu, C.I. & Oprescu, C.G. (1982) *Eur. Polym. J.* **18** 639
52. Goodwin, J.W. & Ottewill, R.H. (1991) *J. Chem. Soc., Faraday Trans.1* **87** 357
53. Osada, Y., Takase, M. & Mizumoto, A. (1984) *J. Chem. Soc. Jpn., Chem. Ind. Chem.* p. 1685

7
Suspension polymerization

7.1 MECHANISM FOR THE FORMATION AND STABILIZATION OF POLYMER PARTICLES IN A SUSPENSION

Several decades have elapsed since the first description of suspension polymerization by Bauer & Lauth [1]. The technology of the preparation of polymers by suspension polymerization has improved remarkably and nowadays is widely applied, e.g. in the preparation of poly(styrene), poly(vinyl chloride), poly(tetrafluoroethylene), poly(methyl methacrylate), poly(ethylene), etc.

A brief description of the process of suspension polymerization appeared as early as 1930 [1, 2]. The authors characterized suspension polymerization as a very unstable process which produces a polymer. Several years later, Crawford [2, 3] and Rohm & Trommsdorff [4] contributed fundamentally to a better understanding of the mechanism for suspension polymerization, clarifying the roles of the continuous aqueous phase, the dispersed monomer droplets, initiator, and stabilizer in the reaction system.

Suspension polymerization uses an aqueous phase where the immiscible monomer or monomers are dispersed by agitation. If the agitation is stopped, the phases often separate the less dense rising to the top of the reactor. The process is also referred to as bead pearl or granule polymerization because of the appearance of the final product. The polymerization takes place entirely in the monomer phase. No mass transfer occurs between the monomer and the aqueous phase. The initiators used are soluble in the monomer phase and are the same as those used in bulk and solution polymerization. The volume ratio of the continuous aqueous phase to the dispersed organic phase [5] varies from 1 : 1 to 6 : 1. Higher ratios are required in rapid polymerizations, where the heat is removed quickly. Suspension polymerization is one of the most important unit processes in the polymer industry, its major advantages lying in the relatively large heat transfer out of the reacting monomer droplets and the ability to control the size of the polymer beads produced. In the practical operation of suspension polymerization, the most important issue is how to control the mean size and the size distribution of the final polymer beads. It is imperative to produce polymer particles of the desired and uniform size. This aim is achieved as follows:

(i) by using the stabilizing effect of inorganic and polymeric species for the suspension polymerizing systems [6–9],

(ii) by forming monomer droplets of uniform diameter at the beginning of the polymerization [10–13],
(iii) by the effect of the geometries of the reactor [14–17] and impeller [6, 18, 19].

The first method is based on controlling the coalescence and break-up of the monomer droplets during polymerization by the addition of many types of chemical species, because the size distribution and mean size of the final polymer particles are functions of the coalescence and break-up of the monomer droplets or monomer/polymer particles.

The second approach is based on completing the polymerization in individual droplets which have a uniform diameter and no history of coalescence and break-up throughout the polymerization process.

The third method is based on controlling the dispersing behaviour of the droplets such as coalescence and break-up by modifying the geometry of the reactor and impeller.

The stabilizer, which is adsorbed on the surface of the monomer droplets and suppresses coalescence of the monomer and polymer particles, plays a particular role in the system. High-RMM water-soluble polymers are most frequently used as stabilizers for polymer suspensions, e.g., gelatine, methyl cellulose, poly(vinyl alcohol), poly(methacrylamide), polyacrylic and salts of poly(methacrylic acid) [20]. Water-soluble copolymers of high RMM are also widely used, e.g., the copolymer of styrene and maleic anhydride. Some salts, e.g., NaCl, also play a partial role in stabilization since they reduce the interfacial tension and water solubility of the monomer and, conversely, they increase the density of the continuous phase. Aluminium hydroxide, barium sulfate, magnesium carbonate, manganese phosphate and kaolin have also found application [21].

Suspension polymerization is, like emulsion polymerization, a reaction which proceeds in the disperse medium. When using a stabilizer in suspension polymerization, the monomer is emulsified by vigorous stirring into the aqueous phase. The polymerization starts, as with homogeneous polymerizations, where an oil-soluble initiator decomposes into initiating radicals. Initiation, propagation and termination proceed in monomer droplets, obeying the laws of block radical polymerization [22]. The stabilizer is a decisive factor, affecting the dispersion of monomer in water, particle formation and the stability of the polymer suspension (as in emulsion polymerization).

The emulsified monomer droplets function as small block reactors, in which the polymerization takes place via radical homogeneous polymerization with the difference that the removal of heat is easily controlled and does not influence the course of polymerization even at high conversions. The aqueous phase surrounding the polymer particles effectively removes the heat of reaction and keeps the temperature constant in the critical phases of the polymerization. The advantages of suspension polymerization also include the low viscosity of the final dispersion over the range of conversions and the use of the cheap nontoxic continuous reaction phase. However, the process has a disadvantage, namely contamination of the final polymer with the stabilizer which, after evaporation, becomes a component of the polymer and adversely affects the product quality. Purification of the product to remove the stabilizer is energetically costly and involves washing of the products with water, organic solvents and drying.

At low conversions, the monomer particles are subject to coalescence, their reverse dispersion into smaller particles being possible. The polymerization conditions, especially stirring, influence the average particle size and distribution at this point and finally also the properties of the polymer suspension. The stage of medium conversion is characterized by increased stickiness of the polymer particles and their greater tendency to coalesce [23]. The coalescence of these particles is irreversible: neither addition of extra stabilizer nor more vigorous stirring lead to dispersion of the associates. As the reaction reaches higher conversions, the polymer particles lose their stickiness and are no longer vulnerable to coalescence.

Poly(vinyl acetate) is a classical representative of the group of stabilizers formed by water-soluble polymers. After dispersing the monomer phase in water, the partially hydrolysed poly(vinyl acetate) is adsorbed onto the monomer particle surface, and the particles then change gradually into polymer particles. It is localized on the surface such that the terminal acetate groups contact the surface of the oil or polymer phase, while the hydroxyl groups are oriented outwards from the interphase into the water (see Fig. 31). The stabilizer, being spread over the particle surface, creates a protective membrane which prevents coalescence of the particles during collision [22]. In this situation, the polymer particles do not coalesce if the repulsive forces prevail over the attractive (Fig. 32). Collisions of these

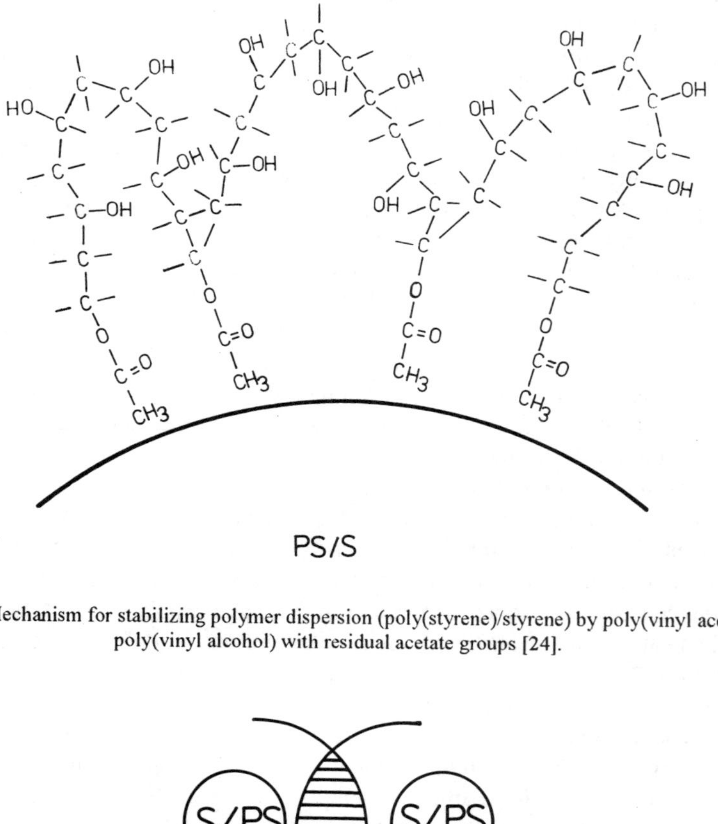

Fig. 31 — Mechanism for stabilizing polymer dispersion (poly(styrene)/styrene) by poly(vinyl acetate) or poly(vinyl alcohol) with residual acetate groups [24].

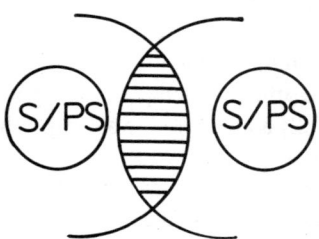

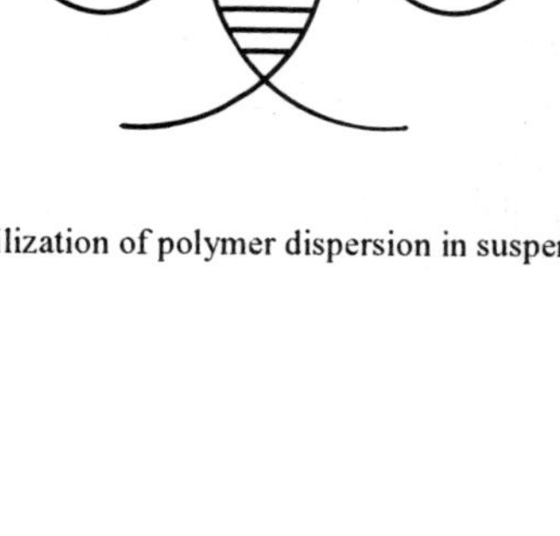

Fig. 32 — Model of steric stabilization of polymer dispersion in suspension polymerization of styrene.

stable particles are accompanied only by overlap of their surface layers; in the zone of mutual overlap the osmotic pressure increases (owing to the flow of water into these regions), which gradually separates the particles one from another [25, 26]. The rate of separation of the suspension particles, whose surface layers mutually overlap after collision, is not only a function of the osmotic pressure [27] but also of the forces of mutual interaction of the stabilizer molecules and of interactions between the stabilizer molecules and polymer or monomer.

Colloidal stability is important in suspension polymerization [28–30]. For colloidal suspensions to be stable, it is essential to provide a repulsive barrier between the particles so that the London and van der Waals attractive energies do not exceed the energy of thermal motion which is $(3/2)\,kT$. Suspensions of neutral colloidal particles flocculate rapidly as a result of the long-range attractive forces. In particle suspensions, the total potential is the sum of the energies of attraction and repulsion. When the potential energy maximum is quite large as compared to the thermal energy, then the corresponding suspensions show long-term stability.

In order to achieve a potential energy maximum, it is necessary to provide a repulsive potential energy between the particles. In suspension polymerization, this is achieved mainly by steric stabilization. If the stabilizer has ionic groups, then electrostatic stabilization can take place.

Interparticle repulsion due to Coulombic forces between two particles is a function of the dielectric constant of the continuous medium. The surface potential develops by adsorption of the relevant ions. If the particles approach a point where the attractive forces overcome the electrostatic forces, then flocculation will occur.

The interpenetration of particles due to Brownian motion results in compression of a nonionic hydrocolloid polymer or its chain segments. The compression produces a change in the free energy, given by the Gibbs–Helmholtz equation:

$$\Delta G_R = \Delta H_R - T \Delta S_R. \tag{1}$$

There are three possibilities regarding the change of free energy: (i) $\Delta G_R = 0$, if no nonionic hydrocolloid is present; (ii) $\Delta G_R = +$, the hydrocolloid stabilizes the particles and does not sensitize them; (iii) $\Delta G_R = -$, the hydrocolloid sensitizes the particles and destabilizes them. In suspension polymerization a positive ΔG_R is important for stability during polymerization. However, a less positive or negative ΔG_R is needed for recovery of the polymer particles and better coagulation.

Some hydrocolloids such as acidic and nonionic polysaccharides, have the advantage that they can stabilize the colloidal system even at relatively low concentrations [31]; such stabilization is due to steric repulsion force.

Yokoyama et al. [32] have found the stability of colloidal dispersions can be controlled by changing the pH. Photon correlation spectroscopy studies have shown that the steric layer thickness increases as the pH decreases. At high ionic strength, the amount of nonionic polysaccharide adsorbed does not change with pH. However, as the pH is lowered, the steric layer thickness expands from 16 to 70 nm. This suggests that a nonionic polysaccharide changes its conformation on the surface because of the formation of loops and tails at lower pH.

By applying high concentrations of stabilizer, mostly stable polymer emulsions are prepared but the quality of the polymer is reduced and, simultaneously, the contribution of emulsion polymerization increases. If lower stabilizer concentrations are used, a higher degree of coalescence (limited coalescence) is observed, which is used for preparing polymer dispersions with larger particles [33].

Inorganic compounds insoluble either in the organic or inorganic phases are effective stabilizers of the suspension in the presence of a low-RMM tenside. The emulsification of $BaSO_4$, styrene and water gives a three-phase dispersion. Addition of a small amount of a low-RMM tenside changes the surface properties of the small particles of barium sulfate by adsorption on their surface [23]. The inorganic substrate modified in that way positively influences the stability of the monomer or polymer particles in suspension.

The mutual interactions of the particles (or their surface layers) cause an increase in the viscosity of the polymer dispersion [34–38]. Increase in the concentration of the stabilizer (low-RMM tensides and high-RMM polymers) produces an increase in viscosity of the system and, in parallel, the stability of the suspension. The increase in stability follows from the formation of an efficient polar protective layer on the particle surface.

At a certain concentration of stabilizer, the mean diameters shift towards larger values and no polymer droplets grow in size. This means that there is a critical concentration of stabilizer at which no coalescence between the monomer droplets or monomer/polymer particles takes place. The critical concentration and the critical surface coverage of the stabilizer, C_{cr} and S_{cr}, and the mean droplet diameter can be related as [14]

$$C_{cr} = \frac{6\Phi}{(1-\Phi)d} S_{cr}, \qquad (2)$$

where Φ is the volume fraction of the dispersed phase.

Tanaka & Hosogai [15] conducted suspension polymerization of styrene using calcium triphosphate as stabilizer in a circular loop reactor and reported that $C_c = 1.6 \times 10^{-6}$ kg m^{-2}.

Leng & Quarderer [6] carried out the suspension polymerization of styrene in the presence of poly(vinyl alcohol) as stabilizer using a conventional stirred tank reactor and reported that $S_{cr} = 8 \times 10^3$ m^2 kg^{-1}.

Borwankar et al. [39] found that polymeric suspension stabilizers, when used in sufficient quantities, can stabilize the dispersions against coalescence. The critical surface coverage of hydroxypropyl methyl cellulose was found to be between 2.3×10^{-6} and 1.2×10^{-5} kg m^{-2}. When concentrations of stabilizer used are such that the surface coverage is above S_{cr}, the oil suspension shows no tendency towards coalescence when the stirring speed is much reduced. By using high concentrations of suspending agents so that the surface coverage exceeds S_{cr}, it is possible to study the breakage mechanism even at high volume fractions of the dispersed phase without further complications arising from coalescence. Study of the breakage mechanism reveals that the volume swept by the turbine impeller is applicable in calculating the power input per unit mass. The correlation between the Sauter mean diameter and the degree of agitation under turbulent conditions was found to be:

$$d_{32} = C_1 (1 + C_2 \Phi)(\delta/\rho_c)^{3/5} N_{IP} d_{IP}^{4/5}, \qquad (3)$$

where $C_1 = 0.053$, $C_2 = 4.6$, δ is the interfacial tension, ρ_c is the density of the continuous phase, N_{IP} is the impeller sped, and d_{IP} is the impeller diameter.

The course of suspension polymerization and the properties of the polymer suspension are controlled by the character and intensity of stirring; this affects the average particle size and size distribution and the surface properties of the particles [40–42]. The rate and duration

Sec. 7.1]   Formation and stabilization of polymer particles   295

of stirring enable the preparation of monomer emulsion with a specific monomer particle size. After the completion of stirring, the particles (in the absence of the stabilizer) rapidly associate into the separating oil phase. Addition of the stabilizer suppresses association of the particles and lowers the rate of coalescence. The size of the final polymer particles depends on the size of the monomer droplets formed by dispersing the oil phase in water in the presence of the stabilizer. The influence of stirring on the mechanism for the formation and stability of the polymer dispersion is rather complex. At lower stirring rates, an increase in the intensity of stirring usually leads to a decrease of the average size of the polymer particles. On exceeding a certain critical rate of stirring, no further increase in the number of particles is observed, but the association of the particles is initiated which causes an increase in particle size [43]. Under these conditions, a larger number of smaller particles with a much greater surface area are produced: a higher concentration of stabilizer is needed to stabilize them. If the system does not contain the required amount of stabilizer, the particles coagulate and a dispersion is formed with a larger particle size (see Fig. 33).

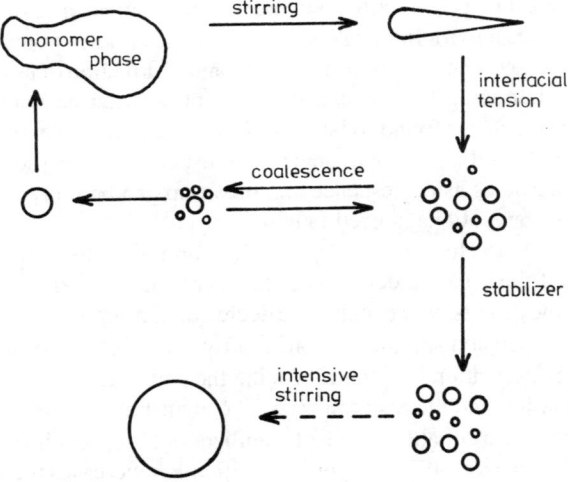

Fig. 33 — Dispersion and coalescence of monomer droplets [37].

The size of the polymer particles is also affected by the reaction conditions, the type of stirrer, and the dimensions of the reactor. Its dependence on these parameters is expressed by semiempirical relationships, e.g. [44, 45]

$$d/d_R = 0.01 \, Re^{0.3} \, Fr^{0.19} \, We^{-1.19} \, (d_R/d_i)^{1.5}, \qquad (4)$$

where d is the particle diameter (cm), d_R is the diameter of reactor (cm), d_i is the impeller diameter (cm), and Re, Fr, and We are the Reynolds, Froud, and Weber numbers, respectively.

Lešek & Eichler [46, 47] followed the suspension copolymerization of methacrylic acid and divinylbenzene and proposed the following semiempirical expression for the dependence of the average size of the polymer particles on the independent parameters

$$d/d_R = 1783\ We^{0.38}\ Re^{0.57}\ Fr^{1.0}. \qquad (5)$$

Matějíček et al. [17] proposed an expression for calculating the variation of the average size of the polymer particles with independent parameters

$$d/d_i = 5.22 \times 10^{-5}\ We^{-0.29}\ Re^{0.24}\ Eu\ (d_i^3/V)^{-1.53}, \qquad (6)$$

where Eu is the Euler number and V is the overall volume of the reaction mixture.

By investigating the influence of the reactor size on the suspension copolymerization of ethyl acrylate, acrylonitrile and divinylbenzene, it was found [17]:

(i) with increasing diameter of the reactor, the width of the polymer particle distribution decreases and
(ii) the amount of coagulate increases with increasing reactor size.

These types of equations give a fairly good prediction of droplet size on the initial mixing. However, they do not account for limiting coalescence, which is an extremely important factor at lower levels of stabilizer. The adsorption behaviour and distribution of the stabilizer are other important factors in determining the bead size which are not taken into account in these equations.

The circular loop reactor is found to be superior for the production of polymer particles of more uniform size [15]. In the loop reactor such dispersing processes of the droplets or monomer/polymer particles as coalescence and break-up can be more easily controlled in comparison with the conventional stirred-tank reactor [14].

The dispersity is used to elucidate the degree of uniformity of the droplet diameters and is defined as the ratio of the standard deviation to the mean diameter, σ/d. This means that the smaller the value of the dispersity, the higher the degree of uniformity of the droplet diameter.

Investigation of suspension polymerization is partly made difficult by the greater or lesser role of emulsion polymerization [48] in influencing the particle size and its distribution, the RMM and its distribution and the resulting rate. The contribution of emulsion polymerization can be suppressed by a suitable choice of stabilizer or by lipophilic modification of the water-soluble stabilizers used [49]. Lipophilic modification increases the surface activity of the stabilizer. The diffusion of the monomer and radical fragments into water and latex particles decrease in parallel. A measure of the retardation of the diffusion of monomer, initiator and other low-RMM fragments into water is proportional to the effective thickness and density of the stabilizer molecules adsorbed on the particle surface. The thickness of the layer is positively influenced by the RMM of the stabilizer and its interaction with monomer and polymer molecules. These effects promoting suspension polymerization are augmented by lipophilic modification of the water-soluble polymer.

The use of lipophilically-modified hydroxyethyl cellulose as stabilizers in suspension polymerization of vinyl monomers has also been studied [50]. This new class of water-soluble polymers showed some distinct advantages over conventional water-soluble polymers used as stabilizers. For comparison, polymerization runs were also conducted with unmodified hydroxyethyl cellulose precursors as stabilizers. The amount of latex by-product was much less than 1% compared to about 5–10% latex formed with unmodified hydroxyethyl cellulose precursors. Increased levels of stabilizer gave higher levels of latex. However, this increase was insignificant in the case of modified hydroxyethyl cellulose. On the other hand, unmodified hydroxyethyl cellulose showed significant increase in the level of latex formation with increasing concentration. Methyl methacrylate, which is more soluble in water than styrene, gave higher levels of latex particles compared with styrene when polymerized with unmodified hydroxyethyl cellulose. Modified hydroxyethyl cellulose showed the opposite trend. This is more likely due to differences in the adsorption behaviour of modified hydroxyethyl cellulose on these two substrates. Similar reduction in the amount of latex particle formation during suspension polymerization was found with modified poly(ethylene oxide) as stabilizer.

The modification of hydroxyethyl cellulose also suppresses, for example, the diffusion of styrene through the interphase formed by the modified stabilizer of hydroxyethyl cellulose; this was indirectly confirmed by the observation that tetrahydrofuran was a poor swelling agent for poly(styrene) particles covered by modified hydroxyethyl cellulose [51]. Intensive dispersion of the monomer into the aqueous phase leads to the formation of a fine monomer dispersion, characterized by readier diffusion of the monomer into water, which increases the contribution of emulsion polymerization. Easier diffusion of the monomer into water is observed in the presence of small, oil-swollen polymer particles [52].

References

1. Bauer, W. & Lauth, H. (1942) *Ger.* 656134 (Röhm und Haas, Darmstadt)
2. Crawford, J.C.W. (1942) *Brit.* 427494 (ICI)
3. Crawford, J.C.W. (1940) *U.S.* 2108044
4. Rohm, O. & Trommsdorff, E. (1943) *Ger.* 735284 (Röhm und Haas, Darmstadt)
5. Munzer, M. & Trommsdorff, E. (1977) In: Schildknecht, C.E. & Skeist, I. (eds) *Polymerization processes*, vol. 29, *High polymers*. J. Wiley & Sons, New York
6. Leng, D.E. & Quarderer, G.J. (1982) *Chem. Eng. Sci.* **14** 177
7. Konno, M., Arai, K. & Saito, S. (1982) *J. Chem. Eng. Jpn.* **15** 131
8. Tanaka, M. & Oshima, E. (1982) *Kagaku Kogaku Ronbunshu* **8** 734
9. Tanaka, M. & Morishima, T. (1986) *Kagaku Kogaku Ronbunshu* **12** 231
10. Edward, A.C. (1982) *U.S.* 57109905
11. Noguchi, H., Noda, S. & Baba, H. (1983) *Japan* 5891701
12. Noguchi, H., Morishita, T. & Baba, H. (1986) *Japan* 61115902
13. Mikos, A.G., Takoudis, C.G. & Peppas, N.A. (1986) *J. Appl. Polym. Sci.* **31** 2647
14. Tanaka, M. & Oshima, E. (1988) *Can. J. Chem. Eng.* **66** 29
15. Tanaka, M. & Hosogai, K. (1990) *J. Appl. Polym. Sci.* **39** 955
16. Langner, F., Moritz, H.U. & Reichert, K.H. (1979) *Ger. Chem. Eng.* **2** 329
17. Matejíček, A., Seidl, J. & Musil, V. (1984) *Angew. Makromol. Chem.* **126** 177
18. Schröder, R. & Piotrowski, B. (1982) *Ger. Chem. Eng.* **5** 139
19. Hoff, H., Lüssi, H. & Hammer, E. (1965) *Makromol. Chem.* **84** 286
20. Crawford, J.C.W. & McGrath, J. (1940) *U.S.* 2191520
21. Rohm, O. & Trommsdorff, E. (1943) *Ger.* 735285
22. Reinhart, H.J. & Thiele, R. (1972) *Plaste Kautsch.* **19** 648
23. Wenning, H. (1956) *Makromol. Chem.* **20** 196
24. Wenning, H. (1958) *Kunstst. Plast.* **5** 328
25. Napper, D.H. (1970) *Ind. Eng. Chem., Prod. Res. Dev.* **9** 467
26. Napper, D.H. (1977) *J. Colloid Interface Sci.* **58** 390
27. Napper, D.H. (1983) *Polymeric stabilization of colloidal dispersions*. Academic Press, London
28. Maadhah, A.G., Amin, M.B. & Usmani, A.M. (1985) *Polym. Bull.* **14** 433
29. Amin, M.B. & Usmani, A.M. (1984) *Polymer. Mater., Sci. Eng.* **51** 660
30. Usmani, A.M. & Salyer, I.O. (1981) *J. Elastomers Plast.* **13** 90
31. Bergenstahl, B., Fogler, H.S. & Stenius, P. (1985) In: Phillips, G.O., Wedlock, D.J. & Williams, P.A. (eds) *Gums and stabilizers for the food industry*, vol. 3. Elsevier, New York, p. 285
32. Yokoyama, A., Srinivasan, K.R. & Fogler, H.S. (1988) *J. Colloid Interface Sci.* **126** 141
33. Wiley, R.M. (1954) *J. Colloid Sci.* **9** 427
34. Davidson, J.A. & Winterhafer, D.E. (1980) *J. Polym. Sci., Polym. Phys. Ed.* **18** 51
35. Mark, M. & Hohenstein, W.P. (1946) *J. Polym. Sci.* **1** 127
36. Schildknecht, C.E. (1957) *Polymer processes*. Interscience, New York, p. 71

37. Winslow, F.H. & Matreyek, W. (1951) *Ind. Eng. Chem.* **43** 1108
38. Black, W. (1968) *Chem. Ind.* (London) **1** 450
39. Borwankar, R.P., Chung, S.I. & Wasan, D.T. (1986) *J. Appl. Polym. Sci.* **32** 5749
40. Hoff, H., Lüssi, H. & Gerspacher, P. (1964) *Makromol. Chem.* **78** 37
41. Hoff, H., Lüssi, H. & Hammer, E. (1965) *Makromol. Chem.* **82** 175
42. Hoff, H., Lüssi, H. & Hammer, E. (1965) *Makromol. Chem.* **84** 274
43. Kelsall, D.G. & Maitland, G.C. (1983) In: Reichert, K.H. & Geiseler, W. (eds) *Polymer reactions engineering.* C. Hanser Verlag, München, p. 131
44. Hill, B.A. (1964) *Br. Chem. Eng.* **6** 104
45. Kundel, N. (1979) M.S. *Thesis.* Akron University, Akron
46. Lešek, F. & Eichler, J. (1975) *Chem.-Ing.-Tech.* **47** 855
47. Lešek, F. & Eichler, J. (1975) *Chem. Prům.* **25** 299
48. Maclay, W.N. (1971) *J. Appl. Polym. Sci.* **15** 867
49. Landoll, L.M. (1982) *J. Polym. Sci., Polym. Chem. Ed.* **20** 443
50. Ahmed, S.M. (1984) *J. Dispersion Sci., Technol.* **5** 421
51. Liang, S. & Fitch, R.M. (1982) *J. Colloid Interface Sci.* **90** 1
52. Higuchi, W.I. & Misra, J. (1962) *J. Pharm. Sci.* **51** 454

7.2 SUSPENSION POLYMERIZATION AND COPOLYMERIZATION OF UNSATURATED MONOMERS

7.2.1 Styrene

The suspension polymerization of styrene is carried out in the presence of protective colloids, e.g. poly(vinyl acetate), polyvinylpyrrolidone, poly(vinyl acetamide), cellulose derivatives and powder colloids such as the phosphates of alkaline earth metals [1–7]. Conventional types of oil-soluble peroxide, e.g., dibenzoyl peroxide, *tert*-butyl perbenzoate, di-*tert*-butyl peroxide, and azo compounds like AIBN are used. The rate of polymerization is a function of the number of particles, of the stabilizer and initiator concentrations, and the temperature. The particle size generally increases with increasing conversion. The particle size is strongly influenced by the concentration of stabilizer. At high concentrations of stabilizer (above 3–5 mass % related to monomer), growth of the polymer particles stops already at lower conversions. At normal stabilizer concentrations (3 mass %), an inflexion point is observed at the end of particle growth at conversions of *ca.* 70% (see Fig. 34). At very low stabilizer concentrations, the volume of the particles increases throughout the conversion range.

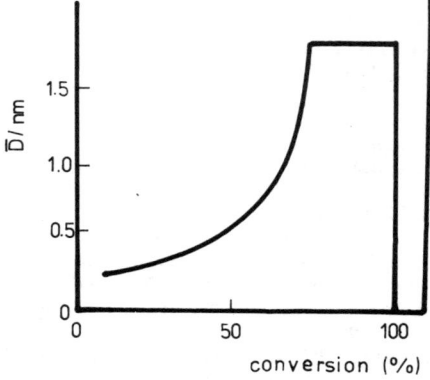

Fig. 34 — Polymer particle growth in suspension polymerization of styrene [10].

A study of the suspension polymerization of divinylbenzene has shown that the size of the polymer particles depends on the RMM of the stabilizer [10]. As the RMM increases, the average size of the particles formed shifts to lower values. The partial hydrolysis of poly(vinyl acetate) leads to the formation of larger particles; the increase is proportional to the fraction of hydroxyl groups in the stabilizer molecule. As the content of hydroxyl groups increases, the lipophilic character of the stabilizer and the strength of interaction between the stabilizer chains and the polymer in the interphase decreases.

As with emulsion polymerization, increase in the stabilizer concentration leads to rate enhancement of suspension polymerization; this is confirmed by the suspension polymerization of styrene and dichlorostyrene in the presence of gelatine or poly(vinyl alcohol) [11].

In suspension polymerizations of styrene [12] and styrene/divinylbenzene [13], the effect of the volume fraction of the dispersed phase on the particle size distribution has been studied at relatively high volume fractions ($\Phi > 0.1$) and high stabilizer concentrations. The authors found that increase in the volume fraction results in larger particles.

Apostolidou & Stamatoudis [14] have investigated the particle size distribution at lower volume fractions of the dispersed phase ($\Phi \leq 0.1$) and low stabilizer concentrations. They found that the poly(styrene) particle size

(i) decreased with increasing stabilizer concentration,
(ii) is affected by the type of impeller used (the pitched open-type impeller is the least effective, thus giving the largest particles),
(iii) is affected by the impeller speed, giving larger sizes at higher speeds, and
(iv) is affected by the volume fraction of the dispersed phase, giving larger sizes at higher volume fractions.

The particle size increased with increasing volume fractions of the monomer or polymer phases regardless of the type of impeller or the stabilizer concentration.

Tanaka & Hosogai [15] have found that the transient droplet diameter distributions and the mean particle diameter shift to larger diameters with reaction time, volume fraction of the dispersed phase and impeller speed. By contrast, on increasing the stabilizer concentration the mean particle size decreases and the particle distribution shifts towards the small particle size. For example, at stabilizer concentration $C_s = 0.6$ mass % (for the aqueous phase), no shift in the diameter size or distribution is observed.

As regards the effect of the impeller speed, the mean diameters are nearly constant in the early stage of polymerization and then start to grow due to coalescence. The growth of monomer/polymer particles due to coalescence becomes very marked at the middle stage of polymerization especially at higher volume fractions of the dispersed phase and higher impeller speeds. An expression correlating the mean size of the polymer particle (d) with the impeller speed (N_{IP}), the volume fraction of the dispersed phase (Φ) and the stabilizer concentration (C_s) was derived, i.e.

$$\ln d = \ln A' + a \ln N_{IP} + 0.3 \ln \Phi - 1.0 \ln C_s , \tag{7}$$

$$A' = 6.5 \times 10^{-3}, a = -0.6 \text{ if } 15 \leq N_{IP} < 30 \text{ s}^{-1},$$

$A' = 9.8 \times 10^{-3}$, $a = 0.3$ if $30 < N_{IP} \leq 50$ s^{-1},

where a circular loop reactor was used.

Similar results, i.e. the shift of the particle distribution and the growth of polymer particles with impeller speed, volume fraction of the dispersed phase and stabilizer concentration were reported by Konno et al. [1] and Hirose and Oshima [9] who used a conventional stirred-tank reactor.

In their study of the suspension polymerization of styrene, chlorostyrene and divinylbenzene, Balakrishnan & Ford [7] have observed that the polymerization process is markedly affected by the ratio between the oil and aqueous phases, the rate of stirring, and the presence of a tenside. As the monomer fraction decreases, the average particle size also decreases and the bimodal particle distribution becomes more marked. Increase in the stirring rate shifts the average particle size to lower values. No notable change in the size or particle size distribution was visible over a wide range of concentrations of stabilizer (gelatine and poly(dimethyl ammonium chloride); about 50%). Addition of the anionic emulsifier, sodium dodecylbenzene sulfonate, in concentrations equal to or below the CMC influenced considerably the average size of the particles and their distribution: it suppressed the production of the large particles of 250 μm diameter (formed mostly in the absence of this emulsifier) and positively affected the formation of polymer particles with a diameter of 70 μm. The nonionic emulsifier (Triton X-100) showed no significant effect on the suspension polymerization. The influence of an anionic emulsifier on suspension polymerization was determined elsewhere [17]. Discussion has shown that the anionic emulsifier increases the role of emulsion polymerization which is seen as suppressing the formation of larger particles and initiating the production of smaller particles.

The influence of oil-soluble initiators on the suspension polymerization of styrene was examined by Hohenstein et al. [18]. These authors evaluated the efficiency of the commonly used peroxides and came to the conclusion that dibenzoyl peroxide was the most effective initiator of the suspension polymerization of styrene. Its rate of polymerization was found to be approximately proportional to the square root of the initiator concentration. The rates and RMMs were higher in suspension polymerization than in the corresponding homogeneous system.

The specific effect of the interphase formed by the adsorbed stabilizer was applied in the suspension copolymerization of styrene with methyl methacrylate initiated by peroxides or azo initiators [19]. When no additives were present, the suspension polymerization proceeded normally.

The addition of soot did not influence the suspension copolymerization of styrene and methacrylates, however, it suppressed the initiating effect of the peroxides completely.

Bakshaee et al. [20, 21] prepared carbon-black polymer particles by the suspension copolymerization of styrene and n-butyl methacrylate. The presence of carbon during polymerization inhibits the free-radical process and the RMMs of the copolymer are generally reduced and their distribution broadened. As a consequence, the glass transition, T_g, is also reduced. The polymer particles of ca. 10 μm diameter were used successfully as toner particles. The polymer particles were impregnated with carbon black during their formation from the liquid comonomers. This procedure eliminates the costly technology associated with conventional melt-mix processes. The triboelectric coefficients of such toners are generally lower than those of conventially-produced materials, as a result of the different nature of the carbon-black suspension prepared. The coefficients pass through a maximum value as the carbon-black content is increased and good correlation with conventionally-produced materials can be achieved in copolymerizations that are made highly alkaline.

Peppas et al. [22] have developed a reaction engineering model for the batch-suspension polymerization of monomers that are either good solvents for their polymers or good swelling agents for their crosslinked polymers, in order to prevent phase separation during polymerization. For example, it can be used to calculate the particle size distribution of crosslinked poly(styrene) beads produced in suspension polymerization. However, it is inapplicable to polymerizations where the separation of phases occurs, as with vinyl chloride.

The particle size distribution can be evaluated from the droplet mass distribution, as shown in the following equation, assuming that the unreacted monomer swells the polymer produced

$$g(v,t) = \frac{\rho_p}{m_s} \bar{f} \ (\rho_p \ v/m_s, \ 10^{-3} \ \lambda \ t) \ , \qquad (8)$$

where $g(v,t)$ is the fraction of particles with volume between v and $v + dv$ at time t, ρ_p is the polymer particle density, m_s is the mass of a stable droplet and λ is the frequency of eddies arriving on the droplet surface.

The density of the polymer particles changes during polymerization according to the equation

$$\frac{d\rho}{dt} = -\left[r_M w \left(\frac{1}{\rho_m} - \frac{1}{\rho_p} \right) \right] \rho \ . \qquad (9)$$

Here, r_M is the rate of monomer reaction, w is the RMM of the monomer, ρ is the density, and the subscripts m and p correspond to monomer and polymer, respectively.

Peppas et al. [22] found that the average size and the standard deviation of the particle size distribution scale to the same quantity, which is a lumped parameter of the monomer density, the impeller speed, the impeller diameter, and the surface tension of the dispersion before the polymerization is initiated.

Some models to calculate the droplet mass distribution of the liquid dispersion have been developed [23–25].

The development of internally-porous polymer particles has been of considerable significance in ion-exchange [26] and gel-permeation chromatography [27]. There are efforts to establish both the mechanism by which the pores are generated and the relationship between the conditions of the suspension polymerization and the properties of the resultant polymer particles or polymer.

Porous polymer particles are prepared by the suspension copolymerization of styrene/divinylbenzene in various ratios. Larger-scale preparations of porous, crosslinked poly(styrene) particles are performed by the method of Millar et al. [28].

Howard & Midgley [29] prepared porous polymer particles by the suspension copolymerization of styrene/divinylbenzene in various ratios together with various quantities of diluents (toluene, decane, xylene, ethylbenzene, dibutyl phthalate, tetrachloroethane) both solvating and nonsolvating. The porosity in the suspension-polymerized styrene/divinylbenzene copolymers is always induced by nonsolvating diluents, whereas those samples made without diluent are nonporous. Solvating diluents occupy an intermediate position. The formation of inhomogeneities during polymerization with nonsolvating diluent was observed.

Bead polymers could not be produced by suspension polymerization beyond certain limits. With solvating diluents, the practical limit at high diluent contents is largely one of polymer yield; however, with nonsolvating diluents, there is a more clearly defined upper bound. At crosslinked volume fractions ca. 0.3, powders rather than beads are formed above 0.6 volume fraction of decane, while at 0.5 volume fraction crosslinked beads could not be made at 0.4 diluent volume fraction.

Trochimczuk & Kolarz [30] prepared a series of terpolymers from acrylonitrile, butyl acrylate and divinylbenzene by suspension polymerization. The acrylonitrile/butyl acrylate/divinylbenzene terpolymers obtained in the presence of inert diluents (heptane, toluene) had a porous structure. Thermal degradation studies showed that the addition of a butyl acrylate unit to the acrylonitrile/divinylbenzene system results in a decrease in the intensity of the exotherm and of the heat of oligomerization. Oligomerization starts at somewhat higher temperatures.

7.2.2 Vinyl chloride

7.2.2.1 Mechanism of particle nucleation

In the first stage of the suspension polymerization of vinyl chloride, liquid monomer is dispersed in water in the presence of a stabilizer into a fine dispersion. The polymerization starts with the decomposition of an oil-soluble initiator to produce initiating radicals which attack the monomer particles. The suspension polymerization of vinyl chloride as regards the insolubility of the polymer in the monomer itself, is characterized by a special mechanism of nucleation and polymerization. After the start of polymerization, the polymer precipitates from the emulsified monomer particles as very small particles (d ca. 10 nm). They are unstable and associate effectively into larger aggregates with ca. 100 nm called 'primary particles'. At low conversions (below 1%), a high concentration of primary particles develops with a low polymer content and which undergo Brownian motion. Their mutual collision leads to the formation of larger particles, the size of which depends on the effect of coagulation. Their coalescence is proportional to the concentration of particles, the stirring rate and temperature, and is also affected by the nature of the polymer particle surface [31, 32].

As polymerization proceeds, the size of the primary particles and the polymer fraction (gel) in the particle increase. By the association of primary particles, a continuous network is formed with a character depending on the size of the particles, their number and the character of their mutual interaction [33]. The network appears at conversions from 10 to 30% and its cohesion affects the properties of the polymer gel phase and the porosity of the resulting polymer.

If the continuous network is formed at lower conversions, the final product shows high porosity. Stirring leads in this case to a particle distribution according to the sizes and their efficient association. If the conditions for growth of the primary polymer particles as individuals are dominant, in the presence of high interparticle repulsive forces, the primary particles grow side by side without aggregation. The product being formed shows low porosity as a consequence of the tight packing of the particles accompanied by a considerable contraction of the monomer [34].

As polymerization proceeds, the particle loses its porous character because the fraction of the polymer phase formed by the primary particles increases. With increasing size of the primary particles and with the generation of new particles, the space between the particles is filled and the original spherical aggregates are deformed and transform to a continuous polymer phase [35].

The formation of the primary particles is thus the combined result of four processes that take place inside a monomer droplet [36, 37]:

(i) the formation of polymer molecules due to the polymerization reactions,
(ii) the separation of the polymer phase from the monomer phase with the formation of insoluble polymer particles (basic particles),
(iii) the aggregation of the particles,
(iv) the growth of the particle due to polymerization in the polymer phase.

The microstructure of polymer particles of vinyl chloride is also significantly influenced by temperature. As the temperature increases, a larger number of primary particles, and particle growth by coalescence are observed. The increase in the number of particles is proportional to the rate of initiation which increases with the decomposition rate of the initiatior. Increase in temperature leads to the formation of a polymer product in a shorter reaction time but does not affect the mechanism of particle formation. Temperature increase sensitizes the rate of coagulation of primary particles which shifts the limiting values of the size of the particles and their number to lower conversions [38]. The growth in size of the polymer particles causes a decrease in the interparticle contacts and a lowering of the toughness of the flocculation network and, conversely, it increases the contraction of the monomer droplets.

With growing size of the primary particles, the porosity of the particle, mainly in its shell, decreases. Davidson & Witenhafer [39] have found that the porosity of particles is proportional to the inverse of the temperature; i.e. by raising the temperature, the fall in the particle porosity is shifted to lower conversions, e.g., at polymerization temperature of 71°C, the polymer particle surface loses its porous character at conversions from 30 to 50% but, at 51°C, it happens at 45–75% conversion.

The temperature also affects the shape of the final polymer granules. If the polymerization takes place near the boiling point of vinyl chloride, the shape of the polymer particles formed differs considerably from spherical [40]. The formation of particles of an irregular shape is explained by the formation of the primary polymer particles preferentially on the monomer particle surface, which produces an isolating layer hindering the transfer of the heat to the water. The reduction in heat transfer is augmented by the layer of adsorbed stabilizer. The impediment to the transfer of heat to water causes a rise in temperature in the particle and gasification of the liquid vinyl chloride. The escaping vinyl chloride deforms both the surface and the spherical shape of the particle. This deformation can be reduced by a suitable choice of the stabilizer which secures a certain access of air to the particle and thus preserves its spherical shape.

Discussion in ref. [41] directed to the structure of the polymer particles shows that the surface particle layer is made of a shell of primary particles formed in the initial steps of polymerization. The formation of a graft copolymer of vinyl chloride and vinyl acetate in the suspension polymerization of vinyl chloride in the presence of a protective colloid of vinyl acetate is a result of polymerization on the surface of the polymer particle [42]. At lower conversions (to 20%), the structure of the particle is mainly influenced by the coagulation of the primary particles. The distinction between the associates disappears as reaction proceeds; and the associates coalesce inside the polymer particles. The porous structure of the surface layer of thickness from 200 to 500 nm is preserved as a function of the reaction conditions up to higher conversions; it disappears, however, at the end of the polymerization.

In suspension polymerization as in emulsion polymerization, the agglomeration of the monomer particles, the mechanism of agglomeration of the primary particles, and the porosity of the polymer are strongly affected by stirring. More vigorous stirring lowers the contraction of the monomer droplets in agglomerates and affects particle growth by the coalescence of the primary and layer particles in a complex way [43].

Investigation of the suspension polymerization of vinyl chloride [41, 43] shows that both the size and the number of primary and monomer/polymer particles are controlled by a protective colloid (e.g. poly(vinyl acetate)). The stabilizer added is adsorbed at the monomer–water interface and controls the size of the final polymer particles and their porosity. In the presence of the stabilizer, a network of primary particles is created at lower conversion and leads to a higher porosity of the final product. Stabilizers and costabilizers are sometimes added to the reaction mixture after a certain conversion solely to stabilize the primary particles more effectively [44] and to fix the network formed by aggregates of the primary particles. The association of stabilizer molecules with a polymer at the monomer–water interface helps to preserve the aggregates and controls the completion of the reaction of the monomer and the deformation of the primary polymer particles directed to the formation of a more highly porous polymer [35].

7.2.2.2 Kinetic model of suspension polymerization

Over the past 20 years, several mathematical models have been developed to describe the two-phase polymerization of vinyl chloride in suspension in bulk reactors. The key feature in all these models is that poly(vinyl chloride) is practically insoluble in its monomer, and polymerization occurs simultaneously in the two phases almost from the start. These models have been summarized in [45, 46].

Talamini *et al.* [47] were the first to derive a two-phase model to predict the time dependence of the conversion of fractional monomer in a suspension PVC reactor. The assumptions made in this model are:

(i) No transfer of radicals between the two phases occurs.
(ii) The initiator concentration does not change with conversion.
(iii) The volume of the reacting mixture remains constant. Abdel-Alim & Hamielec [48] modified Talamini's model by replacing assumptions (ii) and (iii) with assumptions (iv) and (v):
(iv) The initiator concentration *does* change with conversion.
(v) The volume of the reacting mixture changes linearly with conversion.

Moreover, Abdel-Alim–Hamielec's model accounted for the diffusion control of the propagation and termination reactions at high conversions.

In contradistinction of the above two models, the kinetic model of Ugelstad *et al.* [49] accounts for the transfer of radicals between the two phases: they assumed the production of radicals in both phases. These authors suggested that an equilibrium distribution of radicals is quickly established between the two phases. The main difference between Ugelstad's model and that of Kuchanov & Bort [50] is that the latter does not utilize the assumption of an equilibrium distribution of the radicals between the two phases.

Kuchanov & Bort [50] assumed that the radical desorption from the polymer-rich phase can be neglected.

Olaj's model [51] takes into account the formation of precipitated radicals. This model, however, does not take account of the termination of dissolved radicals in the monomer phase. Thiele *et al.* [52] and Kafarov *et al.* [53] proposed a new modified two-phase model. Both models take into account the mass transfer of radicals between the two phases; i.e. the monomer-rich and the polymer-rich phases.

Kelsall & Maitland [54] proposed a kinetic model that takes into account the mass transfer of various species (i.e. monomer, initiator, etc.) between the different phases and the possibility of an inhomogeneous distribution and efficiency of the initiator. It includes also a detailed description of the polymer precipitation and growth processes. The model does not include any expressions related to the formation of anomalous structures observed in PVC.

One of the latest models describing the suspension polymerization of vinyl chloride and the distribution of the polymer and monomer between the individual phases is that of Xie *et al.* [55]. At lower conversions, when $X \leq X_f$ (X_f being the critical fractional conversion at which the pressure decrease of the system is observed), the monomer distribution in the reaction system can be expressed by Eq.

$$M_0 = M_{g1} + M_{w1} + M_1 + M_{p1} + W_p, \tag{10}$$

where $M_0, M_{g1}, M_{w1}, M_1, M_{p1}$, and W_p are the masses of batch monomer at the beginning of the polymerization (the total amount), in the gas phase, dissolved in water, in the monomer phase, in the polymer heterogeneous phase, and of the polymer itself, respectively.

At higher conversions (i.e. $X > X_f$), free monomer particles are no longer present and Eq. (10) takes the form

$$M_0 = M_{g2} + M_{w2} + M_{p2} + W_p. \tag{11}$$

The amount of monomer in the polymer phase can be expressed by the volume fraction of the polymer Φ_2 in the particles

$$M_p = M_0 X \rho_m (1 - \Phi_2) \Phi_2 \rho_p, \tag{12}$$

where ρ_m and ρ_p are the densities of the monomer and polymer, respectively. The polymer content in the system is expressed by the relationship

$$W_p = M_0 X. \tag{13}$$

The volume fraction of the polymer, Φ_2, is obtained by enumerating the Flory–Huggins relationship [56]

$$\ln P_m/P_{m0} = \ln(1 - \Phi_2) + (1 - 1/n)\Phi_2 + \chi \Phi_2^2, \tag{14}$$

where χ is the interaction parameter for the monomer-polymer pair, n is the number of structural units in the polymer chain, and P is the monomer pressure in the reactor. If the effect of monomer mass is neglected, Eq. (14) can be expressed as

$$\ln(\alpha_1/(1-\Phi_2)) - \Phi_2 = \chi\Phi_2, \tag{15}$$

where $\alpha = P_m/P_{m0}$.

Xie *et al.* [55] proposed a semiempirical relationship for calculating the interaction parameter

$$\chi = 1286.4/T - 3.02 \tag{16}$$

on the basis of literature data [57–59].

The solubility of vinyl chloride in water depends on the pressure in the reaction system

$$K = P_{m0}/H, \tag{17}$$

where H (atm) is Henry's constant, or, as the semiempirical Eq.

$$K = 0.0472 - 11.6/T \tag{18}$$

obtained from available data on the temperature dependence of the monomer solubility [55, 60–63]. If the temperature increased to 40, 50, and 60°C, the value of K increased in the order 0.01, 0.011, and 0.13 [55]. The water solubility of vinyl chloride is positively affected by increase in concentration of the stabilizer; a visible increase occurs above a concentration of 0.5 mass % [60].

The model of Xie *et al.* [55] describes the variation of the conversion with temperature and pressure by Eq. (19)

$$X = \frac{M_0 - ((P_m M_m/RT)(1-W_i)V_r + X_f M_0(1/D_m - 1/D_p)D_{g0}/(D_m - D_{g0})) - KW_w P_m/P_{m0}}{M_0 + M_0 D_m(1-\Phi_2)/(\Phi_2 D_p) + P_m M_m M_0(1/D_m - 1/D_p)/(RT)},$$

$$\tag{19}$$

where W_i is the overall reactor feed, V is the volume of the reactor, W_w denotes the water phase feed, and the index 0 the initial state of individual components.

The critical fractional conversion can be obtained from Eq. (20) [56]

$$X_f = \frac{M_0 - (D_{g0}(1.0 - W_i)V_r + KW_w)}{M_0(1.0 + D_{g0}(1/D_m - 1/D_p)/(1.0 - D_{g0}/D_m) + D_m(1.0 - \Phi_2)/(\Phi_2 D_p))}. \tag{20}$$

The applicability of this model has been experimentally confirmed by several investigators [57–59]. These papers have shown that X_f decreases with increasing temperature. For instance, if the temperature increases from 0°C to 40°C or up to 80°C, then X_f decreases from 0.91 to 0.78 or as low as 0.39. Most kinetic data for suspension polymerization and copolymerization have been obtained at about 30°C and therefore, modelling of the polymerization at higher temperatures provides information on the kinetic parameters even for regions for which no experimental data are available.

Increase in conversion is accompanied by a linear decrease of the monomer concentration in the monomer phase and an increase of the monomer concentration in the polymer phase, reaching its maximun at X_f [64]. Under the polymerization reaction conditions $V_r = 5 \, l$, $M_0 = 1200$ g, $W_w = 2500$ g and at 50°C, the monomer is divided into water, gas and monomer phases at zero conversion in the ratio 2.4 : 1.6 : 96%. At the conversion X_f, the ratio of the monomer distribution is 2.4 : 2.2 : 23.2%.

Sidiropoulou & Kiparissides [65] have recently developed a fairly comprehensive but realistic model. The model assumes a two-phase polymerization occurring in three distinct stages. Accordingly, the mass balance equations for initiator, monomer, and the leading moments of the RMM distribution are derived for each stage of polymerization.
Stage 1: polymerization takes place only in the monomer phase.
Mass balance for initiator:

$$dI_1/dt = -k_{d1} I_1 \, ; \, I_1(0) = I_0 .$$

Mass balance for monomer:

$$dM_1/dt = -k_{p1} M_1 [\lambda_{01}] \, ; \, M_1(0) = M_0 . \qquad (22)$$

Finally, the fractional conversion of monomer can be calculated by

$$dx/dt = k_{p1} (1-x)[\lambda_{01}] , \qquad (23)$$

where k_d is the rate constant for decomposition of initiator, k_p is the rate constant for chain propagation, x is the fractional conversion of monomer and λ_{ij} is the moment of living polymer distribution.
Stage 2: $0.1\% < x \leq x_c$.

The polymerization occurs in two separate phases. Therefore, species mass balances are derived for each phase. The total initiator and monomer balances are given by the sum of the corresponding species balances derived for each phase. The two separate phases exist until a fractional conversion x_c has been reached at which the monomer-rich phase disappears.
Total initiator mass balance:

$$dI/dt = -k_{d1} I_1 - k_{d2} I_2 . \qquad (24)$$

Total monomer mass balance:

$$dM/dt = -k_{p1}[M_1]\lambda_{01} - k_{p2}[M_2]\lambda_{02} . \qquad (25)$$

The fractional monomer conversion can be calculated as

$$dx/dt = k_{p1}[\lambda_{01}](1 - x - Ax) + k_{p2}[\lambda_{02}]Ax, \tag{26}$$

where

$$A = (1 - X_c)/X_c.$$

The volumes of the monomer- and polymer-rich phases are:

$$V_1 = V_0 (1 - x - Ax), \tag{27}$$

$$V_2 = V_0 x (A + \rho_m/\rho_p), \tag{28}$$

where V_0 is the volume of the reacting mixture, and ρ_m and ρ_p are the densities of monomer and polymer, respectively.

Stage 3: $X_c < X$.

For conversions greater than X_c, the polymerization takes place only in the polymer-rich phase. During this stage the following mass balance equations can be derived:

Mass balance for initiator:

$$dI_2/dt = -k_{d2} I_2. \tag{29}$$

Total monomer mass balance:

$$dM_2/dt = -k_{p2} M_2 [\lambda_{02}]. \tag{30}$$

The fractional monomer conversion is given by

$$dx/dt = k_{p2} (1 - x)[\lambda_{02}], \tag{31}$$

V_2 is given by

$$V_2 = V_0 (1 - x + x\rho_m/\rho_p). \tag{32}$$

7.2.2.3 Suspension polymerizations and copolymerizations

Poly(vinyl chloride) is one of the most widely commercially produced and used thermoplasts. About 12 million tons of PVC is prepared per annum via suspension polymerization, which is about 80% of the world production.

The suspension polymerization of vinyl chloride, like the emulsion polymerizations of vinyl monomers, is carried out in a batch or semicontinuous way, the difference being that the preparation of the PVC polymers or copolymers requires the use of high-pressure reactors.

Great attention in the vinyl chloride-based preparation of polymers and copolymers is devoted to selection of the stabilizing system. In kinetic studies, about 1–3 mass % of stabilizer (with respect to monomer) is used for stabilizing the polymer particles [41, 43]. With regard to the adverse effect of the stabilizer on the physical properties of the polymer in the preparation of larger amounts of polymer, the use of smaller amounts of stabilizer (around 0.1 mass %) is preferred.

Poly(vinyl acetate) of different RMM is the universally-used stabilizer of the polymer suspension. In addition to its influence on the RMM, poly(vinyl acetate) enables changes in the degree of hydrolysis (by introducing hydroxyl groups) and various arrangements of the acetate groups in the molecule of stabilizer. As the fraction of hydroxyl groups in the stabilizer molecule increases, the porosity of the polymer particles decreases and, in proportion, an increase in their size is observed. If hydrolysed poly(vinyl acetate) with block sequences of acetate groups is used, the particles prepared are smaller than when a stabilizer containing the same number of hydrolysed groups, but with a statistical distribution of acetate groups is used. Modification of the poly(vinyl acetate), especially by hydrolysis enables the preparation of polymer particles of different sizes, size distribution, and porosities [66, 67].

The properties of PVC prepared by suspension polymerization depend mainly on the RMM of the polymer and the particle porosity. The final granulated product contains particles with a diameter of *ca.* 150 μm; these particles have mostly a porous shell [8, 68]. The RMM of the polymer determines its viscosity in the melt and thus also its processability in the mass. It can be affected by the temperature of the polymerization and by a chain-transfer agent [73]. The RMM increases with conversion, especially at lower conversions. The increase in RMM of the polymer is attributed to a reduction of the fraction of termination reactions in the polymer particle gel; this results from the decrease in mobility of the propagating radicals in a high-viscosity medium. The fraction of the polymer gel increases as polymerization proceeds which is why the rate of the bimolecular termination processes decreases in proportion. This reduction of termination in the system leads to an increase in the polydispersity of the RMMs (> 2). The PVC gel contains about 20–30 mass % of monomer on average [70] and is usually characterized by the value of T_g.

Information about the particle porosity is usually obtained from the dependence of the monomer contraction on conversion. PVC has a higher density than its monomer ($\rho_{PVC} = 1.4$; $\rho_{VC} = 0.86$ at 50°C) and therefore the volume of the monomer/polymer particles decreases during polymer formation. The densities of VC and PVC show that, by changing monomer to polymer, the volume of the reaction system should be reduced (this refers only to vinyl chloride) by about 40%. The volume contraction of the monomer droplets is seen to be much smaller and ceases in many reaction systems as early as in the initial step of polymerization [38]. The reduction of the monomer contraction positively affects the porosity of the final polymer.

The suspension polymerization of vinyl chloride has been investigated kinetically by an unconventional calorimetric method [71]. The results obtained showed that the termination rate constant in the PVC phase is diffusion-controlled and very sensitive to the vinyl chloride concentration. The variation in k_t is given by

$$k_t = 10^{(7.3 - 0.3(1/V_f - 7.5))} + K, \tag{33}$$

where K accounts for termination by small radicals and V_f is the free volume for PVC swollen by vinyl chloride calculated from the following expression:

$$V_f = (\text{total volume} - (\text{van der Waals volume})F)/\text{total volume}. \qquad (34)$$

The factor $F = 1.3658$, was calculated assuming the free volume to be 0.025 at the glass transition composition (0.04 g VC per g of swollen PVC phase).

During suspension polymerization of vinyl chloride the pressure begins to drop at a much lower conversion than in emulsion polymerization. Mass balance calculations showed that this is due to entrapment of a large amount of liquid vinyl chloride in the fine capillary system of the suspended particles.

An undesirable phenomenon accompanying the suspension polymerization of vinyl chloride is the formation of particles which are difficult to process. The presence of such particles in PVC powders adversely affects the quality, appearance, and transparency of films from plasticized PVC. These particles are called 'fish eye' particles. It is generally assumed that they contain a large amount of strongly branched or crosslinked PVC macromolecules. The fish eye (FE) particles are not easily transferred or transferred at all into the gelated state, and differ from normal particles of suspension PVC either in their structure at the molecular level or in their morphology or in both. Structural heterogeneities in PVC are caused by the occurrence of 'fish eye' particles [37, 72, 73].

Lukáš et al. [74–76] reported that FE particles are formed from vinyl chloride droplets containing larger amounts of initiator. In other words, the authors presumed an inhomogeneous distribution of initiator in the vinyl chloride droplets during suspension polymerization. The formation of FE particles was discussed in terms of the molecular diffusion of initiator and monomer in the suspension system during the polymerization and its dependence on the size of the drops in which all the components are dispersed.

References [76, 77] report that FE particles contain a higher concentration of conjugated double bonds and are thermally unstable. The lower thermal stability of FE particles results from the side reaction between initiator radicals and the PVC chain. This reaction leads to dehydrochlorination of the PVC under polymerization conditions.

Lukáš et al. [76, 78, 79] discussed the relation between the polymerization process and PVC dehydrochlorination and proposed a new mechanism for the latter. They reported that (i) the defective structures which facilitate the elimination of HCl along the polymer chains have only a transient effect on PVC dehydrochlorination, (ii) cisoid enone, dienone or polyenone structures are permanently active structural defects, which act by an 'enzyme-like' mechanism as catalysts of elimination of HCl from the regular PVC structural units giving rise to chloroallyl structures, (iii) formaldehyde is formed in the polymerization system by the decomposition of polymeric peroxides (arising from the copolymerization of vinyl chloride with oxygen) and creates aldehyde and ketone groups in the polymer chains by transfer reactions, and (iv) following dehydrochlorination, β,β'-dichloroketone groups give rise to cisoid enone, dienone or polyenone structures. The occurrence of such structures in PVC chains is responsible for the permanently increased constant rate of dehydrochlorination of PVC.

Series of suspension copolymerizations of vinyl chloride with various types and concentrations of 1-alkenes were carried out by Mrázek et al. [80, 81]. When estimating the effect of the individual 1-alkenes on the copolymerization, we may conclude, by using the total conversion of this reaction related to the same molar concentrations in the initial mixture, that the total copolymerization rate of the pairs of monomers investigated decreases in the following order:

VC–propene > VC–1-butene ≃ VC–1-pentene . (35)

Propene, 1-butene and 1-pentene were found to be less reactive monomers in the copolymerization than vinyl chloride. Their addition to the propagating polymer radicals proceeds at a lower rate than that of vinyl chloride. 1-Butene and 1-pentene, even when present in the reaction system at low concentrations, slow down the polymerization in a manner comparable with that of the vinyl chloride homopolymerization. The 1-alkenes reduce the intrinsic viscosity of the copolymer solutions compared with the homopolymer of vinyl chloride, and the following dependence may be given:

$$[\eta]_{VC-propene} > [\eta]_{VC-1-butene} > [\eta]_{VC-1-pentene} .$$ (36)

At the same time, with increasing content of 1-alkenes in the initial monomeric mixture, the stability of the suspension decreases.

Mrázek et al. [82, 83] reported that GC enables us to describe the kinetic behaviour of vinyl chloride quickly and reliably. The possibility of investigating comonomer kinetics was found to depend on a sufficiently low boiling point, which should not be higher than ca. $-10°C$, and on its low sorption in the copolymer. Only then is the condition of the sufficiently quick and quantitative desorption of monomers from the sample needed for their quantitative determination fulfilled. Such a conclusion follows from an investigation of the copolymerization of vinyl chloride with 1-butene and 1-pentene, the boiling points of which already exceed the limiting value specified.

7.2.2.4 Methods of kinetic investigation

Few methods are used in PVC chemistry for kinetic investigations. These include, for example, dilatometry, gravimetry, calorimetry, and gas chromatography. Dilatometry is particularly suited for the investigation of polymerization at low conversions. The gravimetric method can be applied to the investigation of polymerization at all conversions. Hitherto, investigation of the suspension polymerization of VC to high conversion has been almost exclusively restricted to gravimetric determination of the polymer formed. Unfortunately this universal method is extremely time-consuming.

Tömell et al. [71, 84] have used a calorimeter to examine the suspension and emulsion polymerization of VC. In this work, by using a calorimetric reactor, which continuously measures the thermal power developed by the process, a more direct measurement of the rate has been attempted. The heat of reaction, H, was calculated from Eq. (37):

$$P_A = H + P_{agit} + P_{ph.tr.} ,$$ (37)

where P_A is the total power developed in the autoclave, P_{agit} the power introduced by the stirrer, and $P_{ph.tr.}$ the thermal power associated with phase transitions. P_{agit} was calculated from the output signal of the torque transducer and the stirrer speed. $P_{ph.tr.}$ was calculated from the material balance and the enthalpy data for the transition:

$$VC_{(gas)} \rightarrow VC_{(liq)}, \ VC_{(liq)} \rightarrow VC_{(aq)}, \ VC_{(liq)} \rightarrow VC(PVC) .$$ (38)

The rate of polymerization was calculated using a value for the heat of polymerization. The total mass balance is given by the equation:

$$\text{Added VC} = \text{PVC} + \text{VC}_{(\text{PVC})} + \text{VC}_{(\text{liq})} + \text{VC}_{(\text{gas})} + \text{VC}_{(\text{aq})} . \quad (39)$$

The amount of PVC formed was calculated from the time integral of the rate of polymerization. The amounts of VC in the PVC, aqueous and gas phases were calculated from the pressure reading using the solubilities [85] and PVT-data [85] for VC as reported elsewhere [85]. By using Eq. (39), the possible presence of small amounts of solubilized monomer in the adsorbed layer of the tenside at the particle surface is overlooked. Surface tension effects on the swelling of PVC were not considered in the analysis of the suspension polymerization experiments. During the short periods of the polymerization experiments in which the VC activity in the PVC phase decreased rapidly with time, the equilibrium between $\text{VC}_{(\text{gas})}$ and $\text{VC}_{(\text{aq})}$ was not always maintained. Under such conditions, Eq. (39) may give negative values for $\text{VC}_{(\text{liq})}$. When this occurred, the solution of Eq. (39) was based on the assumption that $\text{VC}_{(\text{PVC})}$ and $\text{VC}_{(\text{aq})}$ were in equilibrium and that $\text{VC}_{(\text{liq})}$ was zero. This procedure would be incorrect if a phase of liquid VC actually existed. In suspension polymerizations this might occur because of capillary condensation of the monomer in the porous particles.

Mrázek & Lukáš [83] used gas chromatography (GC) for investigation of suspension polymerization of VC at all conversions. The method is based on periodical sampling of small volumes of samples of the reaction mixtures and on the GC determination of the monomers contained therein. The method can be used for determining the conversion curves of the individual monomers, for calculating reaction rates, and for determining copolymerization parameters. Using this method, it is possible to follow the amounts of monomers in the reaction mixture depending on time irrespective of the degree of conversion. Gas chromatography can be used to advantage in the suspension homopolymerization and copolymerization of VC, especially in testing initiation systems and in the investigation of kinetics depending on the composition of the monomer feed and the reaction conditions. Its advantages consist primarily in its speed, reliability, sufficient accuracy and ready automation.

7.2.3 Other lipophilic monomers

The ratio between the water and oil phases is also an important factor affecting the kinetics of suspension polymerization [86, 87]. The influence of the water/oil ratio was investigated during the suspension polymerization of methyl methacrylate at both low and high stabilizer (poly(vinyl alcohol)) concentrations. At low poly(vinyl alcohol) concentrations, change in the water/oil ratio exerted a similar effect on the size of the final polymer particles as in emulsion polymerization. In contrast, if high stabilizer concentrations were used, change in the water/oil ratio did not affect the particle size (which was only proportional to the stabilizer concentration).

Suspension polymerization is also used for preparing poly(vinyl acetate) — a polymer of hydrophilic character. In ref. [40] the problems associated with the preparation of this polymer by the suspension technique are discussed. At polymerization temperatures lower than the b.p. of vinyl acetate, the formation of spherical polymer particles was observed. On raising the polymerization temperature above the b.p. of the monomer, polymer granules were formed in a nonspherical shape. This phenomenon was explained by the formation of primary polymer particles on the surface of monomer droplets, which form a shell to inhibit effective heat transfer from the particles at higher conversions. This resulted in a local rise in temperature in the particle followed by gasification of the monomer. The monomer escaping from the particle disrupts the regularity of its surface and thus affects the resulting shape of the granules.

The effect of the nature and concentration of the stabilizer was studied in the suspension copolymerization of acrylonitrile and vinyl acetate with a high portion of acrylonitrile in the monomer feed [88]. With increase in the stabilizer (polyvinylpyrrolidone, poly(ethylene glycol), etc.) concentration, the polymerization rate increases and the RMM of the copolymer decreases. The relation between the polymerization rate and the reciprocal of the RMM supports the application of the mechanism for homogeneous polymerization. The transfer to stabilizer also contributes to the reduction in RMM. Increased flocculation of the monomer/polymer particles was observed at lower stabilizer concentrations. Higher concentrations of the stabilizer suppressed coagulation to a certain limiting value and led to the production of rather stable suspended particles [89, 90]. The suspension polymerization of acrylonitrile with vinyl acetate can also be controlled by the composition of the water-soluble initiating system ($K_2S_2O_8$–SO_2–$FeSO_4$) [88]. Increase in the overall concentration of initiator led to acceleration of the polymerization and to a decrease in the RMM of the polymer. Reduction in the RMM was observed on raising the SO_2 concentration: e.g. increasing it from 0.17 to 2.5 mass % (with respect to monomer) reduced the viscosity of the copolymer from η_{sp} = 0.75 to 0.19.

The mechanism of the suspension copolymerization of acrylonitrile and styrene was studied by Karaenev et al. [91], who showed that with a higher water solubility of one of the monomers, the reaction mechanism becomes more complex. It is suggested that the process takes place not only in the monomer droplets but also in the aqueous phase. The presence of acrylonitrile increases the solubilities of styrene and the initiator in water. It was, therefore, expected that polymerization proceeds in water where the oligomeric radicals are produced and sorbed or dissolved in the monomer/polymer particles. The copolymerization of the enriched acrylonitrile monomer was suggested to take place on the particle surface as well as at higher conversion.

A decrease in the RMM of the polymer can be affected by adding good transfer agents which suppress the gel effect: this approach is often used in suspension polymerizations. The advantages and disadvantages of using efficient transfer agents (alkyl mercaptan derivatives) in the suspension polymerization of methyl methacrylate initiated by lauroyl peroxide in the presence of sodium methacrylate have been discussed by Giannetti et al. [92]. The dependence of the polymerization rate on conversion exhibits two regions: a region of low rate before reaching the gel effect and one of the high rate afterwards. With increasing fractions of mercaptan, the linear region of the curve is shifted to higher conversions. In the absence of the transfer agent, the gel effect appears at 30–35% conversion; on adding the mercaptan, it becomes shifted to ca. 60% conversion. This shift is associated with a decrease in the RMM of the polymer in proportion to the concentration of transfer agent. In the absence of mercaptan, poly(methyl methacrylate) was formed at lower conversions with a RMM of M_n ca. 2×10^5; at 90% conversion, the RMM was five times higher. The use of mercaptan caused a reduction in the RMM by one order of magnitude on average. Of the two types of mercaptan, n-butyl and n-dodecyl, the latter decreased the RMM more efficiently. If a mercaptan was present, the RMM of the polymer was almost constant throughout the range of conversions, although an increase in rate of polymerization was observed above 60% conversion. The sulfur content (in ppm) varied between 400 and 600, reached a maximum at low conversions, and slightly decreased with further conversion.

Matějíček et al. [93] followed the suspension copolymerization of ethyl acrylate, acrylonitrile and divinylbenzene by the calorimetric method. The kinetics were described by means of the time dependences of conversion and polymerization rate. The conversion curves were sigmoidal with a maximum polymerization rate at conversions between 40 and 60%. The appearance of the gel effect even at low conversion is a characteristic feature of this copolymerization. The polymerization rate was found to increase linearly with temperature. The magnitude of the activation energy of the copolymerization, i.e. $125\,\text{kJ}\,\text{mol}^{-1}$, was determined from the above data. At low conversions, the rate of polymerization increases linearly with initiator concentration. In contrast, at higher conversions, where the gel effect operates, the polymerization rate increases exponentially with increase in the initiator concentration.

7.2.4 Unconventional suspension polymerizations

A development of major importance in the production of hydrophilic polymeric microparticles is the suspension polymerization technique of Mueller & Heiber [94]. This technique is preferred because it provides spherical particles while avoiding the use of potentially toxic organic suspending phases. Briefly, the water-soluble monomer, for example 2-hydroxyethyl methacrylate, is suspended in an aqueous phase that includes dissolved sodium chloride and a magnesium hydroxide precipitate under controlled agitation. Initiation is made by classical free-radical techniques. The presence of the sodium chloride in the aqueous phase reduces the monomer solubility and allows a suspension to be formed. The magnesium hydroxide precipitate acts as a suspending agent.

Of particular interest to Peppas [95] is the production of porous and nonporous microspheres of poly(2-hydroxyethyl methacrylate), which has been shown to be a promising biomedical polymer. A method of producing hydrophilic polymer microparticles of 2-hydroxyethyl methacrylate by aqueous suspension polymerization was investigated by Peppas et al. [96]. The solubility of the monomer in the aqueous phase was reduced as the NaCl concentration increased, indicating a salting-out effect. Quantitative analysis of the salting-out effect revealed that a significant amount of water was present in the monomer phase while very little monomer was present in the aqueous phase. These results suggest that the polymerization reaction is really a combination of a suspension polymerization which takes place in the droplets of dispersed monomer, and a solution polymerization inside the droplets, since there is a significant amount of solvent in the droplets. Since the amount of water is below 40 mass %, the reaction will produce structurally homogeneous networks. The particles formed by this suspension polymerization process were spherical with d (average) ca. 100 μm.

Horák et al. [97, 98] used the suspension polymerization method in decalin to produce poly(2-hydroxyethyl methacrylate) microparticles crosslinked with ethylene glycol dimethacrylate.

Crosslinked hydrophilic polymer particles of diameter both below and above 100 μm have been widely applied in medicine. They are prepared via the copolymerization of a monomer feed containing hydrophilic and tetrafunctional monomer, by classical suspension polymerization [99, 100]. The degree of crosslinking of the hydrophilic particles by a crosslinking agent affects not only the applications properties of a dispersion but also its mechanical stability. The RMM of a polymer segment located between two crosslinked points decreased with increasing particle size. The network density increased with the duration of polymerization and with increase in the particle size, and decreased towards the particle surface.

Procedures for the suspension polymerization of acrylic monomers are described mainly in the patent literature, usually with a lack of detail. Skovby & Kops [101] prepared porous beads by the suspension copolymerization of methacrylic acid, methyl methacrylate, ethyl acrylate, methyl acrylate and divinylbenzene in the presence of a porogen. The porosity varies greatly with the polymer composition, with the amount and type of porogen (toluene, octane, benzene) and with the conditions of preparation. Using the suspension polymerization of acrylic monomers and divinylbenzene, crosslinked porous beads have been obtained which are suitable for enzyme binding of a lipase. It was found that the highest degree of enzyme binding and measured activity was obtained with porous crosslinked poly(methyl methacrylate).

The preparation of porous acrylic polymer particles by suspension polymerization is also described in ref. [102].

Polymers prepared by the suspension polymerization of monomers in the solid phase are represent of a special type of suspension polymer. The initiating radicals are generated by γ-radiation, X-rays and UV radiation [103, 104]. This procedure leads to the preparation of polymers with a tacticity differing from that of polymers prepared by conventional suspension and solution polymerization. N-vinylcarbazole is suitable for preparing polymers and copolymers with such a structure and with the possibility of using a redox initiator [105]. This method enables the preparation of polymers with low RMM (η_{sp}/c from 0.04 to 0.07; $c = 1$; 30°C).

In the suspension polymerization of N-vinyl-*tert*-butyl carbamate [106], the crystals of monomer are suspended in the aqueous phase, a water-soluble polymer being used as stabilizer. The contact of the solid particles of monomer with water depends on the nature of the stabilizer and on the mechanism of stirring. More vigorous stirring improves the contact of the particles with water which is important for the penetration of initiator radicals into the monomer particles. Water-soluble initiators may also be used to start the polymerization. The polymer particle size was, in most polymerizations, equal to that of the original monomer crystal particles. The investigation of the effect of various additives showed that N-hydroxysuccinimide (NHSI) initiated decomposition of the monomer crystals of N-vinyl-*tert*-butyl carbamate during polymerization, the system became turbid, and the crosslinking efficiency of the particles decreased. With increasing NHSI concentration, on the one hand, the RMM of the polymer decreased, and, on the other hand, the crystallinity increased. At higher monomer/initiator ratios, a polymer was formed with a higher RMM but a lower crystallinity. This procedure may produce polymers from other solid monomers (e.g. from vinylnaphthalene).

The polymerization of acrylamide in solution is characterized by a high polymerization rate, a high exothermicity and a highly viscous final reaction mixture. The high rate of polymerization, as well as the marked increase of viscosity, prevent transfer of the reaction heat, even when runs are carried out in small volumes. The uncontrolled increase in temperature causes side reactions, such as intra- and intermolecular imidizations [107] and transfer to polymer, which are made evident by the formation of an insoluble polymer fraction. These difficulties can be avoided if the polymerization of acrylamide takes place in an inverse suspension system. Thus, microsuspensions have lower viscosities, provide easier heat removal and can be used at higher monomer concentrations than the solution process. Furthermore, the inverse latices produced can easily be inverted and added to water so that water-swollen polymer particles dissolve rapidly.

The interest in inverse suspension polymerization is rather recent. Major new applications of certain water-soluble polymers, such as poly(acrylamide), have prompted progress in this field. Data on inverse suspension polymerizations are few and mainly confined to the patent literature [108–112]. By studying various stages of the inverse suspension polymerization of acrylamide, the following reaction steps were identified [113]:

(i) An inverse suspension is formed when the aqueous phase is dispersed into the organic phase. At low or moderate concentrations of the stabilizer, the suspension is unstable and tends to separate without stirring.
(ii) The starting of the polymerization, by addition of initiator, is closely associated with an increased viscosity with gel formation and a sudden inversion of the phases.
(iii) Later in the reaction, when enough polymer is formed, the gel is broken up into small particles by stirring and the organic phase is partially removed from these particles which settle down. The gel breaks up during stirring and the formation of particles probably occurs at the interface where the stabilizer is adsorbed.

There have been several investigations into the kinetics of heterophase suspension polymerization of acrylamide [114–119], most recently and extensively by Hunkeler et al. [120]. They reported that the kinetics of heterophase acrylamide polymerization are a function of the chemical nature of the continuous phase and the solubility of the initiator. When water-soluble initiators are used, all components needed for the polymerization reside in the dispersed droplets. Each particle acts essentially as a small batch reactor and the kinetics are similar to those of solution polymerization [113].

When oil-soluble initiators are used with aromatic continuous phases, the kinetics have been shown to resemble emulsion polymerization [114]. However, when paraffin oil phases are used, the locus of initiation is in the monomer droplets. This polymerization physically and kinetically resembles a suspension, and is referred to as inverse-microsuspension [115, 120].

Hunkeler et al. [120] developed a kinetic model for the inverse-microsuspension polymerization of acrylamide. The model predicts well the conversion, RMM particle size and number for acrylamide polymerizations at various levels of initiator, monomer, stabilizer, temperature and rate of agitation, and for different types of initiators and emulsifiers. The mechanism was characterized as a free-radical polymerization where unimolecular termination with interfacial emulsifier is dominant and the mass transfer of primary radicals and oligoradicals is important. The rate of polymerization was found to depend on the initiator concentration to a power greater than one-half and to be inversely proportional to the concentration of surface emulsifier. Certain classes of stabilizer have been identified as being most suitable for the production of very high RMM polymers.

References

1. Konno, M., Arai, K. & Saito, S. (1982) *J. Chem. Eng. Jpn.* **15** 131
2. Wenning, H. (1956) *Makromol. Chem.* **20** 196
3. Wenning, H. (1958) *Kunstst. Plast.* **5** 328
4. Hoff, H., Lüssi, H. & Hammer, E. (1965) *Makromol. Chem.* **82** 175
5. Mino, G. & Judson, C.M. (1959) *J. Appl. Polym. Sci.* **2** 203
6. Schröder, R. & Piotrowski, B. (1981) *Chem.-Ing.-Tech.* **53** 748
7. Balakrishnan, T. & Ford, W. (1982) *J. Appl. Polym. Sci.* **27** 133
8. Bieringer, H., Flatau, K. & Reese, D. (1984) *Angew. Makromol. Chem.* **123/124** 307
9. Hirose, M. & Oshima, E. (1970) *Kagaku Kogaku* **34** 181

10. Winslow, F.H. & Matreyer, W. (1951) *Ind. Eng. Chem.* **43** 1108
11. Mark, M. & Hohenstein, W.P. (1946) *J. Polym. Sci.* **1** 127
12. Noguchi, H., Morishita, T. & Baba, H. (1986) *Japan* 61-115902
13. Leng, D.E. & Quarderer, G.J. (1982) *Chem. Eng. Sci.* **14** 177
14. Apostolidou, C. & Stamatoudis, M. (1990) *Collect. Czech. Chem. Commun.* **55** 2244
15. Tanaka, M. & Hosogai, K. (1990) *J. Appl. Polym. Sci.* **39** 955
16. Konno, M., Arai, K. & Saito, S. (1982) *J. Chem. Eng. Jpn.* **15** 131
17. Wiley, R.M., Kim, K.S. & Rao, S.P. (1971) *J. Polym. Sci., A-1* **9** 805
18. Hohenstein, W.P., Vingiello, F. & Mark, H. (1944) *India Rubber World* **110** 291
19. Bakshaee, M., Pethrick, R.A., Rashid, H. & Sherrington, D.C. (1985) *Polym. Commun.* **36** 185
20. Bakshaee, M., Pethrick, R.A., Rashid, H. & Sherrington, D.C. (1985) *Polymer* **26** 185
21. Bakshaee, M., Daly, J.H., Hayward, D., Pethrick, R.A., Rashid, H., Roe, S. & Sherrington, D.C. (1987) *Polymer* **28** 1605
22. Mikos, A.G., Takoudis, C.G. & Peppas, N.A. (1986) *J. Appl. Polym. Sci.* **31** 2647
23. Narsimhan, G., Gupta, J.P. & Ramkrishna, D. (1979) *Chem. Eng. Sci.* **34** 257
24. Hulburt, H.M. & Katz, S. (1964) *Chem. Eng. Sci.* **19** 555
25. Valentas, K.J. & Amundson, N.R. (1966) *Ind. Eng. Chem., Fundam.* **5** 533
26. Kun, K.A. & Kunin, R. (1967) *J. Polym. Sci.* **16** 1457
27. Moore, J.C. (1964) *J. Polym. Sci., A-1* **2** 835
28. Millar, J.R., Smith, D.G., Marr, W.E. & Kressman, T.R.E. (1963) *J. Chem. Soc.* p. 218
29. Howard, G.J. & Midgley, C.A. (1981) *J. Appl. Polym. Sci.* **26** 3845
30. Trochimczuk, A. & Kolarz, B.N. (1989) *Polym. Commun.* **30** 369
31. Cooper, W.D., Spiers, R.M., Wilson, J.C. & Zichy, E.L. (1979) *Polymer* **20** 265
32. Willmouth, F.M., Rance, D.G. & Henman, K.M. (1984) *Polymer* **25** 1185
33. Stewart, R.F. & Sutton, D. (1984) *Chem. Ind.* p. 373
34. Buscal, R. & Ottewill, R.H. (1985) In: Buscal, R. *et al.* (eds) *Polymer colloids.* Elsevier, Amsterdam
35. Kendall, K. & Padget, J.C. (1982) *Int. J. Adhesion Adhesives* **2** 149
36. Boissel, J. & Fischer, N. (1977) *J. Macromol. Sci., Chem.* **11** 1249
37. Barclay, L.M. (1976) *Angew. Makromol. Chem.* **52** 1
38. Smallwood, P.V. (1986) *Polymer* **27** 1609
39. Davidson, J.A. & Witenhafer, D.E. (1980) *J. Polym. Sci., Polym. Phys. Ed.* **18** 51
40. Emirova, I.V., Savelyanov, V.P. & Golikov, M.V. (1984) *Vysokomol. Soedin.* **12** 890
41. Zichy, E.L. (1977) *J. Macromol. Sci., Chem.* **11** 1205
42. Weber, J., Thuemmler, W. & Kaltwasser, H. (1983) *Plaste Kautsch.* **30** 606
43. Tornell, B.E. & Uustalu, J.M. (1982) *J. Vinyl Technol.* **4** 53
44. Smallwood, P.V. (1981) *U.S.* 4446287
45. Ugelstad, J. (1977) *J. Macromol. Sci., Chem.* **11** 1281
46. Ugelstad, J., Mork, P.C. & Hansen, F.K. (1981) *Pure Appl. Chem.* **53** 323
47. Crosato-Arnaldi, A., Gasparini, P. & Talamini, G. (1968) *Makromol. Chem.* **117** 140
48. Abdel-Alim, A.H. & Hamielec, A.E. (1972) *J. Appl. Polym. Sci.* **16** 783
49. Ugelstad, J., Flogstad, H., Hertzberg, T. & Sund, E. (1973) *Makromol. Chem.* **164** 171
50. Kuchanov, S.J. & Bort, D.N. (1973) *Polym. Sci. USSR* **15** 2712
51. Olaj, O.F. (1975) *Angew. Makromol. Chem.* **1** 47
52. Thiele, R., Nelles, J. & Rauchstein, D. (1978) *Plaste Kautsch.* **7** 395
53. Kafarov, V.V., Dorokhov, I.N., Dudorov, A.A. & Kusy, P. (1978) *Dokl. Akad. Nauk SSSR* **243** 711
54. Kelsall, D.G. & Maitland, G.C. (1983) In: Reichert, K.H. & Geiseler, W. (eds) *Polymer reactions engineering.* C. Hanser Verlag, München, p. 131
55. Xie, T.Y., Hamielec, A.E., Wood, P.E. & Woods, D.R. (1987) *J. Appl. Polym. Sci.* **34** 1766
56. Flory, P.J. (1953) *Principles of polymer chemistry.* Cornell University Press, Ithaca
57. Berens, A.R. (1975) *Angew. Makromol. Chem.* **47** 97
58. Abdel-Alim, A.H. (1978) *J. Appl. Polym. Sci.* **22** 3697
59. Gerrens, H., Fink, W. & Kohlein, E. (1967) *J. Polym. Sci., C* **16** 2781
60. Hayduk, W. & Laudie, H. (1973) *AIChE J.* **19** 1233
61. Berens, A.R. (1974) *Polym. Prepr.* **15** 197
62. Hayduk, W. & Laudie, H. (1974) *J. Chem. Eng. Data* **19** 253
63. Patel, C.B., Grandin, R.E., Gupta, R., Philips, E.M., Reynolds, C.E. & Chan, R.K.S. (1979) *Polym. J.* **11** 43
64. Chan, R.K.S., Langsam, M. & Hamielec, A.E. (1982) *J. Macromol. Sci., Chem.* **17**, 969
65. Sidiropoulou, E. & Kiparissides, C. (1990) *J. Macromol. Sci., Chem.* **27** 257
66. Barth, H. (1982) *Plastverarbeitung* **33** 1047
67. Semon, W.L. & Stahl, G.A. (1981) *J. Macromol. Sci., Chem.* **15** 1263
68. Clark, M. (1982) In: Burges, R.H. (ed.) *Particulate nature of PVC.* Applied Science Publ., London
69. Burges, R.H. (1982) In: Burges, R.H. (ed.) *Manufacturing and processing of PVC.* Applied Science Publ., London

70. Hause, A.F. (1971) *J. Polym. Sci., C* **33** 1
71. Nilsson, H., Silvergen, C. & Törnell, B. (1983) *Angew. Makromol. Chem.* **112** 125
72. Menges, G. & Berndtsen, N. (1976) *Kunststoffe* **66** 736
73. Pazonyi, T., Pukanszky, B., Kallo, A. & Komaromi, S. (1979) *Angew. Makromol. Chem.* **81** 109
74. Lukáš, R., Tyráčková, R. & Kolínský, M. (1984) *J. Appl. Polym. Sci.* **29** 901
75. Lukáš, R., Michalcová, J. & Tyráčková, V. (1985) *J. Appl. Polym. Sci.* **30** 843
76. Lukáš, R. (1989) *Makromol. Chem., Macromol. Symp.* **29** 21
77. Světlý, J., Lukáš, R., Michalcová, J. & Kolínský, M. (1984) *Makromol. Chem.* **185** 2183
78. Lukáš, R. & Přádová, O. (1986) *Makromol. Chem.* **187** 2111
79. Lukáš, R., Přádová, O., Michalcová, J. & Palečková, V. (1985) *J. Polym. Sci., Polym. Lett. Ed.* **23** 85
80. Mrázek, Z., Jungwirt, A. & Kolínský, M. (1982) *J. Appl. Polym. Sci.* **27** 2079
81. Mrázek, Z., Jungwirt, A. & Kolínský, M. (1987) *J. Appl. Polym. Sci.* **34** 2681
82. Mrázek, Z., Jungwirt, A. & Kolínský, M. (1982) *J. Appl. Polym. Sci.* **27** 1513
83. Mrázek, Z. & Lukáš, R. (1989) *Makromol. Chem., Macromol. Symp.* **29** 155
84. Törnell, B. (1988) *Polym. Plast. Technol. Eng.* **27** 1
85. Nilsson, H., Silvergren, C. & Törnell, B. (1978) *Eur. Polym. J.* **14** 737
86. Hoff, H., Lüssi, H. & Gerspacher, R.P. (1964) *Makromol. Chem.* **78** 37
87. Hoff, H., Lüssi, H. & Hammer, E. (1965) *Makromol. Chem.* **82** 184
88. Gupta, D.C. (1985) *J. Appl. Polym. Sci.* **30** 487
89. Clark, J. & Vincent, B. (1981) *J. Colloid Interface Sci.* **82** 208
90. DeGennes, P.G. (1980) *Macromolecules* **13** 1069
91. Karaenev, S., Mikhailova, M. & Mihna, R. (1988) *Dokl. Bulg. Akad. Nauk* **41** 81
92. Giannetti, E., Mazzochi, R., Fiore, L. & Visani, F. (1986) *J. Polym. Sci., Polym. Chem. Ed.* **24** 2517
93. Matějíček, A., Seidl, J., Vladyka, J. & Krejcar, J. (1980) *Angew. Makromol. Chem.* **88** 113
94. Mueller, K.F. & Heiber, S.J. (1982) *J. Appl. Polym. Sci.* **27** 4043
95. Peppas, N.A. (ed.) *Hydrogels in medicine and pharmacy*, vol. 2. CRC Press, Boca Raton
96. Scranton, A.B., Mikos, A.G., Scranton, L.C. & Peppas, N.A. (1990) *J. Appl. Polym. Sci.* **40** 997
97. Horák, D., Švec, F., Kálal, J., Gumargalieva, K., Adamyan, A., Skuba, N., Titova, M. & Trostenyuk, N. (1986) *Biomaterials* **7** 188
98. Horák, D., Metalová, M., Švec, F., Drobník, J., Kálal, J., Borovička, M., Adamyan, A., Voronkova, O.S. & Gumargalieva, K. (1987) *Biomaterials* **8** 142
99. Ohsuka, Y., Kawaguchi, H. & Yamamoto, T. (1985) *J. Appl. Polym. Sci.* **27** 3279
100. Howell, B.D.B. & Peppas, N.A. (1987) *Eur. Polym. J.* **23** 591
101. Skovby, M.H.B. & Kops, J. (1990) *J. Appl. Polym. Sci.* **39** 169
102. Galina, H. & Kolarz, B.N. (1979) *J. Appl. Polym. Sci.* **23** 3017
103. Nishii, M. & Hayashi, K. (1975) *Annu. Rev. Mater. Sci.* **5** 135
104. Murano, M. & Harwood, H.J. (1970) *Macromolecules* **3** 605
105. Higashimura, T., Matsuda, T. & Okamura, S. (1970) *J. Polym. Sci., A-1* **8** 483
106. Chen, C.C., Milburn, M.V. & Overberger, C.G. (1986) *J. Polym. Sci., Polym. Lett. Ed.* **24** 627
107. Boyadijan, R., Seytre, G., Sage, D. & Berticat, P. (1975) *Eur. Polym. J.* **12** 409
108. Vanderhoff, J.W. & Wiley, R.M. (1966) *U.S.* 3284393
109. Akio, W., Mitsuaki, S., Ryoichi, Y. & Shigenobu, U. (1975) *Japan* 75-107089
110. Kazutaka, J. (1977) *Japan* 77-154888
111. Shibahara, H., Okada, M., Tominaga, Y., Noda, K. & Osuga, Y. (1979) *Japan* 79-74 841
112. Schenk, H.U., Krapf, H. & Oppenlaender, K. (1977) *Ger. Offen.* 2536597
113. Dimonie, M.V., Boghina, C.M., Marinescu, N.N., Marinescu, M.M., Cincu, C.I. & Oprescu, C.G. (1982) *Eur. Polym. J.* **18** 639
114. Graillot, C., Pichot, C., Guyot, A. & El-Aasser, M.S. (1986) *J. Polym. Sci., Polym. Chem. Ed.* **24** 427
115. Singh, P., Huang, P.C. & Reichert, K.H. (1986) In: Reichert, K.H. & Geiseler, W. (eds) *Polymer reaction engineering*. Hüthing and Wepf Verlag, Basel
116. Baade, W. & Reichert, K.H. (1986) *Makromol. Chem., Rapid Commun.* **7** 235
117. Candau, F., Leong, Y.S. & Fitch, R.M. (1985) *J. Polym. Sci., Polym. Chem. Ed.* **23** 193
118. Vanderhoff, J.W., Disteffano, F.V., El-Aasser, M.S., O'Leary, R., Schoffer, O.M. & Visioli, D.L. (1984) *J. Dispersion Sci., Technol.* **5** 323
119. Trubitsyna, S.N., Ismailov, K. & Askarov, M.A. (1978) *Vysokomol. Soedin., A* **20** 2608
120. Hunkeler, D., Hamielec, A.E. & Baade, W. (1989) *Polymer* **30** 127

8

Methods of characterization of polymer dispersions

8.1 METHODS OF PARTICLE SIZE DETERMINATION

The properties of colloidal systems are mainly affected by the particle size and its distribution. A number of approaches and techniques for particle size analysis are available nowadays. According to Scarlett [1], about 400 methods have been reported for determining the size of colloid particles. These methods can be classified into two main categories: fractional and nonfractional.

The nonfractional methods include:

– light scattering: classical, turbidimetric, Fraunhofer, dynamic, etc.;
– neutron scattering; electron and X-ray scattering;
– electron microscopy, etc.

The fractional methods include:

– hydrodynamic chromatography (HDC);
– gel chromatography (GC);
– disk centrifugation (DCF);
– sieving (SIV);
– field-flow fractionation (FFF), which can be sedimentation, steric, flow, thermal, etc.

8.1.1 Classical light scattering

Light scattering can be used for determining particle sizes between 10 and 10 000 nm. It also provides data about the particle shape and interparticle interaction. This is possible because the amplitude of the scattered light is a function of the shape, size and refractive index of the polymer particle. Light scattering is accompanied by unwanted interference due to light scattered by reflection which is minimized by a suitable choice of wavelength of the incident light λ_1. The measurements are carried out with highly dilute colloidal systems (polymer dispersions) containing spherical particles on condition that the particle diameter (d) is smaller than 1/20 of λ_1.

Rayleigh [2, 3] derived a relationship between the reduced intensity of scattered light, I_r, on λ_1 and on the particle size

$$I_r = \frac{N\pi^4 d^6}{S\lambda_1^4} \left(\frac{m^2-1}{m^2+2}\right)^2, \tag{1}$$

where N is the number of particles dispersed in 1 cm^3, m is the relative refractive index, given by the ratio between the refractive indices of the particles and the molecules from the medium, and S is the surface area of the dispersed particles. The Rayleigh law holds only if $\alpha \leq 0.4$ for any value of n, where α equals $\pi d/\lambda$, where λ represents the wavelength of the scattered light. Equation (1) is unsuitable for calculating the particle size of normal latices containing particles with d of ca. 100 or 200 nm.

Debye [4] modified Eq. (1) to calculate the size of spherical particles

$$I_r = \frac{3c\pi^3 d^3 n^4}{4\rho\lambda_1^4} \left(\frac{m^2-1}{m^2+2}\right)^2 \tag{2}$$

and

$$I_r = \frac{c\pi^3 \rho d^3 n_0^2}{3\lambda^4} \left(\frac{\delta n}{\delta c}\right)^2, \tag{3}$$

where ρ is the polymer particle density, n_0 and n are the refractive indices of the medium and particles, respectively, λ is the wavelength of the light *in vacuo*, c is the concentration of the polymer in g cm^{-3}, and $\delta n/\delta c$ is the refractive index increment.

Debye's equations are valid if $m \leq 1.05$. The theories of Rayleigh [2] and Gans [5] can be used for any particle size on condition that m is small enough [5]. The validity of the light-scattering theories for particle size determination has also been verified by other methods, e.g. electron microscopy, on samples of monodisperse polymer latices of styrene and vinyl benzene [7]. Provided $d/\lambda \leq 0.6$, good agreement was observed between the two methods. If $d > \lambda_1$, then I_r decreases rapidly with the scattering angle and only relative data about the particle size are obtained.

If $d > \lambda_1/20$, the reduced intensity of scattering is a function of the angle of observation. This effect is dealt with in Mie's theory [8], but the complexity of the functions involved make it difficult to deal with polydispersity.

The suitability of Mie's theory was confirmed by experimental data. By measuring the light scattering of monodisperse polymer latices with particles smaller than 500 nm and by applying Mie's theory, Marx & Mulholland [9] obtained very precise results.

8.1.2 Photon correlation spectroscopy (PCS)

Photon correlation spectroscopy, also known as dynamic or quasi-elastic light scattering, provides data on the particle size, size distribution, the translational and rotational diffusion of macromolecules, inter- and intramolecular interactions and intramolecular relaxations.

The use of PCS to determine particle size is based on measurements, via autocorrelation of the time dependence of the scattered light, of the diffusion coefficients of suspended particles undergoing Brownian motion [10–13]. The measured autocorrelation function, $G^{(2)}(\tau)$, is given by

Particle size determination

$$G^{(2)}(\tau) = A \left[1 + \beta/g^{(1)}(\tau)^2 \right], \qquad (4)$$

where A is the base line, β is an equipment-related constant and $g^{(1)}(\tau)$ is the first-order autocorrelation function. The base-line constant, A, can be obtained from the long-time asymptote of the measured autocorrelation function or from the square of the average photon flux. The measured autocorrelation function of scattered light intensity as a function of the delay time is easily transformed into the normalized first-order autocorrelation function, $g^{(1)}(\tau)$, which depends upon the delay time, τ, the scattering vector, k, and the particle-size distribution as shown in Eq. (5)

$$g^{(1)}(\tau) = \int_0^\infty F(\Gamma) \exp(-\Gamma\tau) \, d\Gamma, \qquad (5)$$

where $F(\Gamma)$ is the normalized distribution function of the decay constants, Γ, $\Gamma = k^2 D$, $k = 4\pi n/\lambda \sin(\theta/2)$, n is the refractive index of the medium, λ_1 is the wavelength of incident light, θ is the scattering angle, and D is the particle diffusion constant.

When data are analysed by means of Eq. (5), the resulting distribution is the probability density function, $F(\Gamma)$. $F(\Gamma)$ is converted from light intensities to mass fractions by use of the appropriate Mie corrections and from distributions of decay constants to distributions of particle sizes by the Stokes–Einstein expression for the diffusion coefficient, $D = kT/(3\pi\eta d)$, where k is Boltzmann's constant, T is the temperature, d is the particle diameter and η is the viscosity. The experimental problem is to measure the autocorrelation function as accurately as possible.

Measurements are done at multiple scattering angles. There are special problems that arise when d is large relative to λ_1. This is interesting since many latex systems have particles with d from 300 to 1000 nm and a large ratio of particle to fluid refractive index. The usual laser wavelengths used in light scattering are *ca.* 500 nm.

In addition, the refractive index difference between the particle and the medium will also cause a phase shift in the scattered light leaving the particle. Owing to these effects, the intensity of the light scattered by suspended spherical particles depends on d relative to λ_1, θ and the refractive index ratio. In order to account for these effects, one can use Mie's theory to obtain the volume or mass distribution of the latex particles instead of the intensity distribution initially obtained from PCS data.

8.1.3 Turbidimetry

When light passes through a disperse medium not absorbing the light, the intensity of the primary beam is reduced because of light scattering and reflection on the dispersed particles. A measure of this reduction is expressed by the coefficient of turbidity τ [14]. A relationship between the primary beam intensity, I_0, and the beam intensity I after passing through the disperse medium of thickness l and the turbidity coefficient τ, is expressed by Eq. (6)

$$\tau = \frac{1}{l} \ln \frac{I_0}{I}. \qquad (6)$$

Mie's theory introduces the turbidity, τ, into a relationship between the particle number, N, the particle size and the optical constants of the polymer particles:

$$\tau(\lambda) = N l \int_0^\infty \frac{\pi}{4} d^2 Q_s (\alpha,m) f(d) \, \mathrm{d}d , \tag{7}$$

$$\alpha = \frac{\pi d}{\lambda} , \qquad m = n_p/n_m , \tag{8}$$

where $\tau(\lambda)$ is the turbidity at wavelength λ, N is the particle diameter, $f(d)$ is the frequency distribution of the particle size, λ is the wavelength in the medium relative to that *in vacuo*: $\lambda = \lambda_0/n_m$, n_p and n_m are the refractive indices of the particle and medium respectively, and the coefficient, Q_s, characterizes the effectiveness of light scattering by the particles.

The specific turbidity is given by the relationship

$$\frac{\tau}{c} = \frac{A}{cl} 2.303 \tag{9}$$

and Eq. (10)

$$d = \frac{\tau}{c} \frac{\rho \lambda_1^4}{4\pi n^4} \left(\frac{m^2 + 2}{m^2 + 1} \right)^{1/3} \tag{10}$$

enables calculation of the latex particle size, where A stands for absorbance, l is the pathlength, c is the concentration (g cm^{-3}) and ρ is the particle density.

Melik & Fogler [15] proposed a rather simple procedure to estimate the particle size and size distribution in turbidimetric measurements at two wavelengths. It is necessary to carry out measurements in strongly diluted-solutions which eliminate multiple scattering. To evaluate the measurements, knowledge of the refractive indices of the polymer particles and the medium and their dependences on the wavelength and shape of the particle distribution function is required. The particle size distribution can be described by the dependence

$$f(d) = \frac{1}{2\pi\sigma_g d} \exp\left(-\frac{(\ln d - \ln d_g)}{2\sigma_g^2} \right) \tag{11}$$

and is determined by two parameters, the geometrical diameter, d_g, and the geometrical standard deviation, τ_g.

In order to eliminate the dependence of the turbidity on the particle number and to obtain information about the particle size and distribution, the ratio of the turbidities from the two measurements was used.

This method has been partially modified by Kourti *et al.* [16]. To obtain the two crucial parameters, the authors analysed at least three turbidimetric measurements at various wavelengths regardless of the use of the ratio of turbidities or the wavelengths.

The determination of the particle number and the size and distribution of the particles is influenced not only by the mathematical solution of complicated functions but also by the adequacy of the initial physical parameters, such as the refractive indices for the particles and medium. A small change in the refractive index leads to a rather large change in the particle number as has been observed in the characterization of poly(vinyl acetate) latices [17]. A change in the refractive index of the particle by 0.25% led to a 14% change in the particle number.

8.1.4 Stopped-flow spectroscopy

Stopped-flow spectroscopy has proved to be useful as a simple method for particle size determination [18]. This method gives an average particle diameter (d_s) near the mass-average diameter when applied to polydisperse particles. The data required to calculate the particle diameter are simply the rate constant of rapid coagulation (k_r) and the optical factor (the turbidity), both of which are readily available from modified Smoluchowski theory [19, 20] and the Rayleigh–Gans–Debye theory [2–6]. In the stopped-flow method, the particle diameter is determined by measuring not only the turbidity in the steady state but also the change in turbidity with time during coagulation, i.e. τ_0 and $(d\tau/dt)_0$ in the following equation:

$$d_s = \left[\frac{12 w k_r}{\pi \rho (1/\tau_0)(d\tau/dt)_0} \right]^{1/3}, \tag{12}$$

where w is the solid content of the polymer latex and ρ the density of the particles.

The value of $(d\tau/dt)$ is directly proportional to the number of primary polymer particles, N_0, as seen from the equation

$$N_0 = (1/\tau_0)(d\tau/dt)_0 / 2 k_r F, \tag{13}$$

where $F = (C_2/2C_1) - 1$, and C_1 and C_2 are extinction cross-sections of primary and double particles, respectively.

8.1.5 Small-angle neutron and X-ray scattering

Small-angle neutron scattering (SANS) is currently used to determine very small particles with d ca. 2 nm and to detect the structure of colloidal particles. It is also used for analysis of clusters of macromolecules and to determine the number and size distribution of latex particles.

The intensity of the scattered neutron beam (with a scattering vector (q) is given by Eq. [21]

$$I(q) = A (\rho_m - \rho_s)^2 V_m^2 P(q), \tag{14}$$

where A is the constant of the apparatus, ρ_m and ρ_s characterize neutron scattering at the particles and molecules of the medium and their density, V_m is the particle volume, and $P(q)$ is the particle shape factor.

The scattering angle θ is defined as

$$\theta = 4\pi \sin(\theta/2) \lambda, \tag{15}$$

where λ is the wavelength of the incident light.

The function $P(q)$ is given by

$$P(q) = \int_0^{\pi/2} \frac{\sin^2(qH \cos \beta)}{q^2 H^2 \cos^2 \beta} \frac{4 J_1^2(qr \sin \beta)}{q^2 r^2 \sin^2 \beta} \sin \beta \, d\beta, \tag{16}$$

where $J_1(qr \sin \beta)$ is the Bessel function, H is the half length and r is the particle radius. The mathematical solution of Eq. (14) yields the particle size.

Hayter & Penfold [22] used SANS to determine the size of the micelles of the mixed emulsifying system (sodium dodecyl sulfate and dodecyltrimethylammonium halide) and the micellar aggregation number, and to characterize the surface structure and charge of the micelles.

SANS enables us to examine the nature of the interparticle interactions based on micelles or polymer particles [23], e.g. the interaction radius for a concentrated polystyrene latex characterized by an 18 nm particle radius was about 27 nm.

The results from an analysis of the polymer particles formed by poly(methyl methacrylate) and poly(12-hydroxystearic acid) provide an example of the application of SANS in studying the structure of polymer particles [24]. The particle core was formed by poly(methyl methacrylate) with a radius of *ca.* 18 nm; the 6 nm thick surface layer was mostly formed by polyacid and the interaction radius was 38 nm. The poly(12-hydroxystearic acid) chains were adsorbed on the core surface and closely packed.

Small-angle X-ray scattering (SAXS) gives data not only about the size but also about the shape of colloidal particles and is mainly applicable to smaller particles (*d ca.* 10 nm). The intensity of scattered radiation (on the assumption that no multiple scattering and interference of radiation occurs) is expressed by Eq. (17)

$$I(q) = \int_0^\infty N(d) d^6 I_0(q,d) \, dd, \tag{17}$$

where the function $I_0(q,d)$ is the normalized shape factor of a particle with diameter d, $N(d)$ is the dimensional distribution function, $q = (4\pi/\lambda)\sin(\theta/2)$ and θ is the scattering angle.

Several techniques can be employed to determine $N(d)$ from SAXS data. The existing methods can be divided into three groups. Firstly, the size distribution can be found by assuming a simple distribution function for $N(d)$ containing only a few constants such as the log-normal distribution. The actual constants of such a function can be evaluated from the measured intensity. In contrast to this approach the other procedures require no assumption about the form of $N(d)$ to solve Eq. (17). Here one must distinguish between general numerical techniques for solving Eq. (17) and so-called transform solutions.

If no adequate corrections are made for radiation penetration through the polymer particles, the data about the form and the size distribution of the particles may be subject to a certain error. Pleshakov & Nagornyi [26] dealt with the problem, proposing mathematical model for eliminating the effect of radiation penetration through the particles.

Idhii *et al.* [27] used SAXS not only to determine the size of the micelles of a cationic emulsifier (alkyltrimethylammonium chloride) but also to find the distance between two neighbouring micelles. The radii of dodecyl-, tetradecyl-, and hexadecyltrimethylammonium chloride (DTAC, TTAC, HTAC) micelles varied around 1.5 nm, 1.8 nm, and 2.0 nm. The distance between the micelles was *ca.* 5 times greater than the particle radius and decreased with increasing micelle concentration. On raising the concentration of HTAC from 0.02 to 0.05 g cm^{-3}, the interparticle distance of the micelles decreased from 14 nm to 10 nm.

8.1.6 Chromatographic methods

Particle chromatography makes use of various column packings and is used for determining the size distribution of spherical polymer particles of submicron size [28–30]. Depending on the column packing, this technique can be classified into hydrodynamic chromatography (HDC) and gel chromatography (GC).

HDC uses columns packed with nonporous packings; the separation is based on the different rates of particle flow through the column dependent on the particle size. The velocity gradient is produced by the flow of the mobile phase through the interstices among the packing grains. Smaller particles penetrate into smaller interstices, therefore travelling further than larger particles, and hence eluating last. Larger particles travel through larger free spaces only, they are forced into streamlines at higher flow and lower axial velocities, therefore eluting first.

With porous packings (GC), the particle separation follows the steric exclusion mechanism. Polymer particles smaller than the packing grain pores diffuse into the pores; their motion is retarded by possible interaction with the packing or their longer diffusion route and they elute last as in the HDC.

Capillary columns can also be used for particle separation. The particle separation ranges here from 700 to 50 000 nm. The efficiency of capillary columns can be altered by the eluent viscosity, flow rate, and the capillary diameter [31].

The detector response at a retention volume v, $F(v)$ can be expressed as

$$F(v) = \int W(y) G(v,y) \, dy, \tag{18}$$

where $W(y)$ is the actual chromatogram which would be obtained in the absence of axial scattering, $G(v,y)$ is the normalized response of the detector for a monodisperse system with a retention volume y. To solve Eq. (18), we have to know the function $G(v,y)$ and its parameters. To calculate the size distribution, $W(y)$ the relationship between the average retention volume y and the particle diameter d (calibration curve) has to be known. Chromatograms of monodisperse polymer latices are used to construct a calibration curve related to a particle with diameter d and with an average retention volume y

$$\ln d = a + by + cy^2. \tag{19}$$

By injecting a monodisperse polymer latex into the column, a certain chromatographic peak broadening is observed as a result of axial scattering, which requires some correction of the detector response. The corrected detector response at a retention volume, v, relative to the particle number, N, extinction coeffiecient, K, and the particle diameter, d, can be expressed by the relationship

$$W(v) = N(v) \, d^a(v) \, K^b(v), \tag{20}$$

where a and b are the parameters of the detection system and the extinction coefficient K is given by light-scattering theory [8].

The area of the fractogram W_j is related to N_j, the total number of particles [32]

$$N_j = \frac{2.303 W_j}{R_{ext} \, \chi}, \tag{21}$$

where χ is the optical pathlength and R_{ext} is the extinction cross-section which is the sum of the scattering and absorption cross-sections.

The particle diameter, d, is obtained from the power law expression

$$R_{ext} = Kd^{\alpha}, \tag{22}$$

where K and α are constants, e.g. for polystyrene particles, [32] gives the following values for K and α: 2.21×10^{-15} cm^2 nm^{-1} and 2.42 at 220 nm.

The resolution of the chromatogram can be positively influenced by a suitable choice of column packing and wavelength of the detection system [33]. The operation of van der Waals forces in the system also leads to better separation and polymer particle resolution [34].

8.1.7 Field-flow fractionation

The FFF technique is based on influencing the rate of the polymer particle flow through a narrow field channel. The flow rate of a liquid in a narrow channel decreases rapidly from the centre to the wall. By applying a field (see below), smaller particles are located closer to the centre and larger ones nearer the walls, where the flow rate of the fluid is lowest and particle motion will thus be most retarded. The application of a field in the flow of fluid results in the separation of individual polymer particle fractions of different sizes along the channel. Fractions of small particles elute first followed by those of larger particles. The rate of movement of the particles in the channel and a measure of their distribution may be determined continuously by density, particle size or particle mass measurements.

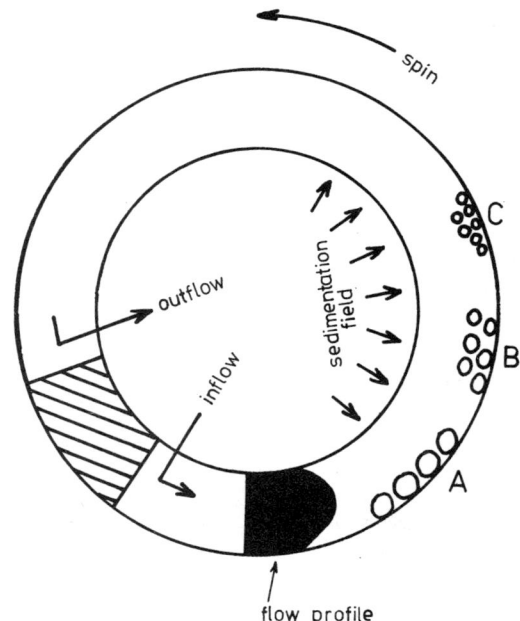

Fig. 35 — Separation of large (A), medium (B) and small (C) particles via sedimentation FFF. Outline represents a side view of 0.025 cm thick channel [36].

FFF can be classified according to the type of applied field into sedimentation, thermal, flow and steric. Sedimentation FFF has been most widely applied in determining particle size distribution with particle dimensions of between 10 and 1000 nm. Steric FFF is suitable for the separation of large particles (1–100 μm) [35].

The measurement of size and size distribution of colloid particles by sedimentation FFF has been reviewed by Giddings *et al.* [36, 37] and Kirkland *et al.* [38]. These various authors described a procedure for obtaining experimental data for particle separation and analysis according to a theory they derived; they described the relationship between the size of the particles and their parameters which affect the mechanism of particle motion in the sedimentation FFF channel. Figure 35 shows particle separation by a sedimentation field generated by a centrifuge. A parabolical cross-section of the channel positively affects the axial separation of the particles according to their size and mass. The theory of sedimentation FFF produces an expression to calculate the particle cross-section

$$d = \left(\frac{6kT}{\Delta\rho\pi Gwa}\right)^{1/3}, \quad (23)$$

where k is Boltzmann's constant, T is the absolute temperature, $\Delta\rho$ is the difference in density of the particle and liquid carrier, a is a retention parameter characterizing the thickness of the particle cloud, w is the channel thickness and G is the field strength.

The particle distribution in an FFF channel follows an exponential law and is given by

$$c(x) = c_0 \, e^{-x/aw}, \quad (24)$$

where $c(x)$ is the concentration at distance x from the wall and c_0 is the particle concentration at the wall ($x = 0$). The rate of particle flow in the liquid carrier stream is expressed by the retention ratio R which is given by

$$R = V^0/V_r = 6a[\cotgh(1/2a) - 2a] = V/v. \quad (25)$$

Here V^0 is the volume of the channel, V_r is the retention volume, V is the velocity of the particle motion, and v is the velocity of carrier motion.

Sedimentation FFF is applicable not only to monodisperse but to polydisperse polymer latices in particular and can provide fractions with a narrow particle size distribution throughout the fractionation. Variation of the detector response with the elution volume (fractogram) illustrates the polydispersity of the colloidal dispersion. The retention volume of each particle volume fraction is used for calculating a, which, on substituting into Eq. (23), gives the value of d.

Giddings *et al.* [39, 40] succeeded in applying the sedimentation method to the characterization of poly(styrene) and poly(vinyl chloride) latices. The authors compared FFF with other techniques, mainly with electron microscopy, and found good agreement. Increase in the field intensity positively influences the resolving power of FFF and enables the resolution of particle distributions with small differences in size and mass [38].

Steric FFF has also been applied industrially, mainly to disperse particles with diameters around 50–100 nm [41]. The difference between this technique and classical FFF consists in the sequence of separation of large and small particles in a thin channel following from the different dependence of the particle retention volume on particle size. For systems using classical FFF, the retention volume is in proportion to the cube of the particle diameter

$$V_r = \frac{GwV^0}{36kT}d^3 = Bd^3 \qquad (26)$$

and, conversely, for steric FFF, it is proportional to the reciprocal of this diameter

$$V_r = \frac{B'}{d}. \qquad (27)$$

These variations show that in normal FFF, the smallest particles move at the highest rate in a thin channel and elute first, whereas in steric FFF the largest particles migrate most rapidly.

FFF is mainly used to characterize the sizes of spherical silica particles as a chromatographic material and to characterize biological material and inorganic colloids [41–43].

8.1.8 Electron microscopy

This is a direct and absolute method for determining polymer particle size distributions. Menold *et al.* [44] were the first to apply electron microscopy (EM) to the problem of particle sizes of polymer suspensions and emulsions. The authors put forward a procedure which involves strong freezing of the colloid system, particle separation from the liquid medium and preparation of latices and is still successfully applied. EM is mainly used for calibration measurements to verify the accuracy and appropriation of new methods for the particle sizing colloidal systems. A disadvantage of EM is its length and laboriousness; in addition, it requires the use of special high-performance vacuum systems.

In sample preparation, the procedure of spraying a very dilute polymer latex on a support followed by evaporation of the liquid medium is commonly used. The procedure is suitable for hard polymer particles which are not subject to deformation and show little tendency to associate. When sizing particles formed by a soft polymer, one has to make additional corrections for the procedure and evaluation with regard to particle deformation and the tendency of the particles to associate.

To reduce the deformability of soft particles, various types of polymer latices into which the particles are inserted, and where they preserve their original shape, have been proposed [45]. Another way to reduce the deformability of the particles measured is to reinforce them before measurement; this is carried out by subjecting the particles to a high temperature [46] or to radiation [47]. Increased hardness of the particles may also be introduced by saturation of the double bonds present by various reactive additives or by slight particle swelling using a monomer which forms a hard polymer (e.g. by styrene). Peltzbauer *et al.* [48] hardened soft poly(chloroprene) particles by bromination prior to determining their size and number by EM. Reduction in the deformability of the particles is achieved by latex sample measurements at lower temperatures, as used by Walter & Bryant [49] and Talmon & Miller [50]. The latter used metal (copper) and also polyimine foils, which enabled measurement of particles smaller than 250 nm.

8.1.9 Centrifugation

The disc centrifuge — photosedimentometer has been used for particle-sizing for a wide variety of dispersed materials [51]. In this device, the particles intercept the light beam of the optical detector near the outer edge of the rotor which produces an electrical signal proportional to the optical density of the fluid in the beam. By using Stokes' law one can calculate the hydrodynamic size of the particles in the beam at any instant.

The centrifugal method is the second absolute technique (EM being the first) for sizing submicron particles. The disc centrifuge with an optical detection system provides quick and reliable data on particle sizes. The data are obtained by evaluating sedimentation curves which depend on the rate of sedimentation and the particle size. The dependence of the particle size (diameter d) on the sedimentation time t is expressed by Eq. (28):

$$d = \frac{6.299 \times 10^6 \, \eta}{t\omega^2 \Delta\rho} \log \frac{R_2}{R_1}, \qquad (28)$$

where η is the solvent viscosity, ω is the speed of rotation, $\Delta\rho$ is the difference between the particle and solvent densities, and R_1 and R_2 are the initial and final sedimentation radii obtained from the dimension of the disc and the fluid volume, respectively [52].

Polymer latices with d ca. 100 to 1000 nm can be sized with the line-start (LIST) and the homogeneous-start methods (HOST) in a disc centrifuge with a broad-spectrum optical detector. In LIST a small amount of sample is injected into the spinning rotor, which contains fluid with an externally-generated density gradient. An alternative to LIST is HOST, where the spin fluid initially contains a uniform concentration of the colloid whose particle size distribution is to be determined [53].

The centrifugal method resolves effectively polymer particles between 10 and 50 000 nm and is used successfully for determining the particle size distribution of polymer dispersions [52–54].

8.1.10 Titration of latices with an aqueous solution of emulsifier

This is a classical but reliable technique used for determining the particle surface area. It was proposed, worked out mathematically, and experimentally verified by Maron *et al.* [55–57]; they based their study on the idea that the original amount of emulsifier in the system does not completely cover the surface area P of the particles. P is given by the product of the amount of emulsifier adsorbed on the particle surface (S_A), the effective surface of the emulsifier molecule (A_S) and Avogadro's number (N_A)

$$P = S_A N_A A_S. \qquad (29)$$

The continuous addition of an aqueous solution of emulsifier to a polymer dispersion will eventually cover the total particle surface area with emulsifier molecules. The attainment of the inflexion point in the titration is detected by a sudden change of physical property of the titrated solution (conductivity, viscosity, surface tension, etc.).

This change results from the formation of emulsifier micelles in the system

$$S_T = S_A + S_{aq}. \qquad (30)$$

The value of S_A is obtained by examining the variation of the overall emulsifier concentration with the concentration of the polymer latex, m,

$$S_T = S_A/m + S_{aq} \tag{31}$$

or from [57]

$$S_T = (S_A/m - I)m + I, \tag{32}$$

where I is related to the amount of micellar emulsifier present in the system. The expression S_A/m denotes the molar content of emulsifier adsorbed per g of polymer; it is constant for a particular latex and varies with the particle size. It is better to express the emulsifier adsorption by the parameter A_S, which does not vary for the particular surface composition of the polymer particle [58, 59]:

$$A_S = \frac{6m}{N_A S_A V}, \tag{33}$$

where V is the particle size. With increasing lipophilicity of the polymer particle surface, the value of A_S decreases owing to the increased number of adsorbed emulsifier molecules per unit area of particle surface. The value of A_S also provides data about the surface properties of the polymer particle.

The accuracy of this method depends on the choice of a suitable emulsifier [60] and detection of the inflexion point of titration [61]. Errors occur, for example, in the preparation of dilute solutions of the basic latex and their titrations as a result of the change in ionic strength and hence the CMC. It is therefore necessary to keep the ionic strength constant; this is done by adding a suitable material [62]. The inflexion point has to be estimated in most cases from wider or narrower concentration ranges of emulsifier; the calculation is thus subjected to 5% error. To minimize the error, various mathematical procedures are available. Linear regression [63] seems to be the best for modelling the experimental titration curve as regards detection of the inflexion point.

The titration method can be applied to polymer dispersions with a particle size > 100 nm. Its great virtue is its speed and cheapness; its disadvantage is that the possible emulsification of the monomer present and any imprecision in the estimate of the inflexion point adversely affects the accuracy of the sizing [64].

References

1. Scarlett, B. (1982) In: Stanleywood, N. & Allen, T. (eds) *Particle size analysis*. J. Wiley & Sons, Chichester, p. 219
2. Rayleigh, J. (1871) *Phil. Mag.* **41** 274
3. Rayleigh, J. (1871) *Phil. Mag.* **41** 447
4. Debye, J. (1947) *J. Phys. Colloid Chem.* **51** 18
5. Rudin, A. (1982) In: *The elements of polymer science and engineering*. Academic Press, San Diego, Chapter 3
6. Gans, R. (1925) *Ann. Phys.* (Leipzig) **76** 29
7. Burnett, G.M., Lehre, R.S., Ovenall, D.W. & Peaker, F.W. (1958) *J. Polym. Sci.* **29** 417
8. Mie, G. (1908) *Ann. Phys.* (Leipzig) **25** 377
9. Marx, E. & Mulholland, G.W. (1983) *J. Res. Natl. Bur. Stand.* (U.S.) **88** 321
10. Koppel, D.E. (1972) *J. Chem. Phys.* **57** 4814
11. Danheke, B.E. (ed.) (1983) *Measurement of suspended particles by quasi-elastic light scattering*. J. Wiley & Sons, New York
12. Morrison, I.D. & Grabowski, E.F. (1985) *Langmuir* **1** 496
13. Bedwell, B., Gulari, E. & Melnik, D. (1985) In: Dahneke, B.E. (ed.) *Measurement of suspended particles by quasi-elastic light scattering*. J. Wiley & Sons, New York
14. Zollars, R.L. (1980) *J. Colloid Interface Sci.* **74** 163
15. Melik, D.H. & Fogler, H.S. (1983) *J. Colloid Interface Sci.* **92** 161

16. Kourti, T., MacGregor, J.F. & Hamielec, A.E. (1987) *J. Colloid Interface Sci.* **120** 292
17. Zollars, R.L. (1981) *J. Dispersion Sci., Technol.* **2** 331
18. Egusa, S. (1982) *J. Colloid Interface Sci.* **86** 135
19. Smoluchowski von, M. (1917) *Z. Phys.* **17** 557
20. Overbeek, J. Th. G. (1977) *J. Colloid Interface Sci.* **58** 408
21. Cebula, D.J. & Ottewill, R.H. (1982) *Colloid Polym. Sci.* **260** 1118
22. Hayter, J.B. & Penfold, J. (1983) *Colloid Polym. Sci.* **261** 1022
23. Alexander, K., Cebula, D.J., Goodwin, J.W., Ottewill, R.H. & Parentlich, A. (1983) *Colloids Surface* **7** 233
24. Cebula, D.J., Goodwin, J.W., Ottewill, R.H., Jenkin, G. & Iabony, J. (1983) *Colloid Polym. Sci.* **261** 555
25. Glatter, O. (1980) *J. Appl. Crystallogr.* **13** 7
26. Pleshakov, V.F. & Nagornyi, V.G. (1983) *Khim. Tverd. Topl.* **5** 130
27. Idhii, Y., Matsuoka, H. & Ise, N. (1986) *Ber. Bunsenges. Phys. Chem.* **9** 50
28. Penlidis, A., Hamielec, A.E. & MacGregor, J.F. (1983) *J. Liq. Chromatogr.* **6** 179
29. Provder, T. & Rosen, E.M. (1970) *Sep. Sci.* **5** 485
30. Hamielec, A.E. (1984) In: Barth, H.G. (ed.) *Modern methods for particle size analysis.* J. Wiley & Sons, New York, p. 251
31. Brough, A.W.J. & Hillman, P.R.W. (1981) *J. Chromatogr.* **208** 175
32. Dos Ramos, J.G. & Silebi, C.A. (1990) *J. Colloid Interface Sci.* **135** 165
33. Nagy, D.J., Silebi, C.A. & McHugh, A.J. (1981) *J. Appl. Polym. Sci.* **26** 1567
34. Dodds, J. (1982) *Analysis* **10** 109
35. Peterson, R.E., Myers, M.N. & Giddings, J.C. (1984) *Sep. Sci., Technol.* **19** 307
36. Giddings, J.C., Karaiskakis, G., Caldwell, K.D. & Myers, M.N. (1983) *J. Colloid Interface Sci.* **92** 66
37. Yang, F.S., Caldwell, K.D. & Giddings, J.C. (1983) *J. Colloid Interface Sci.* **92** 81
38. Kirkland, J.J., Rementer, S.W. & Yau, W.W. (1981) *Anal. Chem.* **53** 1730
39. Giddings, J.C., Myers, M.N. & Caldwell, K.D. (1981) *Sep. Sci., Technol.* **16** 549
40. Giddings, J.C., Lin, H.C., Caldwell, K.D. & Myers, M.N. (1983) *Sep. Sci., Technol.* **18** 293
41. Myers, M.N. & Giddings, J.C. (1982) *Anal. Chem.* **54** 2284
42. Giddings, J.C., Myers, M.N., Caldwell, K.D. & Pav, J.W. (1979) *Chromatographia* **185** 261
43. Dalas, E. & Karaiskakis, G. (1987) *Colloids Surfaces* **28** 169
44. Menold, R., Lutte, B., Kaiser, W. & Schmidt, A. (1972) *Chem.-Ing.-Tech.* **44** 1226
45. Kelsey, R.H. & Hansen, E.E. (1946) *J. Appl. Phys.* **17** 675
46. Wilson, E.A., Miller, J.R. & Rowe, E.H. (1949) *J. Phys. Colloid Chem.* **53** 357
47. Branford, E.B. & Vanderhoff, J.N. (1959) *J. Colloid Sci.* **14** 543
48. Peltzbauer, Z., Hynková, V. Bezděk, M. & Hrabák, F. (1967) *J. Polym. Sci., C* **16** 503
49. Walter, E.R. & Bryant, G.H. (1977) In: Bailey, G.W. (ed.) *Proceedings of the 35th Annual Electron Microscopy Society of America.* Claitor's Publ. Div., Baton Rouge, p. 314
50. Talmon, Y. & Miller, W.G. (1978) *J. Colloid Interface Sci.* **67** 284
51. Langer, G. (1979) *Colloid Polym. Sci.* **257** 522
52. Oppenheimer, L.E. (1983) *J. Colloid Interface Sci.* **92** 350
53. Coll, H. & Searles, C.G. (1987) *J. Colloid Interface Sci.* **115** 121
54. Coll, H. & Searles, C.G. (1986) *J. Colloid Interface Sci.* **110** 65
55. Maron, S.H., Elder, M.E. & Ulevitch, I.N. (1954) *J. Colloid Sci.* **9** 89
56. Maron, S.H., Elder, M.E. & Moore, C. (1954) *J. Colloid Sci.* **9** 104
57. Maron, S.H. & Elder, M.E. (1954) *J. Colloid Sci.* **9** 263
58. Paxton, T.R. (1969) *J. Colloid Interface Sci.* **31** 19
59. Okubo, M., Yamada, A. & Matsumoto, T. (1980) *J. Polym. Sci., Polym. Chem. Ed.* **16** 3219
60. Piirma, I. & Chen, S. (1980) *J. Colloid Interface Sci.* **74** 90
61. Brodnyan, J.G. & Brown, G.L. (1960) *J. Colloid Sci.* **15** 76
62. Hrabák, F., Bezděk, M., Hynková, V. & Peltzbauer, Z. (1967) *J. Polym. Sci., C* **16** 1345
63. Maurice, A.M. (1985) *J. Appl. Polym. Sci.* **30** 473
64. Chatterjee, S.P., Banerjee, M., Bera, B. & Konar, R.S. (1979) *Indian J. Chem., A* **17** 9

8.2 METHODS FOR CHARACTERIZING THE NATURE OF POLYMER PARTICLES

The properties of polymer latices are strongly influenced by the type, distribution and concentration of the functional and polar groups on the surface of the polymer particles and the nature of the polymer. The functional and polar groups affect not only the colloidal but also the usefulness of the polymer dispersion. Our understanding of how the incorporation of functional groups onto the surface or into the particle core depends on the reaction conditions and contributes to the formulation of a detailed mechanism for polymer particle formation.

The polymer latex for which the nature and distribution of the functional groups in polymer particles needs to be determined, is first freed from all low- and high-RMM free and adsorbed products. A low concentration of functional groups bound on the surface or in the particle core requires the use of effective purification procedures. To characterize these groups, the following methods are used: conductometric titration [1], potentiometric titration [2], electrophoresis [3], the use of radioactive components, fluorescence measurement [4], dye distribution [5,6], titration of the latex with an aqueous solution of emulsifier [7], IR spectroscopy [8], equilibrium adsorption procedures [9], etc.

8.2.1 Conductometric and potentiometric titrations

Conductometric titration is used for determining acidic groups mainly localized on the surface [1]. To determine the equivalence point, the change in conductivity caused by titration of the dilute aqueous latex solution by alkaline sodium hydroxide is utilized. The change is particularly pronounced at the end-point. Conductometric titration is the superior, and in many cases the only, method for finding the equivalence point since the classical colour indicators cannot be used because of the turbidity of the dispersion.

The shape of the titration curve and the inflection point provide information on the concentrations and types of acidic groups. The essence of conductometry lies in the exchange of one type of ion by another one with a different mobility and conductivity. The shape of the titration curve can be simulated using the Eq. (1):

$$K = \frac{\lambda_i c_i z_i}{1000}. \tag{1}$$

Here K is the specific conductivity, c_i is the molar concentration of particle i, z_i is the charge of particle i, and λ_i is their equivalent conductivity [10]. The inflexion point is expressed in milliequivalents of H^+ per g of polymer. Equation (1) describes the shape of the titration curve only approximately since it neglects the influence of ionic strength and the structure of the electric double layer. If these factors are considered, the model used for describing the course of titration becomes very complicated.

The density of the surface charge σ ($\mu C\ cm^{-2}$) obtained from the conductometric titration and from equation (35):

$$\sigma = cfF/S, \tag{2}$$

where c is the concentration of titrant needed to neutralize the acid groups present, f is the titration factor, F is Faraday's constant (C mol^{-1}), and S is the overall surface area of the polymer particles, markedly affects the properties of the polymer particles.

Egusa & Makuuchi [11] have applied direct and back titrations in the characterization of the carboxyl groups bound on the surface of polymer particles. They have found that back titration (the excess sodium hydroxide added to neutralize the acid groups of the polymer particles is back titrated by acid) gives a larger number of acid groups on the particle surface than direct titration, a result which led the authors to conclude that hydroxyl groups also penetrate into the inner particle spheres, where further carboxyl groups are neutralized.

Hoy [12] demonstrated the mechanism (Fig. 36) for back titration based on carboxylated polymer particles. The conclusion that back titration neutralizes a greater number of carboxyl groups than direct titration of a latex by sodium hydroxide is similar to that drawn earlier [11]. Direct titration leads to neutralization of the surface-bound carboxyl groups and the particle surface preserves its original contracted form (Fig. 36a). Addition of excess sodium hydroxide leads to neutralization of the surface-bound carboxyl groups and to the production of dissociated groups. Strong electrostatic repulsive forces start to function between the groups, expanding the particle surface and allowing diffusion of other hydroxide ions into the particles, where they neutralize further acid groups (Fig. 36b). By means of back titration of sodium hydroxide, excess NaOH and carboxyl anions localized within the particles or on their surface are titrated. The results of conductometric titrations confirmed that carboxyl groups are mostly bound on the particle surface and form a hydrophilic particle shell; only a small fraction enters the lipophilic core.

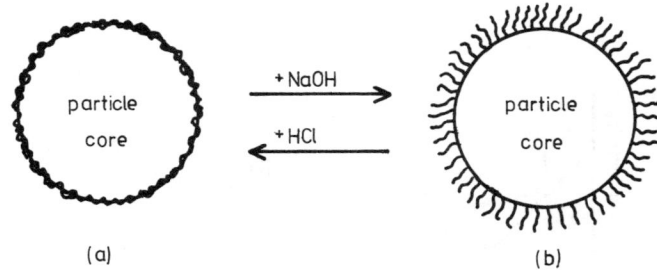

Fig. 36 — Mechanism of effect of titration agent on structure of surface layer of latex particle. Figure shows a particle (a) before and during titration by NaOH, (b) before and during titration by acid.

Expansion of the surface layer of the polymer particle is reached, for example, by addition of 1-propanol [11]. By direct titration of the particles thus swollen, acid groups located inside the particles can also be neutralized [12]. Direct titration of carboxylic latices is used for determining the concentration of surface-bound acid groups while back titration (or expansion or swelling of the particles followed by direct titration) provides data about the concentration of acid groups incorporated into the inner particle spheres [13–16].

The titration curve of the direct titration of carboxylated latices prepared via emulsion copolymerization of lipophilic (styrene) and carboxylic monomers can be divided into three zones (see Fig. 37). The first zone corresponds to neutralization of the protonated sulfate end groups of the polymer molecules, the second zone to the neutralization of carboxyl groups, and the third (ascending) represents the increase in concentration of excess sodium hydroxide [17]. Back titration of these latices indicates as many as four stages on the conductometric curve (see Fig. 38). The first stage corresponds to the neutralization of excess NaOH followed by the neutralization of carboxyl groups on the particle surface and then in the polymerization medium, and finally, the fourth stage of rapid conductivity increase corresponds to increase in concentration of the free acid. If the latex is prepared using peroxodisulfate as initiator, the titration curve is complicated by a further stage representing the conversion of sulfate salts into the acid form; this stage is observed in the initial stage of titration.

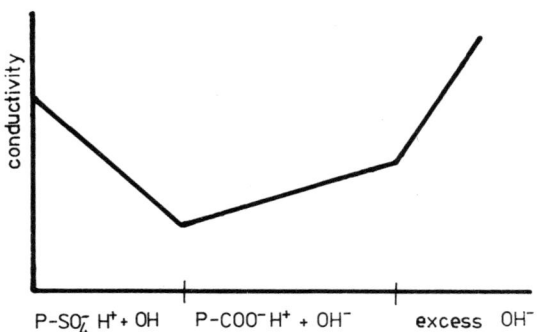

Fig. 37 — Conductometric titration curve obtained by direct titration of carboxyl latex subjected to ion exchange [17].

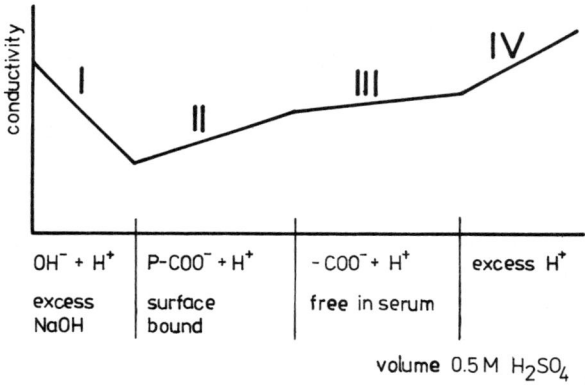

Fig. 38 — Conductometric titration curve of carboxylated poly(styrene) latex obtained by back titration [13].

The concentration of carboxyl groups neutralized by back titration increases with the time of exposure to alkali. This phenomenon follows from the increase in effectiveness of the particle swelling and of the penetration of hydroxyl groups into the particles [13, 18]. With increased contact time, the acid groups, which are also buried in the particle core, are gradually neutralized. Table 47 gives the concentrations of carboxyl groups obtained via direct or back titrations. The results from direct titration show that most carboxylic monomer is positioned through copolymerization on the particle surface. Back titration provides the sum of the surface-bound and buried carboxyl groups. The prolonged effect of NaOH on the latex increases the extent of hydrolysis of the acid groups [19].

Table 47 — Determination of concentration of carboxyl groups in poly(styrene) particles [13]

Latex[a]	μeq. carboxyl group / g latex	
	back titration	direct titration
Styrene / MA	118	83
Styrene / AA	154	130
Styrene / IA	208	164

[a] Latex contained 2% carboxyl comonomer. MA — methacrylic acid, AA — acrylic acid, IA — itaconic acid.

Conductometry can also be used for the indirect determination of hydroxyl groups present in polymer particles. The procedure is as follows: the content of acid groups in a particle is at first determined by conductometric titration. If there are sulfate groups, they must be converted via hydrolysis into hydroxyl, and then, together with the original hydroxyl groups, into carboxyl groups; their total number is determined by titration and the original number of hydroxyl groups is calculated by a simple mathematical procedure [3, 20].

In potentiometric titrations (the pH-dependent suface charge), the equivalence point of the titration is detected by the rapid change of potential of a hydrogen electrode dipped into the titrated solution accompanied by a sudden change of pH [2]. The method is widely applied in the chemistry of polymer dispersions and is used for determining the surface charge of particles stabilized by cationic or anionic emulsifiers [21]. In addition, it can be used, like direct conductometric titration, for determining surface-bound acid groups.

8.2.2 Electrophoresis

Electrophoresis (the pH-dependent mobilities of particles) is a process whereby the electrically-charged particles migrate in the electrical field of the serum phase towards oppositely charged electrodes.

Viscous forces acting on the particle tend to hinder this movement and, when an equilibrium is reached between electrical attraction and viscous drag, the particle moves at constant rate. The motion of these particles in the electric field depends on the type and magnitude of the particle charge and on the particle size. Investigation of the rate and direction of the particle motion thus provides information about the specified parameters. The speed of the particles depends on the strength of the electric field or voltage gradient, the dielectric constant and the viscosity of the medium, and the zeta potential ζ. The particle velocity at unit electric field is referred to as its electrophoretic mobility.

The rate and direction of the electrically charged particles are also influenced by the pH value of the serum phase. Amphoteric polymer latices are particularly sensitive to change of pH. At a particular pH value, the particles are immobile, with a zero value of the isoelectric potential. This pH is a function of the character of the functional groups and their concentration. As the fraction of acid groups increases, the isoelectric point is shifted to the acidic region and, conversely, with an increasing number of basic groups, to higher pH values.

The electrokinetic or zeta potential (ζ) is taken as a measure of the electrical charge of the colloidal systems [22]. It depends on the character of the copolymer, on the type and concentration of initiator and emulsifier, and the reaction conditions used in preparation of the polymer dispersion [23, 24]. The zeta potential is related to the electrophoretic mobility by Eq. (3):

$$U_E = \frac{\varepsilon \zeta f(\kappa r)}{6\pi\eta},\qquad(3)$$

where U_E is the mobility, ε is the dielectric constant, η is the viscosity, r is the particle radius and κ is the Debye–Hückel parameter, which depends on electrolyte concentration.

Electrophoretic determinations of zeta potentials are usually made in aqueous media (high dielectric constant) and moderate-to-high electrolyte concentrations, which results in a large value for κr. Under these conditions, $f(\kappa r) = 1.5$ and Eq. (3) becomes

$$U_E = \frac{\varepsilon \zeta}{4\pi\eta} \qquad(4)$$

which, in water at 25°C reduces to

$$\zeta = 12.85 U_E mV.\qquad(5)$$

Thus the measurement of electrophoretic mobility leads to a simple calculation of the zeta potential.

Polymer dispersions subject to electrophoresis undergo physical changes, e.g. emulsifier molecules are desorbed from the polymer particle surface, ion exchange in the surface layers of the polymer particles and hydrolysis of the functional groups take place. These facts adversely affect the data about the character of the original functional groups as well as the evaluation of kinetic parameters. These effects are particularly evident on the electrophoretic mobility of polymer particles with small charges [3]. The electrophoretic study of particles with a large surface charge is much less sensitive to these effects. However, with regard to the presence of the particles with various sized charges, it is often rather difficult to obtain a precise value of the surface particle charge.

Shortcomings associated with the use of classical electrophoresis can be overcome by more recent developments, e.g. laser Doppler electrophoresis [25]. It is a rather rapid method directly providing a value of the rate of motion of polymer particles in an electrical field. This rate is obtained by investigating and evaluating the intensity of the light scattering and the Doppler shift. The relation between the particle mobility U and the Doppler shift Δf can be expressed as

$$U = \frac{(\lambda/n)\Delta f}{E \sin \theta},\qquad(6)$$

where U is the electrophoretic mobility, E is the electrical field between the electrodes, (λ/n) is the wavelength of radiation in the disperse medium and θ is the angle of light scattering.

The spectra obtained by this method inform not only about the surface particle charge and about the character of the electrical double layer, but also about the sizes of the polymer particles. The value of the ζ potential can be obtained directly from data on the mobility by the procedure proposed by Wiersema et al. [26]. Laser Doppler electrophoresis or microelectrophoresis yield a number of important parameters, e.g. the potential of the polymer particle surface, the potential of the diffusion layer, and the type of functional group localized on the particle surface; these data follow from an investigation of particle mobility as a function of the concentration of various added electrolytes (e.g. NaCl) [27].

8.2.3 Other methods

A number of investigators have used neutron scattering to determine latex particle structures. An early important paper [28] described the preparation of seed latices of poly(methyl methacrylate) using d-styrene or d-methyl methacrylate (MMA). The small-angle scattering data contained sharp maxima and minima in the scattering intensity as a function of angle. The data could be further analysed to yield dimensions in agreement with stoichiometric and other measurements.

Wai et al. [29] studied the structure of polymer particles by SANS. They utilized trideuteromethyl methacrylate as the shell, over a 50/50 poly(styrene-co-methyl methacrylate) seed. While they succeeded in getting maxima and minima in the scattering intensity as a function of angle, the sharpness of the maxima/minima was less pronounced than those obtained by Ramakrishnan & Wignal [28].

Summerfield & Ullman [30] derived a model to obtain the diffusion coefficient from SANS, starting from mixtures of deuterated and protonated particles. The theory suggests an 'exploding star' model, where a deuterated particle is surrounded with protonated particles.

Linne et al. [31] studied the structure and information of polymer chains in latices using SANS. Samples showed abnormally high scattering intensities which were interpreted as due to the development of internal structure in the particles. The author also investigated the geometric effects of supramolecular structure in poly(styrene) latex particles of 200–300 nm diameter.

Sperling et al. [32] used SANS to clarify several aspects of the internal structure of polymer latices and subsequent modes of film formation. Two points have been made clear:

(i) The appearance of core-shell phenomena in latices depends on the size of the polymer chain located in the latex particle; the phenomenon is most marked when the radius of gyration of the chain is about one fifth as big as the latex radius.
(ii) Film formation depends on the extent of interdiffusion of the chains.

^{13}C NMR and mainly ^{19}F NMR spectroscopy are applicable to characterization of the surface and interior of the polymer particles [33, 34]. NMR spectroscopy is used for determining the number of hydroxyl groups localized on the surface of, and buried in the particles and enables elucidation of their origin (e.g. formed by hydrolysis of sulfate groups, etc.). The kinetics of adsorption of acid and basic dyes on the surface of polymer particles also provide information about the character of the surface-bound functional groups [5, 6]. The adsorption curves are not only a function of the character of the organic dye and the surface of the polymer particles but also of the pH. Their course is influenced by the level of electrostatic interaction between the particle and the dye and therefore, for example, dissociation of the surface group changes the character of the given interaction. Hydrogen bonds formed between polymer substrate and additive also contribute to the electrostatic interaction.

Data on the character and concentration of the functional groups on the surface and in the interior of the polymer particles are obtained by the method of gradual dissolution of the polymer particles in alkaline solvents [35]. This method is only suitable for those polymers which are soluble in the solvents used. The polymer particle gradually dissolves, through a peeling off of layers down to its core. Analysis of the solution gives information about the morphology of the polymer particle.

Some aspects of the structure of polymer particles are obtainable from a study of the effect of the degree of neutralization of acid groups on the viscosity [18, 36]. The volume of the particles of carboxylated polymers increases with pH owing to the increase in electrostatic repulsion and solvation of the anionic groups. On the other hand, at lower pH, the volume of the particle may decrease as a result of shielding of the anionic groups by free cations. Changes in the particle volume, and interactions between the functional groups in a particle and between the particles, cause changes in the viscosity of the system. On these grounds, we can then predict the type and concentration of the functional groups in the particles. Expansion of the particles can be measured directly by light-scattering or sedimentation.

A method of mechanical spectroscopy has recently begun to be applied [37]. This presents data about the composition of the copolymer located in the particle interior and about the heterogeneity of the polymer phase in polymer latices prepared by the emulsion polymerization of monomers of different water solubility.

Okubo et al. [38] applied a soap titration method in order to estimate particle morphology indirectly.

Wide-angle light scattering was used by Ford et al. [39] as the principal probe to examine the core-shell structure proposed for certain alkali-swellable acrylic acid acrylate ester copolymer latices.

Beyer et al. [40] have used SAXS to study the core-shell structure of polymer latices. By this method it was possible to determine where the second polymer is located within these two-stage emulsion polymers. The results indicate that (i) the emulsion polymerization process takes place in a small surface layer region of the seed particles, and (ii) a small interfacial layer exists between the core and shell of the particle.

The polarographic method was used to determine poly(acrylic acid), poly(acrylamide) and their corresponding copolymers (or groups) in a polymer matrix by Dunsch et al. [41]. The suppression of the polarographic oxygen maximum was found to be a sensitive and simple method for the determination of acrylic acid and acrylamide groups if the solution contained Ca^{2+} ions.

Okubo [42] reported that transmitted light spectral measurements from a conventional spectrophotometer gave very clear-cut information on the microstructure of a polymer latex in solution and of a latex film. The reliability of this approach was checked by comparing the latice constant obtained from the transmitted light spectrum with that from two other independent techniques, i.e. reflectance spectroscopy and direct ultramicroscopic observation.

References

1. McCarwill, W. & Fitch, R.M. (1977) *J. Colloid Interface Sci.* **64** 403
2. Hull van den, H.J. & Vanderhoff, J.W. (1971) In: Fitch, R.M. (ed.) *Polymer colloids*. Plenum Press, New York, p. 1
3. Ottewill, R.H. & Shaw, J.N. (1970) *Kolloid-Z. Z. Polym.* **218** 34
4. Tsuda, Y. (1961) *J. Appl. Polym. Sci.* **5** 104
5. Roy, G., Mandal, B.M. & Palit, S.R. (1971) In: Fitch, R.M. (ed.) *Polymer colloids*. Plenum Press, New York, p. 49
6. Palit, S.R. & Ghosh, P.J. (1962) *J. Polym. Sci.* **58** 1225

7. Green B.W. (1973) *J. Colloid Interface Sci.* **43** 449
8. Stone-Masui, J.H. & Stone, W.E.G. (1980) In: Fitch, R.M. (ed.) *Polymer colloids*, vol. II. Plenum Press, New York, p. 331
9. Connor, P. & Ottewill, R.H. (1971) *J. Colloid Interface Sci.* **37** 642
10. Harding, I.H. & Healy, T.W. (1982) *J. Colloid Interface Sci.* **89** 185
11. Egusa, S. & Makuuchi, K. (1981) *J. Colloid Interface Sci.* **79** 350
12. Hoy, K.L. (1979) *Coatings Technol.* **51** 27
13. Hen, J. (1974) *J. Colloid Interface Sci.* **49** 425
14. Green, B.W. (1973) *J. Colloid Interface Sci.* **43** 462
15. Furusawa, K., Norde, W. & Lyklema, J. (1972) *Kolloid-Z. Z. Polym.* **250** 908
16. Wilkinson, M.C. & Fairhurst, D. (1981) *J. Colloid Interface Sci.* **79** 272
17. Hull van den, H.J. & Vanderhoff, J.W. (1970) *Br. Polym. J.* **2** 121
18. Wesslau, H. (1963) *Makromol. Chem.* **69** 220
19. Vijayendran, B.R. (1979) *J. Appl. Polym. Sci.* **23** 893
20. Vanderhoff, J.W. & Hull van den, H.J. (1968) *J. Colloid Interface Sci.* **28** 36
21. Davis, J.A., James, R.O. & Leckie, J.O. (1978) *J. Colloid Interface Sci.* **63** 480
22. Healy, T.W. & White, L.R. (1978) *Adv. Colloid Interface Sci.* **9** 303
23. Homola, A. & James, R.O. (1977) *J. Colloid Interface Sci.* **59** 123
24. Harding, H. (1985) *Colloid Polym. Sci.* **263** 58
25. Ware, B.R. (1974) *Adv. Colloid Interface Sci.* **4** 1
26. Wiersema, P.H., Loeb, A.L. & Overbeek, J. Th. G. (1966) *J. Colloid Interface Sci.* **22** 78
27. Goff, J.R. & Luner, P. (1984) *J. Colloid Interface Sci.* **99** 468
28. Ramakrishnan, V. & Wignal, G.D. (1988) *J. Colloid Interface Sci.* **123** 24
29. Wai, M.P., Gelman, R.A., Fatica, M.G., Hoerl, R.H. & Wignal, G.D. (1987) *Polymer* **28** 918
30. Summerfield, G.C. & Ullman, R. (1988) *Macromolecules* **21** 2643
31. Linne, M.A., Klein, A. & Sperling, L.H.J. (1988) *Macromol. Sci., Phys.* **27** 181
32. Sperling, L.H., Klein, A., Yoo, J.N., Kim, K.D. & Mohammadi, N. (1990) *Polym. Adv. Technol.* **1** 263
33. Leader, G.R. (1973) *Anal. Chem.* **42** 16
34. Ho, F.R.L. (1973) *Anal. Chem.* **45** 603
35. Muroi, S., Hashimoto, H. & Hosoi, K. (1984) *J. Polym. Sci., Polym. Chem. Ed.* **22** 1365
36. Muroi, J. (1966) *J. Appl. Polym. Sci.* **10** 713
37. Zosel, A., Heckmann, W., Ley, G. & Machtle, W. (1987) *Colloid Polym. Sci.* **265** 113
38. Okubo, M., Yamada, A. & Matsumoto, T. (1980) *J. Polym. Sci., Polym. Chem. Ed.* **16** 3219
39. Ford, J.R., Rowell, R.L. & Bassett, D.R. (1981) In: Bassett, D.R. & Hamielec, A.E. (eds) *Emulsion polymers and emulsion polymerization*. American Chemical Society, Washington DC, p. 279
40. Beyer, D., Lebek, W., Hergeth, W.D. & Schmutzler, K. (1990) *Colloid Polym. Sci.* **268** 744
41. Dunsch, L., Feist, U. & Morgenstein, J. (1983) *Acta Polym.* **34** 73
42. Okubo, T. (1986) *J. Chem. Soc., Faraday Trans. 1* **82** 3185

8.3 CLEANING OF POLYMER LATICES

The characterization of the surface of polymer latices and their special applications require separation of the polymer particles from unchanged initiator, free emulsifier, additives and reaction products. To get very clean polymer dispersions, various methods are used either individually or in combination. The most commonly applied methods for latex purification are: dialysis, centrifugation-decantation, ultrafiltration, ion exchange, flow ultrafiltration, serum replacement, competitive adsorption, freeze drying steam, stripping, etc.

8.3.1 Dialysis

This is a classical method for obtaining pure polymer dispersions [1] and is rather simple but slow and requires several weeks. In many cases, even extensive or exhaustive dialysis of polymer latices fails to remove completely the emulsifier from the system; thus a minor fraction remains adsorbed on the polymer particles [2–4].

The separation of the polymer particles from low-RMM compounds is based on the different diffusion rates of the individual components and the different effective volumes.

Dialysing membranes trap larger polymer particles and only smaller molecules can flow through the pores. Particles smaller than the pore dimensions freely diffuse into the pure solvent (Fig. 39). Acceleration of the dialysis process is effected by using electric field, by vigorous stirring of the solution of the polymer dispersion and the water phase and by the continuous exchange of the pure water.

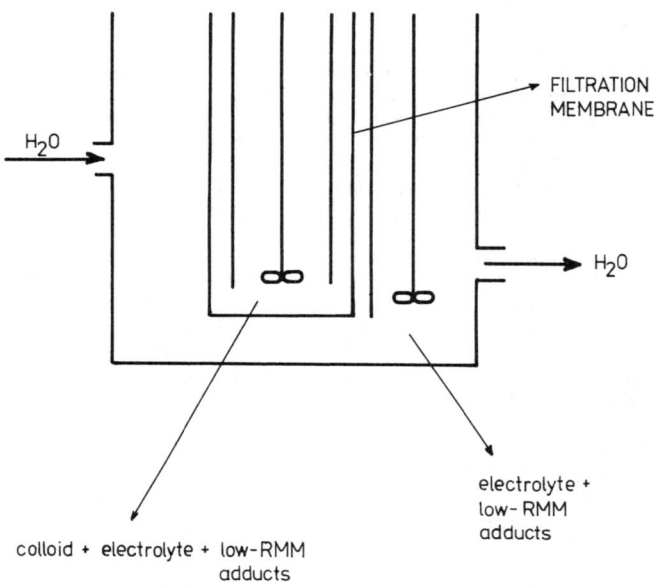

Fig. 39 — Through-flow dialyser.

Commercially available Visking dialysis tubes are used for the dialysis of polymer latices. The dialysis of dilute polymer latices containing maximally 10 mass % of solid polymer lasts two to three weeks on average under continuous exchange of the dialysate. The dialysis time is considerably prolonged in nonflow dialysers. The procedure and effectiveness of dialysis are controlled by spectrophotometry, by measuring the concentration of low-RMM compound in water or, if this compound is dissociated and/or charged, by conductometry.

8.3.2 Ion exchange

The surface of polymer particles can be modified by the exchange of ions which can be achieved by interaction of these particles with a solid substrate bearing functional groups, i.e. ion-exchange resins. The process is rapid but requires high purity of the ion-exchange resins. Commercially available ion exchangers contain a large fraction of low-RMM, and some high-RMM polyelectrolytes which can be only partly removed from the solid substrate. The purification of ion-exchange resins proposed by Vanderhoff *et al.* [3–5] proceeds in several steps and yields a pure solid substrate with a minimum content of unwanted admixtures. This procedure includes washing the resin with aqueous solutions of potassium hydroxide, hydrochloric acid, methanol, and cold or warm water, respectively.

The interaction of a diluted latex containing maximally 10 mass % of solid, with a pure ion exchanger under vigorous stirring for 2–3 hours leads to ion exchange of the polymer particles by ions of the ion exchanger. Depending on the character of the ion exchanger, mostly univalent or bivalent metal cations are exchanged for protons. Highly crosslinked copolymers mainly based on styrene are those most widely applied as ion exchangers.

8.3.3 Ultrafiltration

By the filtration of dilute polymer latices through paper filters with pore dimensions from 1 to 5 μm and through ultrafilters with the pores smaller than 1 μm, the solution is freed from low-RMM compounds and of the fraction of very small polymer particles. A homogeneous, pure solution of polymer particles is obtained in this way. Since there is a high probability of 'clogging' the filter pores, even while stirring the solution, the applicability of the method is reduced. Ultrafiltration is often combined with electrodialysis; the method known as electroultrafiltration affords much better results in the purification of latices than ultrafiltration alone.

Ahmed et al. [6] have used flow ultrafiltration to purify polymer latices. Flow ultrafiltration is a modification of the dialysis process using membranes micron and submicron sized-pores under a continuous flow of solvent through the membrane system. A combination of membranes enables not only removal of low-RMM products but also classification of the polymer particles in terms of their size. In order to prevent clogging of the pores, the latex is vigorously stirred above the membrane; this also assists the flow rate of the solution through the membrane. Clogging is also hindered by using very dilute aqueous solutions of the polymer particles (containing 5 mass % of the polymer portion at maximum). A pure polymer latex is obtained in this way only after extensive washing. Washing of the aqueous polymer dispersion with dilute aqueous hydrochloric acid leads, as with the interaction between an ion-exchange resin and a latex, to the exchange of Ca^{2+}, Na^+, K^+, etc. for protons.

8.3.4 Other methods

Electrodecantation is an important method for preparing pure colloidal systems. The method is based on the diffusion of negatively charged colloid particles directed towards the anodic side of the membrane and positively charged colloid particles towards cathodic side of the membrane. Free, electrically charged low-RMM and high-RMM compounds diffuse in a similar way. The colloid particles become concentrated and microlayers are formed around the membranes. Electrodecantation can thus be used not only for latex purification but also for the concentration of colloidal solutions.

References

1. Shaw, J.N. & Ottewill, R.H. (1965) *Nature* **208** 681
2. Edelhauser, H.A. (1969) *J. Polym. Sci., C* **27** 291
3. Hull van den, H.J. & Vanderhoff, J.W. (1968) *J. Colloid Interface Sci.* **28** 336
4. Hull van den, H.J. & Vanderhoff, J.W. (1970) *Br. Polym. J.* **2** 121
5. McCann, G.D., Brandford, E.B., Hull van den, H.J. & Vanderhoff, J.W. (1971) In: Fitch, R.M. (ed.) *Polymer colloids*. Plenum Press, New York, p. 29
6. Ahmed, S.M., El-Aasser, M.S., Pauli, G.H., Poehlein, G.W. & Vanderhoff, J.W. (1980) *J. Colloid Interface Sci.* **73** 388

9
Future trends in dispersion polymerization

It is perhaps a routine task to predict future trends in the development of a branch of science by extrapolation beyond the present situation until a radically new proposal is made. To achieve a breakthrough, current knowledge can only partly help in stimulating creative imagination. The probability of any fulfilment of one's expectations in, say, just 5 to 10 years is very low, in the light of experience. We often see a breakthrough in the development of a scientific field, which is unforeseen, so that it might seem to be a matter of coincidence or a brainwave. Even here, though, the 'iron' laws of scientific logic are valid, i.e. this jump can only happen at a certain stage of evolution in that branch of science. We can then infer from this that successful prediction of such a major development in a scientific topic will depend on achieving those conditions when our knowledge of the field has matured to a point that a reformulation of our ideas can occur [1]. When considering such future trends, one has to ask whether we are in such a period of maturity.

In a monograph dealing with emulsion polymerization published in 1975, the author [2] raised the question as to which problems needed to be solved in the field of emulsion polymerization. The author reported that the outstanding problems were associated with the reaction mechanism. Was it at all possible to apply a general mechanism for emulsion polymerization, which considers micelles and polymer particles to be the main loci for all emulsion polymerization systems, taking into account differences in monomer reactivity and solubility in the aqueous phase? Or, was it necessary to analyse the individual groups of monomers separately? What is the actual mechanism of particle nucleation? Where is the main propagation site? Which model of emulsion polymerization is most appropriate for a randomly-chosen monomer? To what extent does our experimental knowledge enable preference for one model over another?

As time passed, numerous questions have emerged concerning trends in research which were not predicted beforehand. Many anticipated problems have been resolved, but others await solution. Let us analyse, for example, the question of determining the initiation site of the classical oil-in-water emulsion polymerization. The micellar mechanism, homogeneous nucleation, and other mechanisms for polymer particle formation are discussed in this monograph. Some investigators have attempted to answer the question of the site of initiation in the emulsion polymerization and copolymerization of monomers differing in hydrophilicity (polarity) by applying the known basic principles of radical polymerization to the emulsion polymerization system. This problem might be solved by partitioned polymerization [3]. Application of the basic principles of partitioned polymerization shows that initiation even of weakly polar monomers (styrene, butyl methacrylate) takes place in the aqueous phase, which contradict the Smith–Ewart theory of emulsion polymerization. The problem associated with the initiation site was also examined in ref. [4] on the basis of determination of the composition of the copolymer which arises in the initial stages of copolymerization of a mixture of lipophilic and hydrophilic monomers. The statements about the initiation locus were less unambiguous in this treatment as in partitioned polymerization. We can expect that the problem of the reaction site will stimulate further research in this aspect of emulsion polymerization. The partitioning of the reaction components into different, immiscible phases led investigators to use micellar systems as tools for increasing the concentration and lifetime of free radicals [5].

In the discussion session 'Future trends in heterogeneous free-radical polymerization systems and practical application of the theory' held during the International Microsymposium 'Radical Polymerization in Heterogeneous Systems' in Smolenice [6], attention was concentrated on the problem of obtaining reliable and complete data on factors controlling the particle size and RMM distribution, and the overall kinetics of dispersion polymerization. These properties are also important technologically as they govern the mechanical properties of the latex and polymer formed, not to mention the cost of manufacture. Computer modelling can help in predicting conditions required to produce a polymer product and/or its latex with desired characteristics. At present it is unlikely to give genuinely predictive modelling for copolymer systems although parametrized modelling (using a comparatively small number of strategic measurements) is currently possible and likely to be of increasing use in future. The barrier to increased manufacturing in the field of heterogeneous polymerization is the lack of a widely-applicable, automatic and rapid technique of determining conversion in complex copolymerization systems. Gas chromatography alone cannot be used in multicomponent systems because of the complexities of partitioning between different phases of the dispersion system. Combinations of techniques, including gravimetry and gas chromatography, can be used, but currently cannot give results on a sufficiently short time scale to be useful in on-line systems.

An impediment to qualitative and quantitative understanding of the copolymer systems is our inadequate knowledge of their thermodynamic properties. At present one is unable to predict or rationalize properties such as equilibrium swelling, compatibility, interfacial tension, thermodynamic driving forces for core/shell and/or microdomain morphology. This lack of knowledge is at two levels: inadequacy and incoherence of the present data base and poor understanding of the fundamentals. Even a phenomenological understanding beyond the rudimentary principles currently available would enable dispersions and tailored polymers to be designed and manufactured in a manner vastly superior to those currently available (which still involve a large amount of wasteful trial-and-error, and are much more art than science [7]).

Advances made in our understanding of the degree of organization of systems will enable the preparation of polymers with modified properties in the near future and will provide new possibilities in the field of polymer synthesis. The means of providing polymerized spherical bi- and multilayers will also act as a stimulus. This prediction is justified by the current experiences of their application, e.g. of biomembranes and in emulsion polymerization. We expect that polymerization in organized bilayers will remain an objective of increasing interest [8]. For example, ultrathin polymer films having an organized structure can be used in the fairly near future for optoelectronic devices, and chemi- and biosensor applications [9]. The problem still remains, however, how to fabricate ultrathin films containing a high level of molecular order while possessing the necessary thermal and mechanical stability.

Efforts made in the search for new methods or ways of modifying present procedures for the preparation and applications of polymer dispersions continue. The advantages of disperse systems in various applications are a sufficient motivation for future research. Efforts devoted to the study of the effect of magnetic fields on emulsion polymerization [10] might bring interesting results in the future. A novel method for the preparation of selective composite mebranes from emulsions has been suggested recently [11]. Environmentally more friendly polymer dispersions [12] and concentrated polymer dispersions also deserve greater attention.

Organogels prepared from water-in-oil microemulsions from which the name 'microemulsion gel' was derived have features which are common to all other known gels. However, they also have several distinct structural and physical properties which make them attractive both for basic and applied research [13]. These efforts are also combined with the aim of preparing large monodisperse particles of organic polymers which can be documented by experiments in the gravity-free state [14].

References

1. Moser, F. (1982) *Chimia* **36** 387
2. Blackley, D.C. (1975) *Emulsion polymerization. Theory and practice.* Applied Science Publ., London
3. Bartoň, J., Juraničová, V. & Hloušková, Z. (1988) *Makromol. Chem.* **189** 501
4. Lu, Y.Y., El-Aasser, M.S. & Vanderhoff, J.W. (1988) *J. Polym. Sci., Polym. Phys.* **26** 1187
5. Burkey, T.J. & Griller, D. (1985) *J. Am. Chem. Soc.* **107** 246
6. Gilbert, R.G. (1990) *Makromol. Chem., Macromol. Symp.* **31** 253
7. Vanderhoff, J.W. (1983) In: Poehlein, G.W., Ottewill, R.H. & Goodwin, J.W. (eds) *Science and technology of polymer colloids*, vol. 1. M. Nijhoff Publ., Dordrecht, p. 1
8. Fendler, J. & Tundo, P. (1984) *Acc. Chem. Res.* **17** 3
9. Tredgold, R.H. (1988) *J. Chim. Phys.* **85** 1079
10. Simionescu, C.I., Chiriac, A., Neamtu, I. & Rusan, V. (1989) *Makromol. Chem., Rapid Commun.* **10** 601
11. Ruckenstein, E. (1989) *Colloid Polym. Sci.* **267** 792
12. Palluel, A.L.L., Westby, M.J., Bromley, C.W.A., Davies, S.P. & Backhouse, A.J. (1990) *Makromol. Chem., Macromol. Symp.* **35/36** 509
13. Luisi, P.L., Scartazzini, R., Hearing, G. & Schurtenberger, P. (1990) *Colloid Polym. Sci.* **268** 356
14. Vanderhoff, J.W., El-Aasser, M.S., Micale, F.J., Sudol, E.D., Tseng, C.M., Silwanowicz, A., Kornfeld, D.M. & Vicente, F.A. (1984) *J. Dispersion Sci., Technol.* **5** 231

Subject index

adsorption,
    of coemulsifier, 96, 152
    of dyes, 87, 262, 337
    of emulsifier, 24, 43, 52, 149, 152, 178, 273
    of inhibitor, 155
    of monomer, 162, 241
    of polymers, 185, 220, 221, 256
    of radicals, 31, 43, 241
    reversible, 222, 223, 241, 276

chain length, kinetic, 242, 243, 262
chain transfer, 137
    agent, 50, 52, 61, 131, 132, 134, 135, 214, 219, 245, 263, 315
    constant, 136, 137, 158, 243
    to emulsifier, 126, 145, 152, 157, 159, 212
    to initiator, 157
    to monomer, 50, 52, 129, 136, 137, 157–159, 214, 215
    nondegradative, 211–213
    to polymer, 157, 227
charge density, of particle surface — see surface, charge density
chromatography, 11, 324–328
    gas, 257, 311, 312, 343
    gel, 243, 319, 324
    hydrodynamic, 319, 324, 325
    thin-layer, 260
coefficient,
    diffusion, 53, 55, 92, 180, 321, 337
    distribution, 53, 97, 164
    Einstein, 247
    Hamaker, 181, 219
    Henry, 306
    interaction, 57, 235, 305
    turbidity, 321, 322
coemulsifier, 45, 91, 93, 94, 96, 138, 160, 191, 193, 196, 204
composites,
    dispersion, polymer, 13, 14, 247, 268
    based on organic and inorganic substances, 13, 262–265, 300
    preparation by heterocoagulation, 264, 287
concentration,
    critical coagulation (CCC), 182, 184
    critical micellar (CMC), 31, 32, 40, 43, 46, 78, 81–84, 87, 90, 94, 114, 118, 125, 129, 131, 132, 134, 221, 262, 263, 271, 276, 279, 300, 330
    equilibrium, of monomer, 57, 136, 137
conductivity, 83, 94, 152, 332
    conductometry, 94, 150, 152, 251
    of inverse microemulsion, 91, 191, 202
    of latex,
        equivalent, 83, 332
        specific, 153, 332
    of micellar solution, 83, 150

constant, Flory–Huggins, 277
continuous stirred-tank reactor, 177
conversion, 30, 33, 43, 238, 239
    critical, 33, 305, 306
    curve, 287
    degree of, 43
    fractional, 155, 208, 305, 306
    limiting, 123, 161
copolymer,
    block, 12, 13, 19, 161, 226, 229, 230, 236, 243, 254
    graft, 12, 19, 167, 226–228, 232, 245, 257, 260, 261
    graft, comb, 226, 228, 229, 243
copolymerization,
    emulsion,
        of acrylamide, 172
        of acrylic acid with 2-alkylaminoethyl acrylate, 253
        of acrylonitrile with butyl acrylate, 72, 135, 153, 154, 156–159, 254
        of acrylonitrile with 2-ethylhexyl acrylate, 156
        of butyl acrylate with acrylic acid, 166
        of butyl acrylate with ethyl acrylate, 165
        of butyl acrylate with methacrylic acid, 164, 166, 251
        of butyl acrylate with methyl methacrylate, 164, 165, 254
        of butyl acrylate with vinyl acetate, 57, 68, 159, 160, 171, 175, 254
        of ethyl acrylate with acrylic acid, 166
        of ethyl acrylate with divinylbenzene, 144
        of ethyl acrylate with 1,6-hexamethylene-diacrylate, 144
        of ethyl acrylate with methacrylic acid, 166, 167
        of ethyl acrylate with methyl methacrylate, 150–153
        of methacrylamide, 172
        of methyl methacrylate and butyl acrylate with carboxylic unsaturated monomers, 251
        of methyl methacrylate with acrolein, 160
        of methyl methacrylate with acrylamide, 153
        of methyl methacrylate with 1-naphthylmethyl methacrylate, 245
        of methyl methacrylate with vinyl acetate, 153
        of styrene with acrylamide, 172, 249, 251, 269
        of styrene with acrylic acid, 161–164, 167, 250, 251, 253
        of styrene with acrylonitrile, 72, 134, 254
        of styrene with butadiene, 128
        of styrene with 2-diethylaminoethyl acrylate, 253
        of styrene with N,N-dimethyl acrylamide, 251, 269

of styrene with divinylbenzene, 102, 137, 139
of styrene with glycidyl methacrylate, 270
of styrene with 2-hydroxyethyl methacrylate, 252, 253, 269
of styrene with N-hydroxymethyl acrylamide, 251, 252, 269
of styrene with itaconic acid, 161–163, 250
of styrene with maleic acid, 253
of styrene with methacrylamide, 251, 252
of styrene with methacrylic acid, 161–164, 250, 253, 287
of styrene with methyl acrylate, 135
of styrene with methyl methacrylate, 73, 135, 136
of styrene with α-methylstyrene, 137
of styrene with sodium styryl sulfonate, 252, 271
of styrene with (11-undecanoyl)-1-ethane sodium sulfonate, 270, 271
of styrene with vinyl acetate, 137
of styrene with vinyl benzoate, 137
of styrene with 4-vinylpyridine, 126
of vinyl acetate with dibutyl maleate, 177
of vinyl chloride with acrylamide, 254
inverse emulsion,
of acrylamide with acrylic acid, 199
of methyl methacrylate with acrylic acid, 193, 202
inverse microemulsion,
of acrylamide with N,N-methylenebisacrylamide, 208, 209
of acrylamide with methyl methacrylate, 207, 208
of acrylamide with styrene, 207, 208
seeded emulsion,
of acrylonitrile with vinylidene chloride, in the presence of seed crosslinked particles of poly(butyl acrylate), 258
of styrene with acrylonitrile, 134
in the presence of seed particles of poly(styrene), 257
in the presence of seed particles of poly(butadiene), 134, 262
of styrene with butadiene, in the presence of seed polymer particles of poly(ethyl-co-methyl methacrylate), 259
poly(ethyl acrylate-co-styrene-co-acrylic acid), 259
poly(styrene), 54, 255
of styrene with divinylbenzene, in the presence of seed particles of poly(styrene), 138, 259
seeded, emulsion, batch, 260, 261
seeded, emulsion, semibatch, 260, 261
semicontinuous, 175
suspension,
of acrylonitrile with styrene, 313
of acrylonitrile with vinyl acetate, 313
of ethyl acrylate and acrylonitrile with divinylbenzene, 296, 314
of ethyl acrylate and methyl methacrylate with methacrylic acid, 295, 315
of methacrylic acid with divinylbenzene, 295
of styrene and butyl acrylate with divinylbenzene, 302, 315
of styrene with n-butyl methacrylate, 300
of styrene with divinylbenzene, 299, 301
of styrene with methyl methacrylate, 300
of vinyl chloride with 1-alkenes, 310
critical flocculation volume, 232, 233
crosslinking agent, 258

degree of crosslinking, 314
degree, of dispersion — see polydispersity
degree of grafting, 260
degree of polymerization, 242, 243, 270
dehydrochlorination of PVC, 310
desorption of radical — see radical exit events
dialysis, 339, 340
diffusion, 53, 321
of acrylamide oligomers, 201, 204, 206
of initiator, 201, 239, 296
of monomer, 23, 24, 38, 45, 57, 147, 197, 201, 203, 239, 253, 255, 262, 266, 297
of monomer radical, 43, 53, 159, 212
of radicals, 43, 52–54, 60, 201
dispersant, polymeric, 12, 19, 225, 231, 236
concentration, 221, 236, 238, 247
critical adsorption concentration, 220
desorption, 223
efficiency of stabilization of polymer dispersion, 129, 230, 231
mechanism of stabilization of polymer particles, 216–224, 231, 232, 237, 245
size of micelles, 221
dispersion,
gas in liquid (foam), 18
gas in solid (colloidal inclusion of gas in solid), 18
with heterogeneous particle structure, 249–265
liquid in gas (fog, aerosol), 16, 18
liquid in liquid (emulsion), 17, 18
mixed, 13
modification, modified, 170–177
polymer (see also polymer particles), 9, 10
concentrated, 273, 287, 344
properties and use, 11, 248
rheological properties, 247, 252
solid in gas (smoke, aerosol), 18
solid in liquid (suspension, sol), 16, 18
preparation,
by dispersing method, 16
by condensation method, 16
by condensation-dispersing method, 16, 17
dissociation, of functional groups, 178, 179
double layer, electric, 44, 161, 166, 179–182, 218, 332

electrodecantation, 341
electrolyte, salting-out effect, 86, 199, 314
electron microscopy, 45, 167, 221, 236, 245, 257, 287, 319, 328

Subject index

electrophoresis, 179, 332, 335–337
  laser Doppler, 336
emulsifier, 18, 19, 24, 78
  amphoteric, 78, 96, 130
  anionic, 45, 78, 81, 94, 96, 102, 127, 128, 142, 147, 149, 150, 156, 157, 161, 165, 251, 258, 315
  cationic, 78, 82, 96, 101, 126, 161
  characteristics, 83, 84
  (co) polymerizable — see surface-active monomer
  inorganic, 37
  mixed, 151, 152, 156, 157, 165
  nonionic, 46, 78, 82, 86, 96, 127, 128, 150, 151, 156, 157, 161, 165, 195, 198, 199, 213, 300
epoxidation, 174
equation,
  Brönsted–Bjerrum, 100
  Debye, 320
  Flory–Huggins, 305
  Gibbs–Helmholtz, 293
  Griffin, 86
  Rayleigh, 320
  Smith–Ewart, 25–30, 70
  Stockmeyer, 30
  Stokes–Einstein, 196, 321

Fenton reagent, 133, 141, 142
field-flow fractionation, 319, 326–328
Fuchs ratio, 41, 181

gel, blocking, 200
gel effect, 36, 127, 152, 153, 202, 217, 238, 252, 271, 313, 314, 316
glass transition temperature, 59, 214, 231, 245, 248, 257, 300, 309

Hamaker constant, 219
HLB, 86, 87, 92, 190, 191, 195, 198, 199, 213
hydrolysis,
  of acrylamide groups, 171, 173, 195, 196
  of carbohydrates, 102
  of emulsifier, 161
  of esters, 102
  of functional groups, 171, 172
  of peroxodisulfate, 133, 142
  of phosphate esters, 101
  of sulfate groups, 112, 171, 252, 335
  of vinyl acetate groups, 171, 299, 309
hydroxymethylation, 173

inhibitor, of radical reactions, 154, 155, 159, 271
  2,2-diphenyl-1-picrylhydrazyl (DPPH), 75
  potassium nitrosodisulfonate (Fremy's salt), 75, 76, 204–207
  2,2,6,6-tetramethyl-4-hydroxypiperidinyl-1-oxyl, 75
  2,2,6,6-tetramethyl-4-octadecanoyloxypiperidinyl-1-oxyl (STMPO), 72–75
initiator, 18, 104–118

ammonium peroxodisulfate, 72, 74, 104, 150, 164, 204, 205, 207, 212, 268, 275
azobisisobutyronitrile (AIBN), 71, 72, 74, 75, 109, 138, 147–149, 193, 197–199, 203–206, 211
based on diazoamines, 109
catalysed decomposition, 71, 107, 108, 111–113
cumyl hydroperoxide, 110
decomposition, 49, 50, 66, 71, 72, 104–107, 162, 203, 234, 265, 291, 303
diacetyl peroxide, 252
diazoethers, 109
dibenzoyl peroxide, 109, 203–208, 224, 252, 298, 300
dissociation, 104
Fenton's reagent, 141, 142
high-molecular, 109
hydrogen peroxide, 110
  catalysed decomposition, 110, 111, 113
hydroperoxides, 110
lauroyl peroxide, 252, 313
oil-soluble, 18, 37, 72, 149, 200, 298
peroxodisulfate, 47, 50, 104–109, 111–115, 124, 127, 128, 131, 133, 135, 136, 138, 139, 141, 142, 144, 146–149, 156, 160–163, 167, 193, 197, 198, 203, 249, 251, 252, 254, 261, 263, 269–271, 275, 276, 278, 281, 287, 334
  redox, 104, 160, 251, 263, 313
photo-, 116–118, 139, 160, 163, 196, 210, 211, 264, 276, 286, 288, 315
radioisotope-labelled, 149
redox system, 104, 110–116, 128, 135, 195, 211
surface-active, 109, 126
water-soluble, 18, 23, 25, 30, 37, 71, 75, 104, 109, 126, 150, 194, 200, 212, 243, 252, 256, 278
with surface-active properties, 109, 278
interaction,
  emulsifier–electrolyte, 87, 189
  emulsifier–initiator, 37, 108, 156, 161
  emulsifier–polymer, 149, 152, 184, 219
  emulsifier–water, 89
  micelle–coemulsifier, 96
  particle–dye, 337
  particle–particle, 288, 319, 325
  polymer–solvent, 222, 235
  radical–polymer, 151
  stabilizer–medium, 246
  specific, 222
interpenetrating polymer network (IPN) — see networks, polymer, interpenetrating
interphase, 49, 155, 161, 282, 300
inverse emulsion, 186–214
  kinetics in, 192–209
  preparation, 186–192
inverse microemulsion — see microemulsion, water-in-oil
ion exchangers, 340, 341
ionic strength, 87, 105, 133, 202, 270, 271, 330, 332
isoelectric point — see point, isoelectric

layer,
    Gouy–Chapman, 80, 179
    Stern, 80, 104, 178
light scattering, 81, 88, 95, 189, 191, 196, 202, 210, 211, 221, 236, 319–321, 325, 338
liposomes, transfer of substances, 20, 283
living polymer, 228, 229
locus,
    of initiation, 36, 38, 71, 167, 200, 203–205, 207, 217, 278, 343
    of propagation, 17, 23, 24, 36, 58, 71, 149, 203, 206, 224, 241, 263
loss angle, 258

magnetic field, 153
mechanism,
    adsorption-desorption, 97
    of catalysed reactions, 100
    of coagulation, 44, 181
    coagulative, 44, 154, 201
    of decomposition of potassium peroxodisulfate, 104–109
    of dispersion polymerization in nonaqueous media, 234–244
    of growth of polymer particles, 245
    multistep, 49
    of nucleation of particles in suspension system, 302–304
    of nucleation of polymer particles in emulsion system, 25, 30, 38, 49, 160, 240, 270, 271
    of photoinitiation of polymerization, 116–118
    of polymer particle formation, 199, 303
    of radical polymerization of vinyl monomers, 123–167
    of seeded polymerization, 254
    of stabilization of polymer particles in suspension, 290, 291, 303
    of termination of radicals, 60, 62
method,
    calorimetric, 97, 311, 314
    centrifugal, 328, 329
    centrifugation, 94, 166, 191,196, 211, 328, 329
    chromatographic, 95, 243, 257, 311, 312, 324–326
    conductometric, 94
    dilatometric, 311
    dissolution, 337
    electron microscopic, 45, 166
    fluorescence, 245, 332
    gravimetric, 311, 343
    mechanical spectroscopic, 166
    NMR, 81, 95, 337
    polarographic, 338
    stripping, 95
micelles, 13, 78–90
    aggregation number, 80–82, 85, 88–90, 131
    classical, 13, 20, 186, 187
    core, 99, 222
    cylindrical, 80, 187, 276, 282, 283
    formation, 78, 87, 90, 97
    inverse, lipophilic, 9, 19, 20, 186, 187, 191, 196
    mixed, 89, 150
    reactions, 99–104
    shell, 80, 222
    spherical, 80, 81, 87, 187, 199, 276
    structure, 19, 81, 89, 102, 186, 221, 222
    surface, 99, 127, 142, 167, 324
microemulsion, 17, 19, 90–93, 138, 210
    acrolein/divinylbenzene, 139
    bicontinuous, 92, 93, 188–190
    definition, 17, 19, 93, 186
    inverse – see water-in-oil
    methyl methacrylate/acrolein, 160
    oil-in-water, O/W, 13, 17, 19, 90–92, 139, 188, 190
    poly(styrene), 139, 141
    poly(styrene)/divinylbenzene, 139
    stability, 90, 190, 199
    structure, 19, 90, 92, 189, 199
    styrene/acrylic acid, 168
    water-in-oil, W/O, inverse, 17, 19, 90–92, 186–192, 197–199, 207, 213
        nonpercolating, 202, 204, 207
        percolating, 202, 207
microsuspension, inverse, 201
miniemulsion, 17, 90–92, 94, 138, 139, 159, 160
    stability of, 94, 139
model,
    of Abdel-Alim and Hamielec, 304
    of Asua, 54–56
    of catalysed reactions, 100–103
    coagulative, 44
    of coagulative nucleation, 44, 240, 269
    of diffusion adsorption, 241
    of dispersion polymerization, 175, 201, 217, 240, 242
    of droplet distribution, 301
    first kinetic, 25, 26, 102
    of Frank–Evans, 87
    of Gardon, 268
    of Haber–Weiss decomposition of hydrogen peroxide, 110, 111
    of Kelsall and Maitland, 305
    of Kolthoff, 104, 115
    mass action, 85, 97, 98
    mathematical, 64–71
    of McBain, micellar, 80
    Medvedev's, 36
    micellar, 24, 153
    of micellar block copolymers, 222
    of Nomura, 52–56
    of nucleation from dispersant micelles, 241
    of Olaj, 305
    pseudo-phase, 92, 97
        transfer, 97
    of Rabinovitch, 58
    residual termination, 63
    SCOPE, 67
    of self-nucleation (particles), 40, 240
    Smith–Ewart, 25, 67, 125, 128, 201, 268

of Smoluchowski, 58, 181, 323
of Soh and Sundberg, 58–63
of steric stabilization of particles, 192, 218, 292
of suspension polymerization, 291, 304–308
of two-stage seeded polymerization, 254, 255
of Ugelstad, 30, 304
of Wessling, 268
of Xie and Hamielec, 305
modelling of,
    conversion, 177, 304, 306
    emulsion polymerization, 23, 64, 161
    particle size distribution, 176, 177, 243
monolayers and multilayers, 281–288
    containing polymerizable double bonds, 20, 21, 283
    polymerization of vinyl monomer, 20, 281, 286
    polymerized, 20, 282, 285
    structure, 282–285

networks, polymer, interpenetrating, 13, 258
neutron scattering, 81, 95, 189, 319, 337

oligomer, oligomeric radical — see radical, oligomeric
    critical length, 38, 45, 61, 270
    degree of polymerization, 44, 61, 155

parameters,
    Flory–Huggins, 235
    solubility,
        of emulsifier, 79, 190
        of monomers, 32
        of oil phase, 190
        of polymers, 232
        of solvents, 232, 233, 245
percolation,
    clusters, 202
    threshold, 202
phase inversion, 255
photoinitiator — see initiator, photo-
photon correlation spectroscopy, 320, 321
point,
    cloud, 85, 86, 199
    isoelectric, 172, 253, 335
    Kraft, 83, 84
polydispersity, 11, 17, 52, 131, 253, 270, 296
polymerization,
    dispersion (see also polymer dispersion, polymer particles), 9, 10
    batch method of preparation, 12, 125, 127, 129, 131, 143, 174, 254, 301, 308
    in concentrated emulsion, 287
    continuous method of preparation, 12, 175
        in mixed solvents, 246
        in nonaqueous medium, 216–248
            of acrylamide, 206, 242
            of acrylic acid, 242
            of acrylonitrile, 242
            of lauryl methacrylate, 231
            of methyl methacrylate, 217, 224, 230, 235, 237–239, 242
            of styrene, 224, 237, 243, 245

of vinyl acetate, 242
of 4-vinylpyridine, 229
oil-in-oil type, 12
seeded, two-stage process, 175
semicontinuous process, 159, 175, 240, 249, 308
special methods of preparation, 116, 264, 272, 277,
    oil-in-water, 9, 12, 14, 17, 18, 195
    water-in-oil, 19
emulsion,
    of acrylamide, 272, 287
    of acrylonitrile, 142
    of butyl acrylate, 63, 144–146, 159
    of butyl methacrylate, 63, 72, 74, 147, 149
    of *tert*-butylstyrene, 127, 128
    of divinylbenzene, 126, 138
    without emulsifier, 43, 267–272
        of styrene, 268, 272
        of vinyl acetate, 271, 272
        of vinyl monomers, 272
    of ethyl acrylate, 144, 149
    of 2-ethyl acrylate, 74
    of ethyl methacrylate, 143
    of hexyl acrylate, 149
    of 2-hydroxyethyl methacrylate, 272
    in magnetic field, 153
    mechanism of, 49–76
    of methacrylic acid, 272
    of methyl acrylate, 141, 149
    of methyl methacrylate, 13, 34, 57, 59, 62, 63, 72–75, 118, 143, 150, 264
    of methyl methacrylate, in the presence of fillers, 263, 264
    of octadecyl methacrylate, 150
    with radical scavenger, 72–75
    of styrene, 13, 36, 46, 56–58, 66, 73, 75, 118, 123–140, 246, 272, 287
    of styrene, in the presence of inorganic particles, 263, 264
    thermal, 125
    two-stage model, 131, 254, 255, 258, 260, 268, 269
    of vinyl acetate, 33, 34, 57, 159, 271, 272
inverse emulsion, 14, 186–214, 287
    of acrylamide, 196–199, 201, 203–208, 287
    of styrene, 123, 138, 213, 214
inverse microemulsion, 200
microemulsion,
    of acrylamide, photoinitiated, 210–212
    of methyl methacrylate, 193
    in supercritical fluids, 214
in mono- and multilayers, 20, 281–288
    of vinyl-N-methyl pyridinium methyl sulfate, 282
in monomer droplets, 45–47, 264
of monomer in lamellar mesophase, 213
partitioned, 71–75, 204, 342, 343
    of butyl methacrylate, 72, 73
    of ethylhexyl acrylate, 74

350 Subject index

of methyl methacrylate, 72–76
of styrene, 72, 73, 75, 76
transfer
  of initiator, 71, 74
  of initiator and monomer, 72, 74
  of monomer, 19, 24, 72
seeded, emulsion, 66, 245, 258
mechanism of, 52, 65, 175, 261
  of methyl methacrylate, in the presence of poly(butyl acrylate) particles, 143, 257
  of styrene, in the presence of poly(butyl acrylate) particles, 260
  of styrene, in the presence of poly(ethyl acrylate) particles, 146, 157
  of styrene, in the presence of poly(methyl methacrylate) particles, 259, 261
  of styrene, in the presence of poly(vinyl acetate) particles, 256
  of vinyl acetate, in the presence of poly(styrene) particles, 256
of surface-active initiators, 278–281
of surface-active monomers (surfomers), 273–277
  of allyldimethyldodecylammonium bromide, 276
  of methylacryloylethyldimethylalkyl-ammonium halides, 276
  of nonylphenoxypoly(etheroxy)ethyl acrylate, 277
  di-n-octadecylmethyl(p-vinyl benzyl) ammonium chloride, 276
  sodium 8-nonenoate, 276
  sodium 10-undecenoate, 276
  of vinylimidazolium salts, 274, 276
  of vinylpyridinium salts, 274, 276
suspension,
  of acrylamide, 315, 316
  of chlorostyrene, 300
  of dichlorostyrene, 299
  of divinylbenzene, 299, 300
  of hydroxyethyl methacrylate, 314
  of methyl methacrylate, 290, 297, 313
  of styrene, 52, 53, 66, 131, 290, 292, 294, 298–302
  unconventional, 314, 316
  of vinyl acetate, 312
  of N-vinyl-$tert$-butyl carbamate, 315
  of N-vinylcarbazole, 315
  of vinyl chloride, 290, 301–304, 308–312
polymer particles (see also dispersion polymerization, polymer dispersion), 9, 10
aggregation — see coalescence
association, 182, 184
carriers of catalytically active substances, 11, 14
characterization, 257
  chemical analysis, 257
  by density measurements, 257
  flocculation tests, 257
  microelectrophoresis, 257, 336
  pyrolysis/gas chromatography, 257

solubility experiments, 257, 337
transmission electron microscopy, 257
X-ray photoelectron spectroscopy, 251, 257
charge, 179
coagulation, 20, 43, 44, 69, 218
coagulative nucleation, 44, 69, 218
coalescence, aggregation, flocculation, 31, 45, 126, 142, 143, 149, 156, 176, 182–184, 206, 217, 235, 240, 246, 255, 262, 291–296, 299, 302–304, 313
core-shell type, 12, 38, 143, 156, 159, 162, 166, 252, 254–260, 265, 270, 303, 323, 337, 338, 343
crosslinked, 11, 13, 299, 308, 312
crosslinking, degree, 137, 312, 316
deactivation, 56–64
distribution of radicals in a particle, 40, 258
film formation, 247
  with fish eyes, 310
flocculation — see coalescence
functionalization, 249, 251, 253, 268, 312
growth, 24, 44, 46, 49, 56–64, 140, 162, 201, 245, 294, 298–300, 302
hardening, 328
heterogeneous, 12, 251–265
magnetizable, 14
modification, 10, 170, 250, 309, 340, 343
  of surface, 13, 14, 170, 250, 272
monodisperse, 11, 19, 52, 126, 252, 270, 325
monomer-containing (polymer/monomer), 17, 138, 241, 244, 246, 257, 269, 275
morphology, 175, 245, 253, 255–259, 270, 338
nucleation, 20, 33, 34, 73, 76, 127, 131, 134, 138, 164, 175, 241, 266
  of PVC, 302–304
number, 20, 25, 29, 33, 34, 42, 76, 132–136, 142, 144, 148, 151, 152, 162–165, 240, 256, 268, 269, 271, 272, 303, 304, 322, 323
  of macromolecules in a particle, 197, 198, 200, 211
  of radicals in a particle, 25, 27–29, 33, 35, 41, 43, 54, 60, 65, 70, 131–136, 142, 144, 146, 148–151, 153, 159, 175, 201, 224, 269
porosity, 301, 303, 309
preparation, 170, 217, 249–265, 290
primary, 38, 43, 44, 47, 56, 70, 71, 124, 143, 154–156, 161, 252, 268, 302–304
renucleation, 240
shape, 253, 259, 265, 287, 303, 314, 319, 324
size, 9, 12, 14, 17–19, 31, 52, 59, 126–130, 134, 135, 138–140, 143, 146, 150, 151, 153, 155, 159, 160, 162, 164, 165, 175, 196–200, 202, 203, 210, 211, 216, 235–238, 240, 243–247, 253, 254, 263, 270, 271, 273, 275, 287, 291, 294–296, 298–300, 303, 304, 309, 313, 319, 320, 322, 323, 325, 326, 329, 330
size determination, 319–331
size distribution, 66, 160, 165, 236, 243, 244, 246, 252, 269–271, 280, 291, 294, 300, 309, 321, 322, 325, 327

Subject index

stability, 12, 14, 19, 38, 43, 153, 155, 161, 178–184, 198, 204, 244
structure, 10, 13, 38, 164, 166, 191, 244, 247, 249–265, 303, 337, 338
surface, 26, 36, 149, 150, 152, 159, 162–165, 170, 249, 250, 253, 264, 340
swelling,
 by monomer, 12, 32, 35, 36, 49, 56, 76, 126, 137, 138, 146, 241, 254, 309
 by solvent, 19, 138, 246
volume growth, 166
potential,
 electrokinetic, 179, 181, 335
 interaction, 180

radical,
 alkoxyl and hydroxyl, 110, 118, 142, 163
 average lifetime, 26, 29
 charged, 43
 entry events, 23, 25, 35, 43, 47, 49, 66, 104, 124, 126, 131–133, 148, 161, 163, 212, 252, 261, 263
 exit (desorption) events, 31, 49, 50, 52, 66, 73, 104, 131, 134, 136, 152, 153, 155, 158, 159
 oligomeric, 38–40, 42, 44, 50, 52, 60, 72, 73, 104, 123, 124, 127, 132, 135, 142, 160–163, 206, 235, 240, 241, 252
 primary, 23, 25, 30, 38, 40, 41, 52, 74, 128, 133, 280
 re-entry events, 42, 49, 51, 104, 132
 surface-active, 45, 252
 termination, 201, 206, 212, 226, 262, 305
 transfer, 304
 trapped, 135, 138, 195
 water phase termination, 26, 142
rate,
 of batch emulsion polymerization, 124, 125, 127, 152
 of capture of radicals, 243
 of coagulation, 40, 43, 68, 69, 183, 303
 of copolymerization, 250
 of desorption of radicals, 31, 36, 41, 44, 50, 54, 55, 66, 70, 131
 of dispersion polymerization in nonaqueous media, 238, 242, 243
 of emulsion (co)polymerization, in the presence of surface-active monomers (surfomers), 202
 of emulsion polymerization, 24–26, 28, 29, 33, 37, 44, 268, 271
 in concentrated emulsion, 287
 without emulsifier, 39, 268, 269, 271, 275
 of growth and formation of polymer particles, 39
 of homogeneous nucleation, 42
 of initiation, 241
 of inverse emulsion polymerization, 196, 213
 of inverse microemulsion polymerization, 198
 of micellar nucleation, 42
 of particle formation, 44
 of particle nucleation, 41
 of polymerization in mono- and multilayers, 286
 of production of radicals, 26, 31, 38, 40, 70
 of semicontinuous emulsion polymerization, 176
 of suspension polymerization, 298, 312, 314
 Smith–Ewart, 33, 34
 of termination of radicals, 35, 59, 61, 237
rate constant,
 for chain transfer, 132
 for coagulation, 41, 69, 182
 for decomposition of initiator, 32, 203, 279
 for desorption of radicals, 50, 66, 70
 for entry of radicals into a particle, 25, 39, 49–52, 66, 67, 70
 for exit of radicals from a particle, 132
 for propagation of radical polymerization, 26, 32, 40, 57, 242, 243, 270, 276
 for termination of radicals, 26, 35, 36, 40, 50, 54, 60, 146, 242, 243, 270, 272, 276, 280, 309
reaction,
 bimolecular, 101
 at a double bond, 174, 175, 201, 260, 275
 Hofmann, 172, 173
 Mannich, 173, 174
reaction order of polymerization,
 with respect to emulsifier, 28, 33, 37, 44, 123, 124, 126, 127, 129, 130, 132, 134, 142–144, 146, 147, 149–153, 155–157, 164, 198, 201, 213
 incident light intensity, 117
 initiator, 28, 36, 37, 51, 106, 123, 124, 131, 133, 139, 141, 142, 144, 146, 147, 150, 164, 198, 201, 212, 243, 300
 monomer, 34, 37, 137, 142, 146, 162, 172, 196–198, 211, 212, 242, 243, 271
 particle number, 136, 143, 146, 156

sensitizer, 117, 118, 139, 211
small-angle neutron scattering, 81, 96, 221, 323, 324, 337
small-angle X-ray scattering (SAXS), 221, 323, 324, 338
solubility, of monomers in water, 140, 154, 306
solubilization, 83, 93–99, 150, 153
 ratio, 95, 96
spectroscopy, 95
 mechanical, 166, 338
 NMR, 92, 93, 95
 photon correlation, 320, 321
stability,
 colloidal, 293
 of microemulsions, 90
stabilization,
 of droplets, 159, 268
 of polymer particles, 230–232, 292, 293, 298–300
  effect of temperature, 182, 231
  electrostatic, 181–183, 217, 218, 268, 271, 272, 293
  enthalpic, 219
  entropic, 219
  by polar groups, 163, 182, 293

ratio, 181
steric, 127, 200, 217–220, 225–233, 246, 247, 272
of suspensions, 290, 291
stabilizer, of polymer particles, 223, 224, 291–296, 304, 309
stopped-flow spectroscopy, 323
surface,
- -active initiators, 278–281
- -active monomer (surfomer), 252, 253, 270, 271, 273–277
- charge density, 43, 163, 164, 171, 181, 182, 252, 253, 332
- of emulsifier, 32, 95
- modification, 263
- of polymer particles — see polymer particle, surface
- potential, 293
- tension, 83, 96, 129, 153, 176

termination, of radicals, 50, 59
- biradical, 52, 64, 66, 70
- monomeric radical, 64
- residual, 59–63

terpolymer, 137
terpolymerization,
  emulsion,
    of butyl acrylate, acrylic acid and hydroxyethyl methacrylate, 251
    of butyl acrylate, acrylonitrile and divinylbenzene, 302
    of butyl acrylate, ethyl acrylate and methacrylic acid, 165
    of butyl acrylate, methyl methacrylate and methacrylic acid, 165
    of styrene dimethylaminoethyl methacrylate and methacrylic acid, 253
  inverse emulsion,
    of sodium acrylamidoundecanoate, methyl methacrylate and acrylic acid, 194
test, flocculation, 257
theory,

of coagulative nucleation, 44
collision, 39
of colloid stability, 71, 178–184
DLVO, 44, 71, 178, 180
Debye–Hückel, 179, 336
Flory–Huggins, 235
Frank–Evans, 87
Gardon, 30–36, 268
Gouy–Chapman, 80
Harkins, 23–25
of homogeneous nucleation, 38–46
of light-scattering, 319–321
Medvedev, 36–38
Mie, 320, 321
of mixed films, 92
Romsted, 103
solubilization, 92, 97–99
Stern, 80
Smith–Ewart, 25–30, 135, 198, 201, 343
thermodynamic, 92, 343
theta,
  solvents of polymers, 232, 233
  temperatures, 232, 233
titration, of latex, 329–331
  conductometric, 94, 329, 333
    direct, 332–335
    reverse, 332–335
  potentiometric, 332, 335
transfer reaction, 246, 288
transport, of monomer, 72, 143, 240, 242
  by collisional mechanism, 68
  by interphase diffusion mechanism, 19
turbidimetric titration, 260
turbidimetry, 96, 319, 321–323
turbidity, 95, 199, 203, 252, 323, 332
  specific, 322

ultracentrifugation, 166, 267
ultrafiltration, 341

X-ray photoelectron spectroscopy, 257
X-ray scattering, 189, 221, 319